Principles and Techniques of Gene Manipulation

Principles and Techniques of Gene Manipulation

Editor: Patrick Faraday

www.callistoreference.com

Callisto Reference,
118-35 Queens Blvd., Suite 400,
Forest Hills, NY 11375, USA

Visit us on the World Wide Web at:
www.callistoreference.com

ISBN: 978-1-63239-917-5 (Hardback)

Cataloging-in-Publication Data

Principles and techniques of gene manipulation / edited by Patrick Faraday.
  p. cm.
Includes bibliographical references and index.
ISBN 978-1-63239-917-5
1. Genetic engineering. 2. Genomics. I. Faraday, Patrick.
QH442 .P75 2018
660.65--dc23

Table of Contents

Preface

I am honored to present to you this unique book which encompasses the most up-to-date data in the field. I was extremely pleased to get this opportunity of editing the work of experts from across the globe. I have also written papers in this field and researched the various aspects revolving around the progress of the discipline. I have tried to unify my knowledge along with that of stalwarts from every corner of the world, to produce a text which not only benefits the readers but also facilitates the growth of the field.

Gene manipulation is the modification of genetic elements in plants and animals for commercial purposes. This book on gene manipulation strives to provide a fair idea about this discipline and to help develop a better understanding of the latest advances within this field. Technological development that has greatly affected this field has also been presented. It presents researches and studies performed by experts across the globe that are related to the genome. This book brings forth some of the most innovative concepts and elucidates the unexplored aspects of this field. The readers would gain knowledge that would broaden their perspective about gene manipulation.

Finally, I would like to thank all the contributing authors for their valuable time and contributions. This book would not have been possible without their efforts. I would also like to thank my friends and family for their constant support.

Editor

HLA-B27 and Human β2-Microglobulin Affect the Gut Microbiota of Transgenic Rats

Phoebe Lin[1][*][¶], Mary Bach[3][¶], Mark Asquith[2], Aaron Y. Lee[4], Lakshmi Akileswaran[5], Patrick Stauffer[1], Sean Davin[1], Yuzhen Pan[1], Eric D. Cambronne[9], Martha Dorris[6], Justine W. Debelius[7], Christian L. Lauber[7], Gail Ackermann[7], Yoshiki V. Baeza[7], Tejpal Gill[10], Rob Knight[7,11], Robert A. Colbert[10], Joel D. Taurog[6], Russell N. Van Gelder[5], James T. Rosenbaum[1,2,8]

1 Casey Eye Institute, Oregon Health & Science University, Portland, Oregon, United States of America, 2 Division of Rheumatology, Oregon Health & Science University, Portland, Oregon, United States of America, 3 Division of Rheumatology, University of Washington, VA Medical Center, Seattle, Washington, United States of America, 4 Moorfield's Eye Institute of London, London, United Kingdom, 5 Department of Ophthalmology, University of Washington, Seattle, Washington, United States of America, 6 Department of Rheumatology, University of Texas Southwestern, Dallas, Texas, United States of America, 7 University of Colorado Boulder, Boulder, Colorado, United States of America, 8 Dever's Eye Institute, Portland, Oregon, United States of America, 9 Department of Molecular Microbiology & Immunology, Oregon Health & Science University, Portland, Oregon, United States of America, 10 Pediatric Translational Research Branch, National Institute of Arthritis, Musculoskeletal and Skin Diseases, National Institutes of Health, Baltimore, Maryland, United States of America, 11 Howard Hughes Medical Institute, University of Colorado Boulder, Boulder, Colorado, United States of America

Abstract

The HLA-B27 gene is a major risk factor for clinical diseases including ankylosing spondylitis, acute anterior uveitis, reactive arthritis, and psoriatic arthritis, but its mechanism of risk enhancement is not completely understood. The gut microbiome has recently been shown to influence several HLA-linked diseases. However, the role of HLA-B27 in shaping the gut microbiome has not been previously investigated. In this study, we characterize the differences in the gut microbiota mediated by the presence of the HLA-B27 gene. We identified differences in the cecal microbiota of Lewis rats transgenic for HLA-B27 and human β2-microglobulin (hβ2m), compared with wild-type Lewis rats, using biome representational in situ karyotyping (BRISK) and 16S rRNA gene sequencing. 16S sequencing revealed significant differences between transgenic animals and wild type animals by principal coordinates analysis. Further analysis of the data set revealed an increase in *Prevotella spp.* and a decrease in Rikenellaceae relative abundance in the transgenic animals compared to the wild type animals. By BRISK analysis, species-specific differences included an increase in *Bacteroides vulgatus* abundance in HLA-B27/hβ2m and hβ2m compared to wild type rats. The finding that HLA-B27 is associated with altered cecal microbiota has not been shown before and can potentially provide a better understanding of the clinical diseases associated with this gene.

Editor: Stefan Bereswill, Charité-University Medicine Berlin, Germany

Funding: PL is supported by a Research to Prevent Blindness Career Development Award and an NIH award K08EY022948. JTR is supported by the Stan and Madelle Rosenfeld Family Trust and the William and Mary Bauman Foundation. RK is an HHMI Early Career Scientist. RVG and LA were supported by the Burroughs-Wellcome Translational Scientist Award, an unrestricted award from Research to Prevent Blindness, and by NIH P30 EY001730. MB was supported by NIH T32 AR 007108. TG and RAC were supported by the NIAMS Intramural Research Program Z01 AR041184. The funders had no role in study design, data collection and analysis, decision to publish, or preparation of the manuscript.

Competing Interests: The authors have declared that no competing interests exist.

* Email: linp@ohsu.edu

¶ These authors are co-first authors on this work.

Introduction

The role of the gut microbiota in diseased and healthy states has become increasingly apparent, as evidenced by its importance in a number of conditions including metabolic syndrome, type 1 diabetes, multiple sclerosis, rheumatoid arthritis, and inflammatory bowel disease [1–7]. Many of these diseases are also influenced by the genes of the major histocompatibility complex (MHC), otherwise known as human leukocyte antigens (HLA) in humans. In many conditions, the HLA alleles confer greater disease risk than other genetic factors identified by genome wide association studies [8]. Ankylosing spondylitis (AS), a potentially disabling condition that results in axial and sometimes peripheral arthritis is

highly associated with an MHC class I molecule, HLA-B27, with ~ 90% of patients with AS carrying the HLA-B27 allele. The mechanism by which HLA-B27 predisposes to disease is not well understood, although several hypotheses have been postulated [9–11]. Our study tests the hypothesis that HLA-B27 alters the repertoire of the gut microbiome [11], which may be one mechanism by which it predisposes to disease.

Studies in both humans and rodents support this hypothesis [12]. Bacteria including *Shigella, Salmonella, Yersinia,* and *Cambpylobacter* are well-characterized triggers for HLA-B27-associated reactive arthritis [13–16]. Cross reactivity between monoclonal antibodies to HLA B27 and Gram-negative bacteria has been reported [14,17,18]. There is also sequence homology between a nitrogenase from *Klebsiella* and HLA-B27, although the significance of this finding has been debated [13,19,20]. Raising HLA-B27 transgenic rats in a germ-free environment prevented both gut and joint inflammation that otherwise spontaneously occurs in these rats [21]. Subsequent studies showed that mono-association of HLA-B27 transgenic rats with *Bacteroides vulgatus* in a germ-free environment was sufficient to re-establish colitis in these animals [22,23]. A series of studies by Granfors et al have demonstrated marked effects of HLA-B27 on innate immunity and host defense *in vitro*, including effects on bacterial invasion of cells, intracellular persistence, intracellular signaling, and cytokine production [24–28]. Penttinen et al. showed that HLA-B27-transfected monocytes had enhanced inflammatory responses to lipopolysaccharide found on bacterial cell walls [26,29]. Moreover, disease-prone HLA-B27 transgenic rats exhibit a variety of dendritic cells abnormalities [30,31] including disrupted trafficking of dendritic cells from the gut to the mesenteric lymph nodes [32]. It is plausible that a dysregulated immune response in HLA-B27 individuals alters the composition of the gut microbiome, and that in turn, the altered gut microbiota contributes to disease pathogenesis.

There are various types of animal models of HLA-B27 disease. The HLA-B27 transgenic rats on a Lewis background created by crossing the 21–3 and 283–2 strains express high copy numbers of the HLA-B27 subtype B27:05 and human beta 2 microglobulin (hβ2m) and have a high penetrance and severity of arthritis in the absence of any gastrointestinal inflammation (Table 1) [33,34]. Whereas expression of either transgene alone does not cause disease, expression of both together causes a disease similar to human spondyloarthritis. This Lewis transgenic rat line (21–3×283–2) was chosen for the majority of our studies because 70% develop severe arthritis by 6 months, but the absence of colitis allowed us to avoid the confounding variable of gut inflammation contributing to the changes seen in the gut microbiota. Recent advances in both sequencing technology and our knowledge of the gut microbiome in various mammals including the rat have allowed us for the first time to characterize the differences in the gut microbiota associated with the expression of HLA-B27 and hβ2m.

Materials and Methods

Ethics statement

This study was carried out in strict accordance with the recommendations in the Guide for the Care and Use of

Laboratory Animals of the National Institutes of Health (NIH). The protocols were approved by the Institutional Animal Care and Use Committee at the University of Texas Southwestern in Dallas, Texas, or the National Institute of Arthritis, Musculoskeletal and Skin Diseases (NIAMS) at the NIH.

Animals and sample preparation

The transgenic rat lines bred on a Lewis background expressing human HLA-B27 and hβ2m by crossing the 21–3×283–2 lines have been described previously [33]. Male 21–3 rats develop epididymo-orchitis [35]. Male (21–3×283–2)F1 rats develop epididymo-orchitis, peripheral arthritis, and tail spondylitis [33,35]. None of the males of any of these genotypes develop inflammatory bowel disease [33]. Three main cohorts between 67–70 days old were studied (Table 2): (1) 6 nontransgenic Lewis male rats co-housed with 7 HLA-B27/β2m rats (21–3×283–2 F1 males), and 3 age-matched nontransgenic Lewis males housed separately; (2) 6 nontransgenic Lewis males co-housed with 6 age-matched HLA-B27/hβ2m (21–3×283–3) F1 males; (3) 4 female and 4 male rats of three genotypes: HLA-B27/hβ2m (21–3×283–2 F1), hβ2m (283–2), and nontransgenic, all housed separately by genotype and gender. All of the above rats were produced and housed at the University of Texas Southwestern animal facility under specific pathogen free (spf) conditions. We also investigated a cohort of Lewis rats between 60–180 days old, produced and housed at the NIH in which 28 HLA-B7/ hβ2m rats (line 120–4) were compared to 26 wild type and 28 HLA-B27/hβ2m rats (line 33–3) [36]. All of the 33–3 rats had evidence of colon and cecum inflammation histologically, although the younger animals (8 week old) were clinically normal. The HLA-B7 and wild type animals in this control cohort were phenotypically and histologically normal (Tables 1 and 2).

All specimens from cohorts 1–3 were collected at 67–70 days in age, prior to the onset of arthritis. At this age, rats are expected to have an adult repertoire of gut microbiota, but will not develop arthritis for at least another 40 days, so that we could distinguish genotypic effects from phenotypic effects. Whole cecum samples were collected from the first cohort, and cecal luminal contents and cecal mucosal specimens were separately collected from the other two cohorts as well as the control cohort. Following euthanasia, the cecum was removed under sterile conditions and either the intact cecum (including contents) or the cecal mucosa and cecal contents separately were snap frozen in sterile microfuge tubes. Genomic DNA was then extracted from the samples using a Qiagen DNAeasy kit and then sequenced using biome represen-

Table 1. HLA-B27/hβ2m transgenic rat lines.

Line	Transgene locus zygosity	Transgene scopies		
		HLA-B27	hβ2m	HLA-B7
21–3	hemi-	20	15	–
	homo-	40	30	–
283–2	hemi-	0	35	–
	homo-	0	70	–
21–3x283–2	hemi- x hemi-	20	50	–
33–3	homo-	55	66	–
120–4	homo-	–	10	52

Adapted from [35].

Table 2. Rat cohorts used in experiments.

Cohort	Housing	Gender/Genotype	Samples obtained
1	co-housed	Male/6 WT, 7 (21–3x283–2)F1	whole cecum
	separately housed	Male/3 WT	
2	co-housed	Male/6 WT, 6 (21–3x283–2)F1	cecal lumen, cecal mucosa
3	separately housed by genotype	Male/4 WT, 4 (21–3x283–2)F1, 4 (283–2)	cecal lumen, cecal mucosa
		Female/4 WT, 4 (21–3x283–2)F1, 4 (283–2)	
Control cohort	Depends on age of cohort	Male/15 WT, 15 (33–3), 15 (120–4)	cecal lumen, cecal mucosa
		Female/11 WT, 13 (33–3), 13 (120–4)	

WT: wild type; see Table 1 for genotype designations.

tational in situ karyotyping (BRISK) and/or 16S rRNA gene sequencing techniques (as described below).

BRISK technique

Phi29 amplification was performed on genomic DNA, and BRISK was performed as previously described [37]. Briefly, genomic DNA was digested with BsaXI to yield 33 bp fragments. Following ligation with 'barcoded' Illumina sequencing adapters, these fragments were sequenced at eight samples per lane. Tags are parsed against the reference sequence for rat host tags mapped to the rat genome to provide a karyotype. R^2 values observed for expected tags per chromosome are calculated, and samples were processed further if $R^2>0.95$. Remaining tags were parsed into those associated with known organisms (bacteria, fungi, virus, and parasite). Relative abundance of organisms is then calculated by normalizing the total number of unique tag hits greater than one for that organism by the total number of matched tags or to the total rat host tags present in the specimen when possible. Statistical significance for differences between groups was determined by Monte Carlo analysis, randomly permuting the dataset 100,000 times to establish probabilities for distributions among detected organisms.

16S rRNA gene amplification, taxonomic and diversity analysis

A portion of the 16S rRNA gene was amplified using the 515–806 primers as specified by the EMP (http://www.earthmicrobiome.org/) and sequenced on the Illumina MiSeq platform [38–42]. Data were quality-filtered using QIIME (Quantitative Insights Into Microbial Ecology) [40]. QIIME was used for alignment using Infernal [43], a stochastic context-free grammar aligner, clustering of sequences into operational taxonomic units using uclust [44], phylogenetic reconstruction using the reference + de novo protocol (given the high incidence of likely new clusters), and taxonomy assignment with the RDP classifier [45]. Alpha and beta diversity analyses and visualizations were also performed [46–48]. Statistical significance was assessed using parametric and nonparametric approaches including false discovery rate (FDR) corrections by the Benjamini-Hochberg method.

Quantitative and semi-quantitative PCR

Confirmation of bacterial species differences between the transgenic and wild type animals was obtained using both traditional PCR and then quantified by real time PCR (RT-PCR) using the ΔΔCt relative quantification method as described previously using the 16S rRNA gene as a reference gene [49].

Primer sequences, annealing temperatures, target genes, and product sizes are listed in Table S1.

Results

Expression of the HLA-B27 transgene is associated with alterations in the gut microbiota by 16S rRNA gene sequencing

Using 16S rRNA amplicon sequencing, there were statistically significant differences in bacterial sequences between transgenic and wild-type animals in both cecal lumen and cecal mucosa samples shown by principal coordinates analysis using UniFrac distances [50] (Figure 1a, 1b), with p-values shown in Table 3 by PERMANOVA analysis. While phylum-level differences were not statistically significant (Figure 1c, 1d), genus-level differences between genotypes were more easily distinguishable (Figure 2). Whereas the more abundant bacterial genera (such as *Clostridia* and *Helicobacter*) did not differ between transgenic and wild type animals, some of the rarer genera differed markedly. In both cecal lumen and cecal mucosa samples, an unknown genus of the Rikenellaceae family (dark blue bars) was significantly more abundant in wild type animals than either HLA-B27/hβ2m or hβ2m animals (p = 0.0005 and 0.002, respectively after FDR correction). On the other hand, *Paraprevotella* was higher in transgenic (HLA-B27/β2m or β2m) animals compared to the wild type animals (p = 0.03 in both tissue sites after FDR correction).

In the 3rd cohort of animals, whereas the represented bacterial sequences obtained from the HLA-B27/hβ2m animals from combined cecal lumen and mucosa samples segregated significantly from the wild type animals (pseudo-F 6.6645, p = 0.001) on principal coordinates analysis, there was a much smaller difference between the HLA-B27/hβ2m rats and rats that only expressed only hβ2m (pseudo-F 1.4688, p = 0.039) (Figure 3). Table 3 also illustrates that co-housing abrogated some of the differences seen between transgenic animals and wild type animals (cohort 2). This might be expected because rodents are coprophagic. However, in whole cecum preparations from the first cohort of animals, there remained significant differences in the Unifrac distances between the gut microbiota of wild type and HLA-B27/hβ2m transgenic animals despite co-housing (p = 0.006) (Table 3). HLA-B7, like HLA-B27, is also an MHC I allele, but differs in that it is not associated with arthritis susceptibility and thus serves as an MHC class I control. Significant differences in the gut microbiota were demonstrated in HLA-B7/ hβ2m compared with both wild type controls and HLA-B27/hβ2m rats (pseudo-F 4.6750, p = 0.001 by permanovafor cecal lumen and pseudo-F 5.6001, p = 0.001 by permanova for cecal mucosa). Figure S1 demonstrates the

LUMEN   MUCOSA

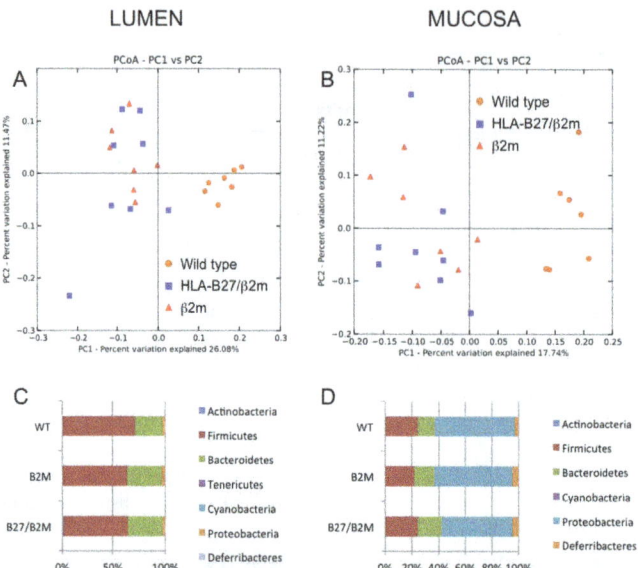

Figure 1. 16S rRNA gene sequencing principal coordinates analysis from non-co-housed rats shows significant differences in cecal microbiota between genotypes. A. Cecal lumen samples from cohort 3; B, Cecal mucosa samples from cohort 3; C, Phylum-level analysis from cecum lumen samples; D, Phylum-level analysis from cecum mucosa samples. WT: wild type rats; B2M: hβ2 microglobulin; B27/B2M: HLA-B27/hβ2 microglobulin transgenic rats.

separation of 16s sequences by genotype using principle coordinates analysis and table S2 shows the p values for genotype group comparisons in this control cohort.

Expression of the HLA-B27 transgene is associated with bacterial species-level alterations in the gut microbiota by BRISK

BRISK is a technique that allows for massively parallel deep sequencing of the gut microbiome generated by a Type IIB

restriction endonuclease. This provides a representation of the specific microorganisms present in a sample on a species level, as well as the abundance of these organisms. In a typical sample, we were able to amplify about 4 million tag, with ~21% of the DNA identified as from the host animal and representing all chromosomes inthe rat genome. Across all samples we identified more than 30,000 tags derived from known bacteria. In total, tags from 909 species were identified in at least one animal. The heat map (Figure 4) shows the relative abundance of specific bacterial

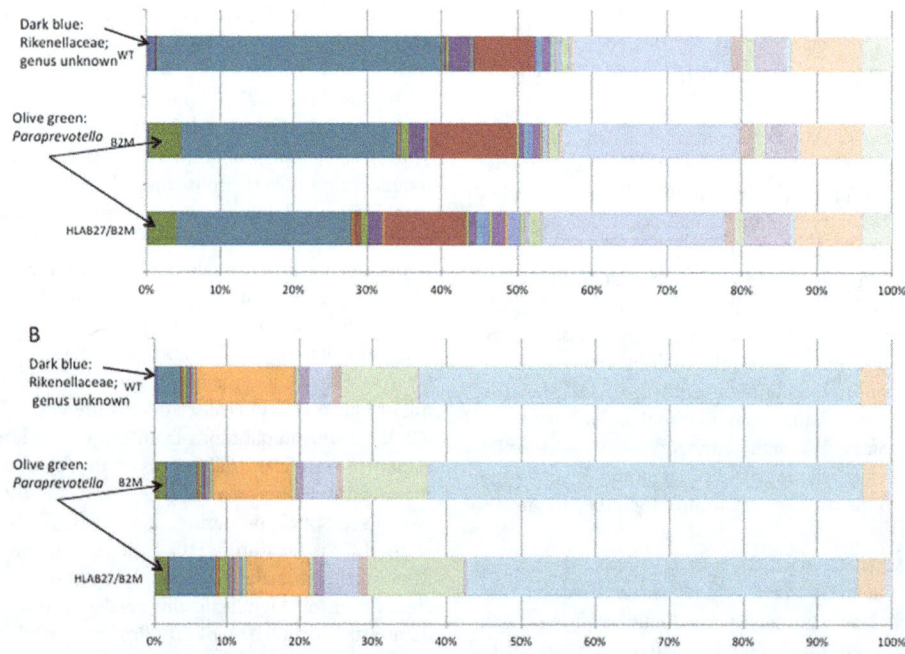

Figure 2. Genus level analysis differences from A, cecal lumen specimens, and B, cecal mucosal samples from cohort 3.

Table 3. HLA-B27 transgenic animals have significant alterations in gut microbiota by 16s rDNA sequencing.

	Animal group comparison	Permanova analysis, p value	n from each group
Co-housed animals	Whole cecum (WT vs B27/B2M)*	1.9066, 0.006	6 vs 7
	Cecal mucosa (WT vs B27/B2M)	1.2121, 0.1688	6 vs 6
	Cecal lumen (WT vs B27/B2M)	1.0113, 0.4266	6 vs 6
Non-co-housed animals	Cecal lumen (WT vs B27/B2M vs B2M)*	3.4031, 0.001	7 vs 8 vs 7
	Cecal lumen (WT vs B27/B2M)*	4.7903, 0.001	7 vs 8
	Cecal lumen (WT vs B2M)*	5.1028, 0.001	7 vs 7
	Cecal lumen (B27/B2M vs B2M)	0.76438, 0.7652	8 vs 7
	Cecal mucosa (WT vs B27/B2M vs B2M)*	2.4418, 0.001	7 vs 8 vs 7
	Cecal mucosa (WT vs B27/B2M)*	3.1632, 0.002	7 vs 8
	Cecal mucosa (WT vs B2M)*	3.0912, 0.001	7 vs 7
	Cecal mucosa (B27/B2M vs B2M)	1.0951, 0.3017	8 vs 7

WT: wild-type Lewis rats; B27/B2M: (21–3×283–2)F1 HLA-B27/hβ2m transgenic Lewis rats; B2M: 283–2 hβ2m expressing transgenic rats; * designates p<0.05.

species from whole cecum preparations of transgenic rats compared to wild type control animals from the first cohort (normalized to amount of host rat DNA), demonstrating an overall difference in the abundance of certain bacterial species. The color code represents the relative abundance of specific bacterial species with each row showing an individual species. Figure 4b quantifies differences between HLA-B27/hβ2m transgenic animals and littermate wild type control animals (co-housed animals). Approximately 30 of these showed significant differences in HLA-B27/β2m transgenic animals compared to wild type animals as determined by iterative Monte Carlo analysis of the dataset. Three of the main differences included alterations in the

abundance of *Faecalibacterium prausnitzii*, *Bacteroides vulgatus*, and *Akkermansia muciniphila*.

Bacteroides vulgatus is more abundant in HLA-B27 transgenic animals than non-co-housed wild type animals

We next attempted to confirm HLA-B27-dependent differences in colonization of these three bacterial species using species-specific real time quantitative PCR. Normalization was performed using 16S rRNA gene PCR. The differences in *F. prausnitzii* could not be confirmed by qPCR. However, we found a marked reduction in *Akkermansia muciniphila* and an increase in

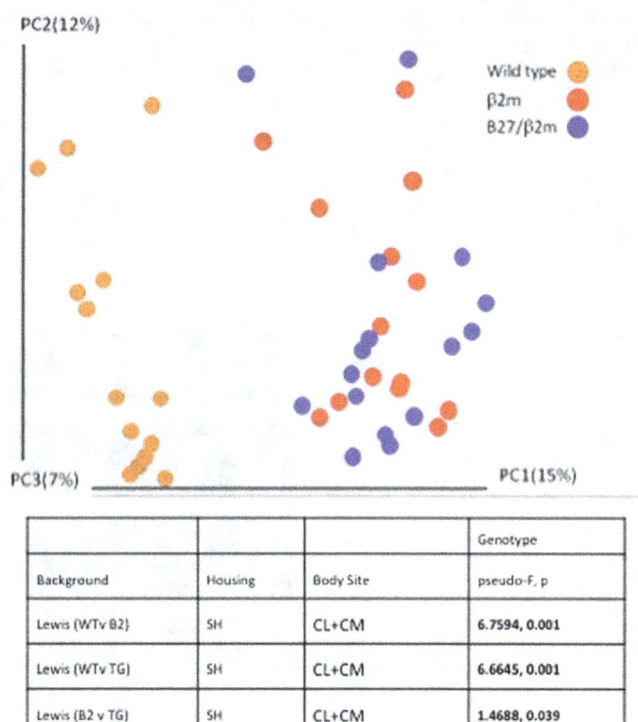

Figure 3. Principal coordinates analysis from combined cecal lumen and cecal mucosa samples from cohort 3.

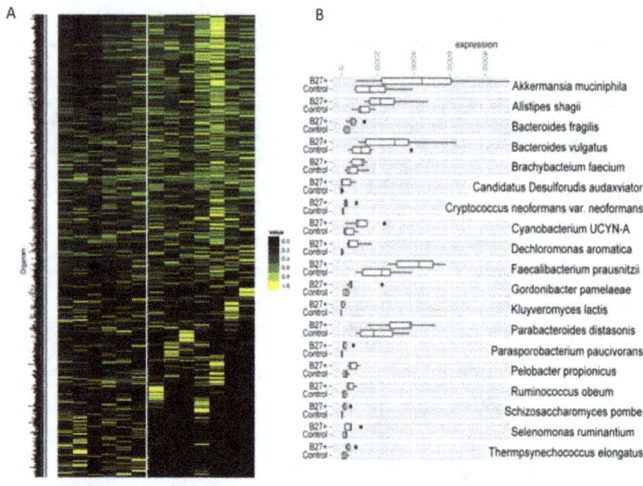

Figure 4. BRISK analysis of whole cecum samples from cohort 1 shows microbial diversity differences on a species level between HLA-B27/hβ2m animals and co-housed wild type animals. A. Each line on the heat map represents a different bacterial species and the color code represents the relative abundance of that organism, with yellow indicating the highest abundance. B, Specific species differences are highlighted, with a black dot indicating statistically significant differences.

Bacteroides vulgatus prevalence in the transgenic HLA-B27/hβ2m animals compared to the wild type rats housed alone (Figure 5). However, the differences were reduced (with levels of these bacteria in wild-type mice comparable to transgenic) when wild type animals were co-housed with the transgenic animals, presumably due to coprophagia. Although there were significant differences in *Bacteroides vulgatus* by BRISK in both co-housed and non-co-housed animals, when confirmatory PCR was performed, a difference was only demonstrated between non-co-housed animals. In cohort 3, *Akkermansia muciniphila* was not present in co-housed and non-co-housed rat cecal samples (neither cecal mucosa nor cecal lumen), suggesting that this species may have colonized this specific group of wild type animals only.

Discussion

To our knowledge this is the first study that shows that expression of the human transgene HLA-B27 (with hβ2m) is associated with alterations in the gut microbiota. The purpose of

this study was to establish the difference seen in the gut microbiota due to HLA-B27, but whether or not specific changes are causative of disease, a result of disease, or unrelated to disease has not yet been established and will require further investigations. Germ-free studies in the 33–3 and 21–4 lines strongly suggest that gut microbiota play an important role in the disease state in this animal model [21–23]. The effect of the germ free state on arthritis in the (21–3×283–2) F1 rats which do not develop colitis remains to be studied.

The rat age in our study (in the first 3 cohorts) was specifically chosen to precede arthritis onset so that the genotype effect could be delineated from the disease effect. At 67–70 days, (21–3×283–2)F1 males show only very localized inflammation in the male genital tract, but no gastrointestinal inflammation or joint inflammation [35]. It is possible that HLA-B27 animals have a distinct T cell selection repertoire resulting in bacterial peptide presentation in a way that affects which microbes are allowed to thrive and/or are eliminated. However, studies that cross a CD8 −/− background to the Lewis 33–3 rat have shown no effect on

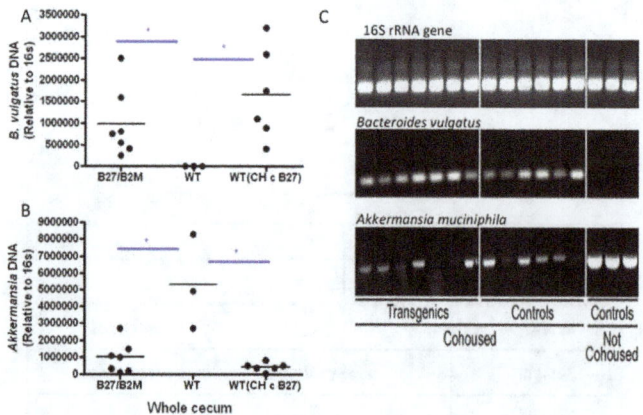

Figure 5. Relative differences in *Bacteroides vulgatus* and *Akkermansia muciniphila* shown by A, quantitative PCR using the 16S rRNA gene as a reference, and B, by traditional PCR using whole cecum samples in cohort 1.

disease prevalence or severity, thus diminishing the validity of this theory [51]. While there is no evidence that the (21–3×283–2)F1animals used in this study have gut inflammation both histologically and by lack of increased cytokine production, the 33–3 rats do have subclinical inflammation from an early age (8 weeks old). Sartor and colleagues have shown that gut inflammation in the Fischer HLA-B27 33–3 transgenic rats which develop clinically-evident inflammatory colitis alters the transcriptional profile of *Bacteroides thetaiotamicron,* which is thought potentially to be a causative organism in colitis, and may result in an augmented adaptive immune response to bacterial antigen [52]. Analogously to the human disease, ankylosing spondylitis, in which 50–60% of patients without GI symptoms have microscopic gut inflammation on histological analyses of biopsy specimens from routine colonoscopy, it is possible that transgenic rats have subclinical gut inflammation that causes changes in the gut microbiota, or alternatively, that transgene expression directly changes the members of the gut microbiota which results in alterations in immunity and/or gut permeability.

We have demonstrated by species-specific sequencing and qPCR confirmation that *Bacteroides vulgatus* is overabundant in the HLA-B27 rat ceca. This organism was found in mono-association studies in germ-free HLA-B27 transgenic Fischer rats to induce colitis [23,53]. While the Lewis transgenic rats used in the first 3 cohorts of our study do not develop colitis, it is possible that *B. vulgatus* is pathogenic in the development of arthritis. Testing this hypothesis would require longitudinal correlation of the presence of *B. vulgatus* with arthritic disease score severity, and/or monoassociation of germ-free rats with *B. vulgatus* to induce arthritis. It is also possible that the overabundance of *B. vulgatus* only reflects more global changes in the microbiota and is not directly related to disease.

We also found that *Paraprevotella* was more abundant in both transgenic animals (HLA-B27/β2m and β2m) by 16s rRNA gene sequencing compared with wild type animals. *Prevotella* species, which are related to *Paraprevotella* have been implicated in the pathogenesis of rheumatoid arthritis in humans [54].

While the components of gut microbiota in different groups of animals can differ depending on the specific micro-environment, as shown by the failure to confirm loss of abundance of *Akkermansia* in our repeat experiments, an overall consistent trend was observed that transgenic animals had different microbiota compared to wild type animals, even in the absence of gut inflammation. Given that individual bacterial species might be inconsistent between groups of transgenic animals, the HLA-B27 disease association might rather be due to broader differences in the metabolic profiles of the bacterial species present, and thus, it is possible that disease and altered immunity may result from the overall milieu of bacterial metabolites interacting with the host rather than an alteration of individual bacterial species.

The transgenic animals that only expressed hβ2m without HLA-B27 exhibited microbiota differences that were more similar to HLA-B27-expressing animals than wild type animals despite being separately housed, although there were some microbiota differences unique to HLA-B27. The hβ2m transgenic animals remain completely healthy in this cohort, while all of HLA-B27/hβ2m rats develop epidiymoorchitis and a majority develop arthritis. One potential explanation for these results could be that the changes in microbiota are at least in part due to the expression of hβ2m rather than HLA-B27, but these alterations are not pathogenic unless HLA-B27 is also present, causing disease either by effecting additional changes in the microbiome, and/or by contributing its effect(s) on the immune response. Another potential explanation is that human β2m may form a functional

complex with rat MHC class I heavy chains, provoking a change in the microbiota that by itself is insufficient to cause disease. There is evidence that overexpression of human HLA-DR*0401 and* 0402, HLA-DR4 subtypes that cause mice to be susceptible or resistant to arthritis, respectively, results in genotype-specific alterations in the gut microbiota [55]. We also show in this study that another MHC I molecule, HLA-B7, can be associated with changes in the gut microbiota differentially from both wild type and HLA-B27 animals. Overall, our results support the concept that changes in MHC expression influence the composition of the gut microbiota. Effects of HLA-B27 that cause disease could turn out to be an unusual specific case within this general principle.

One potential limitation to our study is that random drift, as well as seasonal changes, can contribute to differences detected in the gut microbiome. However, because we compared the microbiota between genotypic groups harvested at the same time, this should not be a large factor in the outcome of our analyses. For instance, within cohort 1, we harvested all genotypes at the same time (as within cohort 2 and 3). Since we have not attempted to pool the three cohorts (which were indeed euthanized at different times) for analyses, random drift should be much less of a factor. It should also be noted that B27+/hβ2m+ rats were littermates of B27-/ hβ2m+ rats in that particular cohort (cohort 3), so these two groups should have the same founder microbiome and differences seen between these two genotypes would argue against drift being the main factor. In terms of differences due to different founder microbiomes, this is theoretically a problem with the wild type rats even though husbandry was similar for all rats at a particular time. While this does not necessarily mitigate the founder effect, we have at least demonstrated that in co-housed rats some genotypic differences exist.

Sequencing techniques have advanced significantly in the last decade to allow for more rapid and in-depth methods of identifying, in a culture-independent fashion, the members of a complex microbial community such as the gut. It is important to note that there are strengths and limitations to the two methods used here. For instance, while the 16s rRNA gene sequencing method can provide better diversity comparisons between two groups in a cohort using principal coordinates analysis, detailed species-specific information was not readily available with the taxa we used in our analysis.It is sensitive to dramatic changes in gut microbiota such as occur with overt clinical gut inflammation, but may fail to find identify potentially crucial differences on a species level. On the other hand, while the BRISK technique is able to identify species-level differences, its sensitivity is limited by identification of specific tags within the database. While 16S sequences are known for many bacteria, BRiSK can generally only identify organisms for which full genome sequences are available. Potentially pathogenic organisms may be among the species that have not been identified or described. Despite these limitations, we have identified substantial microbial biodiversity differences in the gut between wild type and transgenic rats. These findings can potentially direct us to better understand immune-mediated diseases associated with HLA-B27.

Supporting Information

Figure S1   16S rRNA gene sequencing principal coordinates analysis demonstrates significant differences between HLA type in Lewis rats between 2–6 months of age from A, cecal lumen and B, cecal mucosa. B27/β2M: HLA-B27/human β2-microglobulin; B7/β2M: HLA-B7/human β2-microglobulin. Cecal mucosal and lumenal contents from rats in various cohorts were collected using sterile swabs and frozen in sterile 15 ml centrifuge tubes. Samples were shipped to OHSU in dry ice, where genomic DNA was extracted using Qiagen DNAeasy kit and sequenced using 16S rRNA gene sequencing. Control cohorts were handled as follows: 2 month old rats: Weaned at 21 days after birth and singly housed until 2 months of age and sacrificed. 3–4 month old rats: Weaned at 21 days and cohoused with litter mates (random transgenic and wild type) usually 2–3 rats per cage until almost 3 months of age when they were sacrificed. 6 month old rats: Weaned at 21 days and cohoused with litter mates (random transgenic and WT) usually 2–3 rats per cage until almost 3 months of age and then singly housed. Some of these animals were cohoused for mating for a few weeks intermittently.

Acknowledgments

We would like to thank the following people for their help and insight in the preparation of the data and statistical analysis: Christina Metea, Dongseok Choi, Stuart Gardiner, Steve Planck, and Daniel McDonald.

Author Contributions

Conceived and designed the experiments: PL MA MB JDT RK RVG RAC TG JTR. Performed the experiments: PL MA MB AYL LA PS SD YP MD CLL TG GA YVB. Analyzed the data: PL MB MA AYL CLL TG YVB JWD. Contributed reagents/materials/analysis tools: AYL EDC MD JDT RK CLL YVB. Contributed to the writing of the manuscript: PL MB MA RK RVG RAC JTR JWD.

References

1. Atarashi K, Tanoue T, Shima T, Imaoka A, Kuwahara T, et al. (2011) Induction of colonic regulatory T cells by indigenous Clostridium species. Science 331: 337–341.
2. Berer K, Mues M, Koutrolos M, Rasbi ZA, Boziki M, et al. (2011) Commensal microbiota and myelin autoantigen cooperate to trigger autoimmune demyelination. Nature 479: 538–541.
3. Brown CT, Davis-Richardson AG, Giongo A, Gano KA, Crabb DB, et al. (2011) Gut microbiome metagenomics analysis suggests a functional model for the development of autoimmunity for type 1 diabetes. PLoS One 6: e25792.
4. Elinav E, Strowig T, Kau AL, Henao-Mejia J, Thaiss CA, et al. (2011) NLRP6 inflammasome regulates colonic microbial ecology and risk for colitis. Cell 145: 745–757.
5. Kinross JM, Darzi AW, Nicholson JK (2011) Gut microbiome-host interactions in health and disease. Genome Med 3: 14.
6. Scher JU, Abramson SB (2011) The microbiome and rheumatoid arthritis. Nat Rev Rheumatol 7: 569–578.
7. Vijay-Kumar M, Aitken JD, Carvalho FA, Cullender TC, Mwangi S, et al. (2010) Metabolic syndrome and altered gut microbiota in mice lacking Toll-like receptor 5. Science 328: 228–231.
8. Gough SC, Simmonds MJ (2007) The HLA Region and Autoimmune Disease: Associations and Mechanisms of Action. Curr Genomics 8: 453–465.
9. Colbert RA, DeLay ML, Klenk EI, Layh-Schmitt G (2010) From HLA-B27 to spondyloarthritis: a journey through the ER. Immunol Rev 233: 181–202.
10. McHugh K, Bowness P (2012) The link between HLA-B27 and SpA – new ideas on an old problem. Rheumatology (Oxford) 51: 1529–1539.
11. Rosenbaum JT, Davey MP (2011) Time for a gut check: evidence for the hypothesis that HLA-B27 predisposes to ankylosing spondylitis by altering the microbiome. Arthritis Rheum 63: 3195–3198.
12. Biagi E, Nylund L, Candela M, Ostan R, Bucci L, et al. (2010) Through ageing, and beyond: gut microbiota and inflammatory status in seniors and centenarians. PLoS One 5: e10667.
13. Schwimmbeck PL, Yu DT, Oldstone MB (1987) Autoantibodies to HLA B27 in the sera of HLA B27 patients with ankylosing spondylitis and Reiter's syndrome. Molecular mimicry with Klebsiella pneumoniae as potential mechanism of autoimmune disease. J Exp Med 166: 173–181.
14. Scofield RH, Warren WL, Koelsch G, Harley JB (1993) A hypothesis for the HLA-B27 immune dysregulation in spondyloarthropathy: contributions from enteric organisms, B27 structure, peptides bound by B27, and convergent evolution. Proc Natl Acad Sci U S A 90: 9330–9334.
15. van Bohemen CG, Nabbe AJ, Landheer JE, Grumet FC, Mazurkiewicz ES, et al. (1986) HLA-B27M1M2 and high immune responsiveness to Shigella flexneri in post-dysenteric arthritis. Immunol Lett 13: 71–74.
16. Yu D, Kuipers JG (2003) Role of bacteria and HLA-B27 in the pathogenesis of reactive arthritis. Rheum Dis Clin North Am 29: 21–36, v–vi.
17. Lahesmaa R, Skurnik M, Toivanen P (1993) Molecular mimicry: any role in the pathogenesis of spondyloarthropathies? Immunol Res 12: 193–208.
18. Mertz AK, Wu P, Sturniolo T, Stoll D, Rudwaleit M, et al. (2000) Multispecific CD4+ T cell response to a single 12-mer epitope of the immunodominant heat-shock protein 60 of Yersinia enterocolitica in Yersinia-triggered reactive arthritis: overlap with the B27-restricted CD8 epitope, functional properties, and epitope presentation by multiple DR alleles. J Immunol 164: 1529–1537.
19. de Vries DD, Dekker-Saeys AJ, Gyodi E, Bohm U, Ivanyi P (1992) Absence of autoantibodies to peptides shared by HLA-B27.5 and Klebsiella pneumoniae nitrogenase in serum samples from HLA-B27 positive patients with ankylosing spondylitis and Reiter's syndrome. Ann Rheum Dis 51: 783–789.
20. Ewing C, Ebringer R, Tribbick G, Geysen HM (1990) Antibody activity in ankylosing spondylitis sera to two sites on HLA B27.1 at the MHC groove region (within sequence 65–85), and to a Klebsiella pneumoniae nitrogenase reductase peptide (within sequence 181–199). J Exp Med 171: 1635–1647.
21. Taurog JD, Richardson JA, Croft JT, Simmons WA, Zhou M, et al. (1994) The germfree state prevents development of gut and joint inflammatory disease in HLA-B27 transgenic rats. J Exp Med 180: 2359–2364.
22. Hoentjen F, Tonkonogy SL, Qian BF, Liu B, Dieleman LA, et al. (2007) CD4(+) T lymphocytes mediate colitis in HLA-B27 transgenic rats monoassociated with nonpathogenic Bacteroides vulgatus. Inflamm Bowel Dis 13: 317–324.
23. Rath HC, Wilson KH, Sartor RB (1999) Differential induction of colitis and gastritis in HLA-B27 transgenic rats selectively colonized with Bacteroides vulgatus or Escherichia coli. Infect Immun 67: 2969–2974.
24. Laitio P, Virtala M, Salmi M, Pelliniemi LJ, Yu DT, et al. (1997) HLA-B27 modulates intracellular survival of Salmonella enteritidis in human monocytic cells. Eur J Immunol 27: 1331–1338.
25. Penttinen MA, Heiskanen KM, Mohapatra R, DeLay ML, Colbert RA, et al. (2004) Enhanced intracellular replication of Salmonella enteritidis in HLA-B27-expressing human monocytic cells: dependency on glutamic acid at position 45 in the B pocket of HLA-B27. Arthritis Rheum 50: 2255–2263.
26. Penttinen MA, Holmberg CI, Sistonen L, Granfors K (2002) HLA-B27 modulates nuclear factor kappaB activation in human monocytic cells exposed to lipopolysaccharide. Arthritis Rheum 46: 2172–2180.
27. Ruuska M, Sahlberg AS, Granfors K, Penttinen MA (2013) Phosphorylation of STAT-1 serine 727 is prolonged in HLA-B27-expressing human monocytic cells. PLoS One 8: e50684.
28. Sahlberg AS, Penttinen MA, Heiskanen KM, Colbert RA, Sistonen L, et al. (2007) Evidence that the p38 MAP kinase pathway is dysregulated in HLA-B27-expressing human monocytic cells: correlation with HLA-B27 misfolding. Arthritis Rheum 56: 2652–2662.
29. Ruuska M, Sahlberg AS, Colbert RA, Granfors K, Penttinen MA (2012) Enhanced phosphorylation of STAT-1 is dependent on double-stranded RNA-dependent protein kinase signaling in HLA-B27-expressing U937 monocytic cells. Arthritis Rheum 64: 772–777.
30. Dhaenens M, Fert I, Glatigny S, Haerinck S, Poulain C, et al. (2009) Dendritic cells from spondylarthritis-prone HLA-B27-transgenic rats display altered cytoskeletal dynamics, class II major histocompatibility complex expression, and viability. Arthritis Rheum 60: 2622–2632.
31. Fert I, Glatigny S, Poulain C, Satumtira N, Dorris ML, et al. (2008) Correlation between dendritic cell functional defect and spondylarthritis phenotypes in HLA-B27/HUMAN beta2-microglobulin-transgenic rat lines. Arthritis Rheum 58: 3425–3429.
32. Utriainen L, Firmin D, Wright P, Cerovic V, Breban M, et al. (2012) Expression of HLA-B27 causes loss of migratory dendritic cells in a rat model of spondylarthritis. Arthritis Rheum 64: 3199–3209.
33. Tran TM, Dorris ML, Satumtira N, Richardson JA, Hammer RE, et al. (2006) Additional human beta2-microglobulin curbs HLA-B27 misfolding and promotes arthritis and spondylitis without colitis in male HLA-B27-transgenic rats. Arthritis Rheum 54: 1317–1327.
34. Hammer RE, Maika SD, Richardson JA, Tang JP, Taurog JD (1990) Spontaneous inflammatory disease in transgenic rats expressing HLA-B27 and human beta 2m: an animal model of HLA-B27-associated human disorders. Cell 63: 1099–1112.
35. Taurog JD, Rival C, van Duivenvoorde LM, Satumtira N, Dorris ML, et al. (2012) Autoimmune epididymoorchitis is essential to the pathogenesis of male-specific spondylarthritis in HLA-B27-transgenic rats. Arthritis Rheum 64: 2518–2528.
36. Taurog JD, Dorris ML, Satumtira N, Tran TM, Sharma R, et al. (2009) Spondylarthritis in HLA-B27/human beta2-microglobulin-transgenic rats is not prevented by lack of CD8. Arthritis Rheum 60: 1977–1984.

37. Muthappan V, Lee AY, Lamprecht TL, Akileswaran L, Dintzis SM, et al. (2011) Biome representational in silico karyotyping. Genome Res 21: 626–633.

38. Caporaso JG, Lauber CL, Walters WA, Berg-Lyons D, Huntley J, et al. (2012) Ultra-high-throughput microbial community analysis on the Illumina HiSeq and MiSeq platforms. ISME J 6: 1621–1624.

39. Caporaso JG, Lauber CL, Walters WA, Berg-Lyons D, Lozupone CA, et al. (2011) Global patterns of 16S rRNA diversity at a depth of millions of sequences per sample. Proc Natl Acad Sci U S A 108 Suppl 1: 4516–4522.

40. Soergel DA, Dey N, Knight R, Brenner SE (2012) Selection of primers for optimal taxonomic classification of environmental 16S rRNA gene sequences. ISME J 6: 1440–1444.

41. Werner JJ, Zhou D, Caporaso JG, Knight R, Angenent LT (2012) Comparison of Illumina paired-end and single-direction sequencing for microbial 16S rRNA gene amplicon surveys. ISME J 6: 1273–1276.

42. Yatsunenko T, Rey FE, Manary MJ, Trehan I, Dominguez-Bello MG, et al. (2012) Human gut microbiome viewed across age and geography. Nature 486: 222–227.

43. Nawrocki EP, Kolbe DL, Eddy SR (2009) Infernal 1.0: inference of RNA alignments. Bioinformatics 25: 1335–1337.

44. Edgar RC (2010) Search and clustering orders of magnitude faster than BLAST. Bioinformatics 26: 2460–2461.

45. Wang Q, Garrity GM, Tiedje JM, Cole JR (2007) Naive Bayesian classifier for rapid assignment of rRNA sequences into the new bacterial taxonomy. Appl Environ Microbiol 73: 5261–5267.

46. Carvalho FA, Nalbantoglu I, Ortega-Fernandez S, Aitken JD, Su Y, et al. (2012) Interleukin-1beta (IL-1beta) promotes susceptibility of Toll-like receptor 5 (TLR5) deficient mice to colitis. Gut 61: 373–384.

47. Gonzalez A, King A, Robeson MS 2nd, Song S, Shade A, et al. (2012) Characterizing microbial communities through space and time. Curr Opin Biotechnol 23: 431–436.

48. Knights D, Parfrey LW, Zaneveld J, Lozupone C, Knight R (2011) Human-associated microbial signatures: examining their predictive value. Cell Host Microbe 10: 292–296.

49. Navidshad B, Liang JB, Jahromi MF (2012) Correlation coefficients between different methods of expressing bacterial quantification using real time PCR. Int J Mol Sci 13: 2119–2132.

50. Lozupone C, Knight R (2005) UniFrac: a new phylogenetic method for comparing microbial communities. Appl Environ Microbiol 71: 8228–8235.

51. May E, Dorris ML, Satumtira N, Iqbal I, Rehman MI, et al. (2003) CD8 alpha beta T cells are not essential to the pathogenesis of arthritis or colitis in HLA-B27 transgenic rats. J Immunol 170: 1099–1105.

52. Hansen JJ, Huang Y, Peterson DA, Goeser L, Fan TJ, et al. (2012) The colitis-associated transcriptional profile of commensal Bacteroides thetaiotaomicron enhances adaptive immune responses to a bacterial antigen. PLoS One 7: e42645.

53. Hoentjen F, Welling GW, Harmsen HJ, Zhang X, Snart J, et al. (2005) Reduction of colitis by prebiotics in HLA-B27 transgenic rats is associated with microflora changes and immunomodulation. Inflamm Bowel Dis 11: 977–985.

54. Scher JU, Sczesnak A, Longman RS, Segata N, Ubeda C, et al. (2013) Expansion of intestinal Prevotella copri correlates with enhanced susceptibility to arthritis. Elife 2: e01202.

55. Luckey D, Gomez A, Murray J, White B, Taneja V (2013) Bugs & us: The role of the gut in autoimmunity. Indian J Med Res 138: 732–743.

2

Inducible Resistance to Maize Streak Virus

Dionne N. Shepherd[1]*, Benjamin Dugdale[4], Darren P. Martin[2,3], Arvind Varsani[5,6,7], Francisco M. Lakay[1], Marion E. Bezuidenhout[1], Adérito L. Monjane[1,2], Jennifer A. Thomson[1], James Dale[4], Edward P. Rybicki[1,2]

1 Department of Molecular and Cell Biology, University of Cape Town, Rondebosch, Cape Town, South Africa, 2 Institute of Infectious Disease and Molecular Medicine, University of Cape Town, Observatory, Cape Town, South Africa, 3 Centre for High-Performance Computing, Rosebank, Cape Town, South Africa, 4 Centre for Tropical Crops and Biocommodities, Queensland University of Technology (QUT), Brisbane, Queensland, Australia, 5 School of Biological Sciences and Biomolecular Interaction Centre, University of Canterbury, Christchurch, New Zealand, 6 Department of Plant Pathology and Emerging Pathogens Institute, University of Florida, Gainesville, Florida, United States of America, 7 Electron Microscope Unit, Division of Medical Biochemistry, Department of Clinical Laboratory Sciences, University of Cape Town, Observatory, Cape Town, South Africa

Abstract

Maize streak virus (MSV), which causes maize streak disease (MSD), is the major viral pathogenic constraint on maize production in Africa. Type member of the *Mastrevirus* genus in the family *Geminiviridae*, MSV has a 2.7 kb, single-stranded circular DNA genome encoding a coat protein, movement protein, and the two replication-associated proteins Rep and RepA. While we have previously developed MSV-resistant transgenic maize lines constitutively expressing "dominant negative mutant" versions of the MSV Rep, the only transgenes we could use were those that caused no developmental defects during the regeneration of plants in tissue culture. A better transgene expression system would be an inducible one, where resistance-conferring transgenes are expressed only in MSV-infected cells. However, most known inducible transgene expression systems are hampered by background or "leaky" expression in the absence of the inducer. Here we describe an adaptation of the recently developed INPACT system to express MSV-derived resistance genes in cell culture. Split gene cassette constructs (SGCs) were developed containing three different transgenes in combination with three different promoter sequences. In each SGC, the transgene was split such that it would be translatable only in the presence of an infecting MSV's replication associated protein. We used a quantitative real-time PCR assay to show that one of these SGCs (pSPLITrep^III-Rb-Ubi) inducibly inhibits MSV replication as efficiently as does a constitutively expressed transgene that has previously proven effective in protecting transgenic maize from MSV. In addition, in our cell-culture based assay pSPLITrep^III-Rb-Ubi inhibited replication of diverse MSV strains, and even, albeit to a lesser extent, of a different mastrevirus species. The application of this new technology to MSV resistance in maize could allow a better, more acceptable product.

Editor: Baochuan Lin, Naval Research Laboratory, United States of America

Funding: Funding was provided by PANNAR (Pty) LTD, http://www.pannar.com, to DNS, EPR and JAT. The funders had no role in study design, data collection and analysis, decision to publish, or preparation of the manuscript.

* Email: Dionne.shepherd@gmail.com

Introduction

During the past decade a great deal of effort has been spent on the development of crops with transgenic resistance against a number of different economically-important pathogenic single-stranded DNA (ssDNA) viruses in the family *Geminiviridae* [1–4]. Whereas much of the early work focused on pathogen-derived resistance approaches involving the expression of virus-derived genes in plants (see [3,4] for reviews), more recent innovations have seen the application of interfering peptides such as recombinant peptide aptamers [5,6] and zinc finger proteins [7]. All of these approaches have relied on constitutive expression of recombinant proteins, which can have several drawbacks: (1) constitutive expression of resistance genes is redundant when no viral infection occurs and will add unnecessarily to the metabolic load of uninfected transgenic plants; (2) constitutively expressed genes are more likely to be targeted for transgene silencing than inducible genes; (3) constitutive expression limits the types of

transgene that can be used to those whose expression is not detrimental or toxic to plant cells. This last point is particularly pertinent since plants are usually transformed as cells or immature embryos in tissue culture and the expression of toxic gene products can therefore inhibit the regeneration of whole plants.

One way to overcome these problems would be to either delay expression of transgenes until plants have regenerated fully, or, in the case of virus resistance, to make transgene expression inducible only upon viral infection. This has been attempted previously for the geminivirus-induced expression of the cytotoxic ribosome inactivating protein dianthin [8] and the ribonuclease barnase from *Bacillus amyloliquefaciens* [4,9,10]. Both of these proteins are lethal when expressed in plant cells, and therefore can be used to mimic innate hypersensitivity responses to virus infection. However, because of their toxicity such genes need to be "switched off" in the absence of virus infections. In the case of barnase, this was achieved by co-expressing the extracellular barnase with its intracellular inhibitor barstar, which then bind to each other with high affinity [11,12]. If produced at similar levels, barstar inhibits the expression of barnase, resulting in no RNase production. By placing barnase under control of a viral promoter that is activated upon viral infection, and barstar under a viral promoter that is repressed upon viral infection, Zhang et al. [10] surmised that over-expression of barnase relative to barstar would kill virus infected cells, thus preventing further virus spread. The strategy attempted by Hong et al. [8] to express dianthin was a similar one: the gene was placed under control of a viral promoter that is activated by the begomoviral transcriptional activator protein (TrAP).

Despite being promising options for inducible transgene expression, these strategies have certain drawbacks. With both dianthin and barnase, "leaky" or low-level basal expression from the viral promoter occurs in the absence of the viral TrAP ([8,10]. In addition, Hussain et al. [13] have shown with Tomato leaf curl New Delhi virus (ToLCNDV) that the hypersensitive response naturally triggered in *Nicotiana tabacum* and *Lycopersicon esculentum* plants by the ToLCNDV nuclear shuttle protein (NSP) [14] is suppressed by TrAP. If other geminiviruses encode similar anti-hypersensitive response factors it may undermine cell death-inducing resistance mechanisms.

Maize streak disease (MSD), caused by the geminivirus species *Maize streak virus* (genus *Mastrevirus*), results in substantial maize yield reductions throughout sub-Saharan Africa and in some years can cause regional maize crop failures [15]. Throughout the African continent the development of MSD-resistant maize varieties is therefore a prime objective for both conventional maize breeders and biotechnologists. While we have had success in using a constitutively expressed "dominant negative" mutated and truncated replication associated protein (*rep*) transgene to provide resistance to MSV in maize (*rep*$^{1-219Rb-}$ [16]), subsequent research has indicated that far greater degrees of MSV resistance are potentially achievable. In our initial screen of a range of *rep*-derived transgenes, first in a transient expression assay using maize suspension cells, and second in the model plant *Digitaria sanguinalis* [17], we found that a full-length *rep* gene containing mutations in the rolling circle replication (RCR) motif III and retinoblastoma related protein binding domain, pRBR (*rep*$^{III-Rb-}$; Fig. 1A) provided much better resistance against MSV than the truncated version of this gene (all challenged plants were immune); however, we did not progress with this construct because its constitutive expression also led to stunting and infertility in transgenic plants (Fig. 1B).

For MSV-inducible expression of the *rep*-derived transgenes, we developed constructs called "split gene cassettes" (SGCs), based on a novel protein production platform known as INPACT (In Plant Activation [18,19]). These cassettes are arranged such that the gene of interest is split into two exons and the transgene cannot be expressed in the absence of the MSV Rep. Each SGC (Fig. 2) is flanked by two virus-derived long intergenic regions (LIRs), which contain the virion-sense strand origin of replication and Rep binding and nicking sites [20–24], which are in turn embedded within a small synthetic intron termed a syntron [18,19]. The cassettes also include the mastreviral short intergenic region (SIR) which contains the origin of complementary-strand synthesis [25–27]. Upon viral infection the integrated cassette serves as a template for RCR, allowing replicative release [24] and amplification of circular ssDNA forms. Conversion to the dsDNA intermediate form occurs via the SIR using host cell machinery, after which the transgene is transcribed. Removal of the LIR-containing syntron during mRNA processing results in the reconstitution of a translatable in-frame transcript of the gene of interest (Fig. 2).

Replicative release of the integrated construct from the plant genome relies on the specific DNA nicking and joining activities of the MSV Rep, which recognises and binds to sequence-specific repeats known as iterons in the LIR. Because MSD is caused by only one maize-adapted strain, MSV-A [28], Rep-iteron specificity should not be a drawback in terms of obtaining broad resistance to MSD, but will provide an advantage in that functional proteins should only be produced in the presence of mastrevirus Reps that are sufficiently similar to that of MSV-A. This may overcome the problems associated with leaky inducible promoters reported with other systems.

Here we use a cell-culture based assay to demonstrate that, in addition to this inducible transgene expression system being capable of providing particularly high degrees of resistance against MSV-A, it could also potentially provide transgenic maize with broad, albeit less potent, resistance both against diverse grass-adapted MSV strains and other African mastrevirus species such as *Panicum streak virus* (PanSV).

Materials and Methods

Construct Design

Truncation of the MSV Long Intergenic Region and Assaying for Cryptic Splice Sites. The first step in designing the SGCs was to truncate the 3' terminus of the MSV LIR by 70 bp to remove the virion (V) sense promoter region, thus avoiding the possibility of trans-activation of the V-sense promoter and unwanted transcript expression. Primers were designed to amplify a 5'-terminal 241-bp sequence stretch from the LIR of MSV-A4 [ZA-Kom-1989] ([29]; GenBank accession no. AF003952); hereafter referred to as MSV-Kom. This region contains the minimum LIR sequence required for RCR, as determined by Willment et al. [30], and consists of a stem-loop structure and nicking site essential for the initiation of RCR by Rep [22,31], as well as iterons for Rep-binding [32–36]. PacI and SwaI restriction enzyme (RE) sites were incorporated at the 5' terminus of the forward and reverse primers respectively to flank the amplified product for future cloning (Table 1). The PCR product was ligated with pGEMT-Easy (Promega) and sequenced at Macrogen Inc., Korea.

The second step was to test the MSV LIR^{241} sequence for potential intron splice sites, which may cause problems during processing of the functional mRNA when the construct is replicationally released by the viral Rep. To do this, PCR-amplified LIR^{241} was embedded within a synthetic intron (syntron) developed at Queensland University of Technology [18,19]. The LIR-containing syntron was in turn embedded within the GUS reporter gene coding region of a pUC19-based expression cassette (CaMV35S-promoter>GUS>CaMV35S-terminator), thus splitting the coding region into two exons and creating vector p35S-GSLIR^{241} (Fig. 3A). After bombardment of p35S-GSLIR^{241} into Black Mexican sweet (BMS) maize suspension cells using a Bio-Rad PDS-1000/He particle gun (following the methodology of Shepherd et al. [37]), GUS expression was compared with a control construct containing the syntron with no embedded LIR (p35S-GS; Fig. 3A). This was to determine if p35S-GSLIR^{241} expressed the same or similar level of GUS as did the p35S-GS control vector. Lower expression could mean there is a cryptic 3'-terminal splice site in the LIR that interferes with syntron splicing and subsequent GUS translation, while similar expression would indicate no such problem.

Crude protein was extracted from bombarded BMS cells using the GUS extraction buffer from the Marker Gene Technologies (MGT) β-Glucuronidase (GUS) Reporter Gene Activity Detection

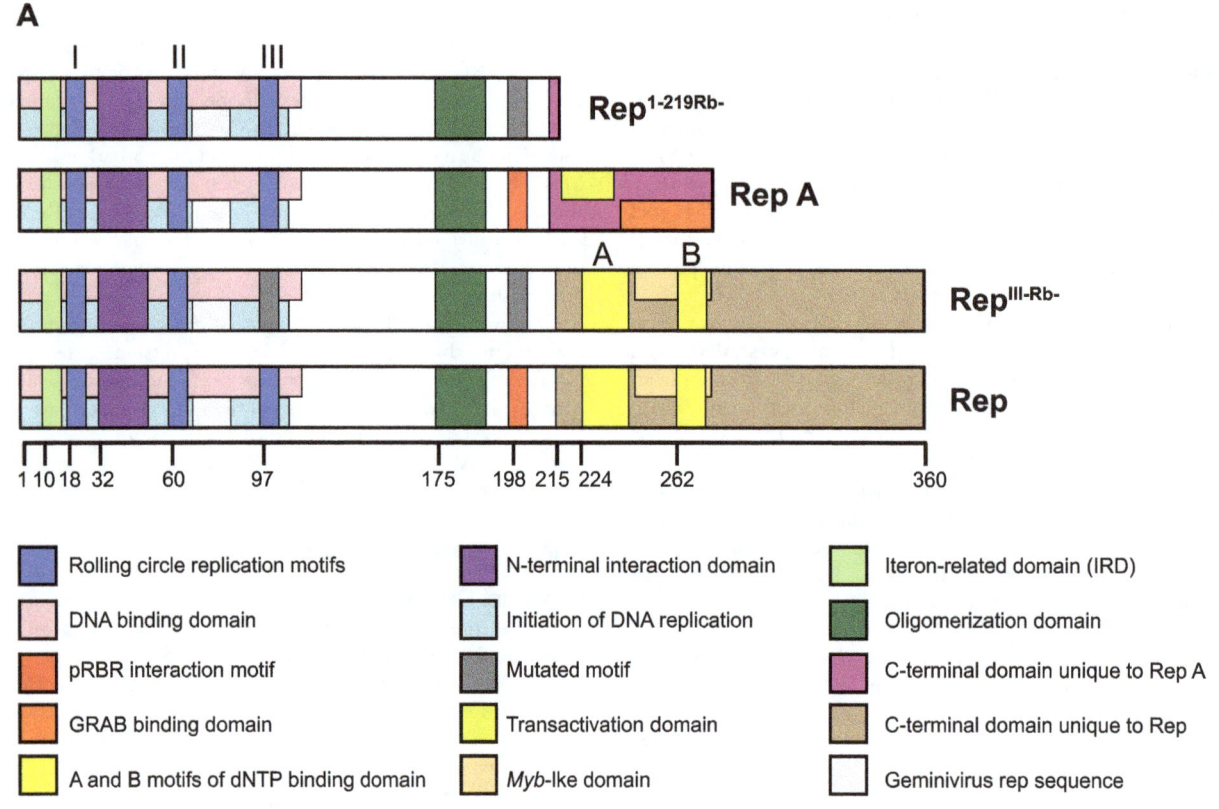

Figure 1. Products of mutated and truncated MSV *rep* genes used in the split gene cassettes, compared with the wild type. A) Known sequence motifs and functional domains present in each gene product are highlighted. Amino acid numbering is relative to the N-terminal methionine. Adapted from Shepherd et al. [17]. B) Three representative *Digitaria sanguinalis* lines constitutively expressing p*rep*[1–219Rb-] (left), p*rep*[III-Rb-] (middle) or Gus and Bar (from pAHC25 [39]; right), illustrate the phenotypic effects of the transgenes. Photo from Shepherd et al. [17].

Kit according to the instruction manual protocol: (http://search.cosmobio.co.jp /cosmo_search_p/search_gate2/docs/MGT_/M0877.20080313.pdf).

Protein in these crude extracts was quantified using the BioRad Protein Assay kit (http://labs.fhcrc.org/fero/Protocols/BioRad_Bradford.pdf) and each sample was diluted to a concentration of 2 mg/ml.

GUS activity was measured using the above-mentioned MGT reporter gene kit according to the kit instructions. The fluorogenic substrate, methylumbelliferyl b-D-glucuronide (4-MUG), was used at a final molarity of 0.04 mM (40 μl of 0.1 mM 4-MUG in 100 μl total volume); while the final concentration of each protein extract (six samples bombarded with p35S-GS; six samples bombarded with p35S-GSLIR[241], and one non-bombarded BMS control sample) was 0.2 mg/ml (10 μl of 2 mg/ml extract in a total

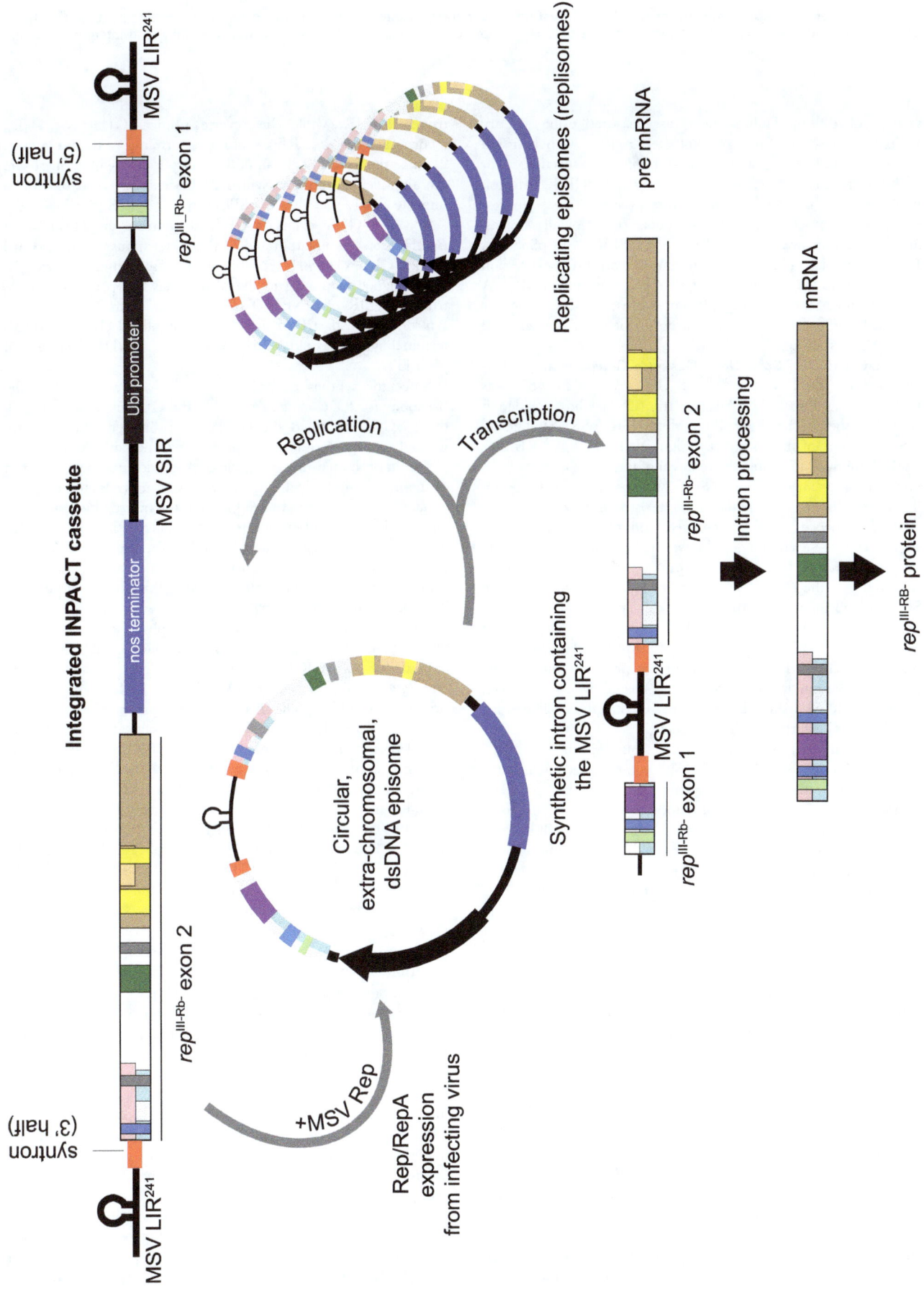

Figure 2. Schematic representation of the INPACT system. MSV-inducible expression from a "split gene cassette" using pSPLITrep$^{\text{III-Rb-}}$-Ubi is used as an example. NosT = nopaline synthase terminator; UbiP = maize ubiquitin promoter. MSVLIR^{241}– = truncated MSV long intergenic region.

volume of 100 µl). Fluorescence was measured using a Cary Eclipse Fluorescence Spectrophotometer (Agilent) with emission and excitation filters set at 455 nm and 365 nm respectively. For each test sample, three replicates and two blanks (GUS extraction buffer in place of protein extract) were assayed. "Test" fluorescence was subtracted from "blank" fluorescence for all samples, and then a ratio was calculated of p35S-GSLIR^{241} to p35S-GS. Ratios below 1 would indicate interference with GUS expression possibly due to the presence of cryptic splice sites in the LIR^{241}. A Mann Whitney test (GraphPad Prism) was used to determine any significant differences in GUS expression between the test and control constructs.

Construction of Split Gene Cassette Constructs. A full-length SGC (pSPLITrep$^{1-219\text{Rb-}}$-35S; Fig. 4A and Fig. S1) was synthesised at Epoch Life Science Inc (USA) who provided it cloned in the SmaI site of pBluescript II SK (pSK; Stratagene, USA). As part of the construct design, NotI and KpnI RE sites flanked the SGC to enable the removal of the entire cassette from pSK (Fig. 4A). The synthesised SGC was designed such that each feature or "module" (e.g. the promoter, terminator, exon 1 or exon 2 sequences) can be removed and replaced with other sequences by restriction digest. However, for downstream cloning purposes some RE sites in the pSK multiple cloning site had to be removed (e.g. the BamHI site, which needed to be unique to the SGC for subcloning of both the promoter and exon 1 sequences; see Fig. 4A). This was achieved by removing the SmaI-cloned SPLITrep$^{1-219\text{Rb-}}$-35S cassette with NotI/KpnI and re-cloning it into the NotI/KpnI sites of pSK, in the process removing the portion of the multiple cloning site that was sandwiched between the KpnI and NotI sites. This was then used as the backbone for the cloning of a further eight constructs.

To generate split exon 1 and exon 2 sequences, the rep$^{1-219\text{Rb-}}$-coding region of pSPLITrep$^{1-219\text{Rb-}}$-35S was designed such that it was split at the first AGGC to create exon 1 (ending in AG at position 155/156, with position 1 being the start codon), and exon 2 (beginning with GC at position 157/158) (See Fig. 4B).

For the cloning of the full-length rep$^{\text{III-Rb-}}$ SGC, the exon 2 fused to the 3′-terminal half of the syntron was also synthesised, with SwaI and SpeI RE sites flanking the fragment (Fig. 4B). Thus, the rep$^{1-219\text{Rb-}}$ exon 2 in pSPLITrep$^{1-219\text{Rb-}}$-35S could be replaced with that of rep$^{\text{III-Rb-}}$ using the SwaI/SpeI RE sites. Exon 1 remained the same in both constructs, since both rep$^{1-219\text{Rb-}}$ and rep$^{\text{III-Rb-}}$ share the same 5′-terminal 295 bp.

For the GUS constructs, a previously made GUS-based SGC (pINPACT-GUS; [18]) was used as template for PCR amplification (see Table 1 for primer sequences) of the 3′-terminal syntron/GUS exon 2 and the GUS exon1/5′-terminal syntron. The design of pINPACT-GUS is essentially the same as for the MSV-based SGC shown in Fig. 4A, except that truncated Tobacco yellow dwarf virus (TYDV) LIRs flank the construct, GUS exon 1 and 2 are in place of MSV rep-derived exon 1 and 2, and some of the RE sites flanking each "module" differ. Also, the GUS coding region is split at the first AGGT to create exon 1 (ending in AG at position 231/232, relative to the GUS start codon) and exon 2 (starting with GC at position 233/234).

For amplification of the 3′-terminal syntron/GUS exon 2 from pINPACT-GUS, the forward primer, GUSex2 (F), was designed to anneal to the last 3′-terminal 21 nucleotides of the TYDV

truncated LIR, while the reverse primer, GUSex2SpeI (R), incorporated an SpeI RE site at the 3′ terminus of the GUS exon 2. Since the 3′-terminal half of the syntron starts with a SwaI site, the amplified 3′-terminal syntron/GUS exon 2 could be cloned into the SwaI/SpeI sites of pSPLITrep$^{1-219\text{Rb-}}$-35S (See Fig. 4A).

For the GUS exon1/5′-terminal syntron amplification, a BamHI site was incorporated at the 5′ terminus of the forward primer, GUSex1BamHI (F), while the reverse primer, GUSex1 (R) was designed to anneal to the 5′-terminal 24 bp of the TYDV truncated LIR of pINPACT-GUS. The 5′-terminal half of the syntron ends with a PacI site; thus the amplified GUS exon1/5′-terminal syntron could be cloned into the BamHI/PacI sites of pSPLITrep$^{1-219\text{Rb-}}$-35S.

Since the synthesised pSPLITrep$^{1-219\text{Rb-}}$-35S was used as the backbone for the cloning of rep$^{\text{III-Rb-}}$ and GUS exons 1 and 2, all three SGCs contained the CaMV35S promoter. However, we wanted to test two additional promoter combinations: the maize ubiquitin promoter (ubi-1) complex, which includes the first intron of the maize ubiquitin-1 gene as well as an untranslated exon for enhanced expression in maize [38] and the maize ubi-1 promoter without the exon and intron. Because splicing of the syntron needs to occur in order to fuse exons 1 and 2 of the transgenes, we were uncertain whether the presence of a second intron (within the promoter region) would interfere with this, hence testing an "intronless" ubi-1 promoter.

Both the ubi-1 promoter complex (simply called Ubi) and the ubi-1 promoter without the exon/intron (called UbiΔI) were PCR amplified from pAHC17 [39] with the addition of flanking AscI and BamHI RE sites (See Fig. 4A and Table 1). The same forward primer (UbiAscI [F]), but different reverse primers (UbiBamHI [R] and UbiΔIBamHI [R]) were used for amplification of Ubi and UbiΔI promoters respectively (see Table 1 for primer sequences).

The CaMV35S promoter in rep$^{1-219\text{Rb-}}$, rep$^{\text{III-Rb-}}$ and GUS-based SGCs was then replaced by (a) Ubi and (b) UbiΔI, resulting in a total of nine SGCs. These were called (1) pSPLITrep$^{1-219\text{Rb-}}$-35S; (2) pSPLITrep$^{1-219\text{Rb-}}$-Ubi; (3) pSPLITrep$^{1-219\text{Rb-}}$-UbiΔI; (4) pSPLITrep$^{\text{III-Rb-}}$-35S; (5) pSPLITrep$^{\text{III-Rb-}}$-Ubi; (6) pSPLITrep$^{\text{III-Rb-}}$-UbiΔI; (7) pSPLITGUS35S; (8) pSPLITGUSUbi; (9) pSPLITGUSUbiΔI. While all constructs were tested initially in a qualitative PCR assay, only Ubi- and UbiΔI-containing SGCs were assayed by quantitative PCR.

Inoculation of Maize Suspension Cells

To rapidly assay the effectiveness of the various SGCs in inhibiting MSV replication, maize suspension cells were bombarded with each SGC and a partial dimer (1.1 mer) of the MSV-Kom genome (pKom602; [40]). MSV-Kom, the isolate from which the rep, LIR and SIR sequences in the SGCs were derived, belongs to an MSV-A subtype known as MSV-A$_4$, which is the most prevalent subtype found in South Africa [41].

BMS suspension-cultured cells were subcultured at a 1:3 dilution three days prior to bombardment. Twenty-four hours before bombardment, 1.0 mL packed volume of actively dividing cells was plated onto solid media.

Different combinations of plasmid DNA (described below) were precipitated onto 1 µm gold particles (50 µl of 60 mg/ml gold suspended in 50% glycerol) according to the protocol of Dunder et al. [42], and these were delivered into the plated BMS cells

Table 1. Primer sequences.

Primer name	[1]Sequence (5'-3')
Primers for SGC cloning	
LIR[241]*Pac*I (F)	TTAATTAAGCCGACGACGGAGGTTGAGG
LIR[241]*Swa*I (R)	ATTTAAATCATACAAAGCAGAACCAGGC
GUSex1*Bam*HI (F)	GGATCCATGGTACGTCCTGTAGAAACCCCAACCCG
GUSex1 (R)	GAGTTTCATCGTACGGTACTTGAG
GUSex2 (F)	GTGCGCCGTAGTTTCCTTTAG
GUSex2*Spe*I (R)	ACTAGTTTATTGGAGATCCTCATTGTTTGC
Ubi*Asc*I (F)	GGCGCGCCAAGCTTGCATGCCTGCAGTGCAG
Ubi*Bam*HI (R)	GGATCCTCTAGAGTCGACCTG
UbiΔI*Bam*HI (R)	GGATCCAGAGGGTGTGGAGGGGGTGTCTATTTATTACG
Real-time PCR Primers	
MSV-Kom Rep (F)	TTGGCTGTCAGAGGGATTTC
MSV-Kom Rep (R)	CCCTGGAGTCATTTCCTTCA
MSV-Kom CP (F)	TAAGCGGGTGCCTAAGAAGA
MSV-Kom CP (R)	TGCTGGAGTGTCTGGATTTG
MSV-VW CP (F)	GGGAGATGATTCGAACTGGA
MSV-VW CP (R)	TGCTGGAGTGTCTGGATCTG
MSV-Set CP (F)	AGTTGTGTCATCGCTTCGTG
MSV-Set CP (R)	TGGTGTATCCGAGCCTATCC
PanSV-Kar CP (F)	CCACACCAACGAGACTCTGA
PanSV-Kar CP (R)	CAACCACATGACACCCACTC
Maize18S (F)	CAGGGATCAGCGGTGTTACT
Maize 18S (R)	GGTAAGTTTCCCCGTGTTGA

[1]Underlined letters highlight engineered restriction enzyme (RE) sites (names of the introduced RE sites are incorporated in the primer names).

using the PDS-1000/He Biolistic particle bombardment delivery system (Bio-Rad) using the method of Shepherd et al. [37]. After bombardment, plates were incubated at 25°C in the dark for four days, after which total DNA was extracted from the BMS cells as described [37].

Initially, each SGC was co-bombarded with pKom602 at a 1:1 weight ratio (as in Owor et al., [43]); i.e. 2 µg of each plasmid per 50 µl gold precipitation. Subsequently, only the Ubi- and UbiΔI-containing SGCs were assayed, this time at SGC:pKom602 weight ratios of 1:1 and 5:1 (2 µg of each plasmid for a 1:1 ratio; 2 µg of pKom602 and 10 µg of SGC for a 1:5 ratio). pSPLIT*rep*[III-Rb]-Ubi was then tested against cloned MSV isolates belonging to the B and C strains of MSV: MSV-B1 [ZA-VW-Triticum-1993] and MSV-C [ZA-Mt Edg-Setaria-1988] (Genbank accession numbers AF239960 and AF007881 respectively), hereafter referred to as MSV-VW and MSV-Set, respectively. In addition, the effectiveness of the SGC was tested against a different species of Mastrevirus, PanSV-A [ZA-Kar-1989] (GenBank accession number L39638), hereafter referred to as PanSV-Kar [40,44,45].

In each bombardment experiment, six plates were bombarded with the infectious mastrevirus clone (MSV-Kom, MSV-Set, MSV-VW or PanSV-Kar) alone, six plates were bombarded with each mastrevirus + pSK (empty vector), and nine plates were bombarded with each mastrevirus + SGC. A non-bombarded BMS plate was always included as a negative control. Each of these experiments was repeated at least twice.

Quantitative Realtime PCR

Quantitative real-time PCR (qPCR), using a Rotor gene RG-3000A device (Qiagen, USA) and SYBR Green I (KAPA SYBR FAST qPCR kit, KAPA Biosystems, South Africa), was performed to determine viral titres in bombarded samples four days post-bombardment. Depending on the level of viral DNA in each sample (initially estimated using replicative-form specific conventional PCR as described by Owor et al. [43]), either 10 ng or 50 ng total DNA was used as template. The realtime PCR was carried out essentially as described [43], except that different primer pairs were used for detection of different viral genotypes (see Table 1). Separate standards were made for each viral genotype, using cloned MSV-Kom, MSV-Set, MSV-VW or PanSV-Kar. In each case, viral plasmid concentrations were 1000, 100, 10, 1, 0.1 and 0.01 pg/ul. As in Owor et al. [43], Maize18S (F) and Maize18S (R) primers were used to amplify a 173 bp product from the *Zea mays* 18S small subunit rRNA gene for normalization of data from different runs. For the 18S standard curve, BMS genomic DNA was gel quantified and diluted to 100, 50, 25, 10, 5 and 1 ng/ul.

Data were analysed using the computer program Rotor-Gene, version 6. Data were used only if amplification efficiencies calculated by the program were above 80% and Pearson's correlation coefficient, r^2, of the standard curves was 0.99 or above. Viral plasmid and 18S standard curves were included in each run, rather than importing a previously performed standard curve.

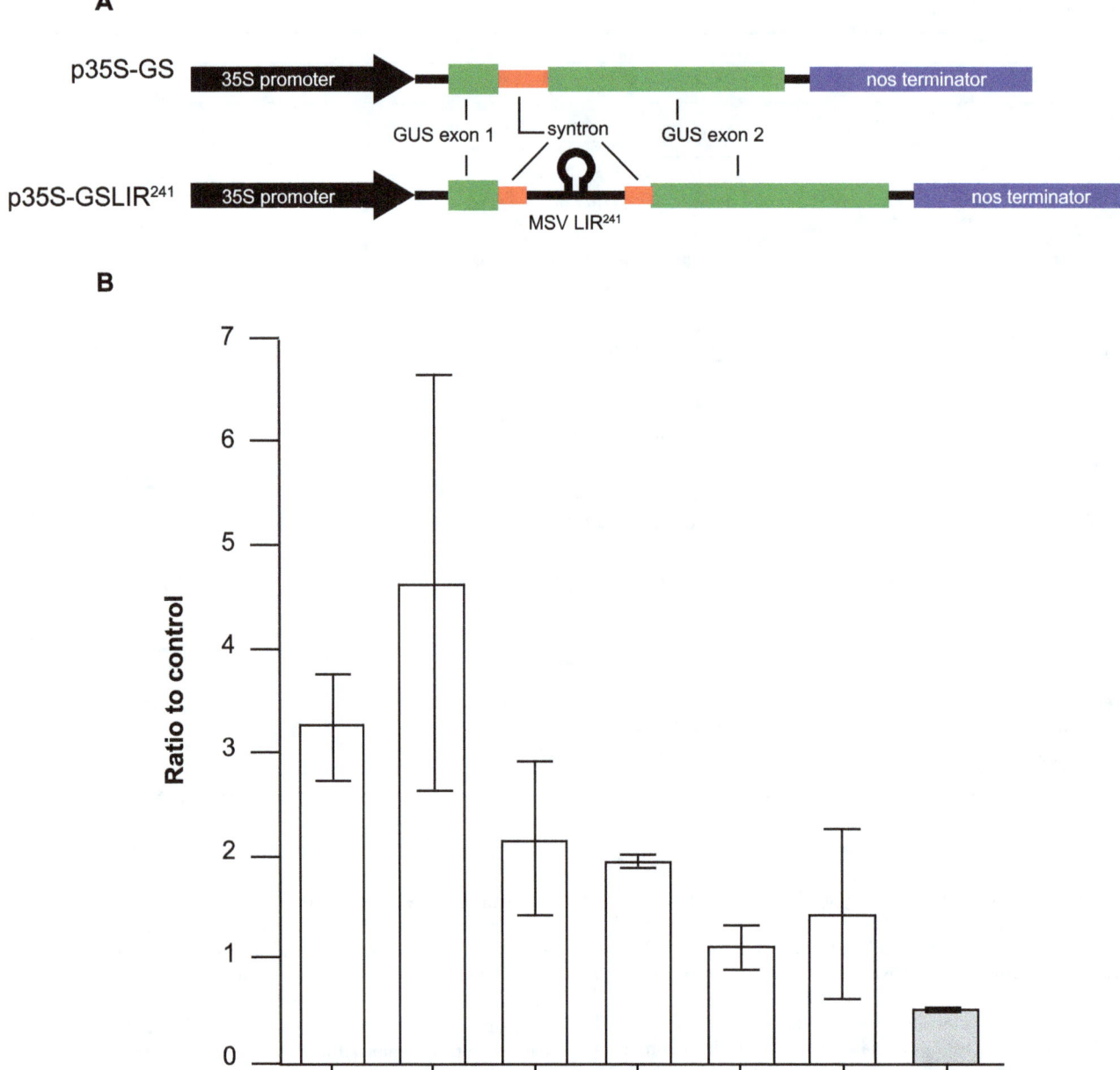

Figure 3. Gus assays to test for cryptic splice sites in the MSV long intergenic region. A) Gus expression cassettes used in the assays. B) Expression of Gus from p35S-GSLIR[241] (test construct) as a ratio to p35S-GS (positive control construct), four days after bombardment. Each bar is an average of three replicates; error bars represent 95% confidence intervals. Negative = negative control (protein extract from a non-bombarded Black Mexican sweet sample).

Statistical Analysis

Real-time PCR data were imported from Rotor-Gene version 6 into Microsoft Excel 2007 for calculation of the virus titres present in each sample. Further statistical analyses (Mann–Whitney tests) were carried out using GraphPad, version 5. Because multiple datasets were compared, a step-down multiple testing correction step was used when calculating P values.

Results and Discussion

Assay for cryptic splice sites within the MSV truncated LIR[241]

The 5′-terminal 241 nucleotides of the MSV-Kom LIR, previously determined to contain all the viral genomic *cis*-acting elements necessary for first strand synthesis [30] were assayed for any potential splice sites that could interfere with splicing of the

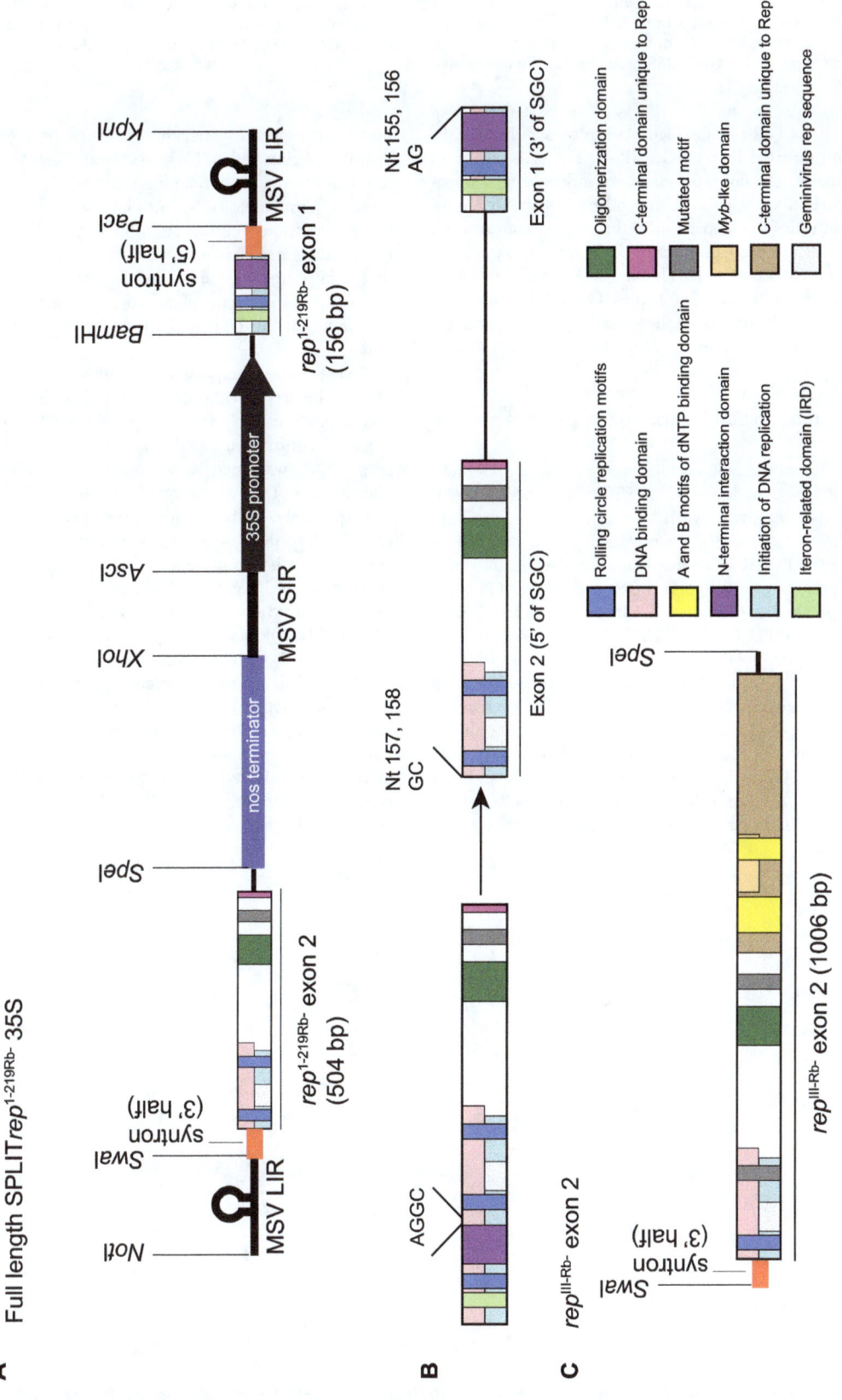

Figure 4. Schematic diagram of synthesised constructs, with restriction enzyme sites incorporated for subsequent cloning. A) pSPLIT*rep*[1-219Rb]-35S containing "modules" that could be removed and replaced with other sequences by restriction digest. B) Illustration showing how the *rep*[1-219Rb] transgene was split at the first AGGC (nucleotides 155, 156, 157 and 158 with respect to the start codon). The exon 2, cloned at the 5' terminus of the split gene cassette in A) therefore began with GC, and the exon 1, cloned at the 3' terminus, ended in AG. C) The synthesised

rep[III-Rb-] (see Fig. 1A for the full-length gene product) exon 2, preceded by the 3'-terminal half of the syntron, flanked by *Swa*I and *Spe*I RE sites. The 3'-terminal syntron/rep[1-219Rb-] exon 2 in pSPLITrep[1-219Rb-]35S was replaced by the 3'-terminal syntron/rep[III-Rb-] exon 2 to create pSPLITrep[III-Rb-]35S. Exon 1 remained the same for both constructs since they share the same 5'-terminal 156 bp. Similarly, other modules were exchanged to create further constructs, such as the CaMV 35S promoter for the maize ubiquitin promoter etc (see text for details).

syntron in which the LIR is embedded. As can be seen in Fig. 3B, GUS expression directed by p35SGSLIR[241] (Fig. 3A), four days post-bombardment, was not reduced in six independent experiments in comparison to the positive control construct p35SGS. All GUS expression ratios in comparison to the positive control were above 1, although this increase in expression was not significant ($p = 0.1143$; Mann Whitney test). This indicated that embedding the LIR[241] within the syntron of the SGCs will not have an appreciable effect on the splicing of the syntron that is required for fusion of the exon 1 and exon 2 sequences prior to their expression.

Viral replication inhibition by split gene cassette constructs

The basic design of the SGCs is illustrated in Figs. 2 and 4 and is described fully by Dugdale et al. [18,19]. Since pSPLITrep[1-219Rb-]35S was the first to be synthesised and was the template upon which the rest of the SGCs were based, we first tested its capacity to inhibit MSV-Kom replication, using a replicative-form specific semi-quantitative end-point PCR assay [17]. Co-bombardment of pKom602 (a cloned 1.1-mer of the MSV-Kom genome; [40]) with pSPLITrep[1-219Rb-]35S was compared with co-bombardment of pKom602 with a construct constitutively expressing Rep[1-219Rb-], which has been shown to inhibit viral replication in transgenic maize plants [16]. As can be seen in Fig. 5, viral titres were reduced in maize suspension cultures bombarded with each of prep[1-219Rb-] and pSPLITrep[1-219Rb-]35S. Confident that the SGC system was working as expected and that expression of rep[1-219Rb-] from the SGC was occurring at a high enough level to inhibit viral replication (indicating that virus-

mediated replicative release, circularisation, transcription and syntron splicing had all likely occurred effectively), we went ahead and constructed the remaining eight SGCs.

Semi-quantitative end-point PCR demonstrated that all but the three GUS-based constructs inhibited viral replication to some extent (data not shown). Of all the promoter combinations, the ubiquitin promoter + ubi-1 exon/intron (Ubi) resulted in the best inhibition. Subsequent quantitative realtime analyses were therefore done on the Ubi- and UbiΔI based constructs only (Figs. 6 and 7).

To control for the possible inhibitory effects on viral replication of both the vector sequences surrounding the SGCs, and the SGC "backbone" itself, pSK and pSPLITGUSUbi were tested as negative controls alongside rep-containing SGCs. In addition, prep[1-219Rb-] was used as a positive control against which the effectiveness of the SGCs could be compared.

Surprisingly, co-bombardment of pKom602 at a 1:1 weight ratio with pSK resulted in a reduction in viral DNA levels compared with bombardment of pKom602 alone (wild type [wt] replication; see Fig. 6A). Average levels of pKom602 co-bombarded with pSK were reduced by 32% when compared with wt ($P < 0.0001$; Wilcoxan signed rank test). Although viral levels were also reduced upon co-bombardment at a 1:1 ratio with pSPLITGUSUbi, the difference from wt was minor (14% reduction; Fig. 6A) and not significant ($P = 0.2783$ Wilcoxan signed rank test). Importantly, the difference between the GUS and pSK datasets was also not significant ($P = 0.0936$; Mann Whitney test). Levels of viral DNA after co-bombardment with pSK were therefore taken as the baseline, and any significant reductions beyond these levels were interpreted as being due to

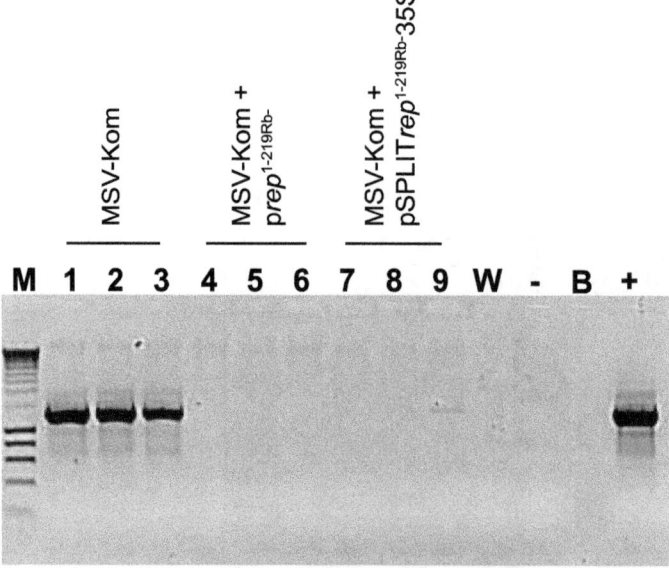

Figure 5. Replicative-form specific end-point PCR assay to test the effectiveness of the synthesised split gene cassette, pSPLITrep[1-219Rb-]35S, in interfering with MSV replication. Black Mexican sweet (BMS) cells were bombarded with an infectious clone of MSV-Kom (pKom602) alone (lanes 1–3); pKom602 and pSPLITrep[1-219Rb-]35S (lanes 7–9), as well as pKom602 and prep[1-219Rb-] (constitutively expressed from the maize ubiquitin promoter [16,17]) for comparative purposes (lanes 4–6). W = water control, − = non-bombarded BMS control, + = positive control (pKom602 plasmid DNA). B = blank. The PCR was performed on total DNA extracted from BMS cells four days post-bombardment.

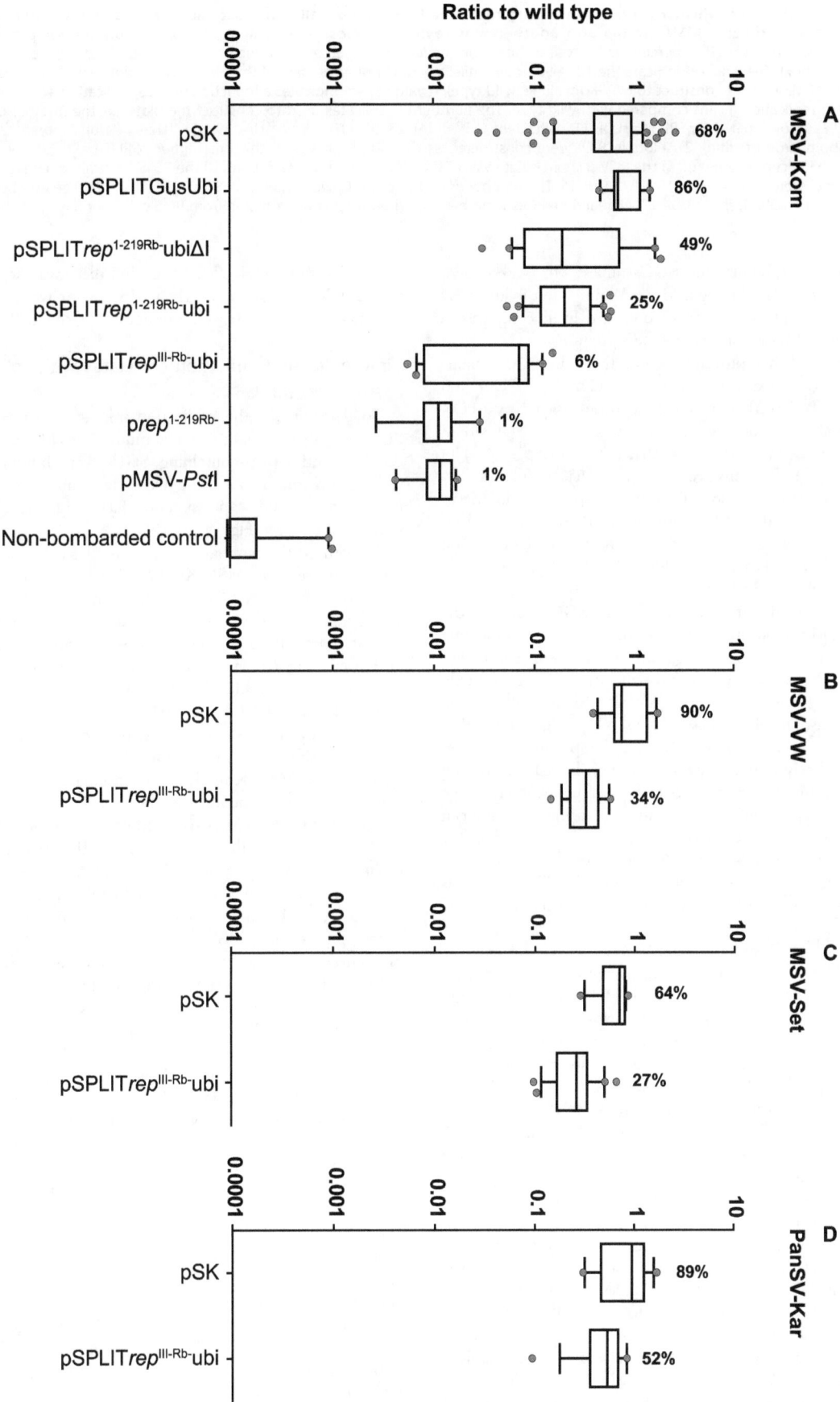

Figure 6. Vertical box-and-whisker plots summarising real-time PCR data on all constructs bombarded at a 1:1 weight ratio with infectious clones of diverse MSV strains and another mastrevirus species. A) MSV-Kom. The plots show the sample minimum and maximum, the lower quartile (25th percentile; bottom of box), the median (50th percentile; horizontal line in box) and the upper quartile (75th percentile; top of box). The whiskers indicate the 10th –90th percentile: any data points outside of this are shown as dots. The y-axis (on a $\log_{10}$ scale) shows the ratio of MSV-Kom + construct to MSV-Kom alone (wild type). A value of <1 indicates a reduction in virus replication. Numbers above each plot are percent replication means compared with wild type. The number of replicates used to construct the plots (i.e. the number of bombarded samples) were as follows: pSK, 75; pSPLITGusUbi, 11; pSPLITrep$^{1-219Rb-}$UbiΔI, 23; pSPLITrep$^{1-219Rb-}$Ubi, 39; pSPLITrep$^{III-Rb-}$Ubi, 21; prep$^{1-219Rb-}$, 9; pMSV-PstI, 14; Non-bombarded control, 21. Plots in B-D) were constructed as described for A), but this time either pSPLITrep$^{III-Rb-}$Ubi or pSK were co-bombarded with infectious clones of: B) the MSV-B strain isolate VW; C) the MSV-C strain isolate Set and D) the PanSV-A strain isolate Kar. The number of replicates for B) were: pSK, 12; pSPLITrep$^{III-Rb-}$Ubi, 15. The number of replicates for C) were: pSK, 17; pSPLITrep$^{III-Rb-}$Ubi, 26. The number of replicates for D) were: pSK, 12; pSPLITrep$^{III-Rb-}$Ubi, 17. All real-time PCRs were performed on total DNA extracted from BMS cells four days post-bombardment.

inhibition by the *rep*-containing SGCs (unless otherwise stated, given P values, calculated using a Mann Whitney test followed by a step down multiple testing correction, are from comparisons between [virus + SGC] and [virus + pSK] datasets).

Since prep$^{1-219Rb-}$ had proven to be effective in transgenic plants when constitutively expressed [17], pSPLITrep$^{1-219Rb-}$Ubi and pSPLITrep$^{1-219Rb-}$UbiΔI, were the first to be assayed by qPCR (Fig. 6A). Viral inhibition was less effective from the SGCs than from the constitutively expressed construct. pSPLITrep$^{1-219Rb-}$Ubi and pSPLITrep$^{1-219Rb-}$UbiΔI resulted in a 75% reduction (P< 0.0011) and a 51% reduction (P = 0.027) respectively in MSV-Kom levels, compared with 99% inhibition (P<0.0011) achieved by prep$^{1-219Rb-}$. This is probably due to the fact that expression from the SGCs first has to be induced by the virus before the protein can inhibit viral replication.

Having determined that the Ubi-containing SGC resulted in the best virus-induced replication inhibition by *rep*$^{1-219Rb-}$ (indicating that the presence of the ubi-1 exon/intron did not interfere with splicing of the SGC syntron), we subsequently further tested only Ubi-containing constructs, starting with pSPLITrep$^{III-Rb-}$Ubi. Previously, constitutively expressed Rep$^{III-Rb-}$, a full-length Rep with mutations in RCR motif III and the pRBR interaction domain, completely inhibited viral replication in cell culture and resulted in immune *D. Sanguinalis* transgenic plants in challenge experiments [17]. However, constitutive expression induced growth and developmental defects in transgenic plants, possibly due to non-mutated motifs in the full-length Rep$^{III-Rb-}$ protein (see Fig. 1) interacting with host regulatory molecules. This gene was therefore an ideal candidate for the virus-induced split gene system.

pSPLITrep$^{III-Rb-}$Ubi inhibited MSV-Kom replication by 94% (P<0.0011) (Fig. 6A), a significant improvement over the truncated *rep* SGC (P<0.0011 between pSPLITrep$^{III-Rb-}$Ubi and pSPLITrep$^{1-219Rb-}$Ubi datasets) and much closer to the replication inhibition (99%) seen by the constitutively expressed prep$^{1-219Rb-}$ that was used to make our MSV-resistant transgenic maize lines. The difference in replication inhibition by prep$^{1-219Rb-}$ and pSPLITrep$^{III-Rb-}$Ubi was not significant (P = 0.0948).

Inhibition by prep$^{1-219Rb-}$ was possibly complete; with the DNA levels detected most likely being due to residual unreplicated plasmid DNA that remains detectable by qPCR four days after bombardment [43,46]. To account for this input DNA, we bombarded maize suspension cells with a cloned replication-deficient MSV mutant, pMSV-*Pst*I (described by Owor et al., [43]) at the same concentration as pKom602. Four days post bombardment, MSV-*Pst*I was detected at levels that were 100-fold lower than MSV-Kom – this was due to the fact that MSV-Kom could be replicationally released [24] from the plasmid in which it was cloned by its wt Rep protein, while a non-functional Rep expressed by MSV-*Pst*I prevented its release from the inoculated plasmid. MSV-Kom DNA levels detectable after co-bombardment with pSPLITrep$^{III-Rb-}$Ubi were only marginally different

(P = 0.055) from MSV-*Pst*I levels, and there was no difference at all from MSV-*Pst*I levels when MSV-Kom was co-bombarded with prep$^{1-219Rb-}$ (both were present at 1% of wt; P = 0.8749).

Viral replication inhibition correlates with split gene cassette dosage

To determine if the replication inhibition observed with the SGCs was a "dominant negative mutant" [47] based mechanism, the Ubi and UbiΔI-containing SGCs were bombarded at 5:1 ratios to pKom602 to achieve over-expression of the mutant Reps relative to viral gene expression (in effect simulating a low-pressure viral infection) (see Fig. 7).

As with a 1:1 bombardment ratio, there was once again a marked decrease in MSV-Kom levels when co-bombarded with five-fold more pSK (Fig. 7A), this time by 69% (a 2.4-fold significant decrease relative to when they were co-bombarded at a 1:1 ratio; P<0.0001). Similarly, five-fold more pSPLITGUSUbi decreased MSV-Kom DNA levels by 49% (a statistically significant 3.5-fold decrease relative to when they were co-bombarded at a 1:1 ratio; P = 0.0075).

Since pSPLITGUSUbi contains two copies of the MSV-Kom LIR and should be replicationally released from the pSK backbone by MSV-Kom Rep, the decrease in viral replication could be due to competition for the viral Rep at Rep binding sites in both the SGC and MSV-Kom LIRs – the former outnumbering the latter. However, this does not explain the greater inhibition of replication seen by pSK at both 1:1 and 5:1 ratios.

To account for this "non-specific" viral inhibition, decreases in viral DNA levels when co-bombarded with *rep*-containing SGCs were compared with MSV-Kom + pSK levels, and only significant differences were taken as being due to the mutant Reps.

Even taking into account the increased inhibition by pSK, viral replication was inhibited significantly more when the SGCs were bombarded at a 5:1 ratio with pKom602 compared with at a 1:1 ratio (Fig. 7A). Compared with a 75% reduction at a 1:1 SGC:virus ratio, viral levels were reduced by 99.6% in the presence of five-fold more pSPLITrep$^{1-219Rb-}$Ubi (P<0.0001). Similarly, reduction in DNA levels by pSPLITrep$^{1-219Rb-}$UbiΔI went from 51% (1:1 ratio) to 96% (5:1 ratio) (p<0.0001). Compared with a 1:1 SGC:virus ratio, the difference between the ubiΔI- and ubi-containing SGCs was more marked at a 5:1 ratio (25-fold vs 236 fold inhibition by pSPLITrep$^{1-219Rb-}$UbiΔI and pSPLITrep$^{1-219Rb-}$Ubi respectively; P<0.0001). This was expected given that the presence of the ubi-1 intron is known to enhance expression from the ubi promoter (intron mediated enhancement [48]).

The effect of bombarding five-fold more transgene DNA was not as great with prep$^{1-219Rb-}$ (99% reduction in viral DNA levels at both 1:1 and a 5:1 ratios; P = 0.4363) or with pSPLITrep$^{III-Rb-}$Ubi (94% reduction at a 1:1 ratio and 97.5% reduction at a 5:1

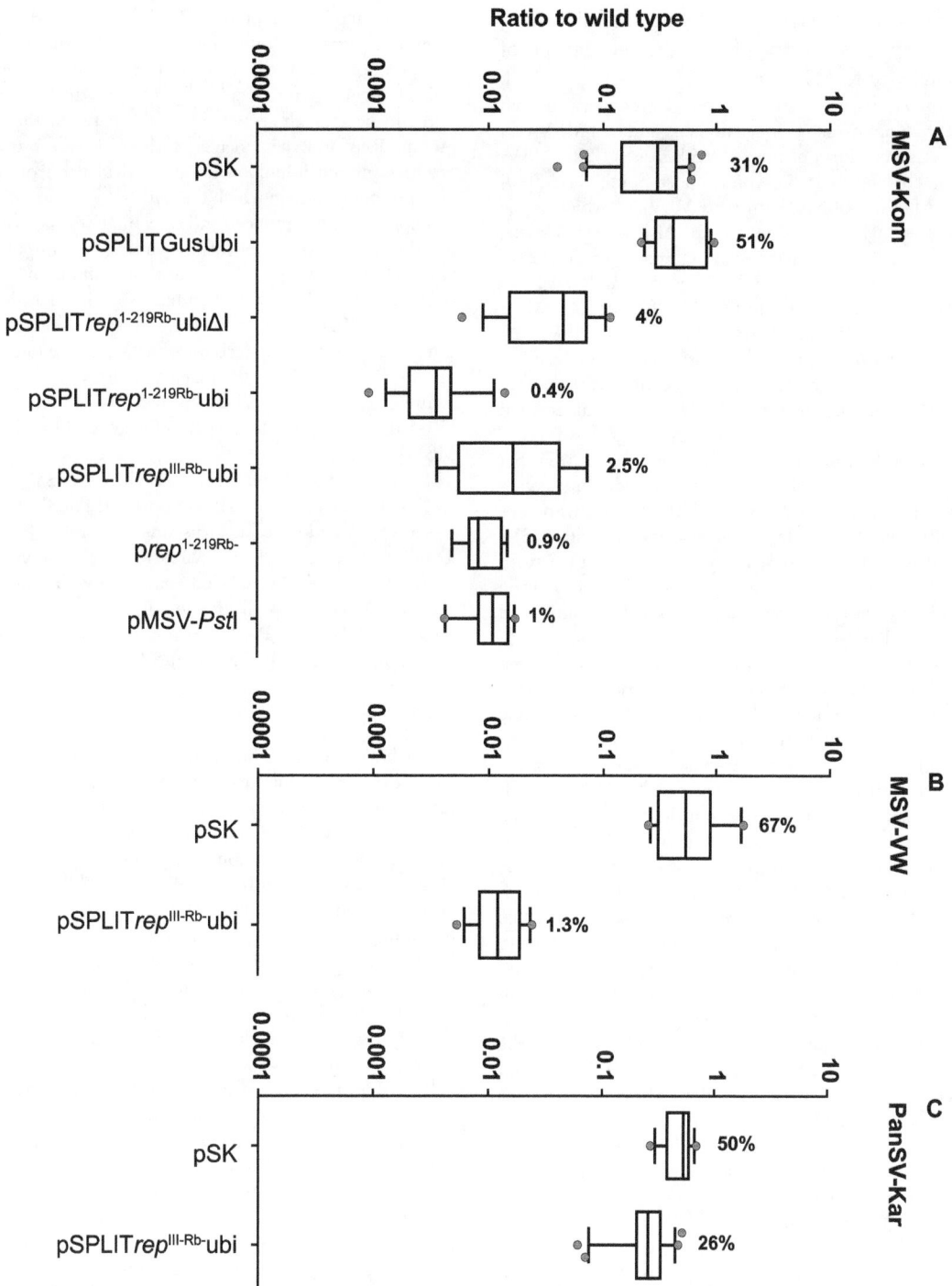

Figure 7. Vertical box-and-whisker plots summarising real-time PCR data on all constructs bombarded at a 5:1 weight ratio with infectious clones of diverse MSV strains and another mastrevirus species. A) MSV-Kom. The plots were constructed as in Figure 6. The number of replicates (i.e. the number of bombarded samples) were as follows: pSK, 34; pSPLITGusUbi, 18; pSPLIT*rep*[1-219Rb-]UbiΔI, 18; pSPLIT*rep*[1-219Rb-]Ubi, 14; pSPLIT*rep*[III-Rb-]Ubi, 8; p*rep*[1-219Rb-], 9; pMSV-*Pst*I, 11. Plots in B-C) were constructed as described for A), but this time either pSPLIT*rep*[III-Rb-]Ubi or pSK were co-bombarded with infectious clones of: B) the MSV-B strain isolate VW; and C) the PanSV strain A isolate Kar. The number of replicates for B) were: pSK, 11; pSPLIT*rep*[III-Rb-]Ubi, 14. The number of replicates for C) were: pSK, 16; pSPLIT*rep*[III-Rb-]Ubi, 25. All real-time PCRs were performed on total DNA extracted from BMS cells four days post-bombardment.

ratio; P = 0.0603). This is understandable for p*rep*[1-219Rb-] considering that at a 1:1 ratio, detected amplicons were probably from input plasmid and not replicated viral DNA, hence there would be no difference in input pKom602 DNA levels whether bombarded

at a 5:1 or 1:1 ratio. For pSPLIT*rep*[III-Rb-]Ubi, the already low levels of viral DNA detected at a 1:1 ratio were reduced so as to make them indistinguishable from input DNA levels.

The effectiveness of pSPLIT*rep*^III-Rb-^Ubi against diverse viral strains or species is associated with their degree of sequence similarity to MSV-Kom

The mutant *rep* transgenes used in the SGCs were derived from the isolate, MSV-Kom, which belongs to the A-strain of MSV that causes the most severe form of maize streak disease [28]. MSV-Kom, first isolated from maize in Komatieport, South Africa [40] belongs to an MSV-A subtype known as MSV-A$_4$, which is the most prevalent subtype found in South Africa [41] Having determined that the most effective SGC in inhibiting replication of MSV-Kom was pSPLIT*rep*^III-Rb-^Ubi (Figs. 6A and 7A), we subsequently decided to test pSPLIT*rep*^III-Rb-^Ubi against MSV isolates belonging to the B and C strains of MSV: (1) MSV-VW, belonging to the MSV-B strain (originally isolated from wheat but representative of viruses normally found infecting *Digitaria* sp. [49]); and (2) MSV-Set, belonging to the MSV-C strain (isolated from a *Setaria* plant and apparently representative of other *Setaria*-adapted MSV isolates [40]). MSV-VW and MSV-Set respectively share 89% and 78% genome-wide nucleotide similarity with MSV-Kom, and 86.5% and 81.4% Rep (with gaps included as a 21st character state) amino acid identity with MSV-Kom Rep. We also tested the construct against a different African streak virus species, PanSV-Kar, which shares 60% genome-wide nucleotide similarity with MSV-Kom and 60.6% Rep amino acid identity with MSV-Kom Rep.

Although MSV-VW, MSV-Set and PanSV-Kar are wild-grass-adapted virus isolates that do not cause serious disease in maize, we deemed it important to test pSPLIT*rep*^III-Rb-^Ubi against these diverse isolates because efficient trans-replication by MSV Rep requires the presence of specific Rep-binding sites (replication specificity determinants, or RSDs) within the LIR [30]. Considering that expression of the transgene from the SGC requires the replicational release of the cassette by the viral Rep which is initiated by the binding of the Rep to these specific sites in the LIR, we sought to determine if this would occur in isolates with non-conserved RSDs. According to Willment et al. [30], while PanSV-Kar is more genetically divergent from MSV-Kom than MSV-Kom is to MSV-Set, the Rep from PanSV-Kar complemented the replication function of a Rep-deficient MSV-Kom genome more efficiently than did MSV-Set's Rep. Efficient trans-replication presumably requires the sharing of RSDs within the LIR: Willment et al. [30] found that the RSDs in the MSV-Kom LIR are indeed more like PanSV-Kar's than MSV-Set's in terms of both spacing and sequence.

As can be seen in Fig. 6B–D, replication inhibition of all three divergent viruses did occur, but to a much lesser extent than inhibition of MSV-Kom. When co-bombarded at a 1:1 SGC:virus ratio, pSPLIT*rep*^III-Rb-^Ubi inhibited MSV-VW by 66%, MSV-Set by 73% and PanSV-Kar by 48%. However, unlike with MSV-Kom and MSV-Set, co-bombardment of pSK did not lead to a reduction in MSV-VW levels (Fig. 6B). Levels of MSV-Kom and MSV-Set DNA were 68% and 64% of wt in the presence of pSK, while levels of VW were 90%. Thus the reduction in MSV-VW DNA levels achieved by pSPLIT*rep*^III-Rb-^Ubi when compared with wt (MSV-VW bombarded alone) should be put into context with viral DNA levels in the presence of pSK. Compared with virus + pSK levels, pSPLIT*rep*^III-Rb-^Ubi reduced MSV-Kom titres by 91% (P<0.0001); MSV-VW by 63% (P<0.0001); MSV-Set by 57% (P<0.0001) and PanSV-Kar by 41% (P=0.0178). This correlates well with the percent identity of each virus' Rep with MSV-Kom. (MSV-Kom>MSV-VW>MSV-Set>PanSV-Kar).

Interestingly, these results do not correlate with the trans-replication efficiencies of MSV-Kom by MSV-Set and PanSV Reps described by Willment et al. [30]. It must be borne in mind

that trans-replication of the MSV-Kom-derived SGC by these divergent viruses is only the first step in the viral replication inhibition process. Once the exon 1 and exon 2 of the *rep*^III-Rb-^ are spliced together and Rep^III-Rb-^ is expressed, presumably there needs to be enough sequence identity between the Kom-derived mutant Rep and the inoculated virus' Reps for trans-dominant negative mutant inhibition to occur. Evidence from challenges of our transgenic lines in which the *rep*^1-219Rb-^ transgene was silenced or expression was reduced indicates that resistance is dependent on the expression level of the transgene (unpublished data). Considering the greater replication inhibition achieved by pSPLIT*rep*^III-Rb-^Ubi when bombarded at a 5:1 rather than a 1:1 ratio with MSV-Kom, it is likely the same would apply for this transgene, and that the resistance mechanism is dependent on the mutant Rep "flooding" the inoculated virus-derived Rep, perhaps forming dysfunctional oligomers with the wt Rep and/or RepA [50–53]) or outcompeting wt Rep for iterated binding sites on the MSV genome.

To test this with the heterologous viruses, MSV-VW (which has the highest identity with MSV-Kom) and PanSV-Kar (which has the lowest identity), each virus was co-bombarded with pSPLIT*rep*^III-Rb-^Ubi at a 5:1 SGC:virus ratio. The replication of both viruses was drastically reduced compared with that observed at a 1:1 ratio: MSV-VW replicated to only 1.3% of wt (P<0.0001), and PanSV to 26% of wt (P<0.0001), representing a 75-fold and 4-fold inhibition respectively (Fig. 7B and C).

Conclusion

We have shown for the first time that the INPACT inducible hyper-expression platform, developed primarily for farming recombinant proteins in plants, can be adapted for virus-inducible resistance to MSV, with potential application in transgenic maize. While replication inhibition was greater with increased dosage of all the SGCs, one construct (pSPLIT*rep*^III-Rb-^Ubi) was extremely effective even at the lower of the two doses that were tested, inhibiting replication of MSV-Kom by 94% when bombarded at a 1:1 ratio with the virus. This construct was also effective against diverse MSV strains and even a different mastrevirus species, although the degree of replication inhibition correlated with the degree of sequence similarity to MSV-Kom (the isolate on which the SGCs were based).

The Rep-inducible nature of the INPACT platform has been demonstrated in transgenic tobacco for the high-level, activatable expression of four different recombinant proteins, including the lethal ribonuclease, barnase [18]. In contrast, it is difficult to definitively prove this using the micro-projectile bombardment transformation system and the SGCs utilised in this study. Firstly, in the absence of MSV Rep protein, *in situ* recombination of the SGC (most likely at the repeated stem-loop sequences) may occur in a small number of cells [20,54–56] and this, in turn, could generate an episomal form of the cassette from which Rep can be expressed. Despite this possibility we were unable to detect SGC-encoded mutant *rep* transcripts in the absence of the MSV Rep protein (data not shown), suggesting these transcript levels are below the limit of detection by qPCR. Secondly, in the presence of MSV Rep protein it is anticipated that the expressed SGC-encoded mutant *rep* gene product inhibits MSV replication, thereby reducing both Rep forms in the system. Similarly, we were unable to detect SGC-encoded mutant *rep* transcripts under these circumstances (data not shown). Ultimately, the practicality of the SGCs described in this study will only be fully realised with the regeneration of phenotypically normal transgenic maize plants engineered to contain the SGC that are resistant/immune to MSV infection. To this end we have regenerated a number of transgenic maize lines containing a SGC capable of expressing the most effective Rep mutant, namely Rep$^{III-Rb-}$. In contrast to lines constitutively expressing this mutant gene, SGC lines have produced T$_2$ generation offspring with normal phenotypes.

Considering that only one strain - MSV-A - causes severe disease in maize throughout the whole geographical range of MSV, and that all isolates so far discovered within this strain have a maximum divergence of only 4.62% at the nucleotide level, it is likely that this novel MSV-inducible resistance construct will be effective against the complete spectrum of severe maize streak disease-causing African MSVs.

References

1. Bosque-Pérez NA (2000) Eight decades of maize streak virus research. Virus Res 71: 107–121.
2. Moffat AS (1999) Geminiviruses emerge as serious crop threat. Science 286: 5446–1835.
3. Shepherd DN, Martin DP, Thomson JA (2009) Transgenic strategies for developing crops resistant to geminiviruses. Plant Sci 176: 1–11.
4. Vanderschuren H, Stupak M, Fütterer J, Gruissem W, Zhang P (2007) Engineering resistance to geminiviruses–review and perspectives. Plant Biotechnol J 5: 207–220.
5. Lopez-Ochoa L, Ramirez-Prado J, Hanley-Bowdoin L (2006) Peptide aptamers that bind to a geminivirus replication protein interfere with viral replication in plant cells. J Virol 80: 5841–5853.
6. Reyes MI, Nash TE, Dallas MM, Ascencio-Ibáñez JT, Hanley-Bowdoin L (2013) Peptide aptamers that bind to geminivirus replication proteins confer a resistance phenotype to tomato yellow leaf curl virus and tomato mottle virus infection in tomato. J Virol 87: 9691–9706.
7. Sera T (2005) Inhibition of virus DNA replication by artificial zinc finger proteins. J Virol 79: 2614–2619.
8. Hong Y, Saunders K, Hartley MR, Stanley J (1996) Resistance to geminivirus infection by virus-induced expression of dianthin in transgenic plants. Virology 220: 119–127.
9. Legg JP, Fauquet CM (2004) Cassava mosaic geminiviruses in Africa. Plant Mol Biol 56: 585–599.
10. Zhang P, Fütterer J, Frey P, Potrykus I, Puonti-Kaerlas J, et al. (2003) Engineering virus-induced African cassava mosaic virus resistance by mimicking a hypersensitive reaction in transgenic cassava. In: Vasil I, editor. Plant Biotechnology 2002 and Beyond, Proceedings of the 10th IAPTC&B Congress June 23–28, 2002 Orlando, Florida, U.S.A. Springer Netherlands. 143–145.
11. Hartley RW (1989) Barnase and barstar: two small proteins to fold and fit together. Trends Biochem Sci.
12. Wang T, Tomic S, Gabdoulline RR, Wade RC (2004) How optimal are the binding energetics of barnase and barstar? Biophys J 87: 1618–1630.
13. Hussain M, Mansoor S, Iram S, Zafar Y, Briddon RW (2007) The hypersensitive response to tomato leaf curl New Delhi virus nuclear shuttle protein is inhibited by transcriptional activator protein. Mol Plant Microbe Interact 20: 1581–1588.
14. Hussain M, Mansoor S, Iram S, Fatima AN, Zafar Y (2005) The nuclear shuttle protein of Tomato leaf curl New Delhi virus is a pathogenicity determinant. J Virol 79: 4434–4439.
15. Martin DP, Shepherd DN (2009) The epidemiology, economic impact and control of maize streak disease. Food Secur 1: 305–315.
16. Shepherd DN, Mangwende T, Martin DP, Bezuidenhout M, Kloppers FJ, et al. (2007) Maize streak virus-resistant transgenic maize: a first for Africa. Plant Biotechnol J 5: 759–767.
17. Shepherd DN, Mangwende T, Martin DP, Bezuidenhout M, Thomson JA, et al. (2007) Inhibition of maize streak virus (MSV) replication by transient and transgenic expression of MSV replication-associated protein mutants. J Gen Virol 88: 325–336.
18. Dugdale B, Mortimer C, Kato M, James T, Harding R, et al. (2013) In Plant Activation (INPACT): an inducible, hyperexpression platform for recombinant protein production in plants. Plant Cell 25: 2429–2443.
19. Dugdale B, Mortimer C, Kato M, James T, Harding R, et al. (2014) Design and construction of an In Plant Activation (INPACT) cassette for transgene expression and recombinant protein production in plants. Nat Protoc In Press.
20. Heyraud F, Matzeit V, Schaefer S, Schell J, Gronenborn B (1993) The conserved nonanucleotide motif of the geminivirus stem-loop sequence promotes replicational release of virus molecules from redundant copies. Biochimie 75: 605–615.
21. Heyraud-Nitschke F, Schumacher S, Laufs J, Schaefer S, Schell J, et al. (1995) Determination of the origin cleavage and joining domain of geminivirus Rep proteins. Nucleic Acids Res 23: 910–916.
22. Laufs J, Traut W, Heyraud F, Matzeit V, Rogers SG, et al. (1995) In vitro cleavage and joining at the viral origin of replication by the replication initiator protein of tomato yellow leaf curl virus. Proc Natl Acad Sci U S A 92: 3879–3883.
23. Stanley J (1995) Analysis of African cassava mosaic virus recombinants suggests strand nicking occurs within the conserved nonanucleotide motif during the initiation of rolling circle DNA replication. Virology 206: 707–712.
24. Stenger DC, Revington GN, Stevenson MC, Bisaro DM (1991) Replicational release of geminivirus genomes from tandemly repeated copies: evidence for rolling-circle replication of a plant viral DNA. Proc Natl Acad Sci U S A 88: 8029–8033.
25. Donson J, Morris-Krsinich BA, Mullineaux PM, Boulton MI, Davies JW (1984) A putative primer for second-strand DNA synthesis of maize streak virus is virion-associated. EMBO J 3: 3069–3073.
26. Hayes R, Macdonald H, Coutts R, Buck K (1988) Priming of complementary DNA synthesis in vitro by small DNA molecules tightly bound to virion DNA of wheat dwarf virus. J Gen Virol 69: 1345–1350.
27. Kammann M, Schalk HJ, Matzeit V, Schaefer S, Schell J, et al. (1991) DNA replication of wheat dwarf virus, a geminivirus, requires two cis-acting signals. Virology 184: 786–790.
28. Martin DP, Willment JA, Billharz R, Velders R, Odhiambo B, et al. (2001) Sequence diversity and virulence in Zea mays of Maize streak virus isolates. Virology 288: 247–255.
29. Muhire B, Martin DP, Brown JK, Navas-Castillo J, Moriones E, et al. (2013) A genome-wide pairwise-identity-based proposal for the classification of viruses in the genus Mastrevirus (family Geminiviridae). Arch Virol 158: 1411–1424.
30. Willment JA, Martin DP, Palmer KE, Schnippenkoetter WH, Shepherd DN, et al. (2007) Identification of long intergenic region sequences involved in maize streak virus replication. J Gen Virol 88: 1831–1841.
31. Laufs J, Jupin I, David C, Schumacher S, Heyraud-Nitschke F, et al. (1995) Geminivirus replication: genetic and biochemical characterization of Rep protein function, a review. Biochimie 77: 765–773.
32. Argüello-Astorga GR, Guevara-González RG, Herrera-Estrella LR, Rivera-Bustamante RF (1994) Geminivirus replication origins have a group-specific organization of iterative elements: a model for replication. Virology 203: 90–100.
33. Argüello-Astorga G, Herrera-Estrella L, Rivera-Bustamante R (1994) Experimental and theoretical definition of geminivirus origin of replication. Plant Mol Biol 26: 553–556.
34. Fontes EP, Luckow VA, Hanley-Bowdoin L (1992) A geminivirus replication protein is a sequence-specific DNA binding protein. Plant Cell 4: 597–608.
35. Fontes EP, Eagle PA, Sipe PS, Luckow VA, Hanley-Bowdoin L (1994) Interaction between a geminivirus replication protein and origin DNA is essential for viral replication. J Biol Chem 269: 8459–8465.
36. Singh DK, Malik PS, Choudhury NR, Mukherjee SK (2008) MYMIV replication initiator protein (Rep): roles at the initiation and elongation steps of MYMIV DNA replication. Virology 380: 75–83.
37. Shepherd DN, Martin DP, McGivern DR, Boulton MI, Thomson JA, et al. (2005) A three-nucleotide mutation altering the Maize streak virus Rep pRBR-interaction motif reduces symptom severity in maize and partially reverts at high frequency without restoring pRBR-Rep binding. J Gen Virol 86: 803–813.
38. Christensen AH, Sharrock RA, Quail PH (1992) Maize polyubiquitin genes: structure, thermal perturbation of expression and transcript splicing, and promoter activity following transfer to protoplasts by electroporation. Plant Mol Biol 18: 675–689.
39. Christensen A, Quail PH (1996) Ubiquitin promoter-based vectors for high-level expression of selectable and/or screenable marker genes in monocotyledonous plants. Transgenic Res 5: 213–218.
40. Schnippenkoetter WH, Martin DP, Hughes FL, Fyvie M, Willment JA, et al. (2001) The relative infectivities and genomic characterisation of three distinct mastreviruses from South Africa. Arch Virol 146: 1075–1088.
41. Shepherd DN, Martin DP, Van Der Walt E, Dent K, Varsani A, et al. (2010) Maize streak virus: an old and complex "emerging" pathogen. Mol Plant Pathol 11: 1–12.

Author Contributions

Conceived and designed the experiments: DNS BD DPM AV JD EPR. Performed the experiments: DNS FML MEB ALM. Analyzed the data: DNS DPM BD. Contributed reagents/materials/analysis tools: DNS EPR JAT. Contributed to the writing of the manuscript: DNS BD DPM AV JAT EPR.

42. Dunder E, Dawson J, Suttie J, Pace G (1995) Maize transformation by microprojectile bombardment of immature embryos. In: Potrykus I, Spangerberg G, editors. Gene Transfer to Plants. Berlin: Springer-Verlag. 127–138.

43. Owor BE, Martin DP, Rybicki EP, Thomson JA, Bezuidenhout ME, et al. (2011) A rep-based hairpin inhibits replication of diverse maize streak virus isolates in a transient assay. J Gen Virol 92: 2458–2465.

44. Varsani A, Oluwafemi S, Windram OP, Shepherd DN, Monjane AL, et al. (2008) Panicum streak virus diversity is similar to that observed for maize streak virus. Arch Virol 153: 601–604.

45. Varsani A, Monjane AL, Donaldson L, Oluwafemi S, Zinga I, et al. (2009) Comparative analysis of Panicum streak virus and Maize streak virus diversity, recombination patterns and phylogeography. Virol J 6: 194.

46. Ruschhaupt M, Martin DP, Lakay F, Bezuidenhout M, Rybicki EP, et al. (2013) Replication modes of Maize streak virus mutants lacking RepA or the RepA-pRBR interaction motif. Virology 442: 173–179.

47. Herskowitz I (1987) Functional inactivation of genes by dominant negative mutations. Nature 329: 219–222.

48. Callis J, Fromm M, Walbot V (1987) Introns increase gene expression in cultured maize cells. Genes Dev 1: 1183–1200.

49. Willment JA, Martin DP, Van der Walt E, Rybicki EP (2002) Biological and Genomic Sequence Characterization of Maize streak virus Isolates from Wheat. Phytopathology 92: 81–86.

50. Horváth G V, Pettkó-Szandtner A, Nikovics K, Bilgin M, Boulton M, et al. (1998) Prediction of functional regions of the maize streak virus replication-associated proteins by protein-protein interaction analysis. Plant Mol Biol 38: 699–712.

51. Orozco BM, Kong LJ, Batts LA, Elledge S, Hanley-Bowdoin L (2000) The multifunctional character of a geminivirus replication protein is reflected by its complex oligomerization properties. J Biol Chem 275: 6114–6122.

52. Missich R, Ramirez-Parra E, Gutierrez C (2000) Relationship of oligomerization to DNA binding of Wheat dwarf virus RepA and Rep proteins. Virology 273: 178–188.

53. Chatterji A, Beachy RN, Fauquet CM (2001) Expression of the oligomerization domain of the replication-associated protein (Rep) of Tomato leaf curl New Delhi virus interferes with DNA accumulation of heterologous geminiviruses. J Biol Chem 276: 25631–25638.

54. Stanley J, Townsend R (1986) Infectious mutants of cassava latent virus generated in vivo from intact recombinant DNA clones containing single copies of the genome. Nucleic Acids Res 14: 5981–5998.

55. Lazarowitz SG, Pinder AJ, Damsteegt VD, Rogers SG (1989) Maize streak virus genes essential for systemic spread and symptom development. EMBO J 8: 1023–1032.

56. Topfer R, Gronenborn B, Schaefer S, Schell J, Steinbiss H-H (1990) Expression of engineered wheat dwarf virus in seed-derived embryos. Physiol Plant 79: 158–162.

3

A Soybean C2H2-Type Zinc Finger Gene *GmZF1* Enhanced Cold Tolerance in Transgenic *Arabidopsis*

Guo-Hong Yu[1✪], Lin-Lin Jiang[1✪], Xue-Feng Ma[1,2], Zhao-Shi Xu[3], Meng-Meng Liu[1], Shu-Guang Shan[1], Xian-Guo Cheng[1]*

1 Key Lab. of Plant Nutrition and Fertilizer, Ministry of Agriculture, Institute of Agricultural Resources and Regional Planning, Chinese Academy of Agricultural Sciences, Beijing, China, 2 Institute of Agro-Products Processing Science and Technology, Chinese Academy of Agricultural Sciences, Beijing, China, 3 Institute of Crop Science, Chinese Academy of Agricultural Sciences (CAAS)/National Key Facility for Crop Gene Resources and Genetic Improvement, Key Laboratory of Biology and Genetic Improvement of Triticeae Crops, Ministry of Agriculture, Beijing, China

Abstract

Zinc finger proteins were involved in response to different environmental stresses in plant species. A typical Cys2/His2-type (C2H2-type) zinc finger gene *GmZF1* from soybean was isolated and was composed of 172 amino acids containing two conserved C2H2-type zinc finger domains. Phylogenetic analysis showed that GmZF1 was clustered on the same branch with six C2H2-type ZFPs from dicotyledonous plants excepting for GsZFP1, and distinguished those from monocotyledon species. The GmZF1 protein was localized at the nucleus, and has specific binding activity with EP1S core sequence, and nucleotide mutation in the core sequence of *EPSPS* promoter changed the binding ability between GmZF1 protein and core DNA element, implying that two amino acid residues, G and C boxed in core sequence TGACAGTGTCA possibly play positive regulation role in recognizing DNA-binding sites in GmZF1 proteins. High accumulation of *GmZF1* mRNA induced by exogenous ABA suggested that *GmZF1* was involved in an ABA-dependent signal transduction pathway. Over-expression of *GmZF1* significantly improved the contents of proline and soluble sugar and decreased the MDA contents in the transgenic lines exposed to cold stress, indicating that transgenic *Arabidopsis* carrying *GmZF1* gene have adaptive mechanisms to cold stress. Over-expression of *GmZF1* also increased the expression of cold-regulated *cor6.6* gene by probably recognizing protein-DNA binding sites, suggesting that *GmZF1* from soybean could enhance the tolerance of *Arabidopsis* to cold stress by regulating expression of cold-regulation gene in the transgenic *Arabidopsis*.

Editor: Fan Chen, Institute of Genetics and Developmental Biology, Chinese Academy of Sciences, China

Funding: This work was supported by the National Key Project for Cultivation of New Varieties of Genetically Modified Organisms (2013ZX08002-005). The funders had no role in study design, data collection and analysis, decision to publish, or preparation of the manuscript.

Competing Interests: The authors have declared that no competing interests exist.

* Email: chengxianguo@caas.cn

✪ These authors contributed equally to this work.

Introduction

Plants are usually exposed to various environmental stress factors affecting plant growth and crop productivity, such as drought, high salt and low temperature. As an adaptive response, the plants have developed a resistance mechanism to abiotic stresses to achieve an optimal adaptation [1]. It was reported that a number of transcription factor gene from different plants was induced or repressed during the responses and acclimations [2]. For example, some typical transcription factors, AP2/ERF, bZIP, NAC, MYB, MYC, WRKY and zinc finger proteins have been well characterized, and confirmed to be involved in stress response via transcriptional regulation modulation [3,4,5], and these transcription factors usually have specific structure and/or conserved domains functioning in the adaptation to abiotic stresses [6,7].

The first zinc finger transcription factor IIIA (TFIIIA) was recognized as a repeated zinc-binding motif in *Xenopus*, and is a classical Cys2/His2-type (C2H2-type) zinc finger gene [6]. The C2H2-type zinc finger has been well characterized in the

eukaryotic transcription factors having specific $CX_{2-4}CX_3$ $FX_5LX_2HX_{3-5}H$ structure (X: any amino acid; number: amino acid amounts), and this conserved motif contains two pairs of specific Cys and His residues which form a tetrahedral structure with a zinc ion [7]. In plants, the C2H2-type zinc finger has a highly conserved QALGGH motif in a putative DNA-contacting surface, which is unique in zinc-finger proteins from plants [8,9]. *In vitro* analysis revealed that the conserved QALGGH motif in plants plays a critical role in DNA binding activity, and each amino acid residue in the motif seems to be essential for maintaining effective binding between DNA and C2H2-type zinc finger proteins [10,11], and these specific domains conferred that zinc-finger proteins play important regulatory role by recognizing the target sequences and regulating expression of target gene in a plant-specific manner [1,8,12].

As a first C2H2-type zinc finger protein in plants, ZPT2-1 was found in petunia (*Petunia hybrid*), and identified to interact with the specific DNA sequences in the promoter region of 5-enolpyruvylshikimate-3-phosphate synthase (EPSPS) [1,13]. Thereafter, several TFIIIA-type zinc finger proteins have been consecutively reported [1,8], and these C2H2-type zinc finger genes paly important regulation roles in responses to various abiotic stresses [8,9,14]. In petunia, the TFIIIA-type zinc finger genes, *ZPT2-2* and *ZPT2-3*, were regulated by cold and/or drought, and over-expression of *ZPT2-3* gene in the transgenic petunia increased the tolerance of plants to drought stress [15,16]. *STZ* is one of the *ZPT2*-related genes in *Arabidopsis*, and was isolated by complementation of the salt-sensitive phenotype with a yeast calcineurin mutant [17], and the expression of *STZ* gene has been identified to be responsive to drought, salt, cold and abscisic acid (ABA), and that constitutive expression of *STZ* resulted in a suppression in growth, and accompanied an enhancement of plants adaptation to drought and osmotic stresses [18,19,20].

To date, few zinc finger proteins from cultivar soybean (*Glycine max*) have been isolated and characterized. *SCOF-1,* a typical C2H2-type zinc finger gene, was isolated from soybean and was confirmed to have a positive role in regulating the expression of cold-regulated gene and enhancing cold tolerance in transgenic plants [21]. *GsZFP1*, a zinc finger transcription factor lacking typical QALGGH motif was isolated from wild soybean (*Glycine soja* L. G07256), and transformed into *Arabidopsis*, and confirmed to play a crucial role in withstanding cold and drought stresses [22], indicating that the presence of QALGGH motifs in the C2H2 zinc finger *GsZFP1* gene from wild soybean seems to be not a crucial element in functioning during adaptation to abitotic stress. In our study, a novel C2H2-type zinc finger gene, *GmZF1* was isolated from cultivar soybean, and identified to have two typical conserved QALGGH motifs, and is significantly different from *GsZFP1* gene in wild soybean [22]. To our knowledge, the understanding on the function of zinc finger C2H2 from cultivar soybean in responses to cold stress is still limited. Therefore, based on the structure analysis and *in vitro* identification of *GmZF1* gene, we transformed the *GmZF1* gene into *Arabidopsis* plants and identified three homozygous lines which were subsequently used to observe the function effects of *GmZF1* gene in responses to cold stress and exogenous ABA. Our data showed that *GmZF1* from cultivar soybean was induced by ABA, and over-expression of *GmZF1* gene significantly enhanced the tolerance of the transgenic *Arabidopsis* to low temperature stress.

Results

Characteristics of the *GmZF1* gene

The *GmZF1* cDNA is composed of 765-bp nucleotides, and eocodes a predicted protein of 172 amino acids with a calculated molecular mass of 19.2 kDa, and flanked by a 97 bp fragment at 5′ end and a 149 bp fragment at 3′ end at two untranslating regions, respectively (accession number DQ055134). Electrophoresis analysis showed that PCR product from the genomic DNA of soybean seedlings has the same size as that from the cDNA length generated by RT-PCR using total RNA as remplate, and sequencing also showed that the generated *GmZF1* clones from gDNA and cDNA have identical sequence, indicating that *GmZF1* gene has no intron in the genomic DNA. A homology search against the GenBank database showed that GmZF1 is a homolog of C2H2-type zinc finger proteins (ZFPs), and has two typical C2/H2 type zinc finger domains ($CX_2CX_3FX_3QALGGHX_3H$) (Fig. 1a). GmZF1 has 57%, 48%, 47% and 46% homology with the C2H2-type ZFPs from *Medicago truncatula*, Petunia, *Arabidopsis* and rice, respectively, and no highly homology was observed between these genes from different plants, suggesting that *GmZF1* belongs to a novel subfamily member of C2H2 zinc finger. Like C2H2 zinc finger genes from four kinds of plants mentioned above, the GmZF1 protein also has plant-specific

QALGGH motifs as well as a conserved motif L-box with more Leu residues, and contains a short hydrophobic region with a highly conserved DLN box near the C-terminus of GmZF1 (Fig. 1a), which may function as a transcription repression domain [16,18]. It is noteworthy that GmZF1 have two conserved domains with QALGGH motifs, and is significantly different from GsZFP1 isolated from wild soybean, the later only has one C2H2 domain lacking QALGGH motif, and the number of amino acid residues between two His (H) residues is composed of four amino acids, which is different from GmZF1 having three amino acids between two His amino acids (as indicated by # and * respectively in Fig. 1a), and no DLN box was found at the C-terminus of GsZFP1 [22].

For profiling the differences in genetic characteristics between C2H2 family members from different plants, a systematic phylogenetic analysis was carried out. The results showed that GmZF1 was clustered on the same branch with six C2H2-type ZFPs from dicotyledonous plants, and distinguished them from monocotyledon species (Fig. 1b), indicating that dicotyledonous plants and monocotyledon plants have significant difference in genetic characteristics of C2H2-type genes, and probably play different regulation roles in plant responses to abiotic stresses. What's worthy reminding is that *GmZF1* from cultivar soybean and *GsZFP1* from wild soybean respectively belong to two different subfamily members in C2H2 zinc finger super family.

Analysis for SDS-PAGE and DNA-binding activity of GmZF1

Analysis of SDS-PAGE indicated that the GmZF1::GST fusion protein with 45.2 kD molecular weight was successfully expressed in *E. coli* strain (Fig. 2a), and the size of expressed protein is identical with the predicted molecular weight of target protein in the fusion vector vector pGEX-4 T-1. As a DNA-binding motif, a C2H2-type zinc-finger domain in the GmZF1 has been identified in many transcription factors. In our study, we observed that the transcription factors containing the zinc finger motif could specifically bind to EP1S core sequence (TGACAGTGTCA), which was originally identified as a *cis*-element within the *EPSPS* gene promoter in petunia [13,23]. For identifying DNA-binding ability of GmZF1, a gel-shift assay was performed *in vitro* using the procedure as described in the method. Data showed that GmZF1 protein was expressed as fusion proteins with GST in *E. coli*. To better understand the molecular mechanism of specific-binding between the GmZF1 proteins and EP1S core sequences, a series of synthesis probes including wild type and mutants was prepared as the following descriptions. Briefly, the EP1S core sequence (wild-type, E1) and two probes (E2: bases substitution, E3: with four bases substitution) in the repeated sequences were prepared (Fig. 2b), and fusion protein was purified for gel shift assay. Assay showed that the GmZF1::GST fusion protein strongly bound to E1 and E2 probes, but weakly binding to E3 probe (Fig. 2c), no complex was found in reaction solution containing GST protein and probes, suggesting that GmZF1 protein could bind specifically to the EP1S core sequence *in vitro*, and two amino acid residues, G and C boxed in core sequence of TGACAGTGTCA probably play key role in recognizing DNA-binding sites in GmZF1 proteins.

Subcellular localization of GmZF1

To identify the subcellular localization of GmZF1, we fused the full length of GmZF1 to GFP vector under control of the constitutive 35S promoter. Both the recombinant DNA constructs encoding a GmZF1::GFP fusion protein and a GFP protein were respectively introduced into *Arabidopsis* protoplast cells.

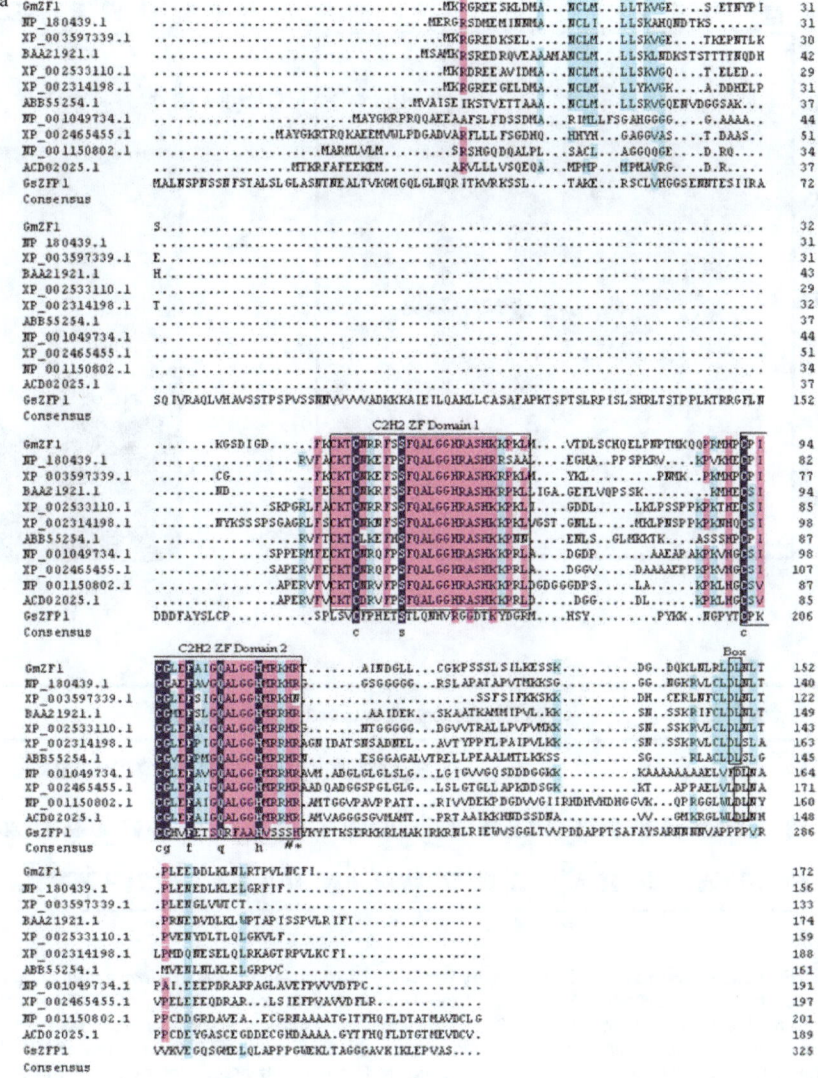

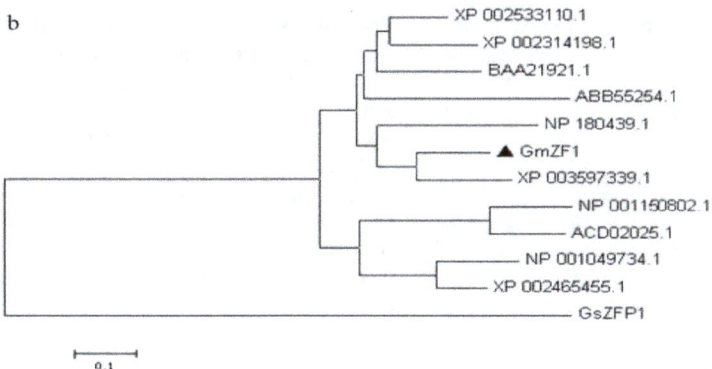

Figure 1. Sequence characteristic of GmZF1 deduced amino acids. a: Multiple alignment of GmZF1 amino acid sequences with XP_003597339.1 (*Medicago truncatula*), BAA21921.1 (*Petunia hybrid*)), NP_180439.1 (*Arabidopsis*) and NP_001049734.1 (*Oryza sativa*). The two zinc finger domains and the conserved DLN amino residues are boxed. b: Phylogenetic analysis of GmZF1 and related proteins. A mid-point rooted neighbor-joining phylogeny was constructed by 12 amino acid sequences from diverse organisms. Excepting for GmZF1, the other C2H2 type zinc finger proteins were demonstrated by XP_002533110.1 (*Ricinus communis*), XP_003597339.1 (*M. truncatula*), BAA21921.1 (*Petunia*), XP_002314198.1 (*Populus*), NP_180439.1 (*Arabidopsis*), ABB55254.1 (*Brassica*), NP_001049734.1 (*Oryza sativa*), XP_002465455.1 (*Broomcorn*), NP_001150802.1 (*Zea mays*), ACD02025.1 (*Triticum aestivum*) and GsZFP1 (*Glycine sojia*. L).

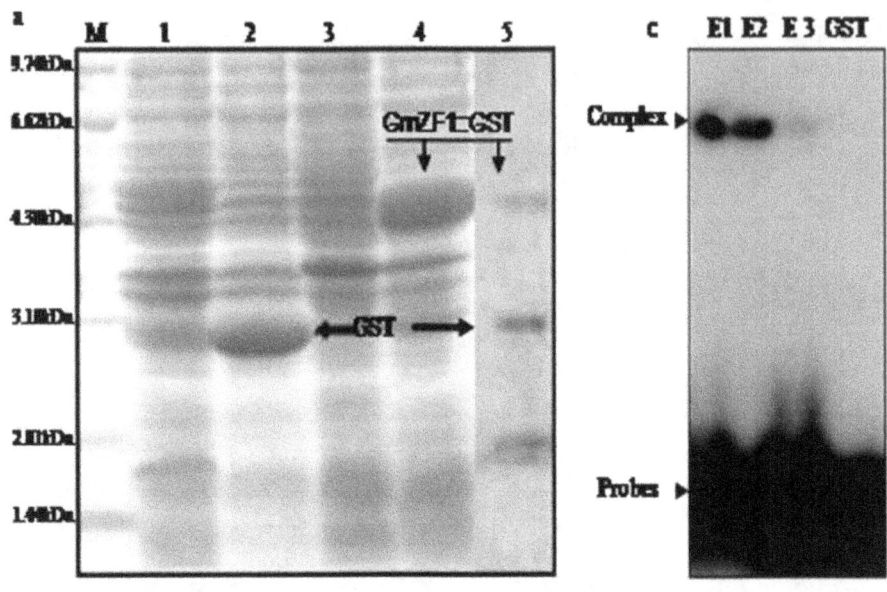

Figure 2. SDS PAGE induction expression and gel shift assay for DNA-binding activities *in vitro*. a: Induction expression and purification of target proteins, GmZF1::GST fusion protein and GST protein. Lane 1-uninduced strains containing pGEX4T-1 vector; Lane 2-induced strains containing pGEX4T-1; Lane-3-uninduced strains containing GmZF1::GST fusion; Lane 4-induced strains containing GmZF1::GST fusion; Lane 5-purified target protein as indicated by arrows. b: Probe design and composition. Nucleotide sequences of EP1S and mutated EP1S (E1, E2 and E3) probes. The nucleotide mutations in the EP1S core motif for each probe are boxed. c: The gel shift assay was performed in a solution containing 0.2 μg GmZF1::GST proteins or GST proteins and the probes ^{32}P-labeled EP1S (E1) or mutant EP1S (E2 and E3), respectively. The GmZF1::EP1S complex and free probes are indicated by arrows.

Localization of the GmZF1::GFP fusion protein was visualized exclusively in the nucleus (Fig. 3b), whereas the control GFP (35S::GFP) was distributed throughout the protoplast cells (Fig. 3a), demonstrating that GmZF1 is a nuclear protein possibly functioning as a transcriptional activator.

Expression of *GmZF1* in response to stress

To examine the transcription profile of *GmZF1* in soybean under cold stress, the expression of *GmZF1* in soybean was detected using semi-quantitative RT-PCR. The result showed that *GmZF1* was weakly expressed in young seedlings, and induced by low temperature and reached a maximum at 5 h after 4°C stress (Fig. 3c). Data also showed that *GmZF1* gene could be induced by exogenous ABA, and seems to have similar expression pattern under cold stress, and maintained at a high expression level when the seedlings were kept at ABA treatment (Fig. 3d), suggesting that *GmZF1* might be involved in plant responses to cold stress via an ABA-dependent pathway.

Phenotype changes of plants exposed to cold stress

To observe the difference in germination and survival rates between the wild type and the transgenic *Arabidopsis*, two independent culture experiments was performed at 22°C and at 4°C, respectively (Fig. 4a, b) Fig. 4a showed that both wild type and transgenic lines almost exhibited identical germination rates under favorable condition, but the survival rate of transgenic line OE1 was significantly higher than that of wild types under cold stress when wild type and transgenic lines of one-week-old seedlings were exposed to 4°C for one week (Fig. 4e). For observing the phenotypic changes of wild type and transgenic *Arabidopsis* under cold stress, the seedlings of two-week-old were transferred onto the MS medium, and cultured at 22°C (Fig. 4c) and 4°C (Fig. 4d) for one week, respectively (note: white bacterial plaques appeared in the mediums because of possible contaminant from culture processes). Data showed that transgenic lines showed a better phenotypic characteristic in withstanding cold stress compared to wild types (Fig. 4d), because both fresh weight and root lengths from the transgenic lines were significantly increased comparing to the wild types (Fig. 4f, g). These data suggested that *GmZF1* in the transgenic *Arabidopsis* could enhance the tolerance

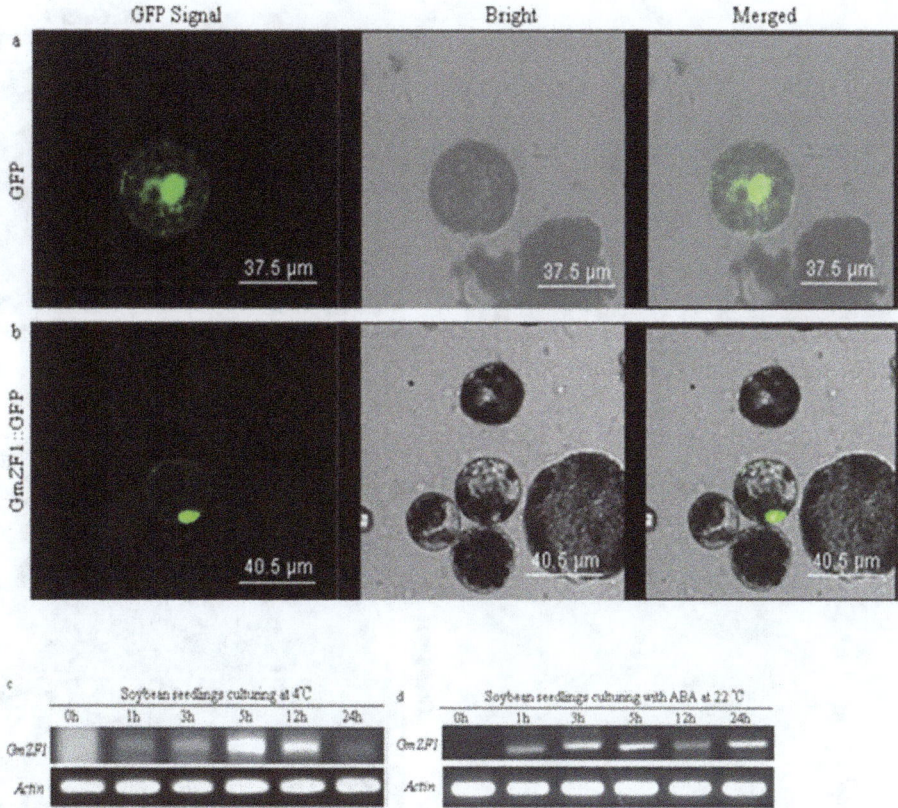

Figure 3. Subcellular localization of the GmZF1 proteins and mRNA accumulation of *GmZF1* gene in soybean seedlings respectively exposed to cold and ABA. a: Images expressing h16318-GFP in protoplast cells; b: images expressing the GmZF1::GFP fusion protein, fluorescent-field illumination was used to examine GFP signal (left); followed by bright-field illumination (middle) and confocal microscopy (right) for an merged image of bright and fluorescent illumination. c: the relative accumulation of *GmZF1* mRNA in soybean seedlings of 10-day-old exposed to 4°C at different time points; d: the accumulation of GmZF1 mRNA in soybean seedlings of 10-day-old exposed to ABA of 200 μM at different time points.

plants to the cold stress. *GmZF1* and *GsZFP1* [22] seems to play similar roles in acclimating the tolerance of plants to cold stress, although these two genes have differences in the structure, especially in the conserved domains of C2H2 zinc finger gene, suggesting that different zinc finger genes in the same super family could demonstrate an uniformity in gene functions even though they have the difference in the genetic characteristics, and this seems to imply that the diversity genetic of genes is an essential factor in regulating the expression of target genes related to abiotic stresses. Therefore, to better elucidate the molecular regulation mechanism of *GmZF1* from cultivar and *GsZFP1* from wild soybean in responding to cold stress, further investigations is necessary.

Over-expression of *GmZF1* led to physiological changes in the transgenic lines under cold stress

To further characterize the function of *GmZF1*, we generated *Arabidopsis* transgenic plants carrying *35S::GmZF1*, in which *GmZF1* was driven by the CaMV35S promoter. Total 22 transformed *Arabidopsis* plants with kanamycin-resistant were obtained, and three independent homozygous transgenic lines (OE1, OE4 and OE7) carrying *GmZF1* were continuously selected by the PCR analysis until T3 transgenic *Arabidopsis* lines. For cold stress, 5-week-old wild-type and transgenic lines were transferred to the incubator at 4°C for two days, and morphological observation showed that the transgenic lines were not significantly inhibited in growth comparing to wild type in spite of being a short

term cold stress, and the growth of the transgenic lines carrying *GmZF1* gene were slightly damaged comparing to wild type (Fig. 5a, b). However, three transgenic lines have remarkable physiological changes in accumulation of the proline, soluble sugar and MDA in the leaves comparing to the wild type after cold stress. Cold stress enhanced accumulation of proline and soluble sugar in the wild type and transgenic lines, but the increase times in the contents of proline in the transgenic lines (OE1, 1.3-fold; OE4, 1.3-fold and OE7, 1.6-fold) were significantly higher than that in the wild type (1.1-fold) (Fig. 5c). Similarly, the increase times in the content of soluble sugar in the leaves from OE1, OE4 and OE7 (1.7-fold, 1.9-fold and 2.2-fold) were also higher than that in the wild type (1.4-fold) after cold stress for two days (Fig. 5d). In contrast, the contents of MDA in the transgenic lines (OE1, 0.012 μmol·g^{-1}; OE4, 0.006 μmol·g^{-1} and OE7, 0.015 μmol·g^{-1}) were decreased comparing to the wild type (0.019 μmol·g^{-1}) after cold stress (Fig. 5e), demonstrating that over-expression of *GmZF1* reduced oxidative damage of membrane lipid peroxidation, and further maintained osmotic balance in the plant cells.

To observe the physiological effects of exogenous ABA supply on the transgenic *Arabidopsis* lines, the leaves from the wild type and the transgenic *Arabidopsis* lines exposed to ABA solution was sampled, and used for the determination in the content changes of praline, soluble sugar and MDA. Data showed that the contents of proline in the wild type were almost identical with that in the transgenic lines before ABA treatment. However, the content of

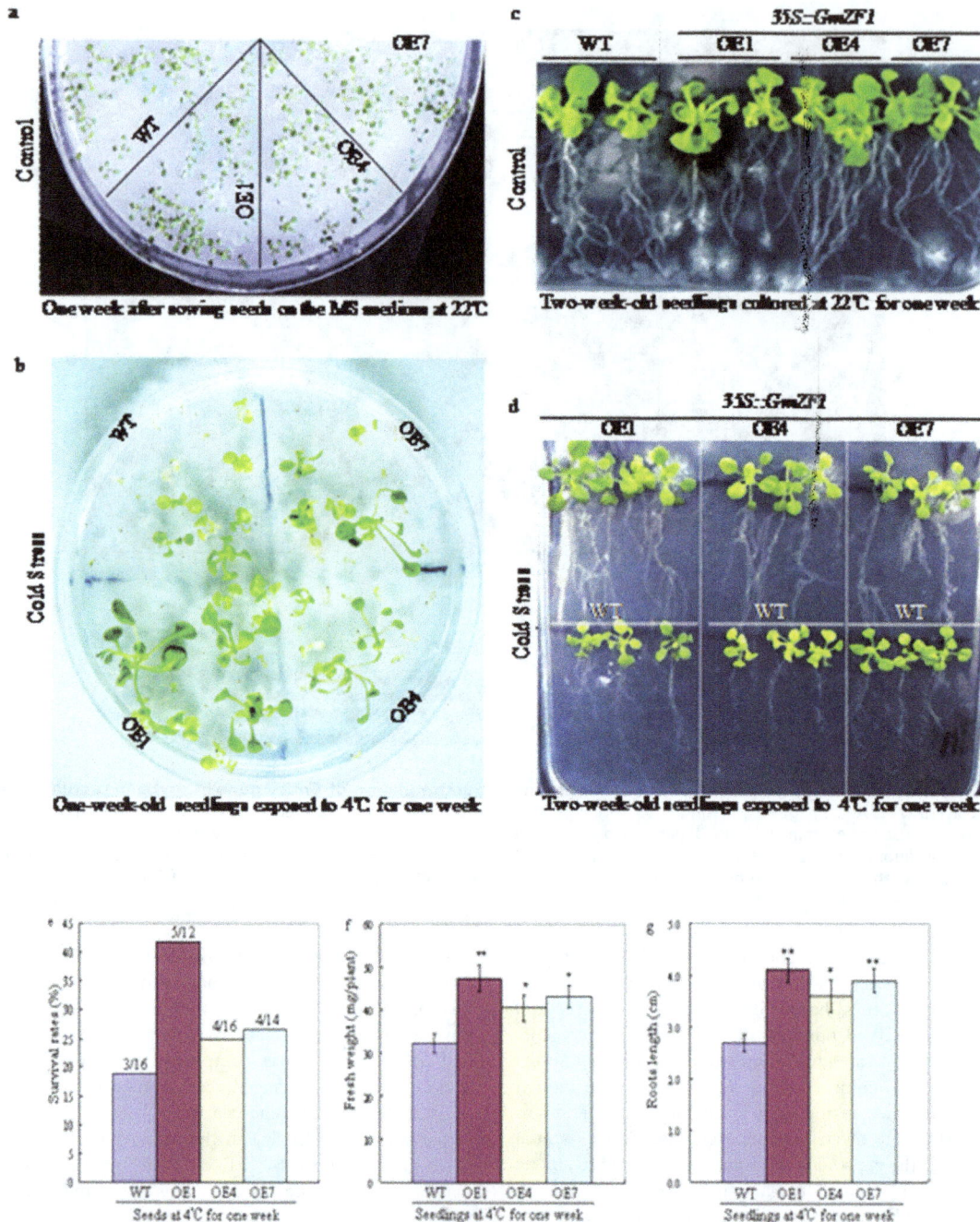

Figure 4. Morphological identification on wild type and transgenic *Arabidopsis* exposed to cold stress at 4°C on the MS medium. a: *Arabidopsis* seedlings of one-week-old after culturing at 22°C; c: *Arabidopsis* seedlings of two-week-old cultured at 22°C for one week; b and d: *Arabidopsis* seedlings of one-week-old cultured at 4°C for one week or two weeks, respectively; e: the survival rates of *Arabidopsis* seedlings exposed to 4°C for one week; f: the fresh weight per plant of two-week-old *Arabidopsis* after culturing at 4°C for one week; g: the roots length of two-week-old *Arabidopsis* seedlings after culturing at 4°C. WT-wild type; OE1, OE4 and OE7- transgenic lines.

proline in the transgenic lines (OE1, 142.4 ng·mg^{-1}; OE4, 427.2 ng·mg^{-1} and OE7, 412.9 ng·mg^{-1}) were significantly higher than that in the wild type (125.7 ng·mg^{-1}) after application of ABA (Fig. 6c). In addition, ABA treatment also enhanced the levels of soluble sugar in the wild type and the transgenic lines, but the increase times in the content of soluble sugar in the transgenic lines (OE1, 1.5-fold; OE4, 2.1-fold and OE7, 2.0-fold) were higher than that in the wild type (1.4-fold) after two days by ABA treatment for (Fig. 6d). The contents of MDA in the transgenic

lines (OE1, 0.012 μmol·g^{-1}; OE4, 0.009 μmol·g^{-1} and OE7, 0.011 μmol·g^{-1}) were reduced comparing to the wild type (0.013 μmol·g^{-1}) after ABA treatment, whereas the contents of MDA in the transgenic lines (OE1, 0.009 μmol·g^{-1}; OE4, 0.009 μmol·g^{-1} and OE7, 0.008 μmol·g^{-1}) were slightly higher than that in the wild type (0.007 μmol·g^{-1}) before ABA treatment (Fig. 6e).

The induction of numerous stress-responsive genes is a hallmark of stress acclimation in plants. To elucidate the molecular

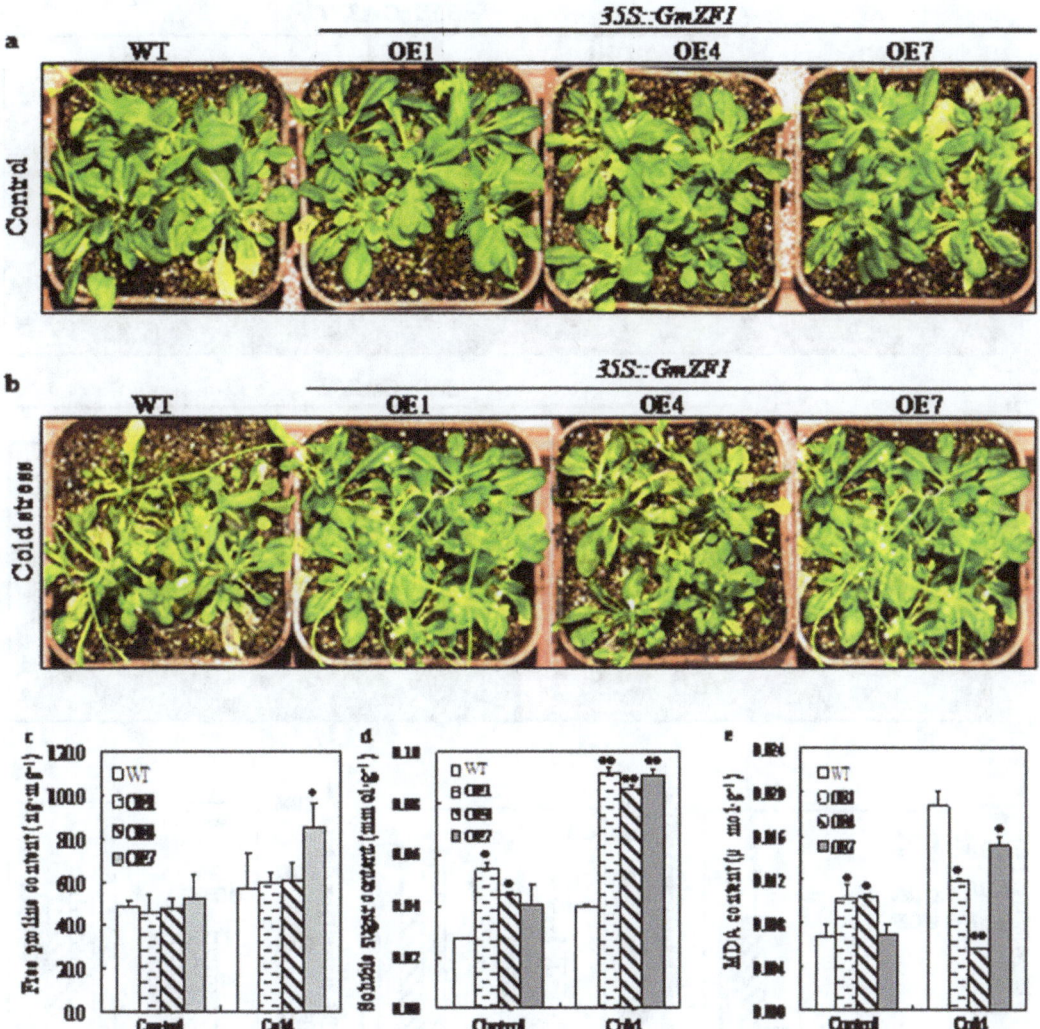

Figure 5. Phenotypic and physiological changes in wild type and transgenic *Arabidopsis* lines under cold stress. a: Phenotypes of wild type and transgenic lines of 5-week-old seedlings before cold stress (Control). b: Phenotypes of wild type and transgenic lines of 5-week-old seedlings exposed to 4°C for one week (Cold stress). c, d and e respectively represent the content changes of free proline, soluble sugar and MDA in the leaves of wild type and in transgenic lines before (Control) and after cold stress (Cold). WT-wild type; OE1, OE4 and OE7-transgenic lines; single * and duble ** respectively means significant difference at 0.05 and 0.01 levels.

mechanism of *GmZF1* in responding to cold, both wild type and transgenic seedlings of *Arabidopsis* were incubated at 4°C or by ABA for 48 hours to examine the expression of cold-responsive genes *cor6.6* at different time points by real-time PCR analysis. As shown in Fig. 7, both wild type and transgenic lines (OE1, OE4 and OE7) exhibited a similar levels in the expression of *cor6.6* before cold treatment, and the OE1 line began to appear significant up-regulations in *cor6.6* expressions in the roots after 12 hours cold treatment, while the OE7 line significantly increased the expression of *cor6.6* gene in the leaves after 12 h cold stress (Fig. 7a, b) When the seedlings of *Arabidopsis* were exposed to exogenous ABA, the expressions of *cor6.6* gene in the roots of the OE1 and OE7 lines were significantly induced after 24 hours (Fig. 7c), and the all transgenic lines (OE1, OE4 and OE7) exhibited remarkable increase in the expression of *cor6.6* gene in the leaves after 12 hours (Fig. 7d). Although both the wild type and transgenic lines at some time points did not exhibit significant differences in the expression of *cor6.6*, but the transgenic lines, OE1, OE4 and OE7 still showed higher transcript levels of *cor6.6*

gene than the wild type under cold stress or exogenous ABA supply (Fig. 7b, c and d), suggesting that *GmZF1* may play an important role in activating cold-resistance genes in the transgenic *Arabidopsis* responding to cold stress through an signal pathway depending ABA.

Fig. 7e, f showed that the expressions of *GmZF1* gene in the roots and in the leaves of transgenic lines significantly increased after exposing the plants to cold stress for 6 hours, and this result seems to be identical with that in soybean seedlings exposed to cold stress or ABA (Fig. 3c, d). Fig. 7e showed that the expression of *GmZF1* gene in the roots of OE4 and OE7 was almost identical, but the OE7 line exhibited obvious increase in the expression of *GmZF1* gene in the leaves after 6 hours cold stress relative to other transgenic lines, and then maintained at relative stable levels at other time points (Fig. 7f). In addition, accumulation pattern of *GmZF1* gene in the transgenic *Arabidopsis* lines treated by exogenous ABA also showed that *GmZF1* gene in the leaves of the transgenic *Arabidopsis* lines was strongly induced by ABA, and the accumulation levels *GmZF1* gene in the roots of OE1 line were

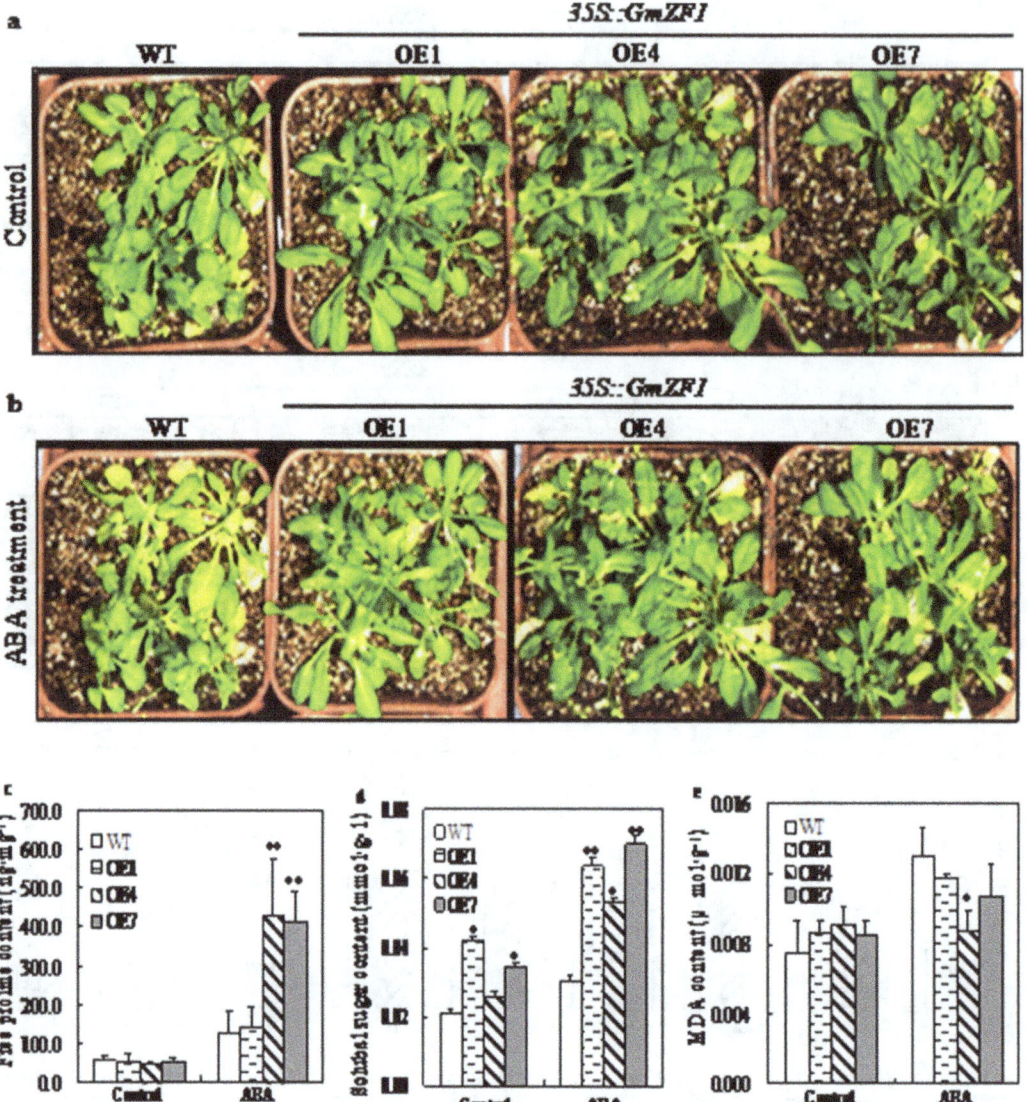

Figure 6. Phenotypic and physiological changes in wild type and transgenic *Arabidopsis* lines exposed to ABA. a: Phenotypes of wild type and transgenic lines of 5-week-old seedlings before ABA treatment. b: Phenotypes of wild type and transgenic lines of 5-week-old seedlings exposed to 200 µM ABA for one week. c, d and e respectively represent the content changes of free proline, soluble sugar and MDA in the leaves of wild type and transgenic lines before (Control) and after ABA treatment (ABA). WT-wild type; OE1, OE4 and OE7-transgenic lines; single * and duble ** respectively means significant difference at 0.05 and at 0.01 levels.

increased significantlyafter exposing to ABA for 24 hours (Fig. 7g) hours, and the accumulation of *GmZF1* mRNA in the leaves of OE the transgenic lines exhibited strong induction by ABA after 6 hours, and only OE7 line has almost identical induction accumulation patterns of *GmZF1* mRNA by ABA, and the induction accumulation seems to be inhibited after 48 hours (Fig. 7h). However, the accumulation of *GmZF1* mRNA in the OE7 transgenic line was little after exposing to ABA supply compared to the other transgenic lines, indicating that different transgenic lines exhibited different accumulation patterns in expressions of *GmZF1* gene in the plant organs, and seems to imply that accumulation patterns of *GmZF1* mRNA in the transgenic plants probably are related to the positions and copy numbers of *GmZF1* gene inserting in the genome of *Arabidopsis*.

Discussion

The C2H2 zinc-finger transcription factors were usually thought to be involved in plant development and have various adaptive responses to the environment stress [10,24]. Although the roles of some C2H2 zinc-finger transcription factors have been identified to be related to stress and developmental processes, the functions of C2H2 ZFPs from soybean involved in stress response are largely unknown [22].

In this study, as a novel C2H2 zinc finger protein gene, *GmZF1* from soybean was characterized. Sequence analysis revealed that the GmZF1 had high identity with other C2H2-type ZFPs, and shared two zinc finger motifs containing a conserved plant-specific QALGGH amino acid sequence which was proved to be critical for DNA-binding activity [11]. Based on the present data, we predicted that the binding activity of GmZF1 to EP1S core elements was probably mediated by the QALGGH sequence.

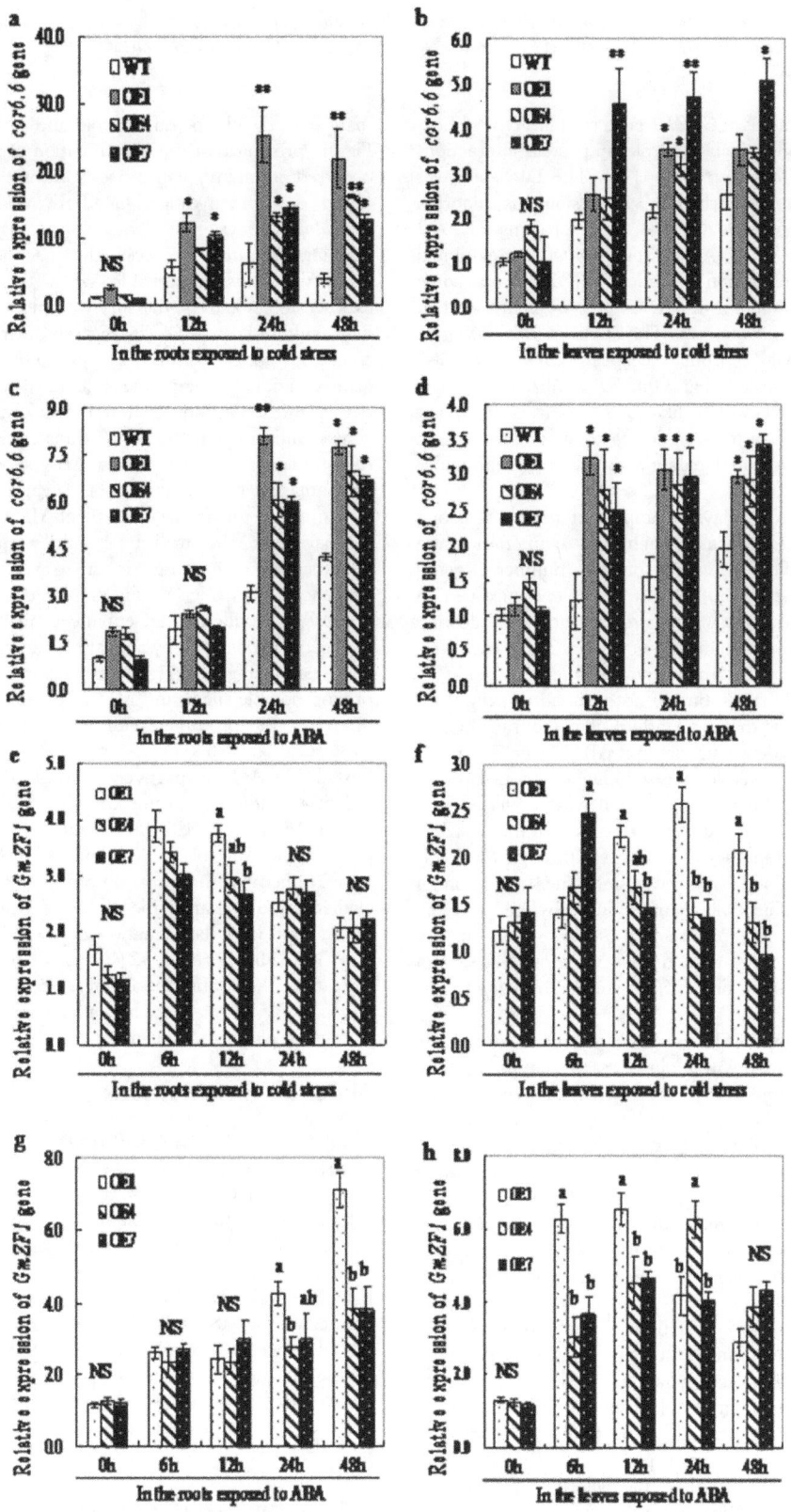

Figure 7. Responsive expression of *cor6.6* **gene and relative accumulation of** *GmZF1* **mRNA in wild type and transgenic lines both under cold stress and ABA supply.** a and b respectively represent the relative expressions of *cor6.6* gene in the roots and leaves of plants exposed to cold stress; c and d respectively represent the relative expressions of *cor6.6* gene in the roots and leaves of plants exposed to ABA, and the expression level of *cor6.6* gene in unstressed wild type was set at 1.0 in the transgenic *Arabidopsis* lines. e: the relative expressions of *GmZF1* gene in the roots; f: the relative expressions of *GmZF1* gene in the leaves of transgenic plants exposed to cold stress; g and h respectively represent the relative expressions of *GmZF1* gene in the roots and leaves of transgenic tolants exposed to ABA. Error bars indicate the standard deviation; NS means

no significant differences; lowercase letters labeled on the columns mean statistical difference at 0.05 level ce at 0.05 level; single * and duble ** respectively means significant difference at 0.05 and at 0.01 levels; WT-wild type; OE1, OE4 and OE7-transgenic lines.

Additionally, the C-terminus of *GmZF1* gene contains typical Leu-rich L-box and DLN-box which play roles in protein interactions or in maintaining the folded structure [19]. The DLN-box was thought to function in transcriptional repression. As reported previously, the DLN-boxes in the zinc finger proteins ZPT2–3 from petunia [16] and STZ/ZAT10 from *Arabidopsis* exhibited repression roles in transcription activities [18]. However, some zinc finger proteins containing the DLN-box were involved in transcriptional activation, such as ThZF1 from *Thellungiella halophila* [25] and CaZF from chickpea [26]. Subcellular localization analysis revealed that GmZF1 localized at nuclei (Fig. 3), implying that GmZF1, like other ZFPs from plant TFIIIA-type, functions as a transcription factor in plant cells and may play an important role in signaling pathway in soybean under abiotic stress.

As a key regulator, ABA plays an important role in signaling pathways under stresses, such as drought, low temperature and osmotic stress, and induces the expressions of a number of genes that respond to abiotic stress [30,31]. For example, the expression of *AZF2* gene from *Arabidopsis* was strongly induced following ABA treatment under drought and salt stress [1,12,18]. However, it was reported that some C2H2-type zinc finger genes are induced by dehydration and cold stress, but do not respond to exogenous ABA [2,32,33], suggesting that two kinds of signaling transduction pathways, ABA-independent pathway and ABA-dependent pathway play different regulation roles in responding to abiotic stresses, repectively, and that the initial stress signal was converted into cellular responses [34]. In this study, our data proved that *GmZF1* was involved in plant responses to cold stress through an ABA-dependent signal transduction pathway since the expression of *GmZF1* in soybean seedlings was clearly induced by ABA (Fig. 4). In addition, accumulation pattern of *GmZF1* gene in the transgenic *Arabidopsis* lines treated by exogenous ABA also showed that over-expression of *GmZF1* gene in *Arabidopsis* was strongly induced by ABA.

Some of the C2H2-type ZFPs from different plant species are confirmed to play regulatory roles in stress responses, such as AZF2 and STZ in Arabidopsis [12,17,27], SCOF-1 in soybean [21], GsZFP1 in wild soybean [22], StZFP1 in potato [9], DST in rice [28] and TaCHP in wheat [29]. Expression analysis revealed that GmZF1 was clearly induced by cold stress (Fig. 4), suggesting that *GmZF1* might be involved in plant responses to cold stress in soybean. Transgenic *Arabidopsis* plants over-expressing *GmZF1* were evaluated for the involvement of *GmZF1* gene in cold tolerance of plants. Although no morphological differences between the transgenic lines and the wild type was observed, the remarkable differences in accumulation of free proline, soluble sugar and MDA were confirmed after cold and ABA treatment (Fig. 5c–e and Fig. 6c–e). Accumulation of proline by stress-induced has been observed in many plant species, and functions as an osmo-protectant coping with stress [32,35,36]. For example, the transgenic tobacco plants over-expressing *GmDREB3* could accumulate much free proline than the wild-type plants after drought stress treatment for 16 d [37]. In plants, soluble sugar has been shown to fulfill a dual role as both metabolites and as signaling molecules [38,39] that may play important roles in the mechanisms of plant responding to the stress [40,41]. In our study, the accumulations of proline and soluble sugar in the wild type and transgenic lines were improved, and increase times in the contents of proline and soluble sugar in the transgenic lines were higher

than that in wild type during cold and ABA treatment (Fig. 5 and Fig. 6), proving that the accumulation of soluble sugars has been occurred in many plant species during cold acclimation [42]. Based on our investigations, both wild type and transgenic *Arabidopsis* lines basically have no significant differences in germination rates. However, the survival rates from these two phenotypes was remarkable when these two seedlings were exposed to cold stress, the survival rates of transgenic *Arabidopsis* lines cultured at 4°C significantly increased after one week comparing to the wild types (Fig. 4 a,b), indicating that *GmZF1* gene could play regulation role during the plant growth and development. Our study showed that over-expression of *GmZF1* gene significantly enhanced the transcription levels of *cor6.6* gene responding to cold stress and ABA supply, and increased the accumulation of proline, soluble sugar in the transgenic *Arabidopsis*, and reduced the content of MDA in the transgenic lines comparing to the wild type, and significantly improved the tolerance of the transgenic lines exposed to cold stress and exogenous ABA (Fig. 5e; Fig. 6e). In conclusion, over-expression of *GmZF1* resulted in an enhancement in cold tolerance of the transgenic *Arabidopsis* by activating transcription of *cor6.6* and/or cold resistance-genes (Fig. 7) as well as accumulation changes of proline, soluble sugar and MDA [43–45]. Bioinformatics analysis fund that the promoter with 1955 bp of *cor6.6* gene contains $7 \times$ GACA and $6 \times$ GTCA repeat in core element TTGA-CAGTGTCAC, respectively (NCBI Reference Sequence in *Arbidopsis thaliana* chromosome 5: NC_003076.8-AT5G15950/ NM_121600.3/NP_197099.1), and these two core elements possibly play key role in recognizing DNA-binding sites in target proteins, because the mutation of G and C in the two boxed core elements could change the ability of GmZf1 protein binding to DNA, especially the mutation of G and C in the GCTA box (Fig. 3c). Therefore, *GmZF1* could play regulation role by activating the transcription of *cor6.6* cold-regulated gene in *Arabidopsis*, and led to an enhancement in the tolerance of transgenic plants to cold stress.

Materials and Methods

Plant materials and growth conditions

Soybean seedlings (*Glycine max*) were grown in a growth chamber at 24°C with 60% relative humidity under 16 h light and 8 h darkness. *Arabidopsis* plants (genotype Colombia) were grown in a controlled environmental chamber at 22°C and 70% humidity with a 14 h light/10 h darkness cycle under normal light intensity (150 $Em^{-2} \cdot s^{-1}$). T_1 seeds were sterilized and planted on the MS medium containing kanamycin of 50 $\mu g \cdot mL^{-1}$ for the selection of transgenic plants. After continuously screening on the MS medium containing kanamycin of 50 $\mu g \cdot mL^{-1}$ for two times and the generated T3 seedlings of transgenic *Arabidopsis* lines were transplanted to pots or MS medium for further investigations.

Isolation and sequence analysis of *GmZF1*

The zinc finger gene *GmZF1* was cloned from soybean based on the expressed sequence tag (EST) database of soybean. Total RNA was extracted from soybean seedlings using RNA Prep. Pure plant kit (Tiangen Biological Company, Beijing), and cDNA synthesis was performed by a reverse transcription kit (TaKaRa Dalian BioCompany). The entire *GmZF1* cDNA of was obtained by PCR using the specific primers (F: 5′-AGAGGAAACTAGCTAGGG-CACTTC-3′ and R: 5′-CCCGAGAACTAAGAAGTTTCG-TATT-3′). The deduced amino acid sequences of soybean GmZF1 zinc finger protein were matched by a protein blast procedure (http://www.ncbi.nlm.nih.gov/BLAST/). BioEdit version 7.0 software was used to multiple sequence alignments. Phylogenetic analysis was carried out by MEGA version 4.0 with adopting position correction distance using a bootstrap replicate number of 1000.

Induction expression of GmZF1 fusion protein and electrophoretic mobility shift assay

An entire 516-bp GmZF1 fragment containing the DNA binding domain was amplified by PCR using a pair of primers F and R (F: 5′-ACAACTCGAGATGAAGAGAGGCAGAGAA-3′ and R: 5′-AGACGAATTCAATGAAACAATTGAGCAC-3′), and was subcloned onto the pGEX4 T-1 vector by inserting at two specific sites, EcoRI and XhoI (Amersham Biosciences), and the recombinant pGEX-4 T-1 plasmids were identified by sequencing, and transferred into Escherichia coli BL21 cells (Amersham Biosciences). The induction of target fusion proteins was performed by adding 0.8 mM IPTG (Isopropyl-β-d-thioga-lactopyranoside) into the culture of E. coli strains, and the cultures carrying GmZF1::GST fusion proteins was incubated for 5 hours at 37°C with 250 rpm The bacterial cells were shattered by ultrasonication in phosphate-buffered saline (PBS) and centrifuged at 11000 g for 10 min to remove insoluble cell debris, and then the supernates were collected, and applied on the gel containing 12% ployacrylamide for electrophoresis detection of fusion protein by the method of SDS-PAGE described in the book [46]. The GST::GmZF1 fusion protein was purified using a glutathione-Sepharose 4B column (Amersham Biosciences) according to the manufacturer's instructions. The 52 bp DNA fragment containing four copies of wild-type or mutant EP1S core sequences were labeled by 25 $\mu Ci \cdot \mu L^{-1}$ of γ-^{32}P-dATP (Amersham Biosciences). The DNA-binding reaction was performed in a 20 μL binding buffer containing ^{32}P-labelled probe, glycerine and purified fusion protein [24], and were subjected to electrophoresis with the gel containing 0.53 Tris-borate-EDTA, 6% polyacrylamide, and the gels were dried and visualized by autoradiography.

Subcellular localization of GmZF1 gene

In brief, an entire GmZF1 cDNA fragment was amplified by PCR procedure with a pair of specific primers (F: 5′-AACACTG-CAGATGAAGAGAGGCAGAGA-3′), and R: 5′-CGGGATC-CAATGAAACAATTGAGCAC-3′.). Based on the specific sites in the MCS of GFP vector, two specific sites, a PstI and BamHI were respectively inserted into the 5′ ends of the primers for subcloning requirement. The digested PCR fragment was inserted into the multiple cloning sites (MCS) in the GFP vector carrying a GFP protein driven by a 35 S promoter, and this construct carrying 35 S::GFP or 35 S::GmZF1::GFP were introduced into Arabi-dopsis protoplast cells, which was prepared by young seedlings of Arabidopsis, and the transformation procedures were performed by the method as described previously [47,48]. The images were captured using a Nikon Eclipse TE2000-U microscope (Nikon).

Transformation of Arabidopsis

A pair of specific primers, the GmZF1 cDNA fragment was generated by PCR procedure using the specific primers (F: 5′-AACACCATGGCCATGAAGAGAGGCAGAGA-3′, and R: 5′-AGACACTAGTAATGAAACAATTGAGCAC-3′. The generat-ed PCR product directly was inserted onto the pCAMBIA1304

vector using a pair of specific restriction sites (NcoI and SpeI), and an over-expression vector p35 S::GmZF1 was constructed, and introduced into Agrobacterium tumefaciens strain C58C1. The flowering Arabidopsis seedlings were used for genetic transforma-tion at 25°C by the method of floral-dip. T3 transgenic Arabidopsis plants were continuously cultured in the medium containing kanamycin of 45 $mg \cdot L^{-1}$ and three positive transgenic Arabidopsis were identified and the T3 seeds of transgenic Arabidopsis lines and wild type Arabidopsis were sowed on MS medium based on the requirements of experiment, and the resulting seedlings were accordingly transferred into the pots and MS medium for cold stress and ABA treatment.

Stress treatments

Cold stress was performed by incubating soybean seedlings of 10-day-old at 4°C for 24 hours and Arabidopsis seedlings of two-week-old seedlings were incubated at 4°C for 48 hours. ABA treatment was respectively carried out by completely spraying 200 $\mu mol \cdot L^{-1}$ ABA onto the 10-day-old soybean seedlings and two-week-old Arabidopsis seedlings Soybean leaves from the same position were respectively sampled at 0, 1, 3, 5, 12 and 24 hours, and Arabidopsis leaves were collected respectively at 0, 6, 12, 24 and 48 hours. Equal amounts of seeds from wild type and transgenic lines were sowed on MS medium at 22°C (Control), and the generated seedlings of one-week-old Arabidopsis were cultured at 22°C and at 4°C (Cold stress), respectively, for one week in two independent growth chambers, and used for investigation in survival rates. For evaluating the changes of Arabidopsis biomass and root development under cold stress, two-week-old seedlings of Arabidopsis from control culture were transplanted on the MS medium, and cultured respectively at 22°C (Control) and at 4°C (Cold stress) for one week in two independent growth chambers, and biological fresh weights per plant and root lengths from two treatments were statistically investigated based on the three replicates.

Gene expression analysis

Based on the equal amounts of leaves RNA from each time point, the expression pattern of GmZF1 gene in soybean was characterized using semi-quantitative RT-PCR by GmF and GmR primers, and accumulation analysis of GmZF1 mRNA in the transgenic Arabidopsis lines was performed using the specific primers (GmF: 5′-ATGAAGAGAGGCAGAGAA-3′ and GmR: 5′-AATGAAACAATTGAGCAC-3′) by quantitative real-time PCR (qPCR) For understanding the expression profile of GmZF1 gene and cold-responsive marker gene cor6.6 in the transgenic Arabidopsis under low temperature stress and exogenous ABA supply, the wild type Arabidopsis and transgenic Arabidopsis lines carrying GmZF1 gene were incubated at 4°C for 48 hours in darkness. All samples including the roots and leaves were collected with three biological replicates at the designated time intervals after cold and ABA treatment, and were quickly frozen in liquid nitrogen and stored at 80°C for total RNA isolation. As a reference gene in Arabidopsis, actin was used as an internal maker, and amplified by the specific primers [(F 5′-AAGTATCCTATT-GAGCATGGTGTTG-3′; R 5′- CTGGCGTACAAGGAGAG-A-3′), (accession number: AEE76148)], and the cor6.6 gene was amplified by the primers (F 5′- ATGTCAGAGACCAACAA-GAATG-3′; R5′- CTTGTTCAGGCCGGTCTTG-3′) (acces-sion number: CAA38894) with qPCR, which was performed using a SYBR premix Ex Taq kit (TaKaRa) according to the manufacturer's instructions on a 7900HT Real-time PCR system.

Accumulation analysis of free proline, soluble sugar and malondialdehyde (MDA)

0.5 g fresh leaves from each treatment of the wild-type and the transgenic *Arabidopsis* lines were respectively collected at different time point, and respectively were treated according to requirements of experiment and used for measurements of free praline, soluble sugar and MDA. The contents of free proline were measured by the previous procedure [40,49,50]. Measurement of MDA content in *Arabidopsis* leaves was determined as described

previously [51,52], and the content of soluble sugar was measured with the method as described by [53].

Author Contributions

Conceived and designed the experiments: XGC ZSX. Performed the experiments: GHY LLJ XFM MML SGS. Analyzed the data: ZSX. Contributed reagents/materials/analysis tools: GHY. Wrote the paper: XGC GHY ZSX.

References

1. Kodaira KS, Qin F, Tran LSP, Maruyama K, Kidokoro S, et al. (2011) *Arabidopsis* Cys2/His2 zinc-finger proteins AZF1 and AZF2 negatively regulate abscisic acid-repressive and auxin-inducible genes under abiotic stress conditions. Plant Physiol 157: 742–756.
2. Yamaguchi-Shinozaki K, Shinozaki K (2006) Transcriptional regulatory networks in cellular responses and tolerance to dehydration and cold stresses. Annu Rev Plant Biol 57: 781–803.
3. Umezawa T, Fujita M, Fujita Y, Yamaguchi-Shinozaki K, Shinozaki K (2006) Engineering drought tolerance in plants: discovering and tailoring genes to unlock the future. Curr Opin Biotechnol 17: 113–122.
4. Zhang G, Chen M, Chen X, Xu Z, Guan S, et al. (2008) Phylogeny, gene structures, and expression patterns of the ERF gene family in soybean (*Glycine max*). J Exp Bot 59: 4095–4107.
5. Xu ZS, Chen M, Li LC, Ma YZ (2011) Functions and application of the AP2/ERF transcription factor family in crop improvement. J Integr Plant Biol 53: 570–585.
6. Laity JH, Lee BM and Wright PE (2001) Zinc finger proteins: new insights into structural and functional diversity. Curr Opin Struct Biol 11: 39–46.
7. CO Pabo, Peisach E, Grant RA (2001) Design and selection of novel Cys2His2 zinc finger proteins. Annu Rev Biochem 70: 313–340.
8. Takatsuji H (1999) Zinc-finger proteins: the classical zinc finger emerges in contemporary plant science. Plant Mol Biol 39: 1073–1078.
9. Tian ZD, Zhang Y, Liu J, Xie CH (2010) Novel potato C2H2-type zinc finger protein gene, *StZFP1*, which responds to biotic and abiotic stress, plays a role in salt tolerance. Plant Biol 12: 689–697.
10. Ciftci-Yilmaz S, Mittler R (2008) The zinc finger network of plants. Cell Mol Life Sci 65: 1150–1160.
11. Kubo K, Sakamoto A, Kobayashi A, Rybka Z, Kanno Y, et al. (1998) Cys2/His2 zinc-finger protein family of petunia: evolution and general mechanism of target-sequence recognition. Nucleic Acids Res 26: 608–615.
12. Chen Y, Sun A, Wang M, Zhu Z, Ouwerkerk PB (2013) Functions of the CCCH type zinc finger protein OsGZF1 in regulation of the seed storage protein GluB-1 from rice. Plant Mol Biol DOI 10.1007/s11103-013-0158-5.
13. Takatsuji H, Mori M, Benfey PN, Ren L, Chua NH (1992) Characterization of a zinc finger DNA-binding protein expressed specifically in Petunia petals and seedlings. EMBO J 11: 241–249.
14. Liu P, Xu ZS, Pan-Pan L, Hu D, Chen M, et al. (2013) A wheat plasma membrane-localized *PI4K* gene possessing threonine autophophorylation activity confers tolerance to drought and salt in *Arabidopsis*. J Exp Bot 64: 2915–2927.
15. Van Der Krol AR, Van Poecke RMP, Vorst OFJ, Voogd C, Van Leeuwen W, et al. (1999) Developmental and wound-, cold-, desiccation-, ultraviolet-b-stress-induced modulations in the expression of the Petunia Zinc Finger Transcription Factor Gene *ZPT2–2*. Plant Physiol 121: 1153–1162.
16. Sugano S, Kaminaka H, Rybka Z, Catala R, Salinas J, et al. (2003) Stress responsive zinc finger gene *ZPT2–3* plays a role in drought tolerance in petunia. Plant J 36: 830–841.
17. Lippuner V, Cyert MS, Gasser CS (1996) Two classes of plant cDNA clones differentially complement yeast calcineurin mutants and increase salt tolerance of wild-type yeast. J Biol Chem 271: 12859–12866.
18. Sakamoto H, Maruyama K, Sakuma Y, Meshi T, Iwabuchi M, et al. (2004) *Arabidopsis* Cys2/His2-type zinc-finger proteins function as transcription repressors under drought, cold, and high-salinity stress conditions. Plant Physiol 136: 2734–2746.
19. Sakamoto H, Araki T, Meshi T, Iwabuchi M (2000) Expression of a subset of the Arabidopsis Cys (2)/His (2)-type zinc-finger protein gene family under water stress. Gene 248: 23–32.
20. Mittler R, Kim YS, Song L, Coutu J, Coutu A, et al. (2006) Gain-and loss-of-function mutations in Zat10 enhance the tolerance of plants to abiotic stress. FEBS Lett 580: 6537–6542.
21. Kim JC, Lee SH, Cheong YH, Yoo CM, Lee SI, et al. (2001) A novel cold-inducible zinc finger protein from soybean, SCOF-1, enhances cold tolerance in transgenic plants. Plant J 25: 247–259.
22. Luo X, Bai X, Zhu D, Li Y, Ji W, et al. (2012) GsZFP1, a new Cys2/His2-type zinc-finger protein, is a positive regulator of plant tolerance to cold and drought stress. Planta 235: 1141–1155.

23. Takatsuji H, Nakamura N, Katsumoto Y (1994) A new family of zinc finger proteins in petunia: structure, DNA sequence recognition, and floral organ-specific expression. Plant Cell 6: 947–958.
24. Lu X, Li Y, Su Y, Liang Q, Meng H, et al. (2012) An *Arabidopsis* gene encoding a C2H2-domain protein with alternatively spliced transcripts is essential for endosperm development. J Exp Bot 63: 5935–5944.
25. Xu SM, Wang XC, Chen J (2007) Zinc finger protein 1 (ThZF1) from salt cress (*Thellungiella halophila*) is a Cys-2/His-2-type transcription factor involved in drought and salt stress. Plant Cell Rep 26: 497–506.
26. Jain D, Roy N, Chattopadhyay D (2009) *CaZF*, a plant transcription factor functions through and parallel to HOG and calcineurin pathways in *Saccharomyces cerevisiae* to provide osmotolerance. PLoS ONE 4: e5154.
27. Nakai Y, Nakahira Y, Sumida H, Takebayashi K, Nagasawa N, et al. (2013) Vascular plant one-zinc-finger protein 1/2 transcription factors regulate abiotic and biotic stress responses in Arabidopsis. Plant J 73: 761–775.
28. Huang XY, Chao DY, Gao JP, Zhu MZ, Shi M, et al. (2009) A previously unknown zinc finger protein, DST, regulates drought and salt tolerance in rice via stomatal aperture control. Genes & Development 23: 1805–1817.
29. Li C, Lv J, Zhao X, Ai X, Zhu X, et al. (2010) *TaCHP*: a wheat zinc finger protein gene down-regulated by abscisic acid and salinity stress plays a positive role in stress tolerance. Plant Physiol 154: 211–221.
30. Xiong L, Schumaker KS, Zhu JK (2002) Cell signaling during cold, drought, and salt stress. Plant Cell 14: 165–183.
31. Finkelstein RR, Gampala SSL, Rock CD (2002) Abscisic acid signaling in seeds and seedlings. Plant Cell 14: 15–45.
32. Davletova S, Schlauch K, Coutu J, Mittler R (2005) The zinc-finger protein Zat12 plays a central role in reactive oxygen and abiotic stress signaling in *Arabidopsis*. Plant Physiol 139: 847–856.
33. Zhu JK (2002) Salt and drought stress signal transduction in plants. Annu Rev Plant Biol 53: 247–273.
34. Shinozaki K, Yamaguchi-Shinozaki K, Seki M (2003) Regulatory network of gene expression in the drought and cold stress responses. Curr Opin Plant Biol 6: 410–417.
35. Igarashi Y, Yoshiba Y, Sanada Y, Yamaguchi-Shinozaki K, Wada K, et al. (1997) Characterization of the gene for 1-pyrroline-5-carboxylate synthetase and correlation between the expression of the gene and salt tolerance in Oryza sativa L. Plant Mol Biol 33: 857–865.
36. Dobrá J, Vanková R, Havlová M, Burman AJ, Libus J, et al. (2011) Tobacco leaves and roots differ in the expression of proline metabolism-related genes in the course of drought stress and subsequent recovery. J Plant Physiol 168: 1588–1597.
37. Chen M, Xu Z, Xia L, Li L, Cheng X, et al. (2009) Cold-induced modulation and functional analyses of the DRE-binding transcription factor gene, *GmDREB3*, in soybean (*Glycine max*). J Exp Bot 60: 121–135.
38. Bolouri Moghaddam MR, Le Roy K, Xiang L, Rolland F, Van den Ende W (2010) Sugar signaling and antioxidant network connections in plant cells. FEBS J 277: 2022–2037.
39. Smeekens S, Ma J, Hanson J, Rolland F (2010) Sugar signals and molecular networks controlling plant growth. Curr Opin Plant Biol 13: 273–278.
40. Sperdouli I, Moustakas M (2012) Interaction of proline, sugars, and anthocyanins during photosynthetic acclimation of *Arabidopsis thaliana* to drought stress. J Plant Physiol 169: 577–585.
41. Ramel F, Sulmon C, Gouesbet G, Couée1 I (2009) Natural variation reveals relationships between pre-stress carbohydrate nutritional status and subsequent responses to xenobiotic and oxidative stress in *Arabidopsis thaliana*. Anna Bot 104: 1323–1337.
42. Guy CL (1990) Cold Acclimation and freezing stress tolerance: Role of protein metabolism. Annu Rev Plant Biol 41: 187–223.
43. Giannakoula A, Moustakas M, Mylona P, Papadakis I, Yupsanis T (2007) Aluminium tolerance in maize is correlated with increased levels of mineral nutrients, carbohydrates and proline and decreased levels of lipid peroxidation and Al accumulation. J Plant Physiol 165: 385–396.
44. Zhan X, Wang B, Li H, Liu R, Kalia RK, et al. (2012) *Arabidopsis* proline-rich protein important for development and abiotic stress tolerance is involved in microRNA biogenesis. Proc Natl Acad Sci USA 109: 18198–203.
45. Xu ZS, Xia LQ, Chen M, Cheng XG, Zhang RY, et al. (2007) Isolation and molecular characterization of the *Triticum aestivum* L. Ethylene-responsive

factor 1 (*TaERF1*) that increases multiple stress tolerance. Plant Mol Biol 65: 719–732.

46. Davis LG, Kuehl WM, Battey JF (1994) Section, 11-3: Acidic guanidine isothiocyanate/phenol/chlororm extraction for isolation of RNA; Section, 17-1: Electrophoresis of protein on sodium dodecy sulfate polyacrylamide gels. In Basic methods in Molecular Biology, 2nd ed.1994, 335–338, 661–668, Appleton & Lange, USA.

47. Han MJ, Jung KH, Yi G, Lee DY, An G (2006) Rice Immature Pollen 1 (RIP1) is a regulator of late pollen development. Plant Cell Physiol. 47: 1457–1472.

48. Liu XM, Nguyen XC, Kim KE, Han HJ, Yoo J, et al. (2013) Phosphorylation of the zinc finger transcriptional regulator *ZAT6* by *MPK6* regulates *Arabidopsis* seed germination under salt and osmotic stress. Biochem Biophys Res Commun 430: 1054–1059.

49. Giannakoula A, Moustakas M, Syros T, Yupsanis T (2012) Aluminium stress induces upregulation of an efficient antioxidant system in the Al-tolerant maize line but not in the Al-sensitive line. Environ Exp Bot 67: 487–494.

50. Zhan X, Wang B, Li H, Liu R, Kalia RK, et al. (2012) Arabidopsis proline-rich protein important for development and abiotic stress tolerance is involved in microRNA biogenesis. Proc Natl Acad Sci USA 109: 18198–203.

51. Yue Y, Zhang M, Zhang J, Tian XL, Duan LS, et al. (2012) Overexpression of the *AtLOS5* gene increased abscisic acid level and drought tolerance in transgenic cotton. J Exp Bot 63: 3741–3748.

52. Ai L, Li ZH, Xie ZX, Tian XL, Eneji AE, et al. (2008) Coronatine alleviates polyethylene Glycol-induced water stress in Two Rice (*Oryza sativa L.*) cultivars. J Agronomy and Crop Sci 194: 360–368.

53. Taji T, Ohsumi C, Iuchi S, Seki M, Kasuga M, et al. (2002) Important roles of drought-and cold-inducible genes for galactinol synthase in stress tolerance in *Arabidopsis thaliana*. Plant J 29: 417–426.

4

Towards the Generation of B-Cell Receptor Retrogenic Mice

Jenny Freitag[1¤a], Sylvia Heink[1¤b], Edith Roth[2], Jürgen Wittmann[2], Hans-Martin Jäck[2], Thomas Kamradt[1]*

1 Department of Immunology, University Hospital Jena, Jena, Germany, **2** Division of Molecular Immunology, Department of Internal Medicine III, Nikolaus-Fiebiger-Center, University of Erlangen-Nürnberg, Erlangen, Germany

Abstract

Transgenic expression of B- and T-cell receptors (BCRs and TCRs, respectively) has been a standard tool to study lymphocyte development and function *in vivo*. The generation of transgenic mice is time-consuming and, therefore, a faster method to study the biology of defined lymphocyte receptors *in vivo* would be highly welcome. Using 2A peptide-linked multicistronic retroviral vectors to transduce stem cells, TCRs can be expressed rapidly in mice of any background. We aimed at adopting this *retrogenic* technology to the *in vivo* expression of BCRs. Using a well characterised BCR specific for hen egg lysozyme (HEL), we achieved surface expression of the retrogenically encoded BCR in a *Rag*-deficient pro B-cell line *in vitro*. *In vivo*, retrogenic BCRs were detectable only intracellularly but not on the surface of B cells from wild type or *Rag2*-deficient mice. This data, together with the fact that no BCR retrogenic mouse model has been published in the 7 years since the method was originally published for TCRs, strongly suggests that achieving BCR-expression *in vivo* with retrogenic technology is highly challenging if not impossible.

Editor: Ari Waisman, University Medical Center of the Johannes Gutenberg University of Mainz, Germany

Funding: This work is supported by KA 755/5–1, www.dfg.de, Deutsche Forschungsgemeinschaft (TK, JF). The funders had no role in study design, data collection and analysis, decision to publish, or preparation of the manuscript.

Competing Interests: The authors have declared that no competing interests exist.

* Email: thomas.kamradt@med.uni-jena.de

¤a Current address: Institute for Infection Immunology, TWINCORE, Centre for Experimental and Clinical Infection Research, a joint venture between the Medical School Hannover (MHH) and the Helmholtz Centre for Infection Research (HZI), Hannover, Germany
¤b Current address: Klinikum rechts der Isar der TU München, Experimentelle Neuroimmunologie, München, Germany

Introduction

Over the last three decades transgenic mice have been valuable tools to study the biology of lymphocytes. Prominent examples include mice that expressing transgenic B-cell receptor (BCR) recognising (neo) self-antigens, which served to identify tolerance mechanisms in B cells [1–3].

Breeding transgenic mice onto different backgrounds either by conventional back-crossing or the speed congenic approach is time consuming and expensive. To overcome these major limitations, a new technique to express TCRα and TCRβ chains from a 2A peptide-linked bicistronic retroviral vector using retroviral-mediated stem cell gene transfer was developed and published in 2006 [4–6]. These mice were designated '*retrogenic*' ('*retro*' from retrovirus and '*genic*' from transgenic; rg). The original publication describes the generation of retrogenic mice expressing either the OTI- or OTII-OVA-specific TCR. Holst and colleagues detected frequencies of OTI[+]- or OTII[+]-T cells in the retrogenic mice that were similar to the one observed in respective OTI or OTII transgenic control animals [4,5]. Subsequently, several other groups published the generation of mice expressing retrogenic TCRs, e.g. a MOG-peptide specific TCR [7]. Altogether, 64 different TCR retrogenic mice were generated in the past seven years [8]. In striking contrast, not a single BCRrg mouse has been published to date.

The generation of retrogenic mice offers several potential advantages compared with the generation of transgenic mice. First, retrogenic mice can be generated using any background strain. Second, the generation of retrogenic mice is faster than the generation of transgenic mice, since there is no need to backcross the retrogenic mice. Third, multiple proteins can be analysed simultaneously.

However, retrogenic mice cannot be propagated by breeding, because there is no germline transduction and the analysis is limited to the hematopoietic system [4].

Another major advantage of the retrogenic approach is the usage of so-called 2A peptides to link two or more target proteins instead of an IRES. Mechanistically, the 2A sequence induces the "skipping" of the ribosome thereby preventing it from covalently linking newly inserted amino acids and letting it continue translation. Therefore, 2A sequences are referred to as CHYSEL (*cis*-acting hydrolase element) sequences. This allows for the stoichiometric expression of the spliced proteins, which is crucial for heterodimeric molecules such as the BCR. Although these 2A like sequences were first discovered in +ssRNA and dsRNA viruses, the ribosomal "skipping" functions also *in vitro* and *in vivo* in all tested eukaryotic systems [9–12].

To demonstrate that the generation of BCR retrogenic mice is feasible per se, we chose the well-characterised Hen-Egg-Lysozyme (HEL)-specific BCR, MD4. A MD4 BCR-transgenic

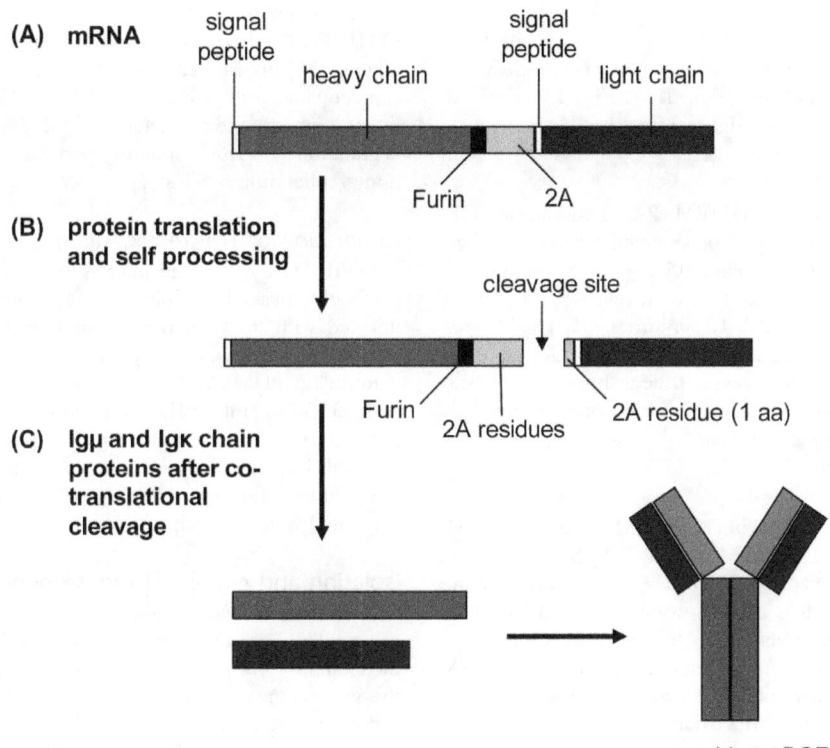

Figure 1. Full-length recombinant HEL-IgM-BCR expression cassette. (A) The HEL-specific Ig chains are linked by the Furin–FMDV-2A self-cleaving sequences. Both Ig chains have a signal peptide sequence, but only the downstream Igκ chain possesses a stop codon. (B) During translation, the first cleavage event occurs at the FMDV-2A peptide sequence; the 2A peptide remains attached as remnant to the C-terminus of the upstream Igμ chain. (C) A second cleavage event, initiated at the Furin site, finally yields both Ig chains without further attachments that can be assembled to build the HEL-IgM-BCR.

line was generated in the 1980s by Goodnow and co-workers and was used as control in our experiments [1].

We show for the first time the expression of a recombinant, membrane IgM-BCR *in vitro* using the pro-B cell line R5B, which is deficient for endogenous Ig chains. We also detected the recombinant αHEL IgM-BCR intracellularly when analysing these retrogenic mice, but to our surprise, we failed to demonstrate the surface expression of the recombinant αHEL IgM-BCR *in vivo*.

Materials and Methods

Mice

$Rag2^{-/-}$ [13], OTII TCR transgenic [14] and C57BL/6 wt mice were bred and maintained under SPF conditions at the animal facility of the University Hospital Jena. HEL-IgM-BCR transgenic mice (MD4) [1] were kindly provided by R.J. Cornall (Oxford, UK).

Ethics statement

All animal experiments were approved by the appropriate governmental authority (Thüringer Landesamt für Lebensmittelsicherheit und Verbraucherschutz; Registered Number 02–038/06) and conducted in accordance with institutional and state guidelines.

Injection of 5-Fluorouracil

Donor mice were injected with 0.15 mg/g bodyweight 5-Fluorouracil (5-FU) intraperitoneally to induce proliferation of hematopoietic stem cells. After 72 hrs, bone marrow was isolated from femur and tibiae (see below).

Irradiation of recipient mice

$Rag2^{-/-}$ (4.5 Gy) or C57BL/6 wildtype (9 Gy) recipient mice were irradiated at the Leibniz Institute for Age Research/Fritz-Lipmann-Institute (FLI), Jena, using a *Gammacell40 Exactor* (Caesium source). Drinking water was supplemented with Sulfamethoxazol/Trimethoprim (Cotrim, 40 mg/ml; changed twice a week; Hexal, Holzkirchen, Germany) before irradiation and after reconstitution.

Cloning of HEL-Igμ$_m$ BCR

Total RNA from sorted transgenic HEL-Igμ^{a+} B cells was isolated using the High Pure RNA Isolation Kit (Roche Applied Science, Penzberg, Germany), reverse transcribed using Oligo(dT) and cDNA was subsequently used as template for cloning of the HEL-IgH and IgL chains. Full length sequence information for HEL-IgH and IgL chain genes were obtained through 5′ and 3′RACE (GeneRACER Life Technologies, Darmstadt, Germany). Using the below mentioned oligos, the FMDV-2A peptide sequence as well as restriction sites were added to the full length sequence clones for either HEL-specific Igμ as well as Igκ chain: forw_Igμ: 5′ GGGACCGGTGCCGCCACCATGATGGTGT-TAAGTCTTCTGTAC; rev_Igμ: 5′GCC GGCAAGCTTCAG-CAGGTCGAAGTTCAGGGTCTGCTTCACGGGGGGCCCG-CCGCCGCCGTTTCACCTTGAACAGGGTGACG; forw_Igκ: 5′CTGCTGAAGCTTGCCGGC GACGTGGAGAGCAACCC-CGGCCCCATGGTTTTCACACCTCAGATACTT; rev_Igκ: 5′TCCCCGCGGGGACTAACACTCATTCCTGTTGAAGCT. Oligos were synthesized by BioTeZ (Berlin, Germany). Sequences of selected clones were analysed (Agowa/LGC Genomics, Berlin, Germany) at appropriate check points. The recombinant construct was subcloned into the retroviral based target vector (pRMYs-eGFP) via the restriction sites AgeI and SacII (see **Fig. 1** for detailed structure of the recombinant HEL-Igμ construct).

Transfections

CHO-K1 cells were grown in DMEM (PAA Laboratories/GE Healthcare, Freiburg, Germany) supplemented with 10% FCS, 100 mM Hepes, 20 U/ml Penicillin, 0.1 mg/ml Streptomycin, 50 μM 2-Mercaptoethanol. For transfection, cells were seeded at 1.5×10^5 cells/ml in RPMI (PAA Laboratories/GE Healthcare, Freiburg) w/o antibiotics. After 24 hrs, cells were transfected using Lipofectamine2000 transfection reagent (Life technologies, Darmstadt) and incubated for further 24 hrs. Transfection efficiency was determined as% GFP$^+$ using FACS.

Western blot

Cells were lysed with lysis buffer (20 mM Hepes, 2.5 mM MgCl$_2$, 1% Triton X-100, 100 mM EGTA, 1 M β-Glycerophosphat, 100 mM ortho-Vanadat, 1 M DTT, 10 mg/ml Aprotinin, 1 mg/ml Leupeptin). Protein extracts were separated on 10% SDS-Laemmli gels and transferred by electroblotting onto nitrocellulose membranes. Membranes were blocked with BSA and incubated with primary antibody (goat anti-mouse Igμ, goat anti-mouse Igκ, both Santa Cruz Biotechnology, Santa Cruz, USA). Membranes were washed in 0.1% Tween/TBS and incubated with the HRP-conjugated secondary Ab (donkey anti-goat-IgG-HRP; Santa Cruz Biotechnology). Detection was performed using ECL reagent (Pierce).

Producer cell generation and determination of viral titers

PhoenixEco cells [15] were grown in DMEM supplemented with 10% FCS, 100 mM Hepes, 20 U/ml Penicillin, 0.1 mg/ml Streptomycin, 50 μM 2-Mercaptoethanol. One day before transfection, cells were seeded at 4×10^6 cells per 100 mm petri dish. Using calcium phosphate precipitation (2.5 M CaCl$_2$, 2xHBS, 25 mM Chloroquine), cells were transfected with 20 μg plasmid DNA. Six to eight hours post-transfection, media was replaced. After 2 days, virus-containing supernatants were collected, sterile-filtered (0.45 μm) and kept at $-80°C$ until further usage. Percentage of GFP$^+$ cells was analysed using FACS.

For determination of viral titers, NIH/3T3 cells were seeded at 2×10^4 cells/ml. At the next day, the cell culture media was removed and cells were centrifuged for 3.5 hrs at 33°C and 3300 rpm with virus-containing supernatants. To allow better transduction of the cells, polybrene (2 μg/ml, Sigma-Aldrich, Taufkirchen) was added. After 24 hrs the percentage of GFP$^+$ cells was determined by flow cytometry. Viral supernatants with>45% GFP$^+$ were used for infection of murine donor stem cells.

Maintenance and transduction of cell lines

The WEHI-231 cell line, expressing surface Igμ [16], as well as the BCR-deficient pro-B cell line R5B [17] were maintained in RPMI supplemented with 10% FCS, 100 mM Hepes, 20 U/ml Penicillin, 0.1 mg/ml Streptomycin, 50 μM 2-Mercaptoethanol. For transduction of the R5B cell line, 5×10^5 cells were centrifuged with virus-containing supernatant plus Polybrene (4 μg/ml) for 3.5 hrs at 3300 rpm and 33°C. After centrifugation, fresh media was added and cells were incubated for 48 hrs at 37°C and 5% CO$_2$. After 2 days, cells were analysed for the surface expression of Igμ and Igκ as well as for the transduction rate (% GFP$^+$) using flow cytometry.

The GP+E86 OTII producer cells used for the generation of OTII TCR retrogenic mice were kindly provided by D.A.A. Vignali (Memphis, Tennessee, USA) and maintained in DMEM supplemented with 10% FCS, 100 mM Hepes, 20 U/ml Penicillin, 0.1 mg/ml Streptomycin, 50 μM 2-Mercaptoethanol 2 mM L-glutamine, 1 mM sodium pyruvate and 100 mM MEM nonessential amino acids [4].

Generation of TCR retrogenic mice

OTII TCR retrogenic mice were generated according to the protocol published by Holst et al. [4]. Therefore, donor mice were injected with 0.15 mg per g body weight 5-Fluorouracil. After 48 hrs, bone marrow was extracted from femur and tibiae and cultured in cDMEM (20% FCS) supplemented with 20 ng/ml rmIL-3, 50 ng/ml rmIL-6 and 50 ng/ml SCF. GP+E86 OTII producer cells were irradiated with 12 Gy 24 hrs after the donor BM cells had been isolated. Transduction was allowed for 48 hrs. A minimum of 4×10^6 cells was used for reconstitution of lethally irradiated wildtype recipient mice.

Isolation and culture of mouse bone marrow-stem cells

Bone marrow was flushed from femur and tibiae of donor mice using a 23xG needle. Cells were seeded at $1.5–2 \times 10^6$ cells/ml and incubated for 48–72 hrs at 37°C +5% CO$_2$ prior to infection. For the generation of BCR rg mice, cell culture media was further supplemented with 20 ng/ml IL-3 (BPV supernatant [18],), 50 ng/ml rmIL-6, 10 ng/ml rmIL-7, 20 ng/ml SCF and 100 ng/ml Flt3-L (all from Miltenyi Biotec, Bergisch-Gladbach, Germany).

Transduction of murine bone marrow-stem cells

For transduction of the donor cells, cell culture plates were coated overnight with RetroNectin (40 μg/ml; TaKaRa Bio, Saint-Germain-en-Laye, France) at 4°C. The next day, the plates were washed twice with PBS and blocked with 2% BSA/PBS. Virus-containing supernatants were added and plates were centrifuged twice for 2 hrs at $2000 \times g$ and 32°C, while the supernatants were replaced after the first 2 hrs of centrifugation. Donor BM cells were placed onto these virus-loaded plates at 2×10^6 cells/ml and incubated at 37°C+5% CO$_2$. After 24 hrs, cells were harvested and intensively washed. Transduction efficiency (% GFP$^+$ cells) was determined using flow cytometry. Transduced cells were resuspended in 2% FCS/PBS and used for reconstitution of irradiated recipient mice (minimum 4×10^6 cells/mouse).

Flow cytometry and cell sorting

Transduced cell lines (WEHI-231, R5B) or single-cell suspensions (prepared from mouse spleens, lymph nodes and bone marrow) wer eincubated with anti-CD16/32 (2.4G2/75; 10 μg/mL) and rat IgG (10 μg/mL; Dianova, Hamburg, Germany) to prevent unspecific binding. Cells were stained with anti-Igμ^a-biotin (clone DS-1), anti-Igμ^b-PE (clone AF 6–78; both BD Biosciences, Heidelberg, Germany), anti-mouse IgM-Cy5 (μ-chain specific), anti-mouse Igκ-PE (both Southern Biotech/Biozol, Eching, Germany), anti-mouse CD19-Alexa Fluor 647 (clone 1D3; eBioscience, Frankfurt/Main, Germany). Streptavidin-APC-eFluo780 (eBioscience) was used as secondary antibody; Hen-Egg-Lysozyme was coupled to Alexa Fluor 647 (kindly provided by René Riedel, Deutsches Rheumaforschungszentrum Berlin, Berlin, Germany). For intracellular stainings, cells were fixed with 2% PFA for 20 min on ice and subsequently permeabilised by washing with saponin-containing buffer. Again, unspecific binding of antibody was blocked by incubation with anti-CD16/32 (2.4G2/75; 10 μg/mL) and rat IgG (10 μg/mL; see above). 1 000 000 events were acquired for each sample using a LSRII cell cytometer (BD Biosciences). Data were acquired using the DiVa software; data analysis was performed using FlowJo software (TreeStar). For the isolation of HEL-specific, Igμ^{a+} B cells, splenocytes of transgenic HEL-IgM-BCR mice were first depleted of T cells by magnetic cell sorting (AutoMACS; Miltenyi Biotec) using CD90-Microbeads (mouse, Miltenyi Biotec). Thereafter CD90-negative cells were further stained with an allotype-specific antibody (anti-Igμ^{a}-FITC, clone DS-1; BD Biosciences) and FACS-sorted (ARIA, BD Biosciences).

Results

Cloning of the recombinant HEL-IgM B-cell receptor

To establish the generation of BCR retrogenic mice we decided to use the Hen-Egg-Lysozyme (HEL)-specific membrane form of μ heavy chain (Igμ) and the κL chain (Igκ) of the BCR MD4. A corresponding transgenic line, expressing HEL-specific IgM as well as IgD antigen receptors, was generated by Goodnow and co-workers. The MD4 μ and δ genes were derived from BALB/c mice (IgH^{a} allotype) and can, therefore, be distinguished from the endogenous C57BL/6 BCRs (IgH^{b} allotype) [1]. We purified the HEL-specific B lymphocytes from transgenic MD4 splenocytes by FACS (purity>95%), using an allotype-specific antibody (anti-Igμ^{a}) (see Figure S1). Total RNA was isolated from these cells, reverse transcribed and used as template. First, we obtained full length sequence information for both Ig chains by performing 5′ and 3′ RACE. A 2A-peptide sequence was added to the full length cDNA sequences of the HEL-specific μ- and κ-light chain by PCR. The recombinant construct was generated through ligation employing the introduced restriction sites. The expression cassette comprises 5′LTR–HEL-Igμ–furin–2A peptide–Igκ–IRES–eGFP–3′LTR (Fig. 1).

To link the two HEL-specific Ig chains (resulting in the recombinant BCR) we chose the Foot-and-Mouth-Disease-Virus-(FMDV-) derived 2A peptide sequence. We used this particular 2A peptide sequence (APVKQTLNFDLLKLAGDVESNPGP) because cleavage efficiencies of>90% were reported using this 2A peptide sequence [19–21]. The DNA sequence of selected clones was confirmed at appropriate time points. One clone, carrying the recombinant HEL-IgM-BCR, was selected for all further experiments. Initially, we also included a different construct (with the Igκ-chain being the upstream protein). However, we did not succeed in generating the recombinant construct for this variant of the HEL-Igμ BCR due to the formation of tertiary structures. As a consequence, we performed all subsequent experiments with the expression cassette shown in Fig. 1.

Verification of the FMDV-2A peptide-mediated cleavage *in vitro*

The 2A peptide-mediated cleavage event occurs during the translation process. Thereafter, the 2A peptide remains attached as remnant to the polypeptide encoded by the sequence upstream of the 2A peptide. To date, no deleterious effects of the remaining 2A peptide have been observed in retrogenic mice [8]. Here, the remnant is attached to the C-terminus of the membrane form of Igμ. Since the correct sequence of the C-terminus is essential for surface expression of the BCR, we decided to include also a Furin cleavage site (RRRR) [22–24] upstream of the FMDV-2A to ensure removal of 2A peptide remnant and thus to obtain a more native Igμ chain by post-translational processing (Fig. 1).

To verify the cleavage efficiency of the FMDV-2A peptide as well as the upstream Furin-cleavage site, we transfected Chinese Hamster Ovary cells (CHO-K1) with the retroviral construct and analysed lysates by western blot (Fig. S2). Lysates of either empty vector-transfected cells or of hybridomas expressing either Igμ or Igκ [25,26] served as controls. Using anti-mouse Igμ- and Igκ-specific antibodies, specific products for both Ig chains (~28 kDa for Igκ; ~70 kDa for Igμ) could be detected in the lysates of HEL-IgM-BCR transfected cells, indicating cleavage of the recombinant protein.

However, with both primary antibodies another high-molecular-weight band was detected. Its apparent molecular weight of ~100 kDa corresponds to the expected size of the uncleaved, recombinant protein providing evidence for incomplete processing. Since the intensities of the Igμ- and the Igκ-specific bands were similar to that of the uncleaved product detected with anti-Igμ and anti-Igκ, respectively, we concluded that the amounts of cleaved versus uncleaved protein were nearly equal. Hence we proceeded with the *in vitro* surface expression of the recombinant BCR.

Expression of the recombinant HEL-IgM-BCR in a BCR-deficient cell line

To examine the production and assembly of the recombinant HEL-Igμ$_M$-BCR *in vitro*, we transduced the IL-7-dependent pro-B cell line R5B (*Rag2$^{-/-}$*) [17,27] with virus-containing supernatant. R5B cells do not express endogenous Ig chains, but do express the surrogate light chain. Supernatants of either empty vector transfected or MOCK-transfected cells as well as the IgM(κ)-producing cell line WEHI-231 [16] were used as controls in this experiment. 48 hours after the transduction, the cells were stained with either fluorochrome-conjugated Igμ- or Igκ-specific antisera and analysed by flow cytometry (Fig. 2A). As expected, neither Igμ nor Igκ could be detected on the cell surface of the non-transduced or empty vector transduced samples. Specific staining for Igμ as well as Igκ was shown for the WEHI-231 control. Furthermore, all cells that were transduced with retrovirus encoding for the recombinant BCR (and thus were GFP$^+$), expressed the recombinant BCR (~28% Igκ$^+$Igμ$^+$). The two Ig chains were always co-expressed. The Ig chains could also be detected intracellularly (Fig. 2B). Again, co-expression of the Igμ and Igκ chain was observed for cells that were transduced with virus encoding for the recombinant BCR (~27% Igκ$^+$Igμ$^+$). Interestingly, the frequencies of μH$^+$Igκ$^+$ cells were equal for surface as well as for intracellular stainings, suggesting that all transduced cells transported the recombinant BCR to their cell surface.

Stoichiometric expression of the two Ig chains could also be shown at the protein level by western blot analysis. Using anti-mouse Igμ- and Igκ-specific antibodies, specific products for both Ig chains (~70 kDa for μH, ~28 kDa for Igκ) could be detected in the lysates of cells infected with supernatant encoding for the recombinant BCR but not in the non-transduced cells and cells transduced with empty vector controls (Fig. 2C).

Generation of retrogenic mice expressing HEL-specific IgM-BCRs

Whereas the generation of many different TCR retrogenic mice has been published; no BCR rg mice have been published to date. Therefore, we decided first to generate TCR rg mice to establish the procedure in our lab. We generated OTII TCR (TCRVα2–P2A–TCRVβ5–IRES-eGFP) rg mice according to the publication of Holst et al. and compared them with the corresponding classical OTII TCR transgenic mice [5]. Flow cytometric analysis of the

(A) <u>surface</u>

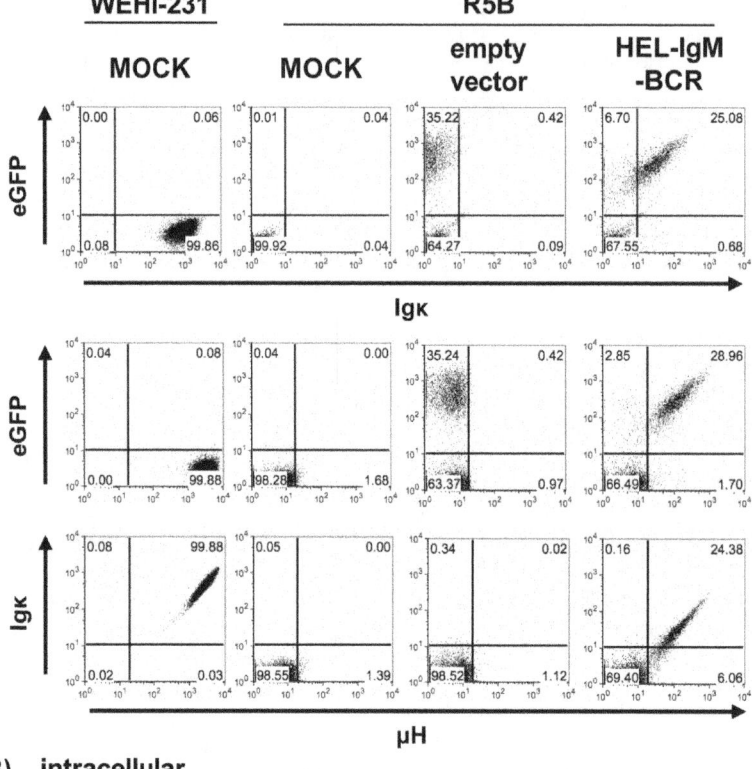

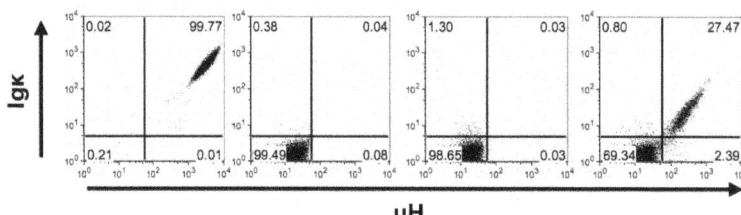

(B) <u>intracellular</u>

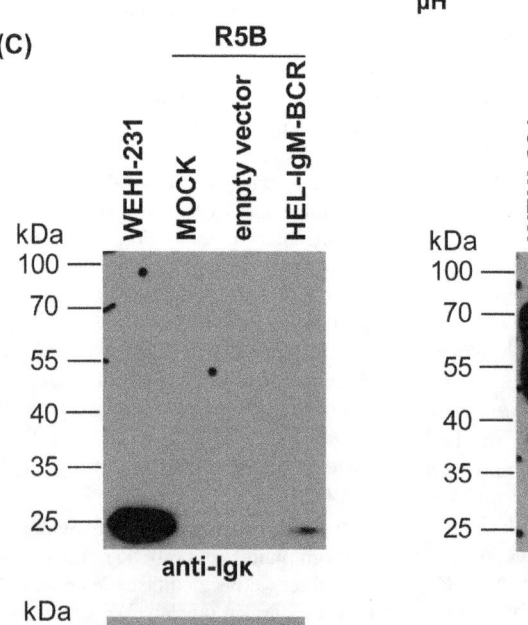

(C)

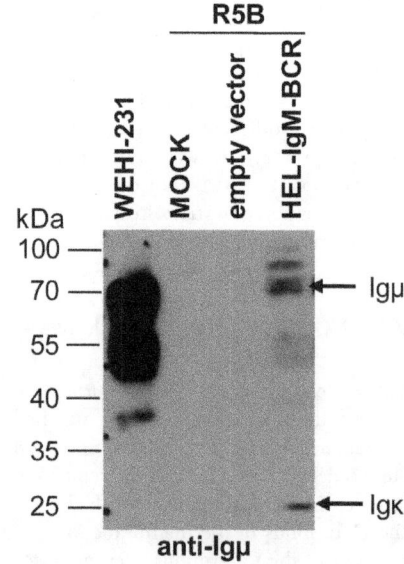

Figure 2. *In vitro* surface expression of the recombinant HEL-IgM-BCR. The pro-B cell line R5B, deficient for endogenous BCR, was infected with virus-containing supernatant encoding for the recombinant HEL-specific IgM-BCR. 24 h post-infection, cells were either analysed by flow-cytometry or western blot analysis. The WEHI-231 B-cell lymphoma served as positive control. Staining for Igμ and Igκ showed surface (A) as well as intracellular expression (B) of the recombinant BCR. Numbers in quadrants indicate the percentage of cells expressing the respective marker. (C) Lysates of infected R5B cells were analysed for κ- and μH-chain expression. Expression of the Ig chains was only detected in lysates of cells infected with the recombinant HEL-IgM-BCR. Incubation with an anti-GFP antibody was used as internal control. Results are representative of three independent experiments.

TCR retrogenic mice at 8 weeks post-reconstitution showed that the ovalbumin-specific OTII TCR was expressed in all lymphatic organs analysed albeit with a lower frequency as compared to the transgenic controls (Fig. S3). Of note, already Holst et al. described in their original publication in 2006 lower frequencies (numbers) in the retrogenic as compared to the transgenic system [4].

Having established the generation of TCR retrogenic mice, we started to modify the culture conditions of the donor cells for the differentiation of B cells rather than T cells (see Materials and Methods section). An adapted protocol was used to transduce donor cells (as compared to [4]) giving better transduction efficiencies in our hands. Figure S4 exemplary shows the flow cytometric analysis of HEL-IgM-BCR transduced donor stem cells at either 24 hours or 5 days post-transduction. Transduction rate was ~6% (as expressed by GFP$^+$ cells) at 24 hrs post-infection. The percentage of GFP$^+$ cells was even increased after further 4 days in culture along with the percentage of GFP$^+$CD19$^+$ cells, indicating the stem cells would at least partly develop into the B cell lineage. Transduced cells were subsequently used to reconstitute lethally irradiated wildtype recipients.

Starting 2 weeks post-reconstitution, mice were bled regularly to track the repopulation of the immune system (i.e. B cells), thereby identifying the optimal time point for analysis. Around 8 weeks post-reconstitution, IgM-BCR retrogenic mice were sacrificed, single cell suspensions of spleen, lymph node, bone marrow as well as blood were prepared, stained with fluorochrome-conjugated antibodies against B cell marker and analysed by flow cytometry. Figure 3 shows exemplary results for splenocytes from HEL-IgM-BCR retrogenic mice. The percentage of GFP$^+$ (and therefore transduced) cells was ~10%. Of note, only those transduced cells will express the retrogenic BCR on their surface that co-express the signaling components of the BCR complex (Igα and Igβ) and therefore will develop into B cells. With the help of the TCR retrogenic mice it was already shown that only cells co-expressing the TCR-co-complex molecule CD3 would express the rg TCR on their cell surface [4].

Analysis of the HEL-IgM-BCR retrogenic mice further revealed GFP$^+$ cells expressing the endogenous Igμ of b allotype at levels comparable to the wildtype control group (42.81% and 42.77% CD19$^+$Igμ^{b+} respectively). However, surface staining demonstrated no expression of the retrogene, since no specific binding of the antigen HEL coupled to a fluorochrome was observed. Specific HEL-binding was observed, as expected in the HEL-IgM-BCR transgenic mice that were carried as control (Figure S5A, middle panel). Of importance, the fluorochrome-conjugated HEL used in these experiments exhibited unspecific binding to a certain degree. Nevertheless, the BCR signaling components Igα and Igβ were expressed by all groups analysed (data not shown).

In another experiment, we reconstituted irradiated wildtype mice with transgenic MD4 bone marrow cells (no retrovirally transduced cells were used in this experiment). Analysis performed at 6–8 weeks post-reconstitution showed that ~13% of splenic as well as ~6% of lymph node cells were positive for the transgene (by staining for CD19, Igμ^a and HEL; see Figure S6).

Whereas the retrogenic mice did not express the HEL-specific IgM-BCR on their surface in detectable amounts, intracellular

staining revealed high expression of the retrogene; (Figure S5B). In contrast to HEL-BCR transgenic mice, B cells from the retrogenic mice expressed both intracellular HEL-IgM-BCR (with Igμ^a), as shown by staining with the antigen HEL coupled to a fluorochrome and the endogenous Igμ (Igμ^b; Figure 3B and Figure S5B). Surface staining for lineage markers other than B cells revealed no major differences between the analysed groups (Figure 3C). Solely the percentages of CD4$^+$ and CD8$^+$ (and therefore also CD3$^+$) T cells were slightly increased in the retrogenic compared with the transgenic mice (27.50% to 21.46%, 8.59% to 6.63% and 31.46% to 28.40%).

In a final attempt, we reconstituted *Rag*2$^{-/-}$ mice to rule out possible expulsion by endogenous, Igμ^{b+} cells. Again, we were not able to detect the retrogenic HEL-IgM-BCR in the analysed mice, although we could identify the transgenic HEL-IgM-BCR in mice reconstituted with transgenic bone marrow cells (as shown in Figure S5C). Furthermore, by staining for the BCR signaling components Igα and Igβ we could not show expression of these molecules for the HEL-IgM-BCR retrogenic mice but for both the transgenic and wildtype reconstituted controls that were carried in this experiment (figure 3D). However, the frequency of retrogenically transduced cells in this experiment was low (~2% GFP$^+$). This should be taken into consideration when interpreting this set of data.

In summary, we were able to generate mice expressing the retrogenic BCR albeit only intracellularly. However, the BCR was not transported to the cell surface.

Discussion

The generation of retrogenic mice was originally described in 2006 by the group of D.A.A. Vignali for TCRs. They described retrovirally-mediated stem cell gene transfer to express TCRα and TCRβ chains from a 2A peptide-linked multicistronic retroviral vector in mice [4,5]. Since then several other groups successfully generated TCR retrogenic mice (for examples see ref. [8]). The number of publications describing the generation of TCR retrogenic mice has exceeded the number of 60, with the retrogenic TCRs being specific for model antigens (as OVA, male antigen) or with the TCR's specificity being relevant during autoimmunity (e.g. insulin, MOG) or host defence (influenza). In contrast and surprisingly, we are unaware of a single publication on the generation of BCR retrogenic mice. We reasoned that this approach should be amenable to the production of BCR retrogenic mice expressing IgH and IgL chains from a 2A peptide-linked multicistronic retroviral vector. Among the early publications on retrogenic technology one also speculated that generating BCR retrogenic mice should be possible [6]. After having produced TCR-retrogenic mice (using the well charac-terised OTII TCR that recognises an ovalbumine peptide presented by I-A^b) we set out to produce BCR retrogenic mice.

As a model BCR we chose the well characterized Hen-Egg-Lysozyme-specific BCR MD4 [1–3]. Using the *Rag*-deficient pro-B cell line R5B [17] we were able to show the surface expression of the retrogenic HEL-IgM-BCR *in vitro*. However flow cytometric analyses of our BCR retrogenic C57BL/6 mice revealed that the

(A) <u>surface</u>

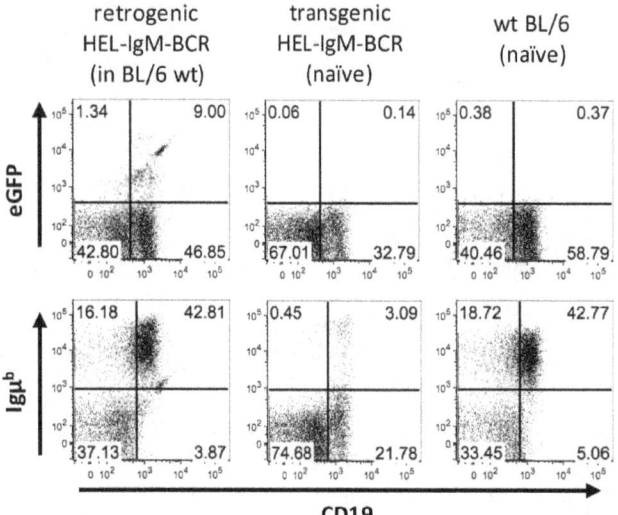

(B) <u>intracellular</u>

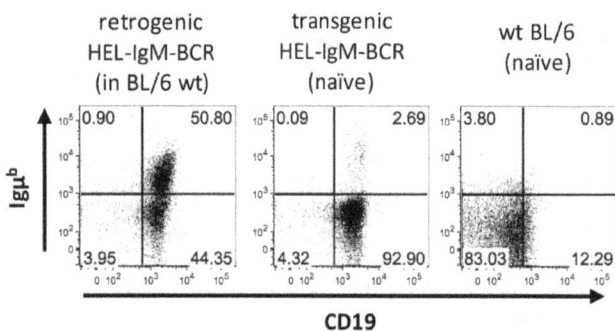

(C) <u>surface</u>

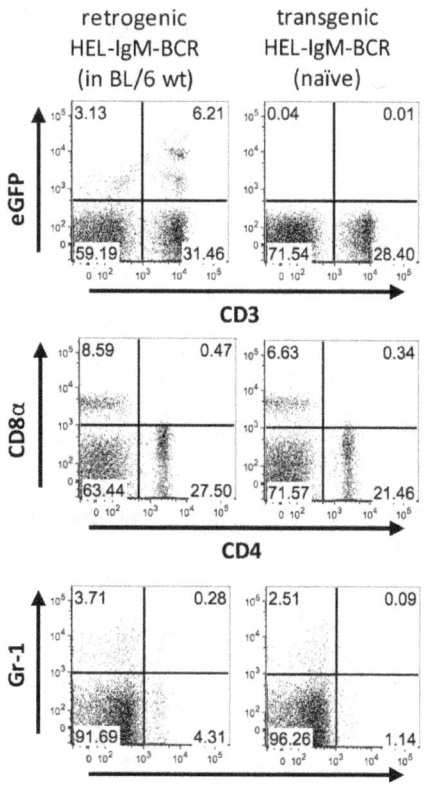

(D) <u>surface</u>

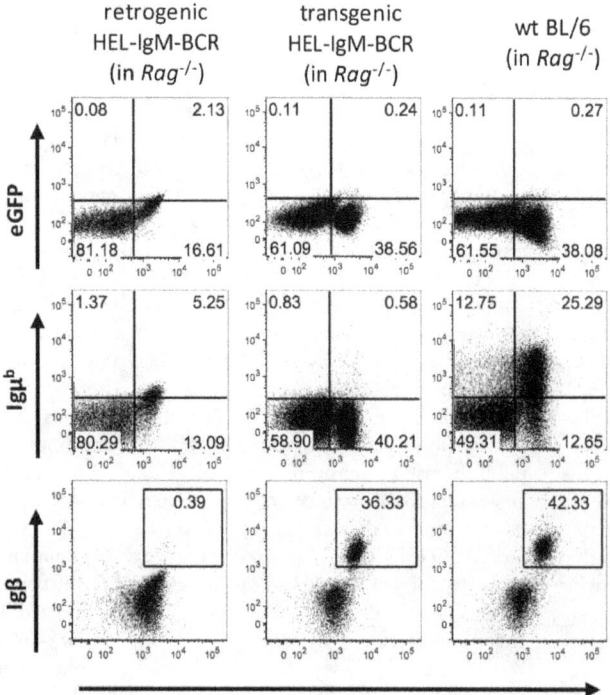

Figure 3. Analysis of HEL-IgM retrogenic mice showed only weak expression of the retrogene. Retrogenic mice were generated as described in the Materials and Methods section. Analysis was performed 6- to 8-weeks post-reconstitution. Single cell suspensions of spleen and lymph nodes were stained with fluorochrome-conjugated antibodies and analysed by flow cytometry. Transgenic HEL-IgM-BCR as well as wildtype C57BL/6 mice served as controls. Results for splenocytes are displayed. (A) Representative surface staining for CD19 and Igμb. Transduced cells are GFP+. Numbers in quadrants indicate the percentage of cells expressing the respective marker. Dead cells were excluded from the analysis. (B) Intracellular staining for the same parameters as in (A). Cells are gated on CD19+ cells. (C) Surface staining for surface markers other than B cells (CD3, CD4, CD8, CD11c, Gr-1). (D) Rag2-/- mice served as donors and recipients instead of C57BL/6 wt mice. Flow cytometric analysis for same parameters as in (A) as well as for Igβ. Results shown are representative of two independent experiments.

HEL-IgM-BCR was expressed exclusively intracellularly but not on the cell surface. To rule out competition by endogenous BCRs we used $Rag2^{-/-}$ mice. Even in these mice the retrogenic HEL-IgM-BCR was detectable only intracellularly (data not shown) but not as membrane protein suggesting that the processing of the BCR for transport to the cell surface in R5B cells differs from the requirements *in vivo*. By contrast, although highest *in vitro* cleavage efficacies were shown for 2A peptide derived from Foot-and-Mouth-Disease-Virus [19–21], Kim et al. showed highest *in vivo* cleavage efficacies for porcine teschovirus-derived 2A sequence [28]. Therefore, it is likely that the cleavage events mediated by the FMDV-derived 2A peptide or furin were improper *in vivo*, thus preventing the correct insertion of the Igμ chain into the cell membrane. At the same time, this could also explain why a transgenic BCR but not a retrogenic BCR is transported to and expressed at the cell surface.

In general, the surface expression of a BCR requires the expression of the BCR signaling components Igα and Igβ in addition to both Ig chains. All four components are required to assemble a transport-competent BCR [29]. Using fluorochrome-conjugated antibodies specific for Igα and Igβ we were able to show the expression of these molecules on the surface of the B lymphocytes of the analysed retrogenic mice. Consequently, the failed surface expression of the retrogenic anti-HEL-IgM-BCR cannot be accounted for the lack of expression of Igα and Igβ. Finally, the GFP expression is not representative of HEL-specific BCR but of transduced cells. Since bone marrow stem cells are used for the transduction, they may differentiate into cells other than B cells (e.g. T cells, as shown by GFP+CD3+ cells in Fig. 3C). This phenomenon was already described by Holst et al. as they detected GFP+ cells that were CD3- and therefore did not express the retrogenic TCR and presumably represented cells other than T cells [4,5].

One important point to consider in our efforts to produce BCR retrogenic mice is the seemingly low rate of transduction. In our experiments we detected transduced (and therefore GFP+) cells with a frequency of ~10%. This is in line with the frequencies described by Holst et al. in their original publication on the generation of retrogenic mice [4]. Therefore, the low rate of transduction is unlikely to explain the lack of surface expression. Of note and as shown in figure S4, when we analyzed our transduced cells 5 days post-infection, the percentage of GFP+ along with the percentage of GFP+CD19+ cells was increased as compared to the 24 hrs post-infection time point arguing the stem cells would at least partly develop into the B cell lineage. Hypothetically, the lower frequency observed in our scenario could be due to the particular receptor used. Holst et al. reported that the results obtained with the retrogenic approach will vary depending on the specificity of the TCR to be investigated. However all retrogenically analysed TCRs were expressed [4]. Whereas we did not formally rule out this possibility for our BCR retrogenic system we consider it unlikely.

One final and important question is whether the failure to produce retrogenic mice with one specific BCR reflects a principle problem to express such a construct in B cells, or whether it is due

technical or other problems particular to the BCR, the vector used or other non-generalisable factors. Whilst this cannot be answered definitively with one, two or several negative attempts, we know of one other group who used a different BCR and a different vector construct and also failed to produce retrogenic mice despite intensive efforts (F. du Pré and L.M. Sollid, U. Oslo, personal communication). The two groups' similarly negative results together with our positive results regarding the production of TCR rg mice, support the interpretation that the problem is general (for retrogenic BCRs) and not particular for one (or two) receptor constructs. Therefore, the failure to generate BCR retrogenic mice points towards an inherent and important difference between retrogenic BCR and TCR constructs.

In summary we used and optimised an approach that has proven successful for the generation of TCR retrogenic mice in several laboratories including ours to produce BCR retrogenic mice. Expression of the retrogenically encoded BCR was routinely detectable intracellularly. In contrast, surface expression of the retrogenic BCR was only achieved *in vitro* in the *Rag*-deficient pro-B cell line R5B but neither in $Rag2^{-/-}$ mice nor in wild type mice *in vivo*.

Based on our results and the absence of any reports on BCR retrogenic mice in the literature we conclude that the retrogenic technology, while very useful to study other receptors including TCR, is not the obvious choice for expressing a BCR *in vivo*. Our findings are further corroborated by experiments performed by another group, who also failed in generating BCR retrogenic mice (M.F. du Pré and L.M. Sollid, University of Oslo, personal communication). Still, these combined negative results together with the absence of any reports on BCR retrogenic mice in the literature do not definitively rule out the potential use of BCR retrogenic expression.

Acknowledgments

The authors thank Karin Müller and Olga Rudeschko for excellent technical assistance as well as Annett Krause for her invaluable help, and Birgit Meißner, Petra Schroth and Regina Zapfe for expert mouse care. We also thank Dr. D.A.A. Vignali for advice and for kindly providing the OTII_pMIG GP+E86 producer cells and Dominique Galendo (FLI Jena) for assisting with irradiation of mice. We thank Drs. M.F. du Pré and L.M. Sollid, University of Oslo, to quote their unpublished results. The present work forms part of the PhD thesis of JF.

Author Contributions

Conceived and designed the experiments: JF SH HMJ TK. Performed the experiments: JF SH ER JW. Analyzed the data: JF SH HMJ TK. Wrote the paper: JF HMJ TK.

References

1. Goodnow CC, Crosbie J, Adelstein S, Lavoie TB, Smith-Gill SJ, et al. (1988) Altered immunoglobulin expression and functional silencing of self-reactive B lymphocytes in transgenic mice. Nature 334: 676–682.
2. Hartley SB, Crosbie J, Brink R, Kantor AB, Basten A, et al. (1991) Elimination from peripheral lymphoid tissues of self-reactive B lymphocytes recognizing membrane-bound antigens. Nature 353: 765–769.
3. Mason DY, Jones M, Goodnow CC (1992) Development and follicular localization of tolerant B lymphocytes in lysozyme/anti-lysozyme IgM/IgD transgenic mice. Int Immunol 4: 163–175.
4. Holst J, Szymczak-Workman AL, Vignali KM, Burton AR, Workman CJ, et al. (2006) Generation of T-cell receptor retrogenic mice. Nat Protoc 1: 406–417.
5. Holst J, Vignali KM, Burton AR, Vignali DA (2006) Rapid analysis of T-cell selection in vivo using T cell-receptor retrogenic mice. Nat Methods 3: 191–197.
6. Nakagawa R, Mason S, Michie A (2006) Determining the role of specific signaling molecules during lymphocyte development *in vivo*: instant transgenesis. Nature Protocols 1: 1185–1193.
7. Alli R, Nguyen P, Geiger TL (2008) Retrogenic modeling of experimental allergic encephalomyelitis associates T cell frequency but not TCR functional affinity with pathogenicity. J Immunol 181: 136–145.
8. Bettini M, Bettini M, Vignali D (2012) T-cell receptor retrogenic mice: a rapid, flexible alternative to T-cell receptor transgenic mice. Immunology 136: 265–272.
9. Donnelly ML, Luke G, Mehrotra A, Li X, Hughes LE, et al. (2001) Analysis of the aphthovirus 2A/2B polyprotein 'cleavage' mechanism indicates not a proteolytic reaction, but a novel translational effect: a putative ribosomal 'skip'. J Gen Virol 82: 1013–1025.
10. de Felipe P (2004) Skipping the co-expression problem: the new 2A "CHYSEL" technology. Genet Vaccines Ther 2: 13.
11. Donnelly ML, Gani D, Flint M, Monaghan S, Ryan MD (1997) The cleavage activities of aphthovirus and cardiovirus 2A proteins. J Gen Virol 78 (Pt 1): 13–21.
12. Halpin C, Cooke S, Barakate A, El Amrani A, Ryan M (1999) Self-processing 2A-polyproteins–a system for co-ordinate expression of multiple proteins in transgenic plants. Plant J 17: 453–459.
13. Shinkai Y, Rathbun G, Lam K, Oltz E, Stewart V, et al. (1991) RAG-2-deficient mice lack mature lymphocytes owing to inability to initiate V(D)J rearrangement. Cell 68: 855–867.
14. Barnden M, Allison J, Heath W, Carbone F (1998) Defective TCR expression in transgenic mice constructed using cDNA-based alpha- and beta-chain genes under the control of heterologous regulatory elements. Immunol Cell Biol 76: 34–40.
15. Swift S, Lorens J, Achacoso P, Nolan G (2001) Rapid production of retroviruses for efficient gene delivery to mammalian cells using 293T cell-based systems. Curr Protoc Immunol: Chapter 10:Unit 10.17C.
16. Boyd AW, Schrader JW (1981) The regulation of growth and differentiation of a murine B cell lymphoma. II. The inhibition of WEHI 231 by anti-immunoglobulin antibodies. J Immunol 126: 2466–2469.
17. Corfe SA, Gray AP, Paige CJ (2007) Generation and characterization of stromal cell independent IL-7 dependent B cell lines. J Immunol Methods 325: 9–19.
18. Haan C, Hermanns H, Heinrich P, Behrmann I (2000) A single amino acid substitution (Trp(666)–>Ala) in the interbox1/2 region of the interleukin-6 signal transducer gp130 abrogates binding of JAK1, and dominantly impairs signal transduction. Biochem J 349: 261–266.
19. de Felipe P, Hughes LE, Ryan MD, Brown JD (2003) Co-translational, intraribosomal cleavage of polypeptides by the foot-and-mouth disease virus 2A peptide. J Biol Chem 278: 11441–11448.
20. Ryan MD, Drew J (1994) Foot-and-mouth disease virus 2A oligopeptide mediated cleavage of an artificial polyprotein. EMBO J 13: 928–933.
21. Ryan MD, King AM, Thomas GP (1991) Cleavage of foot-and-mouth disease virus polyprotein is mediated by residues located within a 19 amino acid sequence. J Gen Virol 72 (Pt 11): 2727–2732.
22. Nakayama K (1997) Furin: a mammalian subtilisin/Kex2p-like endoprotease involved in processing of a wide variety of precursor proteins. Biochem J 327 (Pt 3): 625–635.
23. Roebroek AJ, Creemers JW, Ayoubi TA, van de Ven WJ (1994) Furin-mediated proprotein processing activity: involvement of negatively charged amino acid residues in the substrate binding region. Biochimie 76: 210–216.
24. van de Ven WJ, Voorberg J, Fontijn R, Pannekoek H, van den Ouweland AM, et al. (1990) Furin is a subtilisin-like proprotein processing enzyme in higher eukaryotes. Mol Biol Rep 14: 265–275.
25. Smith-Gill SJ, Lavoie TB, Mainhart CR (1984) Antigenic regions defined by monoclonal antibodies correspond to structural domains of avian lysozyme. J Immunol 133: 384–393.
26. Smith-Gill SJ, Mainhart CR, Lavoie TB, Rudikoff S, Potter M (1984) VL-VH expression by monoclonal antibodies recognizing avian lysozyme. J Immunol 132: 963–967.
27. Milne CD, Corfe SA, Paige CJ (2008) Heparan sulfate and heparin enhance ERK phosphorylation and mediate preBCR-dependent events during B lymphopoiesis. J Immunol 180: 2839–2847.
28. Kim J, Lee S, Li L, Park H, Park J, et al. (2011) High cleavage efficiency of a 2A peptide derived from porcine teschovirus-1 in human cell lines, zebrafish and mice. PLoS One 6.
29. Dylke J, Lopes J, Dang-Lawson M, Machtaler S, Matsuuchi L (2007) Role of the extracellular and transmembrane domain of Ig-alpha/beta in assembly of the B cell antigen receptor (BCR). Immunol Lett 112: 47–57.

5

Increased Drought Tolerance through the Suppression of *ESKMO1* Gene and Overexpression of *CBF*-Related Genes in Arabidopsis

Fuhui Xu[1], Zhixue Liu[1], Hongyan Xie[1], Jian Zhu[1], Juren Zhang[3], Josef Kraus[2], Tasja Blaschnig[2], Reinhard Nehls[2], Hong Wang[1,2]*

1 School of Life Sciences and Technology, Tongji University, Shanghai, China, **2** KWS SAAT AG, Einbeck, Germany, **3** School of Life Science, Shandong University, Shandong, China

Abstract

Improved drought tolerance is always a highly desired trait for agricultural plants. Significantly increased drought tolerance in *Arabidopsis thaliana* (Columbia-0) has been achieved in our work through the suppression of *ESKMO1 (ESK1)* gene expression with small-interfering RNA (siRNA) and overexpression of *CBF* genes with constitutive gene expression. *ESK1* has been identified as a gene linked to normal development of the plant vascular system, which is assumed directly related to plant drought response. By using siRNA that specifically targets *ESK1*, the gene expression has been reduced and drought tolerance of the plant has been enhanced dramatically in the work. However, the plant response to external abscisic acid application has not been changed. *ICE1*, *CBF1*, and *CBF3* are genes involved in a well-characterized plant stress response pathway, overexpression of them in the plant has demonstrated capable to increase drought tolerance. By overexpression of these genes combining together with suppression of *ESK1* gene, the significant increase of plant drought tolerance has been achieved in comparison to single gene manipulation, although the effect is not in an additive way. Accompanying the increase of drought tolerance via suppression of *ESK1* gene expression, the negative effect has been observed in seeds yield of transgenic plants in normal watering conditions comparing with wide type plant.

Editor: Haibing Yang, Purdue University, United States of America

Funding: The research fund was provided by KWS SAAT AG (www.kws.com) to FX as her PhD project. KWS participated in the study design, data analysis, decision to publish and preparation of manuscript via HW.

Competing Interests: The work is financially supported by KWS SAAT AG, the employer of JK, TB, RN, and HW. KWS SAAT AG participated in the study design, data analysis, decision to publish and preparation of manuscript via HW. There are no patents, products in development or marketed products to declare.

* Email: hong.wang@kws.com

Introduction

Drought stress is a major limiting factor for crop production worldwide [1]. In 2012, a severe drought in the United States caused heavy losses in crop production, especially in corn, and farmers produced less than three-fourths of the corn that the U.S. Department of Agriculture anticipated [2]. In China, around 20 million hectares of land are at risk of drought each year [3]. Globally, estimated crop losses due to water limitation exceed $10 billion annually [4]. Improving yield under drought, therefore, is a continuous challenge for agriculture, especially for modern breeding.

The development of modern plant biotechnology provides new hope for generating crops with increased drought tolerance. Understanding the response of plants to drought stress is the first step for development of stress tolerance plant through plant biotechnology. Gene expression experiments have identified several hundred genes that are either induced or repressed during drought. Some of those genes encode proteins that play important role in protecting cells from dehydration, such as the enzymes required for biosynthesis of various osmoprotectants, late-embryo-genesis–abundant (LEA) proteins, antifreeze proteins, chaperones, and detoxification enzymes [5]. Some others are responsible for gene products including transcription factors, protein kinases, and enzymes involved in phosphoinositide metabolism. C-repeat/dehydration–responsive element binding factors (*CBF*s) are AP2/ERF-type transcription factors, which make up a critical gene cluster of the second group. During the stress condition, *CBF* genes are rapidly induced in response to abiotic stress, such as dehydration and cold [6,7]. The CBF proteins in turn activate expression of a set of target effector genes by binding to a core sequence in their promoter, C-repeat (CRT) / dehydration response element (DRE) [8,9,10,11]. *CBF* genes appear to be ubiquitous in plant species and almost always present as a gene family [12,13]. In *Arabidopsis*, the three characterized *CBF* genes are *CBF1*, *CBF2*, and *CBF3*, which are organized in tandem on chromosome 4 [14]. *CBF1* and *CBF3* are positive regulators whereas *CBF2* has a negative regulatory role [15]. *CBF* transcription factor genes are induced by the constitutively expressed inducer of *CBF* expression (*ICE1*) by binding to the *CBF* promoter [16,17]. The *ICE1-CBF* cold response pathway is conserved in diverse plant species [14,17,18]. Constitutive overexpression of *CBF* transcription factors in transgenic plants has increased the plant tolerance to freezing, salt, and drought stresses [8,19,20,21,22,23]. This functional conservation has suggested the *ICE1-CBF* genes are important targets for crop improvement for drought tolerance through genetic engineering [24].

ESK1 is a newly discovered member of the second group. It was initially identified as conferring freezing tolerance; a significantly high proline content accumulates in *esk1* mutants [25]. The gene product of *ESK1* belongs to an uncharacterized plant-specific protein family containing 48 members [26]. Bioinformatics analysis of genes whose expression modified by the *eskimo1* mutation showed that a large number of genes were previously reported linking to plant response to salt, osmotic stress, and the stress hormone abscisic acid (ABA) [25,26]. Later work showed that the mutant has a clear advantage in response to drought and salt stress: In standard and drought conditions, transpiration rate of mutant is lower than in wild type (WT) [27]. A biologically relevant parameter is the water required per biomass unit, and with this measure, the *esk1* mutants clearly have shown a higher water use efficiency and photosynthetic rate compared to WT [27]. This higher water use efficiency was independent of stomata closure through ABA biosynthesis. Measurement of root hydraulic conductivity suggests that the *esk1* vegetative apparatus suffers water deficit because of a defect in water transport system [27,28]. *ESK1* promoter-driven reporter gene expression has been observed in xylem and fibers, the vascular tissue which is responsible for the transport of water and mineral nutrients from the soil to the shoots, via the roots. Moreover, in cross sections of hypocotyls, roots, and stems, collapsed xylem vessel has been observed in *esk1* mutant [28]. The *ESK1* gene, therefore, was inferred to play a major role in whole plant water economy. *ESK1* has homologues in numerous species, and it is reasonable to hypothesize that manipulation of *ESK1* in crops could improve water use efficiency.

With an understanding of the molecular mechanism, several gene manipulation approaches have been employed to increase plant drought tolerance. The manipulated genes include those encoding enzymes required for the biosynthesis of various osmoprotectants or enzymes for modifying membrane lipids, LEA protein, and detoxification enzyme [29,30,31]. To date, the *CBF* genes are the most explored genes for improving crop stress tolerance because of the ability of this transcription factor to regulate an entire set of genes in a stress-response pathway [8,32,33]. When rice *OsDREB1A* [34] or corn *ZmDREB1A* is constitutively overexpressed in *Arabidopsis*, the downstream target genes regulated by the *Arabidopsis DREB1* (e.g., *RD29A*) are induced, resulting in desiccation tolerance under 15% humidity [35]. In another study, 35S: *CBF1* transgenic tomato plants are more resistant to water-deficit stress by showing less plant wilting and leaf curling than WT controls after 21-d water withdrawal in the same pot [36]. Constitutive overexpression of two wheat *CBF* factors in barley substantially improves survival under severe drought or cold. In addition, expression of *DREB* factors in wheat and barley under the control of drought-inducible promoters allows for normal development, together with significantly improved survival under severe drought [37].

In spite of the extensive evaluation of *CBF* factors, only a few studies have shown a clear improvement in drought tolerance in crops under field conditions [38]. Considering the complexity of the plant stress response, it has been assumed that better drought tolerance might be obtained if multiple genes involved in different stress response pathways could be manipulated together through

molecular stacking. In this work, we selected *ICE1* as well as the *CBF1* and *CBF3* genes involved in one stress-activated pathway and the *ESK1* gene from another stress-regulating pathway as targets of gene manipulation. Here, *ICE1* or *CBF1* and *CBF3* genes were overexpressed under the control of a constitutive promoter whereas the *ESK1* gene was suppressed with siRNA technique. Using gene stacking, we combined the two expression cassettes into one transformation vector for *Agrobacterium*-mediated plant transformation. The results obtained by testing the concept in the model plant *Arabidopsis* showed a significant increase in drought tolerance compared to non-transgenics, indicating that multiple gene manipulation might be a promising strategy for improving stress response and especially drought tolerance in plants.

Results

Manipulation of target gene expression through knock-down and overexpression

To knock down gene expression, a siRNA targeting specifically the *ESK1* gene was designed, and to overexpress the desired *ICE1* or *CBF* gene, the 35S promoter was placed in front of the gene for constitutive expression. The gene suppression and overexpression cassettes were integrated into the *Arabidopsis* genome, either alone or combined, via the flora dipping method. In analysis of gene expression by qRT-PCR, the plant transformed with siRNA cassette that suppresses *ESK1* showed clear knock-down of *ESK1* expression. Although the level of suppression varied from line to line, most plants showed about or more than 50% reduction in gene expression (Fig. 1A). In comparison, the transgenic lines derived via using only the pGPTV-*ESKi* vector showed a better suppression of *ESK1* gene expression than those obtained through use of the vector with the stacked overexpression and suppression cassettes (Figs. 1A, C, D).

Analysis of either *ICE1* or *CBF* overexpression showed the transgenic lines with the relevant transgene cassette had significantly higher *ICE1* or *CBF* expression compared to WT (Figs. 1B–E). For further evaluation of the concept, the best-performing plants based on qRT-PCR results were selected and their T2 seeds produced following the protocol described in the Methods part.

The response of transgenic *Arabidopsis* to osmotic stress *in vitro*

A distinct difference emerged in *in vitro* stress response between transgenic and WT plants. When growing on medium without any applied external stress, the majority of transgenic lines showed an almost identical phenotype to WT plants. However, when growing on medium with 30% polyethylene glycol (PEG), the transgenic plants with suppressed *ESK1* had much better root system growth than WT plants: The main root was longer, and the number of lateral roots was greater (Fig. 2). Leaf growth also differed between the transgenic and WT plants although not as much as the root system: All transgenic lines had eight leaves after 18 days of growth on medium with PEG whereas WT had only six. 15 *ESKi* transgenic lines were tested and the result of response to PEG was similar (data not shown).

Compared to the transgenic lines with the suppressed *ESK1* alone, the plants obtained through using stacking vector that suppress *ESK1* and overexpress either *ICE1* (14 lines) or *CBF* (3 lines) presented a similar performance on medium with PEG. After 14 days of growth with PEG, more and larger leaves were observed in transgenic lines compared to WT (Fig. 2). The growth

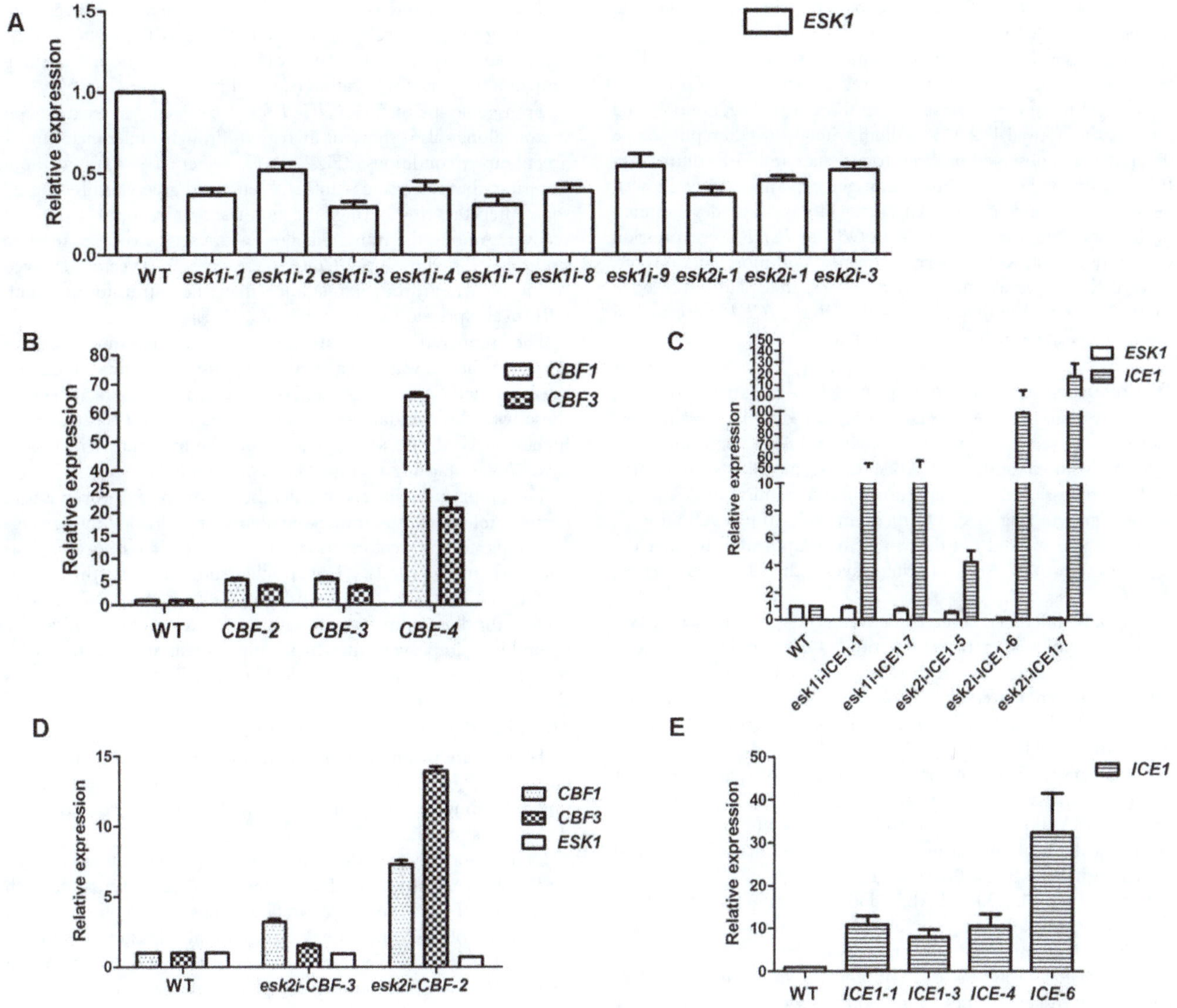

Figure 1. qRT-PCR results show *ESK1*, *ICE1*, *CBF1*, and *CBF3* expression in different transgenic lines. (**A**) *ESK1* was tested in two-week-old *ESKi* transgenic lines. (**B**) *CBF1* and *CBF3* were tested in two-week-old *CBF* transgenic lines. (**C**) *ESK1* and *ICE1* were tested in two-week-old *eski-ICE1* transgenic lines. (**D**) *ESK1*, *CBF1* and *CBF3* were tested in two-week-old *esk2i-CBF* transgenic lines. (**E**) *ICE1* was tested in two-week-old *ICE1* transgenic lines. Values represent means ±SD (error bar) of three replicates.

Figure 2. Response of transgenic lines to osmotic stress. Seedlings of different transgenic lines were subjected to osmotic stress: WT, *esk1i*-3, *esk1i-ICE1*-7, *esk2i-CBF*-2, *CBF*-3, *ICE1*-3. Four-day-old WT or transgenic seedlings were transferred to 1/2 Murashige and Skoog (MS) medium previously infused with 30% PEG for 14 days. Experiments were repeated at least three times with similar results. At least 30 seedlings per genotype were measured in each replicate.

of the root system had the same pattern with transgenic lines showing less effect from PEG treatment.

The responses to *in vitro* osmotic stress observed from transgenic plants with only the overexpression for *CBF* (5 lines), or *ICE1* (4 lines) were also similar to those with *ESK1* suppressed (Fig. 2). On normal 1/2 MS medium without PEG, no phenotype difference was observed between transgenics and WT control. On PEG medium, however, most transgenic plants clearly showed better growth than control plants after 14 days: The development of the root system from plants with *CBF* or *ICE1* overexpression was more robust, with a longer main root and more lateral roots. Nevertheless, several lines of *ICE1* transgenic plants showed a stress phenotype like that of WT, although the *ICE1* expression in those plants was much higher than in WT.

The response of transgenic plants to ABA stress *in vitro*

Suppression of *ESK1* gene expression did not significantly influence the sensitivity of plants to external ABA application. On medium without additional ABA, transgenic plants containing *ESK1* suppression cassette appeared almost identical to WT plants in their growth (Fig. 3A). On medium with 20 μM ABA for 14 days, the two still did not differ, and both appeared to suffer the effects of growth on ABA, showing leaves with yellow-brown stress symptoms (Fig. 3B).

The only difference observed between transgenic lines and WT plants was in seed germination time. On medium containing 0.5 μM ABA (Fig. 4B), the transgenic seeds showed delayed germination compared to WT control whereas on medium without ABA (Fig. 4A), the transgenic seeds germinated at the same time as WT.

Similar results were obtained for plant overexpression of *CBF* or *ICE1* in response to ABA stress *in vitro* (Figs. 3, 4). On medium without ABA, both transgenic and control plants showed similar growth with no abnormal phenotype observed in the majority of plants during the first 18 days of growth.

On medium with 20 μM ABA, both transgenic and WT plants were negatively affected. The elongation of the main root was inhibited, the number of lateral roots was reduced, and the leaves showed yellow-brown stress symptoms. No clear differences between transgenic lines and non-transgenic controls were observed (Fig. 3).

The performance of greenhouse transgenic plants in response to drought

To mimic most completely the interaction of plants and their natural environment, the transgenic plants showing the potential to increase drought tolerance via the *in vitro* osmotic test were transferred to greenhouse for evaluation of drought tolerance. The testing protocol was established based on the water-loss rate in greenhouse and the survival rate of plants in the protocol of water withdrawal and re-watering. Plants growing 2 weeks after germination in the described conditions were withdrawn from water for 14 days, so that the soil water content reached around 20%, which is considered a serious drought condition [27]. Following this threshold point, the plants were then fully watered and grown under a normal watering program for two more weeks before survival rates were evaluated.

Non-transgenic plants usually died completely under this protocol; however, transgenic plants obtained by the vector pGPTV-*ESKi* alone showed dramatic improvement in drought tolerance in contrast to the WT control plant (Fig. 5A). After 14 days without water, the transgenic plants still appeared green although with some level of dryness (Figs. 5A, B). Two weeks after re-watering, many of the transgenic plants had re-gained growth,

and the survival rate of the best-performing line *esk1i-3* reached 80% (Fig. 5C). In comparison, almost all WT plants became yellow and dry after 14 days without water, seldom regaining growth after water restoration (survival rate 1.25%, Fig. 5C).

Transgenic plants with *ICE1* or *CBF* gene overexpression vector alone also showed increased drought tolerance under greenhouse conditions (Figs. 5B–E). After 14 days without watering, the transgenic plants still remained green and recovered soon after the re-watering; the non-transgenic controls did not recover. Among the transgenic lines with various vectors, the line transformed with the *ICE1* overexpression vector showed relatively better drought tolerance than those transformed with *CBF* overexpression vector (Figs. 5C, D, E).

The improved drought tolerance test response under greenhouse conditions was also achieved in transgenic plants containing the combined *ESKi* cassette and overexpression cassette, especially those with the combination of the *ESKi* and *ICE1* cassettes; the highest survival rate was obtained from the line transformed with *ESKi-ICE1* stacking vector at 85.25% (Fig. 5C).

To compare the effects of different vectors on improvement of drought tolerance, the transgenic plants were divided into groups according to the vectors used, and the survival rate data was analyzed using *t*-tests. The results showed that the plants performing best were those transformed with *ESKi-ICE1* stacking vector; the lines transformed with the *ESKi* vector showed the second-best improvement while improvement was weakest with overexpression of the *CBF* genes (Figs. 5D, E).

The effect of transgenes on seed biomass

To evaluate potential applications to agriculture, we investigated the effect of transgenes on plant biomass, especially seed mass, under greenhouse conditions, either normal growth conditions or drought test conditions.

Under normal growth conditions in the greenhouse, some *ESK1* siRNA transgenic plants showed a dwarf phenotype with dark leaves (Fig. 6Aa); a delay in flowering was also observed, but most of the plants, otherwise, looked normal. Measuring the mass of seed harvested under normal greenhouse conditions revealed that the *ESK1* siRNA transgenic plants produced less seed than non-transgenic controls. However, upon harvest of the seed from plants that survived the drought test procedure, a clear contrast was evident between the seed yield of transgenic plants and controls, given that most of the non-transgenic plants died of drought. Comparing the seed yield of transgenic plants under drought and normal conditions, a significant reduction was found under drought, with less than 10% of the yield under normal conditions (Fig. 6B).

The impact of transgenes on seed yield showed a similar trend in plants transformed with *ICE1* or *CBF* vector alone and in those transformed with stacking vectors (Fig. 6B). Some dwarf phenotype development was observed during growth under normal conditions whereas in the drought condition, a higher stress tolerance compared to WT was observed, although accompanied by severe seed yield reduction.

Discussion

Plant genetic engineering has presented great potential for improving the drought tolerance of crops, especially since the discovery of *CBF* genes and their functions in stress tolerance [8,19,20,21,22,23]. *ESK1* was recently found to be involved in a stress tolerance pathway that is separate from the *CBF* pathway [25,26]. The *ESK1* pathway originally was identified as being related to plant freezing tolerance, but Bouchabke-Coussa *et al.*

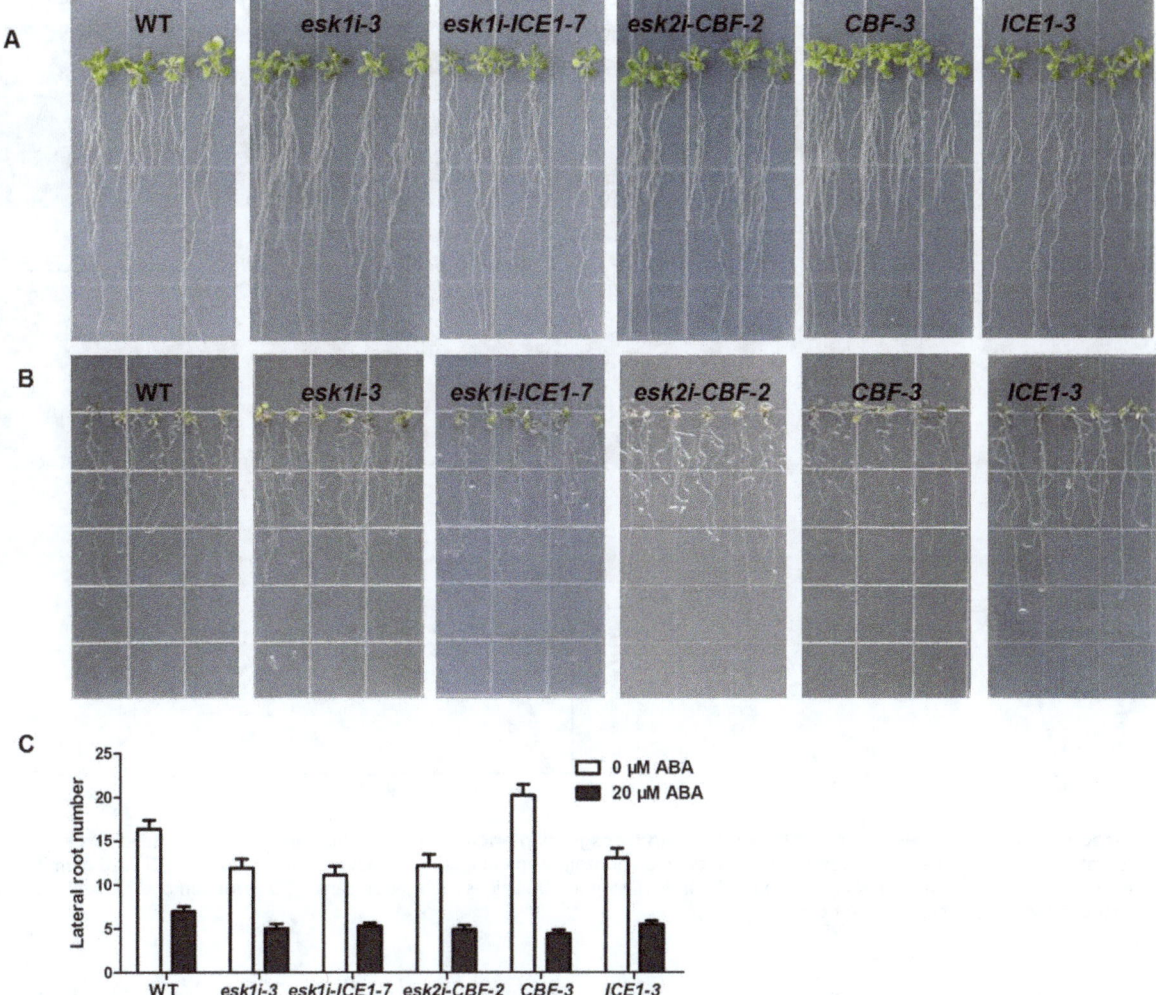

Figure 3. Seedling growth in response to ABA for WT and transgenic plants. (**A**) Four-day-old seedlings grown on 1/2 MS medium were transferred to 1/2 MS medium without ABA. (**B**) Sensitivity of seedlings to ABA. Four-day-old seedlings grown without ABA were transferred to 1/2 MS medium with 20 μM ABA. The photographs were taken 14 d after the transfer. (**C**) Quantification of lateral root number. 30 seedlings were measured in each experiment. Values represent means ±SD (error bar) of three replicates. At least 30 seedlings per genotype were measured in each replicate.

[27] later observed reduced respiration capacity in the *eskl* mutant of *Arabidopsis*, indicating a potential role of *ESK1* in drought tolerance. In our experiment, siRNA specifically targeting *ESK1* was employed to suppress its expression in *Arabidopsis* and the role of *ESK1* was confirmed in drought tolerance: Transgenic plants with efficient *ESK1* expression knock-down showed an obvious increase in osmotic stress tolerance *in vitro* and little effect on root system development by PEG treatment *in vitro* (Fig. 2). Further evaluation confirmed that suppression of the *ESK1* gene significantly enhanced plant drought tolerance (Fig. 5). In ABA *vitro* assay, it has been observed that *ESK1* knock-down transgenic plants has the same sensitivity to ABA change as control (Figs. 3B and 4B), which indicates that acquired stress tolerance of the transgenic plant may not be closely linked to the ABA-regulated network.

The *ESK1* knock-down plants appeared phenotypically normal comparing to non-transgenic plants when grown for 18 days: Plant size, leaf color, and shape were basically same like WT under normal growth conditions (Fig. 3A). Morphological observation also revealed that the number of stomata in the leaves remained unchanged relative to controls (data not shown). The question that

arises is how *ESK1* knock-down confers higher tolerance to drought or other stress. One explanation is that the disruption in *ESK1* function may block the normal development of the vascular system. Lefebvre *et al.* [28] observed a kind of collapsed xylem structure in cross sections of hypocotyls, roots, and stems of an *Arabidopsis esk1* mutant. In the current study, we also observed that the vascular system appeared abnormal in both leaf and stem (data not shown). These observations may explain why the respiration rate of a transgenic plant is lower compared to WT and why drought tolerance is higher than in controls.

Measuring the weight of total seeds showed that knocking down the *ESK1* gene can negatively affect seed production under normal growth conditions but that under the drought condition, the yield is much less affected, in contrast to complete loss in the non-transgenic control (Fig. 6B). Linking the yield penalty and the vascular distortion of the *ESK1* knock-down plant, we can infer that manipulation of *ESK1* gene expression can indeed increase the drought tolerance of plants, but expecting no reduction in biomass is not realistic considering the physical distortion of the vascular structure. Adoption of an inducible promoter or tissue-specific promoter may alleviate some negative consequences, but

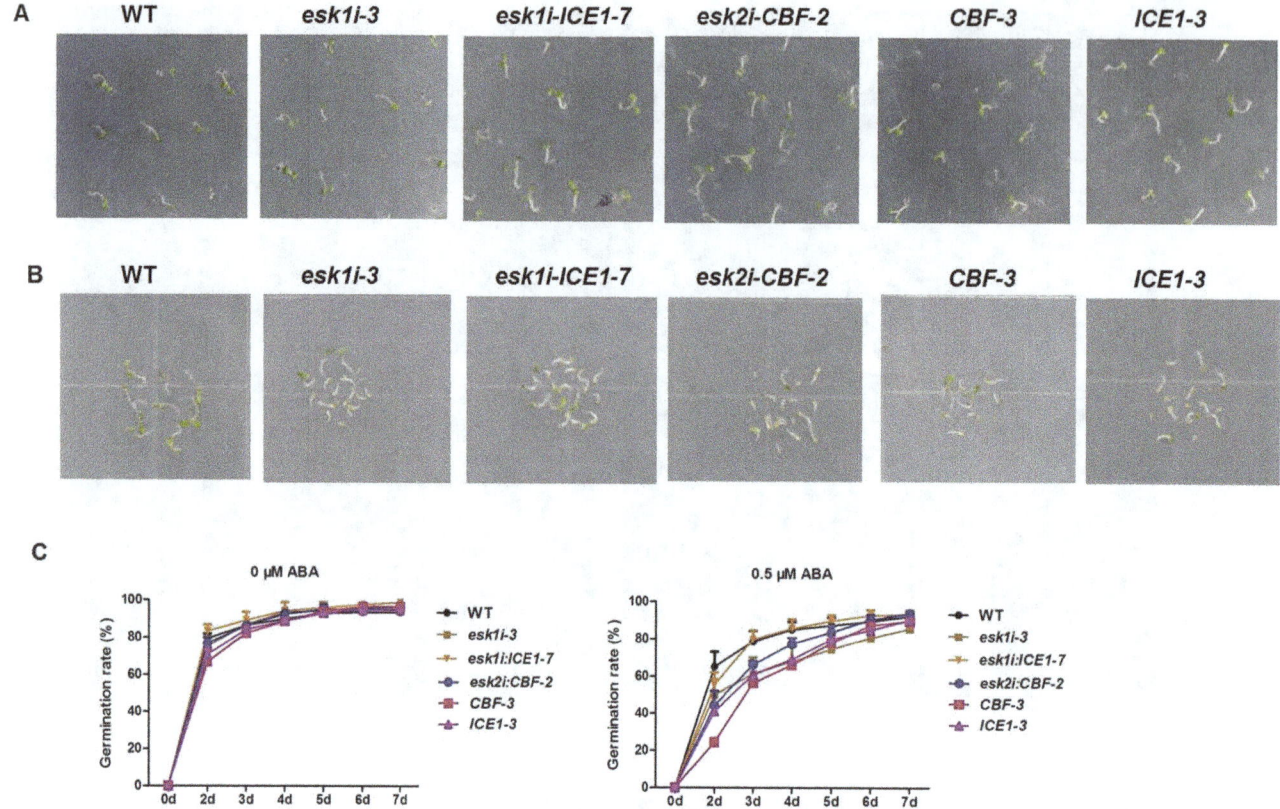

Figure 4. Seed germination in response to ABA for WT and transgenic plants. (**A**) and (**B**) Photographs of young seedlings at 5 d after the end of stratification. Seeds were germinated and allowed to grow on horizontal agar medium containing 0 or 0.5 μM ABA. (**C**) Seed germination time course of the six genotypes grown on medium without ABA or with 0.5 μM ABA. Values represent means ±SD (error bar) of three replicates. At least 100 seeds per genotype were measured in each replicate.

the enhancement of drought tolerance may be compromised at the same time. Therefore, use of *ESK1* for stress tolerance should be carefully balanced. One potential application may be in improvement of ornamental plants, for which biomass is not important.

Many studies have illustrated the potential of manipulating *CBF/DREB* genes to confer improved drought tolerance [20,21,37,39]. For example, overexpression of *CBF1/DREB1B* from *Arabidopsis* improves tolerance to water-deficit stress in tomato, but few plants clearly show enhanced drought tolerance under natural conditions [36]. We speculate that the complexity of the stress response pathway could be the reason and that modification of a single gene in a complicated pathway might not be sufficient to alter plant drought tolerance dramatically. Gene discovery and functional genomics projects have revealed many mechanisms and gene families that confer improved adaptation to abiotic stresses. These gene families can be manipulated into novel combinations, expressed ectopically, or transferred to species in which they do not naturally occur or vary [40,41,42]. Therefore, we have designed a gene stacking strategy by combining manipulation of the *CBF* and *ESK1* genes, hoping to obtain at least an additive effect on drought tolerance.

The result showed that gene stacking can indeed further improve drought tolerance in *Arabidopsis*, but an additive effect was not observed. One possibility for the lack of additive effect is that genes stacked in the same transformation vector may not work as efficiently as those in completely independent transformation vectors. qRT-PCR analysis in our work showed that the suppression effect of siRNA targeting *ESK1* was less significant

when the same vector was used for the stacked RNAi cassette and the overexpression cassette. In the future, the effect of combined gene manipulation will be evaluated through stacking the gene cassette by crossing the transgenic lines with individual vectors.

Materials and Methods

Plant materials and growth conditions

Arabidopsis thaliana ecotype Columbia (Col-0) and transgenic plants in Col-0 background were grown under long-day conditions (16 h light/8 h dark) with light intensity at 100 μE m^{-2} s^{-1}.

For plate-grown plants, *Arabidopsis thaliana* seeds were surface sterilized with 20% (v/v) bleach and sown on medium containing half-strength Murashige and Skoog (MS) salts [43], 1% sucrose, and 0.8% agar. After stratification for 3 d at 4°C, the plates were kept in a growth incubator under a long-day photoperiod (16 h light, 8 h darkness) at 24°C for 10 d. For transgenic seeds, medium was supplemented with 50 μg/mL kanamycin sulfate.

Gene isolation and binary vector construct

The leaf of 2-week-old *Arabidopsis* Col-0 was used for DNA and RNA extraction. The Col-0 cDNA sequences were obtained from GenBank (http://www.ncbi.nlm.nih.gov/). Total DNA was extracted from the samples using the DNeasy Plant Mini Kit (Qiagen, Germany), and total RNA was extracted using the RNeasy Plant Kit (Qiagen, Germany). cDNA was synthesized as described in the Thermo Scientific protocol (#K1631).

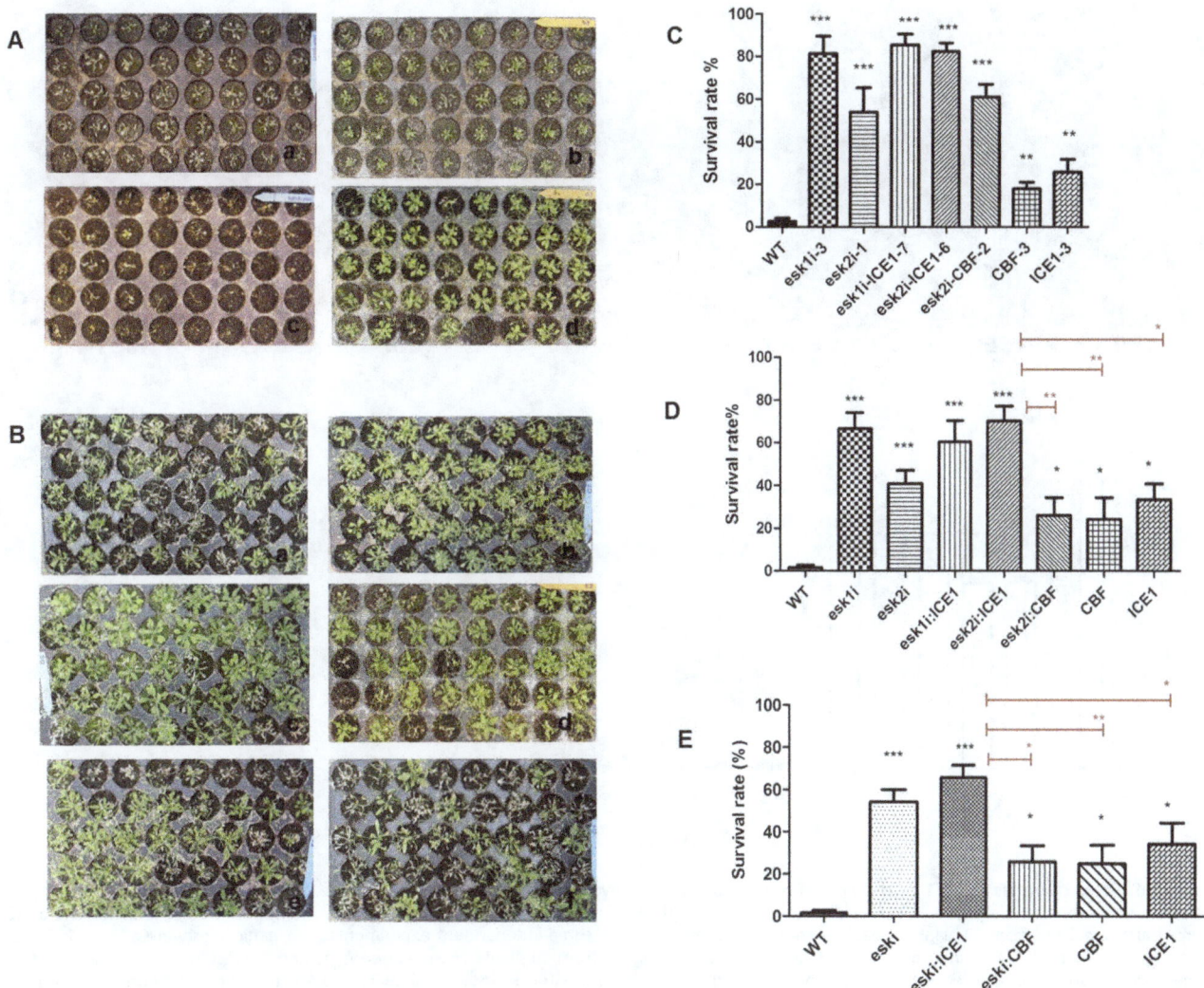

Figure 5. Drought responses of WT and transgenic plants. (A) and **(B)** Drought tolerance assay. 7-d seedlings were transferred to soil for another 1 week, subjected to drought by water withholding for 14 d, and then re-watered for 7 d. WT (Aa and Ac), *esk1i*-3 (Ab and Ad), *esk2i*-1 (Ba), *esk1i-ICE1*-7 (Bb), *esk2i-ICE1*-6 (Bc), *esk2i-CBF-2* (Bd), *CBF*-3 (Be), and *ICE1*-3 (Bf) plants. **(C)** Survival rate of plants from (A) and (B) after re-watering. **(D)** Survival rate of plants from three individual transgenic lines of each vector. **(E)** Survival rate of plants from all transgenic lines as description in methods and materials. SD (error bars) was calculated from results of three independent experiments (n>30 for each experiment). Asterisks indicate significant differences from the corresponding WT values as determined by Student's *t*-tests (*0.01≤P≤0.05, **P≤0.01, ***P≤0.001). Experiments were repeated at least three times with similar results.

The *AtCBF1* (AT4G25490) and *AtCBF3* (AT4G25480) were amplified from *Arabidopsis* Col-0 DNA by PCR. *AtCBF1* was amplified with a forward primer containing a BamHI restriction site at the 5′ end (5′-CGGGATCCCTCTGATCAATGAACT-CATT-3′) and a reverse primer containing a SacI site at the 3′ end (5′-GCGAGCTCTTAGTAACTCCAAAGCGACA-3′). *AtCBF3* was amplified with a forward primer containing an ApaI site at the 5′ end (5′-GGGCCCGATCAATGAACT-CATTTTCTGC-3′) and a reverse primer containing a XbaI site at the 3′ end (5′-GCTCTAGATTAATAACTCCATAACGA-TACGTCG-3′).

The *AtICE1* (AT3G26744) and *AtESK1* (AT3G55990) cDNA fragments were amplified from *Arabidopsis* Col-0 RNA by RT-PCR. *AtICE1* was amplified with a forward primer containing a XhoI site at the 5′ end (5′-GCCTCGAGGCGATGGGTCTT-GACGGAAACAATGGTG-3′) and a reverse primer containing a

XbaI site at the 3′ end (5′-GCTCTAGATCAGATCATACCAG-CATACCCTGCTGTATCG-3′).

For the RNA interference (RNAi) construct, for a 341-bp specific fragment of *ESK1*, *esk1i*, the sense fragment was amplified by PCR using a forward primer (5′-GCCTCGAGTTGCTAG-CATGTCTCCTCTT-3′) and a reverse primer (5′-GCGAGCT-CATTCCACGTGTCAGGTAAAC-3′), as was the antisense fragment (forward primer: 5′-CGCCCGGGATTCCACGTGT-CAGGTAAAC-3′; reverse: 5′-CGGTCGACTTGCTAG-CATGTCTCCTCTT-3′). For a 284-bp specific fragment of *ESK1*, *esk2i*, the sense fragment also was amplified using forward and reverse primers (5′-GCCTCGAGTCAAGTGTGCATTA-GAGACG-3′ and 5′-GCGAGCTCATTCCACGTGTCAGG-TAAC-3′, respectively), as was the antisense fragment (forward: 5′-CGCCCGGGATTCCACGTGTCAGGTAAAC-3′; reverse: 5′-CGGTCGACTCAAGTGTGCATTAGAGACG-3′).   The

A

B

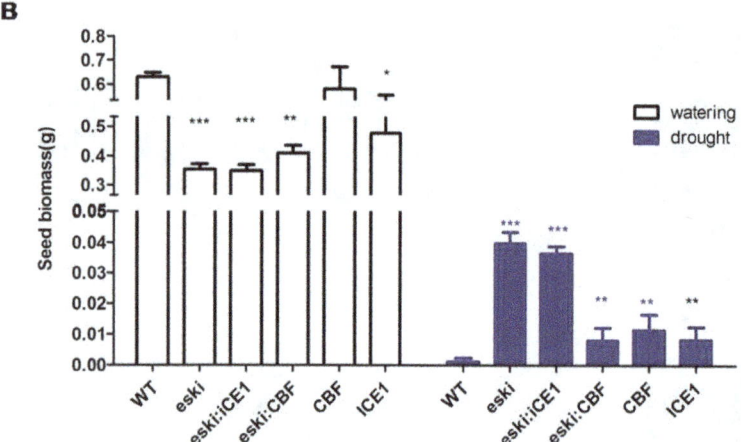

Figure 6. Seed biomass for WT and transgenic plants. (**A**) Phenotype of WT, *esk1i*-3, *esk1i-ICE1*-7, *esk2i-ICE1*-6, *CBF*-3, and *ICE1*-3 in normal watering environment. a, eight-week-old seedlings, b, six-week-old seedlings. (**B**) Average seed biomass. In the normal watering environment, SD (error bars) calculated from 30 plants of each phenotype and the results of three independent experiments, transgenic lines were as description in methods and materials. After drought treatment, SD (error bars) calculated from 10 plants of each phenotype and the results of three independent experiments, transgenic lines were as description in methods and materials. Asterisks indicate significant differences from the corresponding WT values determined by Student's *t*-tests (*$0.01 \leq P \leq 0.05$, **$P \leq 0.01$, ***$P \leq 0.001$). Experiments were repeated at least three times with similar results.

PCR products were sequenced to ensure that they encoded the expected gene products.

The pGPTV-SdaI vector was used to construct a binary expression vector. The pRNAi-vector was used to construct an intron-spliced hairpin RNA (RNAi construct), and the isolated gene or gene fragment was constructed into the relevant vector to yield pGPTV-*CBF1-CBF3* (*CBF*), pGPTV-*ICE1* (*ICE1*), pGPTV-*esk1i* (*esk1i*), pGPTV-*esk2i* (*esk2i*), pGPTV-*esk2i-CBF1-CBF3* (*esk2i-CBF*), pGPTV-*esk1i-ICE1* (*esk1i-ICE1*), and pGPTV-*esk2i-ICE1* (*esk2i-ICE1*). The transcription of each gene or gene fragment was under the control of the 35S promoter and 35S terminator.

Agrobacterium-mediated gene transformation

The floral dip method [44] was applied to stably transform the *Arabidopsis* plant by using the *Agrobacterium* strain ATHV containing the designed binary vector. Seeds from treated plants were germinated on media containing 1/2 MS salts, 1% sucrose, 0.8% agar and 50 μg/mL kanamycin. Resistant seedlings were transferred to soil and grown under 16 h light /8 h dark at 24°C in a growth chamber. After PCR confirmation, positive seedlings were used to produce T2 and T3 seeds, which were always subjected to kanamycin selection and PCR confirmation; only the

positive plants were used for seed production of the next generation. The kanamycin-tolerant and PCR-positive T3 or T4 plants were used in all experiments. Ecotype Col-0 served as control.

Quantitative real-time PCR (qRT-PCR)

Total RNA from different *Arabidopsis* plants was isolated using the RNeasy plant mini kit (Qiagen, Germany) according to the manufacturer's instructions. To eliminate any residual genomic DNA, total RNA was treated with ribonuclease-free DNase I (Thermo Scientific). Two micrograms of the total RNA was used as template to synthesize cDNA employing the RevertAid H Minus First Strand cDNA Synthesis Kit (Thermo Scientific). Real-time PCR was performed according to Kant *et al.* [45]. The PCR reaction was performed with three replicates and repeated with three biological samples. Relative quantification values for each target gene were calculated by the 2 (-Delta Delta C(T)) method [46]. For normalizing the amount of total RNA in all *Arabidopsis* samples, *Actin2* (AT3G18780) and *beta tubulin* (AT5G23860) were used as internal reference genes to compare data from different PCR runs or cDNA samples. The GenScript online tool was used to design the primers (https://www.genscript.com/), and primers used in q-PCR are listed in Supplemental Table S1.

In vitro assay for plant response to ABA and PEG

For germination, seeds were planted on the plate containing 1/2 MS salts, 1% sucrose, and 0.8% agar with 0 or 0.5 μM ABA as indicated. Plates were chilled at 4°C in the dark for 3 d and moved to 24°C with a 16 h light/8 h dark cycle. The percentage of seed germination was scored at the indicated times. Germination was defined as an obvious emergence of the radicle through the seed coat [47].

To study the inhibition effect of ABA in post-germinative growth, seeds were sown on 1/2 MS medium with 50 μg/mL kanamycin for 4 d after 3 d stratification, then transferred to 1/2 MS medium containing 0 or 20 μM ABA.

For response to *in vitro* dehydration, 3-day-old seedlings were transferred to 1/2 MS medium previously infused with 30% PEG8000 for 14 d [48]. After growing for 14 d on the treatment medium vertically, seedlings were photographed with a digital camera.

15 *ESKi* transgenic lines, 14 *eski-ICE1* transgenic lines, 3 *eski-CBF* transgenic lines, 5 *CBF* transgenic lines and 5 *ICE1* transgenic lines were used in both PEG and ABA *in vitro* assay. For each treatment, 30 plants of each line have been subjected to the treatment. Each test has been repeated three times. The influence of the tested element was concluded based the result pooled together from three repeats.

Greenhouse test for drought tolerance

15 *ESKi* transgenic lines, 14 *eski-ICE1* transgenic lines, 3 *eski-CBF* transgenic lines, 5 *CBF* transgenic lines and 5 *ICE1* transgenic lines, which showed desired gene manipulation result in qRT-PCR, were used for drought tolerance test. All seeds from transgenic lines and Col-0 together were sown in 40 5 cm×5 cm pots. Plant growth conditions were as described [49]. All seeds were cold-treated (4°C) for 3 d immediately after planting to ensure uniform germination. Plants were grown in controlled environment chambers at 24°C under sunshine illumination or supplementary illumination from cool-white fluorescent lights ($100–150$ μmol m^{-2} s^{-1}) and irrigated with water three times a week, ensuring soil saturation; the last watering was on day 14 after germination. Two-week-old seedlings were used for drought treatment, in which watering was withdrawn for two weeks. Afterwards, all plants were re-watered according to the pre-drought treatment. Tolerance to drought was determined by the capacity of plants to resume growth after 7 d of re-watering. The test was repeated at least three times. The data obtained was subjected to statistical analysis described as below.

Measurement of drought effect on seed biomass

14 *ESKi* transgenic lines, 12 *eski-ICE1* transgenic lines, 3 *eski-CBF* transgenic lines, 4 *CBF* transgenic lines and 3 *ICE1* transgenic lines were used for the test. At least 30 plants from each transgenic line and wild type were exposed to both control and drought treatment described as above. After the treatment, the survived plants were grown in normal water condition for seed setting. The seeds were harvested by bag that had been covered to plant in reproductive period. The harvested seeds were cleaned and then dried at room temperature. The weight of seed was measured after 2 weeks drying. The average weight (total harvested seed divided by the number of tested plant) was used as result to compare the drought effect on each line of plant. The data was analyzed based the statistical methods described below.

Statistics

The results shown are representative of three independent experiments, and within each experiment, treatments were replicated three times. Data were statistically analyzed using SigmaPlot 10.0 and GraphPad Prism 5.

Acknowledgments

We thank Professor Juren Zhang (School of Life Science, Shandong University) for thoughtful suggestions. This research was supported by KWS SAAT AG.

Author Contributions

Conceived and designed the experiments: HW FX ZL RN. Performed the experiments: FX TB HX. Analyzed the data: FX HW ZL JZ JZ JK. Contributed reagents/materials/analysis tools: FX TB. Contributed to the writing of the manuscript: HW FX.

References

1. Bohnert HJ, Nelson DE, Jensen RG (1995) Adaptations to Environmental Stresses. Plant Cell 7: 1099–1111.
2. U.S. Drought 2012: Farm and Food Impacts. Available: http://www.ers.usda.gov/topics/in-the-news/us-drought-2012-farm-and-food-impacts.aspx. Accessed 2013 July 26.
3. Jenifer Huang McBeath, Jerry McBeath (2010) Environmental Change and Food Security in China. New York: Springer.pp. 112.
4. Sergey M, Govorushko (2012) Natrual Processes and Human Impacts: Interactions between Humanity and the Environment. New York: Springer.pp. 113.
5. Umezawa T, Fujita M, Fujita Y, Yamaguchi-Shinozaki K, Shinozaki K (2006) Engineering drought tolerance in plants: discovering and tailoring genes to unlock the future. Curr Opin biotechnol 17: 113–122.
6. Shinozaki K, Yamaguchi-Shinozaki K (2007) Gene networks involved in drought stress response and tolerance. J Exp Bot 58: 221–227.
7. Zhang X, Fowler SG, Cheng H, Lou Y, Rhee SY, et al. (2004) Freezing-sensitive tomato has a functional CBF cold response pathway, but a CBF regulon that differs from that of freezing-tolerant Arabidopsis. Plant J 39: 905–919.
8. Liu Q, Kasuga M, Sakuma Y, Abe H, Miura S, et al. (1998) Two transcription factors, DREB1 and DREB2, with an EREBP/AP2 DNA binding domain separate two cellular signal transduction pathways in drought- and low-temperature-responsive gene expression, respectively, in Arabidopsis. Plant Cell 10: 1391–1406.
9. Miller AK, Galiba G, Dubcovsky J (2006) A cluster of 11 CBF transcription factors is located at the frost tolerance locus Fr-A^{m}2 in Triticum monococcum. Mol Genet Genomics 275: 193–203.
10. Lata C, Bhutty S, Bahadur RP, Majee M, Prasad M (2011) Association of an SNP in a novel DREB2-like gene SiDREB2 with stress tolerance in foxtail millet [Setaria italica (L.)]. J Exp Bot 62: 3387–3401.
11. Roychoudhury A, Paul S, Basu S (2013) Cross-talk between abscisic acid-dependent and abscisic acid-independent pathways during abiotic stress. Plant Cell Rep 32: 985–1006.
12. Skinner JS, von Zitzewitz J, Szucs P, Marquez-Cedillo L, Filichkin T, et al. (2005) Structural, functional, and phylogenetic characterization of a large CBF gene family in barley. Plant Mol Biol 59: 533–551.
13. El Kayal W, Navarro M, Marque G, Keller G, Marque C, et al. (2006) Expression profile of CBF-like transcriptional factor genes from Eucalyptus in response to cold. J Exp Bot 57: 2455–2469.
14. Medina J, Bargues M, Terol J, Pérez-Alonso M, Salinas J (1999) The Arabidopsis CBF gene family is composed of three genes encoding AP2 domain-containing proteins whose expression is regulated by low temperature but not by abscisic acid or dehydration. Plant Physiol 119: 463–470.
15. Novillo F, Medina J, Salinas J (2007) Arabidopsis CBF1 and CBF3 have a different function than CBF2 in cold acclimation and define different gene classes in the CBF regulon. Proc Natl Acad Sci USA 104: 21002–21007.
16. Zarka DG, Vogel JT, Cook D, Thomashow MF (2003) Cold induction of Arabidopsis CBF genes involves multiple ICE (inducer of CBF expression)

promoter elements and a cold-regulatory circuit that is desensitized by low temperature. Plant Physiol 133: 910–918.

17. Chinnusamy V, Ohta M, Kanrar S, Lee BH, Hong X, et al. (2003) ICE1: a regulator of cold-induced transcriptome and freezing tolerance in Arabidopsis. Genes Dev 17: 1043–1054.

18. Chinnusamy V, Zhu J, Zhu JK (2007) Cold stress regulation of gene expression in plants. Trends Plant Sci 12: 444–451.

19. Jaglo-Ottosen KR, Gilmour SJ, Zarka DG, Schabenberger O, Thomashow MF (1998) Arabidopsis CBF1 overexpression induces COR genes and enhances freezing tolerance. Science 280: 104–106.

20. Haake V, Cook D, Riechmann JL, Pineda O, Thomashow MF, et al. (2002) Transcription factor CBF4 is a regulator of drought adaptation in Arabidopsis. Plant Physiol 130: 639–648.

21. Siddiqua M, Nassuth A (2011) Vitis CBF1 and Vitis CBF4 differ in their effect on Arabidopsis abiotic stress tolerance, development and gene expression. Plant Cell Environ 34: 1345–1359.

22. Orabya H, Ahmad R (2012) Physiological and biochemical changes of CBF3 transgenic oat in response to salinity stress. Plant Sci 185–186: 331–339.

23. Zhang L, Li Z, Li J, Wang A (2013) Ectopic overexpression of SsCBF1, a CRT/DRE-binding factor from the nightshade plant Solanum lycopersicoides, confers freezing and salt tolerance in transgenic Arabidopsis. PLoS One 8: e61810.

24. Boyer JS (1982) Plant productivity and environment. Science 218: 443–448.

25. Xin Z, Browse J (1998) Eskimo1 mutants of Arabidopsis are constitutively freezing tolerant. Proc Natl Acad Sci U S A 95: 7799–7804.

26. Xin Z, Mandaokar A, Chen J, Last RL, Browse J (2007) Arabidopsis ESK1 encodes a novel regulator of freezing tolerance. Plant J 49: 786–799.

27. Bouchabke-Coussa O, Quashie ML, Seoane-Redondo J, Fortabat MN, Gery C, et al. (2008) ESKIMO1 is a key gene involved in water economy as well as cold acclimation and salt tolerance. BMC Plant Biol 8: 125.

28. Lefebvre V, Fortabat MN, Ducamp A, North HM, Maia-Grondard A, et al. (2011) ESKIMO1 disruption in Arabidopsis alters vascular tissue and impairs water transport. PLoS One 6: e16645.

29. Zhang M, Barg R, Yin M, Gueta-Dahan Y, Leikin-Frenkel A, et al. (2005) Modulated fatty acid desaturation via overexpression of two distinct omega-3 desaturases differentially alters tolerance to various abiotic stresses in transgenic tobacco cells and plants. Plant J 44: 361–371.

30. Duan J, Cai W (2012) OsLEA3-2, an abiotic stress induced gene of rice plays a key role in salt and drought tolerance. PLoS One 7: e45117.

31. Badawi GH, Kawano N, Yamauchi Y, Shimada E, Sasaki R, et al. (2004) Overexpression of ascorbate peroxidase in tobacco chloroplasts enhances the tolerance to salt stress and water deficit. Physiol Plant 121: 231–238.

32. Stockinger EJ, Gilmour SJ, Thomashow MF (1997) Arabidopsis thaliana CBF1 encodes an AP2 domain-containing transcriptional activator that binds to the C-repeat/DRE, a cis-acting DNA regulatory element that stimulates transcription in response to low temperature and water deficit. Proc Natl Acad Sci U S A 94: 1035–1040.

33. Nakashima K, Ito Y, Yamaguchi-Shinozaki K (2009) Transcriptional regulatory networks in response to abiotic stresses in Arabidopsis and grasses. Plant Physiol 149: 88–95.

34. Dubouzet JG, Sakuma Y, Ito Y, Kasuga M, Dubouzet EG, et al. (2003) OsDREB genes in rice, Oryza sativa L., encode transcription activators that function in drought-, high-salt- and cold-responsive gene expression. Plant J 33: 751–763.

35. Qin F, Sakuma Y, Li J, Liu Q, Li YQ, et al. (2004) Cloning and functional analysis of a novel DREB1/CBF transcription factor involved in cold-responsive gene expression in Zea mays L. Plant Cell Physiol 45: 1042–1052.

36. Hsieh TH, Lee JT, Charng YY, Chan MT (2002) Tomato plants ectopically expressing Arabidopsis CBF1 show enhanced resistance to water deficit stress. Plant Physiol 130: 618–626.

37. Morran S, Eini O, Pyvovarenko T, Parent B, Singh R, et al. (2011) Improvement of stress tolerance of wheat and barley by modulation of expression of DREB/CBF factors. Plant Biotechnol J 9: 230–249.

38. Xiao BZ, Chen X, Xiang CB, Tang N, Zhang QF, et al. (2009) Evaluation of seven function-known candidate genes for their effects on improving drought resistance of transgenic rice under field conditions. Mol Plant 2: 73–83.

39. Gutha LR, Reddy AR (2008) Rice DREB1B promoter shows distinct stress-specific responses, and the overexpression of cDNA in tobacco confers abiotic and biotic stress tolerance. Plant Mol Bio 68: 533–555.

40. Halpin C (2005) Gene stacking in transgenic plants-the challenge for 21st century plant biotechnology. Plant Biotechnol J 3: 141–155.

41. Ramana Rao MV, Parameswari C, Sripriya R, Veluthambi K (2011) Transgene stacking and marker elimination in transgenic rice by sequential Agrobacterium-mediated co-transformation with the same selectable marker gene. Plant Cell Rep 30: 1241–1252.

42. Ye X, Al-Babili S, Klöti A, Zhang J, Lucca P, et al. (2000) Engineering the provitamin A (beta-carotene) biosynthetic pathway into (carotenoid-free) rice endosperm. Science 287: 303–305.

43. Murashige T, Skoog F (1962) A revised medium for rapid growth and bio-assays with tobacco tissue cultures. Physiol Plant 15: 473–497.

44. Clough SJ, Bent AF (1998) Floral dip: a simplified method for Agrobacterium-mediated transformation of Arabidopsis thaliana. Plant J 16: 735–743.

45. Kant S, Kant P, Raveh E, Barak S (2006) Evidence that differential gene expression between the halophyte, Thellungiella halophila, and Arabidopsis thaliana is responsible for higher levels of the compatible osmolyte proline and tight control of Na$^+$ uptake in T. Halophila. Plant Cell Environ 29: 1220–1234.

46. Livak KJ, Schmittgen TD (2001) Analysis of relative gene expression data using real-time quantitative PCR and the 2 (-Delta Delta C(T)) method. Methods 25: 402–408.

47. Li H, Jiang H, Bu Q, Zhao Q, Sun J, et al. (2011) The Arabidopsis RING finger E3 ligase RHA2b acts additively with RHA2a in regulating abscisic acid signaling and drought response. Plant Physiol 156: 550–563.

48. van der Weele CM, Spollen WG, Sharp RE, Baskin TI (2000) Growth of Arabidopsis thaliana seedlings under water deficit studied by control of water potential in nutrient-agar media. J Exp Bot 51: 1555–1562.

49. Novillo F, Alonso JM, Ecker JR, Salinas J (2004) CBF2/DREB1C is a negative regulator of CBF1/DREB1B and CBF3/DREB1A expression and plays a central role in stress tolerance in Arabidopsis. Proc Natl Acad Sci U S A 101: 3985–3990.

6

Rice *ORMDL* Controls Sphingolipid Homeostasis Affecting Fertility Resulting from Abnormal Pollen Development

Chutharat Chueasiri[1], Ketsuwan Chunthong[1], Keasinee Pitnjam[1], Sriprapai Chakhonkaen[1], Numphet Sangarwut[1], Kanidta Sangsawang[1], Malinee Suksangpanomrung[1], Louise V. Michaelson[2], Johnathan A. Napier[2], Amorntip Muangprom[1]*

1 National Center for Genetic Engineering and Biotechnology, Thailand Science Park, Klong Luang, Pathumthani, Thailand, **2** Biological Chemistry Department, Rothamsted Research, Harpenden, Hertfordshire, United Kingdom

Abstract

The orosomucoids (ORM) are ER-resistent polypeptides encoded by *ORM* and *ORMDL* (ORM-like) genes. In humans, ORMDL3 was reported as genetic risk factor associated to asthma. In yeast, ORM proteins act as negative regulators of sphingolipid synthesis. Sphingolipids are important molecules regulating several processes including stress responses and apoptosis. However, the function of *ORM/ORMDL* genes in plants has not yet been reported. Previously, we found that temperature sensitive genetic male sterility (TGMS) rice lines controlled by *tms2* contain a deletion of about 70 kb in chromosome 7. We identified four genes expressed in panicles, including an *ORMDL* ortholog, as candidates for *tms2*. In this report, we quantified expression of the only two candidate genes normally expressed in anthers of wild type plants grown in controlled growth rooms for fertile and sterile conditions. We found that only the *ORMDL* gene (*LOC_Os07g26940*) showed differential expression under these conditions. To better understand the function of rice *ORMDL* genes, we generated RNAi transgenic rice plants suppressing either *LOC_Os07g26940*, or all three *ORMDL* genes present in rice. We found that the RNAi transgenic plants with low expression of either *LOC_Os07g26940* alone or all three *ORMDL* genes were sterile, having abnormal pollen morphology and staining. In addition, we found that both sphingolipid metabolism and expression of genes involved in sphingolipid synthesis were perturbed in the *tms2* mutant, analogous to the role of ORMs in yeast. Our results indicated that plant ORMDL proteins influence sphingolipid homeostasis, and deletion of this gene affected fertility resulting from abnormal pollen development.

Editor: Carl Ng, University College Dublin, Ireland

Funding: This work was supported by National Science and Technology Development Agency; grant numbers P-00-20101. The funders had no role in study design, data collection and analysis, decision to publish, or preparation of the manuscript.

Competing Interests: The authors have declared that no competing interests exist.

* Email: amorntip.mua@biotec.or.th

Introduction

Orosomucoid (ORM) family proteins are ER proteins encoded by *ORM* and *ORMDL* (ORM-like) genes. These proteins are highly conserved from yeasts to plants to human. Recently, human ORMDL3 was reported as a genetic risk factor associated with asthma in diverse populations [1–3]. In addition, the expression of *ORMDL3* gene was shown to be associated with the disease symptoms [4,5]. ORMDL3 is involved in pro-inflammatory diseases by binding and inhibiting sarco-endoplasmic reticulum Ca^{2+} pump (SERCA), which reduce ER Ca^{2+} concentration and increase unfolded-protein response (UPR), a process believed to induce inflammation [6].

In yeast, deletion of the ORM proteins showed susceptibility to agents that increase protein-misfolding in the ER [7]. In addition, deletion of the ORM proteins increased UPR and slowed ER-to-Golgi transport of the tested proteins though these effects were suppressed by high temperatures [8]. After treating with agents that increase protein-misfolding, *ORM2* gene expression was increased significantly, indicating that it is up-regulated by UPR

[8]. In addition, genetic interaction profiles of *orm2Δ* deletion mutants showed inverse correlation with the interaction patterns of yeast mutants with reduced *LCB1* and *LCB2* expression. Furthermore, yeast cells with over-expression of *ORM1* or *ORM2* showed genetic interaction profiles highly correlated with those seen in yeasts with reduced *LCB1* and *LCB2* expression, indicating that increased expression of ORM reduced LCB 1/2 activity [9]. *LCB1* and *LCB2* encode serine palmitoyltransferase, the first and rate-limiting enzyme in sphingolipid biosynthesis. Over-expression of *ORM1* or *ORM2* resulted in reduced LCB levels, while cells with deleted *ORM1/2* had highly elevated levels of LCBs [9]. Thus, it was proposed that ORM proteins were negative regulators of sphingolipid synthesis [9]. Yeast cells with *ORM1/2* deletions increased flux throughout the sphingolipid pathway, resulting in growth defects [9] though the precise metabolic impact of loss of ORM function on sphingolipid synthesis in yeast is currently ambiguous, since *orm1Δ/orm2Δ* mutants have been reported to have reduced [8] or elevated [9] levels of ceramides. In addition, alteration in *ORM* gene expression or mutations to their phosphorylation sites perturbed

sphingolipid metabolism, leading to the hypothesis that ORMs play a central role in lipid homeostasis [9,10].

Sphingolipids are key cellular membranes and signaling molecules involved in several cellular activities, such as cell proliferation, cell differentiation, apoptosis, and stress responses [11–14]. In plants, sphingolipids are important for cell growth and the establishment of cell polarity through their contribution to the functional organization of the endomembrane system [15]. Accordingly, sphingolipids were reported to be involved in a trafficking pathway with specific endomembrane compartments and polar auxin transport protein [16]. Sphingolipids are also reported to be involved in responses to ABA and cold temperature [17-19], also regulated mineral ion homeostasis and programmed cell death [20,21].

Microsporogenesis has been reported to be controlled by sphingolipids in several plant species [22–24]. In Arabidopsis, the study of *fbr11-2/lcb1-1* mutants showed that the *fbr11-2* mutant, an allele of *lcb1-1*, was transmitted only through female gametophytes, and initiated apoptotic cell death in bi-nucleated microspores [23]. The *FBR11/LCB1* expression was confined in microspores during microgametogenesis. These results suggested that SPT modulated programmed cell death plays an important role in the regulation of male gametophyte development [23]. In addition, Arabidopsis *LCB2* loss-of-function mutant demonstrated that sphingolipids are important for gametophytic development by alterations in the endomembrane system of pollen [22]. Furthermore, sterile or severely reduced in fertility were observed in antisense or RNAi transgenic plants with repressed expression of *dihydrosphingosine C4 hydroxylase 1(DSH 1)*, a key constituent of sphingolipids, which supported the role of sphingolipids on fertility in plant [25].

Male sterility facilitates hybrid seed production. Rice hybrids showed 20% higher yields than the best inbred varieties [26]. Hybrid rice is a promising alternative to increase food production to meet the demand for future need. Previously, we found that *temperature-sensitive genetic male-sterile 2* (*tms2*) rice line (TGMS) lacked a 70Kb section of DNA that encoded 4 expressed genes, one of which is an *ORMDL* ortholog [27]. Although there are three *ORMDL* genes in rice, only one (*LOC_Os07g26940*) is located in the deleted region. In this report, we found that the *LOC_Os07g26940* gene showed strong expression in rice panicles and anthers. In addition, *LOC_Os07g26940* showed differential expression under low and high temperatures, corresponding to fertile and sterile conditions, respectively. When RNAi was used to suppress *ORMDL* genes, either in the deletion region or all *ORMDL* three genes in rice, the resulting transgenic plants showed abnormal pollen development that resulted in male sterility. In addition, we showed that sphingolipids metabolism and expression of genes in sphingolipids biosynthesis were perturbed in the *tms2* mutants. Similar to yeast, our results indicated that the plant *ORMDL* gene modulates sphingolipid homeostasis, affecting fertility by controlling pollen development. Therefore, understanding the processes that control plant fertility could be useful in generating high yield crops by hybrid or by genetic engineering.

Results

Expression analysis of candidate genes

Seven genes located in the 70 Kb loci associated with the TGMS *tms2* phenotype were annotated to encode expressed proteins, four of which expressed in panicles [27]. In this study, we examined the expression of these four genes (*putative cytochrome P450 77A3 LOC_Os07g26870, LOC_Os07g26930, ORMDL*

LOC_Os07g26940, and *LOC_Os 07g26974*) in anthers of wild type plants by RT-PCR, confirming the identity of the amplicons by sequencing. Our results indicated that only *LOC_Os07g26930* and *ORMDL LOC_Os07g26940* were expressed in anthers (data not shown). To determine whether *LOC_Os07g26930* and *ORMDL LOC_Os07g26940* were differentially expressed at low and high temperatures corresponding to fertile and sterile conditions, the expression of these genes was monitored in panicles and anthers of wild type plants grown in controlled growth room at 26 and 32°C, using qRT-PCR. Information from GRAMENE rice database indicated that LOC_Os07g26930 and *ORMDL LOC_Os07g26940* have 1 and 3 splicing forms respectively. Primers specific for all these forms were designed and used for expression analysis by qRT-PCR, compared to *Actin1* as internal control. The results showed that expression of *LOC_Os07g26930* was constant at low and high temperatures in panicle or in anther of wild type plants (Fig. 1A), inconsistent with a role in TGMS. In the case of *ORMDL LOC_Os07g26940*, we detected all three splicing forms by qRT-PCR. The results showed that expression levels of *ORMDL LOC_Os07g26940.1* was similar at low and high temperatures in the panicle. Interestingly in the anther, this splicing form showed much higher expression at low temperature (fertile condition) than that at high temperature (sterile condition) (Fig. 1A). *ORMDL LOC_Os07g26940.2* and *ORMDL LOC_Os07g26940.3* showed similar levels of expression at low and high temperatures in panicle or in anther, with a little higher at high temperatures in panicle for *ORMDL LOC_Os07g26940.2*. While *ORMDL LOC_Os07g26940.3* showed a little higher expression in the anther at high temperature (sterile condition) than at low temperature (fertile condition) (Fig. 1A). To further study tissue-specific expression of *ORMDL LOC_Os07g26940*, multiple different rice tissues were used for qRT-PCR analyses including leaves, stems, roots, anthers, small panicles, medium panicles, and large panicles. Using *elongation factor 1 α, EF-1α* as a reference gene, the results indicated that all the three transcripts derived from *ORMDL LOC_Os 07g26940* were detected in most of the tested tissues, except in root where we could not detect expression of *ORMDL LOC_Os 07g26940.2*. The three forms were predominantly accumulated in the panicle and anther (Fig. 1B).

Phylogenetic Analysis

LOC_Os07g26940 is a member of the ORMDL protein family, which is highly conserved in eukaryotes. Information from the GRAMENE rice database indicated that *ORMDL LOC_Os07g26940* has three splicing forms, and the results from sequence alignments of these transcripts indicated that the three transcripts of *ORMDL LOC_Os07g26940* shared the same sequence for about half of the gene starting from the beginning of the transcripts (Fig. S1). Previously, a BLAST search using the predicted amino acid sequences of *ORMDL LOC_Os07g26940* revealed two other sequences (encoded by *LOC_Os04g47970* and *LOC_Os02g45180)* with high percent sequence similarity to *ORMDL LOC_Os07g26940*, suggesting that at least three *ORMDL* genes are present in rice [27]. *ORMDL LOC_Os04g47970* has two splicing forms but *ORMDL LOC_Os02g45180* has only one. Similarly, results from comparing the two transcripts of *ORMDL LOC_Os04g47970* showed that they share about 75% of the sequences starting from the beginning of the transcripts (Fig. S2). The rice transcripts of *ORMDL* showed high homology in the coding sequences (Fig. S3). To understand the evolutionary relationships of this gene family in rice and other species across kingdoms, phylogenetic analysis was constructed, using deduced amino acid sequences from the longest transcript of

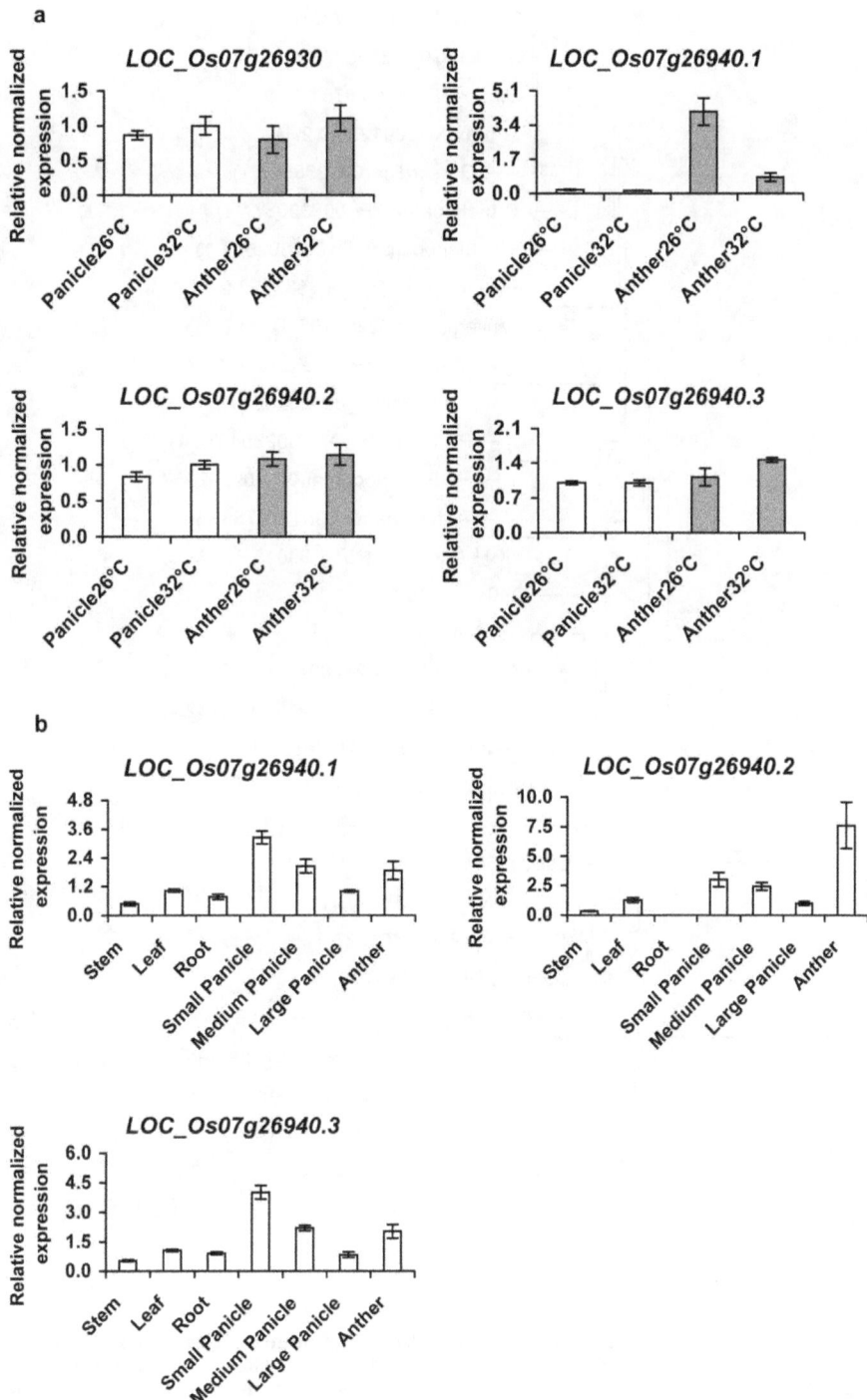

Figure 1. Expression analysis by quantitative RT-PCR. (A) qRT-PCR analysis of *LOC_Os07g26930* and all three splicing forms of *ORMDL LOC_Os07g26940* in panicles and anthers of wild type rice plants grown in control growth room at 26°C and 32°C. The expression level of *Actin1* (*Os05g36290*) was used as internal control. (B) qRT-PCR analysis of all three splicing forms of *LOC_Os07g26940* in wild type rice tissues, including stems, leaves, roots, small panicles (1–2 cm), medium panicles (3–14 cm), large panicles (15–25 cm) and anthers. The expression level of *EF-1α* (*Os03g08020*) was used as internal control. All data are representative from at least two biological repeats, each based on three technical replicates; similar results were obtained in the repeated experiments. Bars indicate the standard error (n = 3), which was calculated from technical replicates. Each biological sample was a mixture of 3 plants.

each rice ORMDL. This analysis showed the phylogenetic relationship between rice ORMDL and other ORMDL from different species selected from NCBI Reference Sequence database. The results showed 4 main groups of genes from human, drosophila, yeast, and plants. As expected, rice ORMDL is closer to *Z. mays* and to *S. bicolor* than to *A. thaliana*. Interestingly, ORMDL LOC_Os07g26940 is in a separate clade from the other two rice ORMDL (Fig. 2).

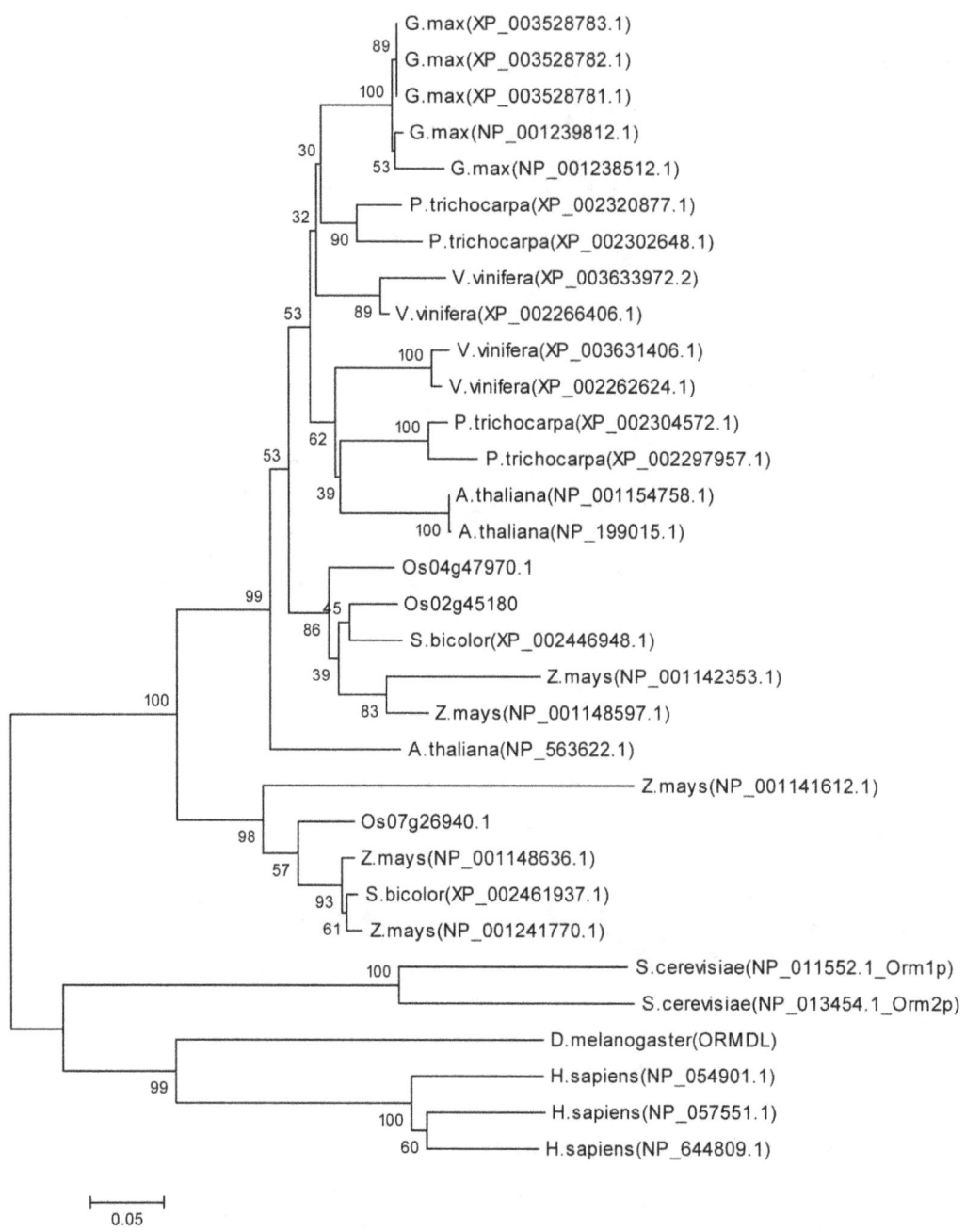

Figure 2. Phylogenetic relationship of rice ORMDL and other ORMDL from different species across kingdom. An un-rooted neighbour joining phylogenetic tree was constructed from the protein sequences of rice ORMDL obtained from GRAMENE rice database and ORMDL from other species selected from NCBI Reference Sequence database. Multiple sequence alignment was performed using the ClustalW in MEGA5 and the tree was generated using MEGA5. The numbers for interior branches indicate the bootstrap values (%) for 1000 replications.

Expression analysis of rice *ORMDL* genes

To determine the patterns of expressions of rice *ORMDL* genes in *tms2* and wild type rice plants, expression of these genes by qRT-PCR was monitored in anthers of *tms2* and wild type plants grown in controlled growth room at 26 and 32°C. Using *EF-1α* as the reference gene, the results indicated that all the three *ORMDL* genes in rice showed similar patterns of expression by having higher expression at low temperature (26°C) than that at high temperature (32°C) conditions, and these genes showed higher expression in *tms2* than in wild type plants in both conditions. As

expected, expression of *ORMDL LOC_Os 07g26940.1*, used as a representative splicing form of *ORMDL LOC_Os 07g26940*, was not detected in the *tms2* mutant rice plants (Fig. 3).

RNAi transgenic rice plants

To investigate the role of *LOC_Os07g26930* and *ORMDL LOC_Os07g26940* on pollen development, RNAi technology was used to down-regulate *LOC_Os07g26930* or *ORMDL LOC_Os07g26940* expression in rice. For the gene-specific *ORMDL LOC_Os07g26940* RNAi construct, sequence present only in the

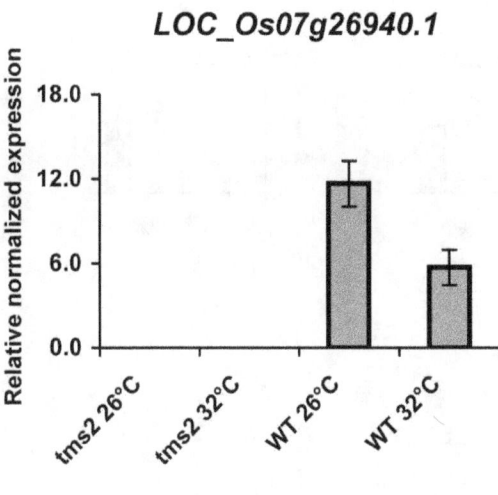

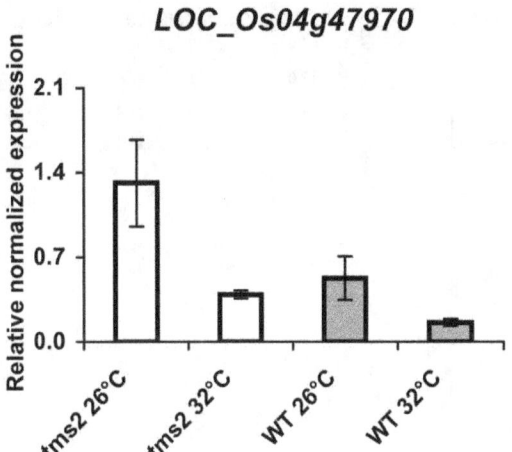
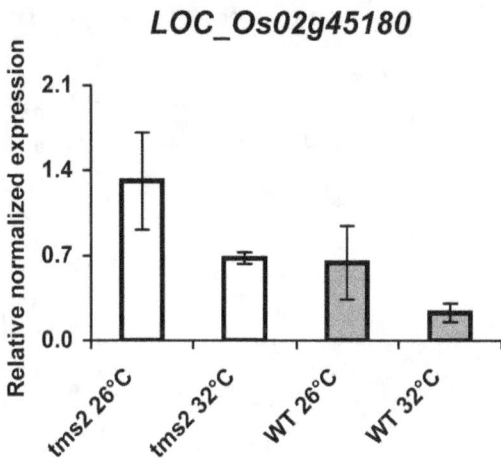

Figure 3. Expression of rice *ORMDL* genes in *tms2* and wild type rice plants. qRT-PCR analysis of rice *ORMDL* genes in anthers of *tms2* and wild type rice plants grown in control growth room at 26°C and 32°C. The expression level of *EF-1α* was used as internal control. All data are representative from at least two biological repeats, each based on three technical replicates; similar results were obtained in the repeated experiments. Bars indicate the standard error (n = 3), which was calculated from technical replicates. Each biological sample was a mixture of 3 plants.

transcripts of *ORMDL LOC_Os07g26940*, but not in *ORDML* genes (*LOC_Os04g47970* and *LOC_Os02g45180)* were used (Fig. S3). As this sequence is different from the other two rice *ORMDL* genes, we predicted that the RNAi only affects *ORMDL LOC_Os07g26940* expression. However, because three *ORMDL* genes are present in rice, knocking down only the *ORMDL LOC_Os07g26940* expression may not reveal all phenotypic effects. Thus, we generated an another RNAi construct using conserved sequences present in all rice *ORDML* genes (Fig. S3), to universally down-regulated all the *ORDML* gene transcripts. These RNAi sequences were under the control of ubiquitin promoter [28]. These RNAi constructs were transformed in to wild type (WT) *Nipponbare* plants. We obtained 27, 8 and 6 transgenic lines for *LOC_Os07g26930*-specificRNAi, *LOC_Os07g26940*-specificRNAi, and *ORMDL*-specific RNAi, respectively. In our experience, it was much more difficult to obtain *LOC_Os07g26940*-specificRNAi, and *ORMDL*-specific RNAi, transgenic lines than that from *LOC_Os07g26930*-specificRNAi construct. In natural condition, at flowering, none of transgenic plants for *LOC_Os07g26930*-specificRNAi were sterile. For,

LOC_Os07g26940-specific RNAi, or *ORMDL*-specific RNAi, two independent transformed lines from each construct were selected for further studies on the basis of low levels of expression of the target genes and a single copy of the transgene construct. As expected, expressions of all three splicing forms of *ORMDL LOC_Os07g26940* were reduced in *LOC_Os07g26940*-specific RNAi transgenic plants, while expression of the other two ORMDL were not lower than that of wild type plants (Fig. 4A). For *ORMDL*-specific RNAi, expressions of all three *ORMDL* were reduced (Figure 4B). At flowering, the pollen of these plants was different from that of wild type rice plants in both morphology and staining. The wild type pollen was round, showing the presence of starch granules, while pollen of *LOC_Os07g26940*-specificRNAi, or *ORMDL*-specific RNAi, was abnormal showing small pear-shaped pollen grains lacking starch (Fig. 4C). In addition, these T0 transgenic plants were sterile when they were grown under glass house conditions. Thus, they were ratooned from one T0 plant to at least 6 plants from each of the original T0 transgenic plant. The plants were grown under glass house conditions for approximately one month. Subsequently, three

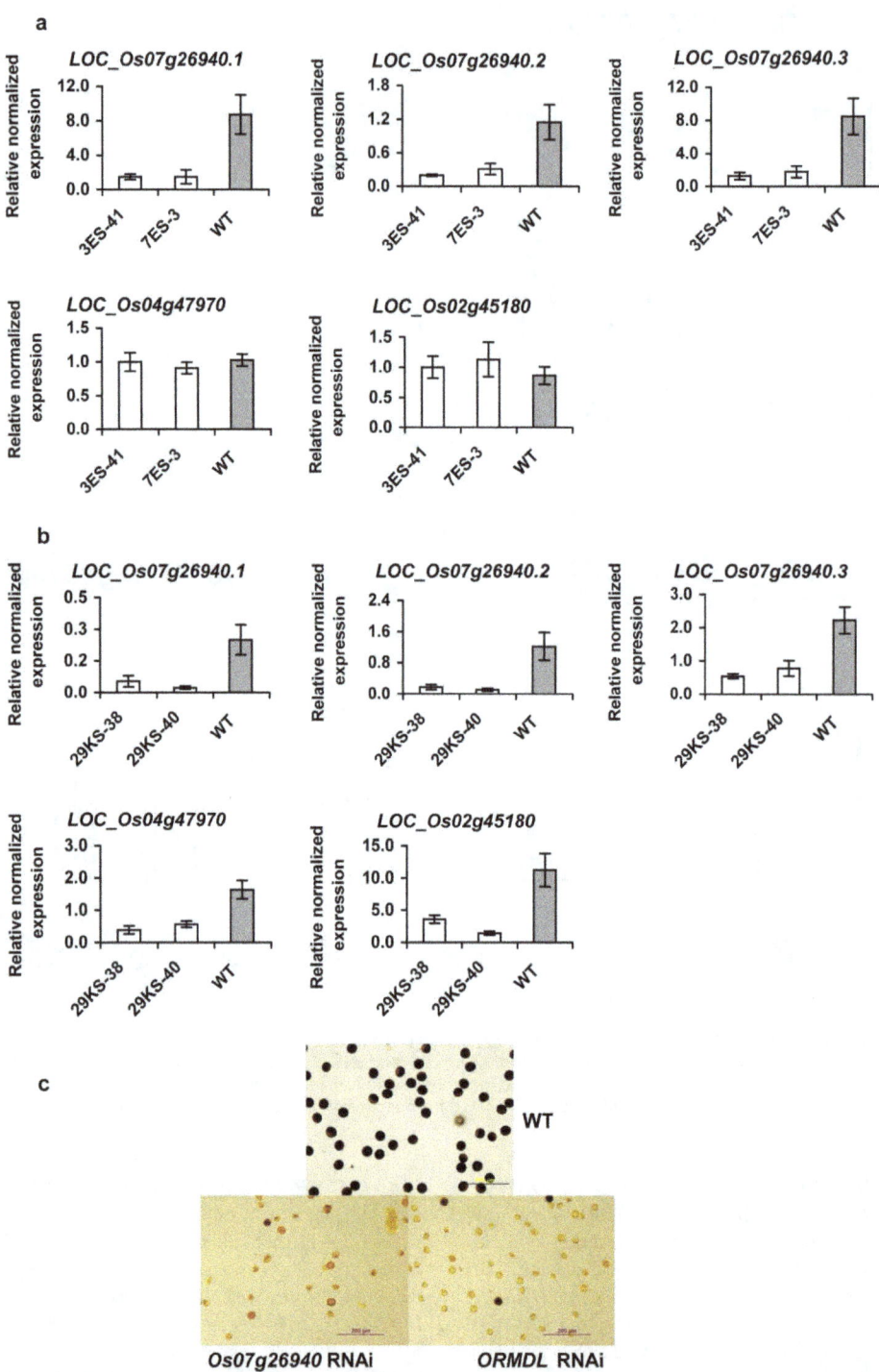

Figure 4. Expression analysis and pollen staining of RNAi transgenic rice plants. A) qRT-PCR analysis of *ORMDL LOC_Os07g26940* and other *ORMDL* gene expression in leaves of *LOC_Os07g26940* RNAi (3ES-41 and 7ES-3). B) qRT-PCR analysis of *ORMDL LOC_Os07g26940* and other *ORMDL* gene expression in leaves of *ORMDL* RNAi (29KS38 and 29KS-40) transgenic rice plants. 3ES-41 and 7ES-3 or 29KS38 and 29KS-40 are two independent transgenic lines from the same construct. The expression level of *EF-1α* was used as internal control. All data are representative from at least two biological repeats, each based on three technical replicates; similar results were obtained in the repeated experiments. Bars indicate the standard error (n = 3). which was calculated form technical replicates. Each biological sample was a mixture of 3 plants. C) Pollen staining of wild type, *LOC_Os07g26940* RNAi, and *ORMDL* RNAi transgenic rice plants. Similar results were obtained from the two tested transgenic lines form each construct. The pictures are a representative from each construct.

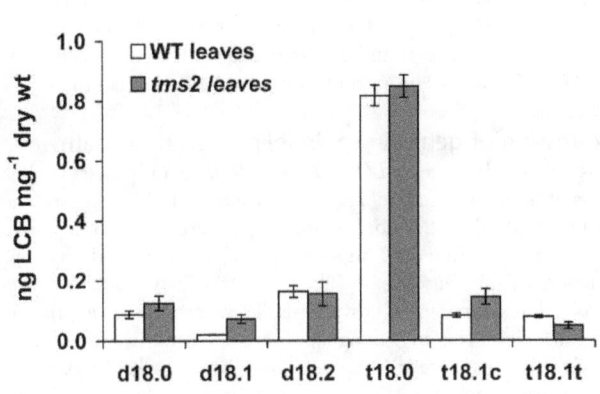

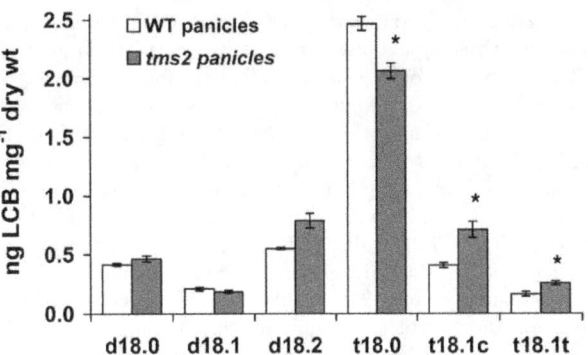

Figure 5. Sphingolipid long-chain base composition ng LCB mg^{-1} dry weight in wild type and *tms2* mutant rice plants in leaf tissue(A) and in panicles(B).

plants from each line were moved to grow in control growth room at 26 and 32°C. At flowering, these plants were sterile, with no plants capable of setting seed. The panicle tissue of the T0 was not used for direct measurement as they were retained to observe seed setting and to allow for T1 seed production.

Sphingolipid metabolism is perturbed in the *tms2* mutant and in RNAi rice lines

Recently, yeast ORM proteins were reported as negative regulators of sphingolipid synthesis [8,9]. Since one *ORMDL* gene was deleted in the rice *tms2* mutant, one scenario is that this may affect or alter sphingolipid synthesis by releasing LCB synthesis from homeostatic regulation. To investigate whether the *tms2* mutant affects sphingolipid composition, total sphingolipid long-chain bases (LCBs) were analyzed from leafs and panicles of wild-type and *tms2* mutant rice plants. Sphingolipids comprise an LCB and an amide-linked a fatty acid moiety, usually also decorated with a sugar head group. The LCB component of sphingolipids is unique to these lipids, and its synthesis in yeast has been shown to be modulated by ORM1/2. Analysis of total LCBs showed that no significant difference in sphingolipid LCB composition was observed in leaves of *tms2* mutant and wild type rice plants (Fig. 5A). Importantly, panicles of *tms2* mutant rice plants had altered sphingolipid LCB composition, with higher levels of d18:2, t18:1c, and t18:1t but lower levels of t18:0, compared with wild type rice plant(Fig. 5B). The other LCB species were similar in *tms2* mutant and wild type rice plants (Fig. 5B). In view of the alteration of LCB composition in the *tms2* mutant, and also previous studies in yeast which revealed altered ceramide levels in *orm1Δ/orm2 Δ* mutants [8,9], more targeted analysis was carried out, to measure the ceramide levels in this mutant and also RNAi-suppressed lines. The results showed that both RNAi-suppressed lines and *tms2* mutant plants showed a reduced accumulation of ceramides, relative to the WT (Fig. 6A–6B; note that in the case of the RNAi lines, analysis was carried out

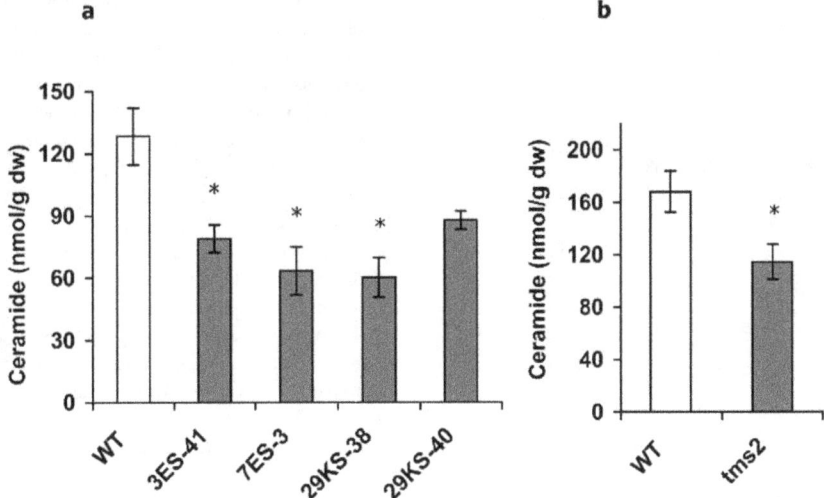

Figure 6. Total ceramide analysis. 6A) Total ceramide analysis in leafs of RNAi rice lines and WT rice plants. 6B) Total ceramide analysis in panicles of tms2 mutant and WT rice plants. Ceramide levels were determined from the leaves of *LOC_Os07g26940* RNAi (3ES-41 and 7ES-3) or *ORMDL* RNAi (29KS38 and 29KS-40) transgenic rice plants, and compared with equivalent tissue from WT. Results were expressed as absolute levels of ceramides (nmol/g dw). Similar analysis was performed on panicles of either WT or tms2 mutant rice, and data is presented as absolute levels of ceramides (nmol/g dw). Absolute levels (A, B) represent the mean of 6 technical replicates for each independent line. Bar indicates the standard error (n = 6). * indicates statistical significance at p-value of 0.05 as determined by t-test. 3ES-41 and 7ES-3 or 29KS38 and 29KS-40 are two independent transgenic lines from the same construct.

Table 1. Ceramide composition in panicles of *Indica* WT and *tms2* rice lines.

		c16:0	c18:0	c20:0	c20:1	c22:0	c22:1	c24:0	c24:1	c26:0	c26:1
WT	d18:0	0.41±0.09	0.01±0.00	0.13±0.01	0.14±0.02	0.04±0.01	0.02±0.01	0.15±0.01	0.02±0.00	0.23±0.02	0.02±0.01
	d18:1	0.13±0.03	0.03±0.02	0.28±0.02	8.86±1.69	0.23±0.07	1.01±0.10	0.60±0.24	0.60±0.13	0.19±0.10	0.10±0.03
	d18:2	1.61±0.50	0.38±0.03	4.29±0.31	77.85±4.52	0.69±0.07	13.29±1.03	0.52±0.08	11.19±1.37	0.21±0.07	2.25±0.47
	t18:0	0.62±0.06	0.40±0.08	2.52±0.28	0.13±0.02	5.29±0.92	0.46±0.08	18.46±8.21	0.88±0.16	5.95±2.77	1.39±0.29
	t18:1	0.13±0.01	0.07±0.04	0.10±0.01	0.72±0.08	0.15±0.03	1.56±0.11	0.94±0.19	2.23±0.22	0.51±0.11	0.30±0.05
tms	d18:0	0.40±0.19	0.01±0.00	0.07±0.01	0.07±0.03	0.02±0.00	0.01±0.00	0.09±0.02	0.02±0.01	0.17±0.04	0.01±0.00
	d18:1	0.07±0.02	0.01±0.00	0.13±0.03	4.39±1.61	0.16±0.04	0.54±0.13	0.47±0.17	0.35±0.10	0.19±0.10	0.06±0.10
	d18:2	1.59±0.59	0.24±0.06	2.85±0.58	54.81±10.10	0.44±0.09	8.46±1.45	0.35±0.06	7.26±1.33	0.09±0.02	1.48±0.29
	t18:0	0.49±0.10	0.30±0.09	1.31±0.33	0.11±0.03	3.36±0.81	0.25±0.04	13.78±6.70	0.55±0.18	4.46±2.39	0.64±0.13
	t18:1	0.13±0.03	0.02±0.00	0.05±0.01	0.45±0.10	0.13±0.04	0.87±0.20	0.61±0.13	1.67±0.46	0.33±0.10	0.28±0.07

Unit: nmol g dw^{-1}, N: a minimum of 3± SE.

on leaf material). This was true for RNAi which suppressed all rice ORMDL transcripts, or specifically just the transcript derived from *LOC_Os07g26940* and absent in *tms2*. Breakdown of these data is shown in table 1 and 2. These data confirm the specific role for *LOC_Os07g26940* in modulating ceramide levels in rice.

Expression of genes in sphingolipid synthesis pathway

Rice *ORMDL* gene *LOC_Os07g26940* is proposed, on the basis of homology, to be a regulator of sphingolipid synthesis, and in line with this, deletion of this gene perturbed sphingolipid metabolism in the *tms2* mutant. Thus, we reasoned that the deletion might consequently also affect the expression of genes involved in sphingolipid biosynthesis. To determine the expression of genes involved in sphingolipid metabolism, we used genes involved in early steps of sphingolipid synthesis in Arabidopsis, which were homologous to those genes in *S. cerevisiae* [29], and then searched for homologs of these genes in rice using Arabidopsis and rice database (http://www.arabidopsis.org/; http://www.gramene.org/). A total of 12 rice genes encoded for four enzymes involved in early steps of sphingolipid synthesis were identified (Table 3) and tested for expression analysis in anthers.

The expression of 9 genes encoding for four enzymes were detected (Fig. 7A–7E), while the primers specific for the other three genes including *serine palmitoyltransferase LOC_Os 01g70370*, *serine palmitoyltransferase LOC_Os 01g70380*, and *ceramide synthase LOC_Os 02g49590* were unable to provide good amplification. Two rice orthologs (*LOC_Os 02g56300* and *LOC_Os 10g11200*) of the *LCB1* subunit of serine palmitoyltransferase (SPT), showed higher expression at the non-permissive (i.e. sterile) temperature (32°C) in the *tms2* mutant compared with the permissive (fertile) temperature (26°C), and these *LCB1* transcripts were higher than those of wild type plants. Another rice ortholog of the *LCB1* (*serine palmitoyltransferase LOC_Os 03g14800*) showed similar expression levels in 26°C and 32°C both in wild type and mutant rice plants, and similar levels of expression in wild type and mutant rice plants (Fig. 7A). Interestingly, a rice ortholog (*serine palmitoyltransferase LOC_Os 11g31640*) of the *LCB2* showed higher expression at permissive (fertile) temperature (26°C) compared with the non-permissive (i.e. sterile) temperature (32°C) in both *tms2* mutant and in the wild type plants, and expressions of this gene were much higher in the mutant than those in the wild type plants at both conditions (Fig. 7B). A rice ortholog (*3-ketosphinganine reductase LOC_Os02g47350*) of *TSC10*, showed similar levels of expression at permissive (fertile) temperature (26°C) compared with the non-permissive (i.e. sterile) temperature (32°C) in *tms2* mutant, but this gene showed higher expression at permissive (fertile) temperature (26°C) compared with the non-permissive (i.e. sterile) temperature (32°C) in wild type plants (Fig. 7C). A rice ortholog (*sphingolipid-C4-hydroxylase LOC_Os06g12250*) of *SUR2* showed higher expression at the non-permissive (i.e. sterile) temperature (32°C) compared with the permissive (fertile) temperature (26°C) in the mutant, but in wild type, this gene showed higher expression at permissive (fertile) temperature (26°C) compared with the non-permissive (i.e. sterile) temperature (32°C). Surprisingly, the other tested rice ortholog (*sphingolipid-C4-hydroxylase LOC_Os02g51150*) of *SUR2* showed higher expression at permissive (fertile) temperature (26°C) compared with the non-permissive (i.e. sterile) temperature (32°C) in both *tms2* mutant and in the wild type plants but expression of this gene were much higher in the mutant than those in the wild type plants at both conditions (Fig. 7D). For *ceramide synthase*, a rice ortholog (*LOC_Os02g37080*) of LAC1/LAG1, showed higher expression at the non-permissive (i.e. sterile) temperature (32°C) in the *tms2*

Table 2. Ceramide composition in leaves of *Japonica* WT and RNAi rice lines.

		c16:0	c18:0	c20:0	c20:1	c22:0	c22:1	c24:0	c24:1	c26:0	c26:1
WT	d18:0	0.02±0.00	0.00±0.00	0.02±0.00	0.02±0.00	0.02±0.00	0.02±0.01	0.20±0.06	0.00±0.00	0.20±0.05	0.01±0.00
	d18:1	0.04±0.00	2.02±1.02	0.02±0.00	1.88±0.15	0.15±0.02	0.51±0.07	0.98±0.27	0.22±0.02	0.14±0.01	0.07±0.00
	d18:2	0.01±0.01	0.14±0.20	2.13±0.79	55.09±12.86	0.54±0.20	14.40±3.31	0.17±0.05	5.05±1.16	0.02±0.00	0.61±0.14
	t18:0	0.21±0.07	0.11±0.02	0.50±0.08	0.05±2.86	4.76±1.13	0.27±0.68	28.87±7.07	0.91±0.09	4.09±1.01	0.22±0.03
	t18:1	0.02±0.00	0.01±0.00	0.05±0.01	0.48±0.53	0.06±0.24	1.41±0.21	0.36±1.50	1.04±0.22	0.08±0.21	0.11±0.02
3ES-41	d18:0	0.03±0.01	0.01±0.00	0.02±0.00	0.02±0.00	0.02±0.00	0.01±0.00	0.25±0.03	0.01±0.00	0.21±0.01	0.01±0.01
	d18:1	0.02±0.00	0.01±0.00	0.01±0.00	1.12±0.07	0.09±0.01	0.55±0.07	0.46±0.05	0.16±0.02	0.05±0.01	0.01±0.00
	d18:2	0.01±0.00	0.10±0.01	2.09±0.26	38.81±4.01	0.55±0.07	10.52±1.17	0.16±0.02	3.28±0.38	0.01±0.00	0.35±0.07
	t18:0	0.10±0.02	0.06±0.02	0.32±0.02	0.05±0.01	2.26±0.36	0.17±0.02	12.26±2.07	0.43±0.06	1.64±0.23	0.08±0.01
	t18:1	0.01±0.00	0.01±0.00	0.05±0.00	0.38±0.06	0.04±0.01	1.11±0.20	0.19±0.04	0.73±0.08	0.04±0.01	0.09±0.01
7ES-4	d18:0	0.03±0.00	0.02±0.01	0.02±0.01	0.03±0.01	0.03±0.00	0.03±0.02	0.45±0.07	0.02±0.00	0.41±0.11	0.02±0.00
	d18:1	0.04±0.01	0.02±0.00	0.07±0.01	0.69±0.17	0.12±0.04	0.28±0.07	0.61±0.19	0.24±0.05	0.11±0.04	0.03±0.01
	d18:2	0.03±0.00	0.08±0.01	0.57±0.23	21.46±8.24	0.17±0.07	7.44±2.99	0.03±0.01	3.31±1.16	0.05±0.01	0.30±0.09
	t18:0	0.25±0.05	0.09±0.05	0.47±0.13	0.18±0.03	3.55±0.75	0.53±0.11	14.25±4.58	1.12±0.00	2.27±0.00	0.23±0.00
	t18:1	0.04±0.01	0.04±0.01	0.21±0.11	0.37±0.18	0.08±0.04	1.37±0.43	0.38±0.09	0.99±0.49	0.07±0.02	0.14±0.05
29KS-38	d18:0	0.05±0.02	0.01±0.00	0.03±0.00	0.02±0.00	0.03±0.01	0.01±0.00	0.60±0.06	0.01±0.00	0.52±0.13	0.07±0.03
	d18:1	0.03±0.01	1.10±1.09	0.04±0.01	1.02±0.20	0.29±0.10	0.42±0.12	0.45±0.06	0.22±0.09	0.10±0.02	0.02±0.01
	d18:2	0.01±0.00	0.09±0.02	0.92±0.49	14.39±6.71	0.03±0.01	8.41±2.66	0.02±0.01	3.67±1.35	0.02±0.01	0.42±0.17
	t18:0	0.39±0.08	0.22±0.10	0.94±0.28	0.12±0.02	6.93±2.25	0.88±0.28	11.22±1.61	0.88±0.13	2.03±0.40	0.16±0.01
	t18:1	0.04±0.01	0.02±0.00	0.03±0.01	0.27±0.04	0.13±0.05	1.26±0.43	0.54±0.20	0.92±0.42	0.12±0.04	0.08±0.03
29KS-40	d18:0	0.03±0.01	0.00±0.00	0.02±0.00	0.03±0.01	0.04±0.00	0.01±0.00	0.31±0.01	0.01±0.00	0.27±0.01	0.01±0.00
	d18:1	0.02±0.00	0.73±0.72	0.04±0.01	1.74±0.05	0.12±0.02	0.54±0.03	0.31±0.05	0.12±0.00	0.05±0.01	0.01±0.00
	d18:2	0.01±0.00	0.14±0.00	0.13±0.00	46.73±0.00	0.03±0.00	12.50±0.00	0.01±0.00	4.02±0.00	0.01±0.00	0.45±0.00
	t18:0	0.13±0.01	0.10±0.05	0.50±0.02	0.05±0.00	3.02±0.09	0.18±0.00	10.70±3.06	0.33±0.07	1.25±0.18	0.09±0.02
	t18:1	0.01±0.00	0.02±0.00	0.02±0.01	0.48±0.04	0.03±0.00	1.30±0.08	0.21±0.01	0.90±0.01	0.05±0.00	0.10±0.00

Unit: nmol g dw^{-1}, N: a minimum of 3± SE.

Table 3. Identification of genes involved in early steps of sphingolipid biosynthesis in rice using homologous genes from *A. thaliana* and *S. cerevisiae*.

Gene	S. cerevisiae	A. thaliana	O. sativa
Serine palmitoyltransferase	LCB1	At4g36480	LOC_Os02g56300, LOC_Os10g11200, LOC_Os03g14800
	LCB2	At5g23670	LOC_Os11g31640, LOC_Os01g70380, LOC_Os01g70370
		At3g48780	LOC_Os11g31640, LOC_Os01g70380, LOC_Os01g70370
3-ketosphinganine reductase	TSC10	At3g06060	LOC_Os02g47350
		At5g19200	LOC_Os02g47350
Sphingolipid-C4-hydroxylase	SUR2	At1g14290	LOC_Os06g12250, LOC_Os02g51150
		At1g69640	LOC_Os06g12250, LOC_Os02g51150
Ceramide synthase	LAC1/LAG1	At3g25540	LOC_Os02g37080, LOC_Os02g49590
		At3g19260	LOC_Os03g15750
		At1g13580	-

mutant compared with the permissive (fertile) temperature (26°C), and this expression was much higher than that of wild type plants. In wild type, this gene showed similar levels of expression, with very low level of expression at both conditions. The other tested rice ortholog (*ceramide synthase LOC_Os02g15750*) of LAC1/LAG1 showed higher expression at permissive (fertile) temperature (26°C) compared with the non-permissive (i.e. sterile) temperature (32°C) in both *tms2* mutant and in the wild type plants, and this gene showed similar levels of expressions in *tms2* mutant and in the wild type plants for each condition (Fig. 7E).

Discussion

The temperature-sensitive genetic male sterility (TGMS) system provides a powerful tool for the production of hybrid rice, in turn leading to greater yields of this vital crop. However, the molecular mechanism(s) of TGMS are currently unknown, precluding the facile incorporation of this trait into germplasm. A previous study reported a TGMS phenotype of *Ugp1*-cosuppressed plants resulting from aberrant transcripts, which undergo temperature-sensitive splicing in florets [30]. A rice *ORMDL*, *LOC_Os07g26940* was predicted to have three splicing variants (http://www.gramene.org). Accordingly, we detected all the three transcripts by qRT-PCR. Interestingly, one splice variant showed strongly differential patterns of expression in anther at low and high temperatures, implying a role for *ORMDL LOC_Os07g26940* in response to temperatures in anther. Importantly, this gene showed high expression in panicle and anthers which, in addition to modulation by temperature, coincides with the TGMS phenotype of *tms2*.

The TGMS phenotype of *tms2* mutant rice plants was reported to be controlled by a recessive gene located in chromosome 7. Previously, the *tms2* mutant plants were reported to contain at least 70 kb deletion including seven genes annotated as expressed proteins [27]. However, only two of these genes (*LOC_Os07g26930 and ORMDL LOC_Os07g26940*) were strongly expressed in panicles and anthers and only the *ORMDL* gene *LOC_Os07g26940* showed differential expression in fertile and sterile conditions in anthers, suggesting a function in this tissue and in TGMS. Accordingly, knock down function of this gene or all *ORMDL* genes including *LOC_Os07g26940*, *LOC_Os04g47970*, and *LOC_Os02g45180*, by RNAi, affected pollen development resulting in male sterility. Reduction of only *ORMDL* gene expression in chromosome 7 (*LOC_Os07g26940*) or all *ORMDL*

genes affected pollen development in both morphology and staining. Pollen grains of *LOC_Os07g26940* RNAi and *ORMDL* RNAi were different in sizes, shapes and staining, with none of these plants setting seeds. Interestingly, the male sterility of these RNAi plants was not modulated by temperature, with *LOC_Os07g26940* RNAi and *ORMDL* RNAi transgenic plants sterile at both low and high temperatures, suggesting that at least one other factor is needed for the conditional male sterility observed in the TGMS phenotype of *tms2*. Thus, deletion of *ORMDL LOC_Os07g26940* and this (currently unknown) factor are critical for the TGMS phenotype. Although genetic analysis indicates that the TGMS phenotype of *tms2* mutant plants was controlled by a single gene [31,32,27], it is possible that this phenotype is controlled by two closely linked genes not resolved by the current mapping population. Thus, a bigger population with greater recombination maybe required to further study this at the genetic level. Alternatively, the apparent inability of RNAi suppression of ORMDL to completely phenocopy the *tms2* TGMS may indicate subtle requirements for the total absence of this gene to deliver conditional fertility.

Recently, *ORMDL* was reported as a negative regulator of sphingolipid synthesis by inhibiting *serine palmitoyltransferase* (*LCB1/LCB2/TSC3*) [9]. Over expression of ORM1 and ORM2 in yeast reduced LCB and ceramide levels. Conversely, deletion of ORM1/2 elevated the levels of LCB and ceramides [9], although very similar studies have also reported decreased ceramides (in conjunction with elevated LCBs) in yeast mutants lacking ORMs [8]. Since a rice *ORMDL* gene is deleted in the *tms2* mutant, an alteration to sphingolipid homeostasis might be predicted. Accordingly, we found that sphingolipid synthesis was perturbed in the *tms2* mutant. Panicles of *tms2* mutant rice plants had higher d18:2, t18.1c, and t18:1t but lower t18:0 than the wild type rice plants, although the overall level of LCBs was not significantly altered. Closer examination of the discrete ceramide species present in the panicles of WT and *tms2* mutant confirmed a concomitant decrease in ceramides levels, in agreement with total ceramide analysis, but also showed some discrete variation in the accumulation of individual ceramides. For example, the predominant C20:1/d18:2 ceramide species is reduced by ~30% in the *tms2* mutant. This is particularly noteworthy, given the total LCB analysis of the *tms2* mutant revealed a moderate increase in the accumulation of this sphingoid base. However, since it is currently unclear if LCB modifying enzymes such as the sphingolipid D8-desaturase and the C4-hydroxylase utilise free LCBs and/or

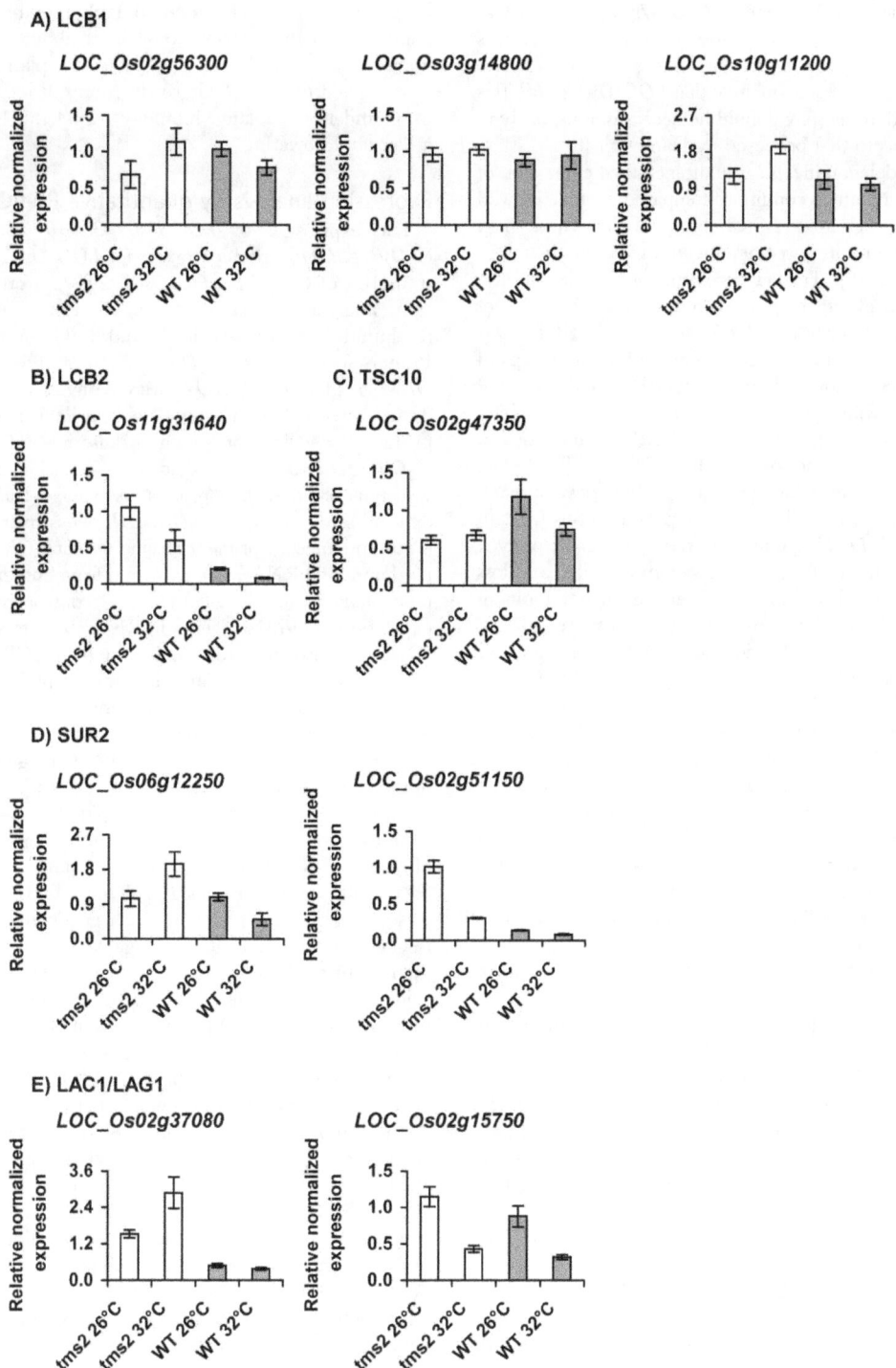

Figure 7. Expression analysis of genes involved in sphingolipid synthesis by quantitative RT-PCR. Expression of genes involved in early steps of sphingolipid synthesis including *LCB1*(A), *LCB2*(B), *TSC10* (C), *SUR2*(D), *LAC1/LAG1*(E) in anthers of *tms2* and wild type rice plants grown in control growth room at 26°C and 32°C. All data are representative from at least two biological repeats, each based on three technical replicates; similar results were obtained in the repeated experiments. Bars indicate the standard error (n = 3), which was calculated from technical replicates. Each biological sample was a mixture of 3 plants.

ceramides as substrates, it is not possible to make causal associations between these two different metabolic pools, or how these are impacted by the *tms2* mutation. It is interesting to note that the rice LAG1/LAC1 ceramide synthase ortholog *LOC_Os02g37080* showed upregulation in *tms2*, and although the substrate-specificity of this enzyme remains to be determined, it may play a role in the synthesis of a discrete subset of ceramides as seen in Arabidopsis [16]. More strikingly, and in clear agreement with the yeast studies of Han et al [8], we observed a significant reduction in the ceramide levels in rice plants with perturbation to the expression of the ORMDL gene *LOC_Os07g26940* – this was true for both the *tms2* mutant and also lines in which RNAi was used to suppress transcript accumulation. On the basis of these two independent interventions, we conclude that *LOC_Os07g26940* is specifically involved in the sphingolipid homeostasis, most likely in a manner analogous to that proposed by Han et al. [8].

In addition, the deletion in *tms2* mutant increased expression of *LCB1* in high temperature condition compared to that in wild type plants, and increased *LCB2* both in low and high temperature conditions in agreement with its previously established role in yeast [9]. Furthermore, the *tms2* deletion also affected expression of other genes involved in early steps of sphingolipid synthesis such as *SUR2* and *LAC1/LAG1*. Our results indicated that most of the genes involved in early steps of sphingolipid synthesis showed higher expression in the mutant than that in the wild type plants, particularly under high temperature treatment. In Arabidopsis, *LBC1* and *LCB2*, encoding subunits of serine palmitoyltransferase (SPT), were reported as important genes for male gametophyte development, involved in programmed cell death during pollen mitosis [23]. In addition, Arabidopsis *LCB2* null mutants resulted in gametophytic lethality by altering the endomembrane system of pollen and loss of pollen viability, which occurred early in pollen development during transition from the uni-nucleate microspore to the bicellular pollen grain [22]. Deletion of *ORMDL* gene in *tms2* mutant rice plants and RNAi suppression of *ORMDL* gene expression affected pollen development, resulting in sterility. The precise mechanism by which ORMDL *LOC_Os07g26940* modulates aspects of male fertility still remain to be resolved – based on our data, it appears that, like their yeast orthologs, higher plant ORMs are involved in sphingolipid homeostasis. However, the consequences of release from ORM regulation (reduced ceramides, and an altered LCB profile) is distinct from that observed in yeast, pointing the way for future studies. Given the significantly larger sphingolipidome of plant compared with *S. cerevisiae*, it is perhaps unsurprising that higher plants have evolved additional tiers of regulation for the control of these bioactive lipids.

In conclusion, our results indicate that *tms2* mutant plants lacking one member of the rice *ORMDL* gene family has altered sphingolipid composition. In addition, seven out of the nine expressed genes involved in sphingolipid synthesis pathway showed higher expression in the mutant compared to wild type plants, in agreement with the role of ORM as a homeostatic regulator of sphingolipid biosynthesis [9]. Our results indicate that *tms2* mutant plants lacking ORMDL *LOC_Os07g26940* affected the expression of several genes involved in early step of sphingolipid synthesis pathway, although the tested genes (orthologous for *LCB1, SUR2*, and *LAC1/LAG1*) had discrete expression patterns, suggesting an aspect of sub-functionalization most likely associated with plant-specific sphingolipid homeostasis. Our findings demonstrate an emerging role for *ORMDL* genes in plants as regulators of ceramide accumulation and also provide support for the important role of sphingolipids on pollen development.

Materials and Methods

Ethics Statement
N/A

Plant materials and growth conditions
Wild type rice (*Oryza sativa* ssp *japonica* cv. Nipponbare, and *indica* ssp cv. Pathumthanee1) and *tms2* mutant (in *indica*

background and backcrossed to Pathumthanee1, BC2F4) were planted in natural green house, and about one month before flowering some of these plants (at least 5 plants/lines/condition) were moved to controlled growth rooms at temperature 26°C or 32°C under 80% relative humidity, 12 h light/12 h darkness with a day until flowering.

Expression analysis by quantitative RT-PCR
For expression of *LOC_Os07g26930* and *ORMDL LOC_Os07g26940*, specific primers for *LOC_Os07g26930* and each transcript of *ORMDL LOC_Os07g26940* were used for expression analysis in panicle and anther of wild type plants (Pathumthanee 1) grown in 24 and 32°C control growth rooms. Primers specific for *LOC_Os07g26930* and for each transcript of *ORMDL LOC_Os07g26940* were designed using Primer-BLAST (http://www.ncbi.nlm.nih.gov/tools/primer-blast/). In addition, primers specific for each transcript of *ORMDL LOC_Os07g26940* were used to test expression in other tissues of wild type plants such as root, stem, panicle, leaf, and anther. For expression of all the three *ORMDL* genes in *tms2* and wild type (Pathumthanee 1), primers specific for each *ORMDL* genes, were used and total RNAs were extracted from anther of *tms2* and wild type plants grown in 24 and 32°C control growth rooms. For expression of *ORMDL* genes in RNAi transgenic rice plants, total RNA were isolated from leaf, and qRT-PCR were performed using specific primers for each transcript of *Os07g26940* or primers specific for each of the other two *ORMDL* genes. For expression of genes involved in early steps of sphingolipid synthesis, primers specific for 12 rice genes identified as homologous genes of Arabidopsis, which are homologues to known genes involved in sphingolipid synthesis in *S. cerevisiae* were used for qRT-PCR using total RNAs extracted from anther of *tms2* and wild type plants (Pathumthanee 1) grown in 24 and 32°C control growth rooms. For all the data presented, one sample was a mixture of 3 plants for each genotype. Total RNAs were isolated using TRIZOL^{TM} reagent (Invitrogen) following the manufacture's instructions. First-strand cDNA was synthesized using cDNA synthesis kit (Fermentas) and cDNAs transcribed from the total RNA (20 μl reaction volume). Gene specific primer were designed for amplicons about 200 bp. qRT-PCR experiments were performed on the BIO-RAD CFX 96 real time PCR systems, using SsoFast EvaGreen Supermix (Bio-Rad, Singapore). At the end of each experiment a melting curve was determined for each primer pair at a temperature stage from 60°C to 95°C to check the specificity of annealing. Primers targeting Actin1 (Os05g36290) were used to normalize the expression data of each gene in wild type plants. Because tms2 mutation affects expression of Actin1, thus as recommended [33] primers targeting Elongation Factor 1 α, EF-1α (Os03g08020) were used to normalize the expression data of genes compared expression in the mutant and wild type plants. For each gene, at least two biological replicates were performed, each with three technical replicated. Similar results were obtained in the repeated experiments. The average Ct values and standard deviation were calculated from three technical replicates. The quantification of gene expression was performed using the relative quantification method (2−ΔΔCT) [34]. All the tested primers were listed in Table S1.

Phylogenetic Analysis
Phylogenetic Analysis was performed using protein sequences encoded by the longest splicing forms of each rice ORMDL obtained from GRAMENE rice database and ORMDL from other species selected from NCBI Reference Sequence database. The tree was generated by neighbour-joining (NJ) algorithm with

p-distance method and pairwise deletion of gap, employing MEGA version 5 [35]. A bootstrap statistic analysis was performed with 1000 replicates to test the phylogeny.

Generation of *Os07g26940-* RNAi and *ORMDL*-RNAi transgenic rice plants

To generate RNAi construct to suppress expression of *ORMDL LOC_Os07g26940*, all *ORMDL* genes, or *Os07g26930*, a 156-bp fragment specific only for *Os07g26940* cDNA was PCR amplified with primer *Os07g26940*-RNAi-F (5′-ggggtaccactagt CCACT-CACCACGCGCCAC-3′, *Kpn*I, *Spe*I) and *Os07g26940*-RNAi-R (5′-ggggatccgagctc ACGTAGTAGGGGTAGGAC-3′, *Bam*HI, *Sac*I), a 250-bp fragment specific for all *ORMDL* cDNA was PCR amplified with primer *ORMDL*-RNAi-F(5′-ggggtaccac-tagtGTGAACAAGAACACGGAGT-3′, *Kpn*I, *Spe*I) and *ORMD* -RNAi-R (5′-ggggatccgagctcTGGTCATCAGCAGCAAAT -3′, *Bam*HI, *Sac*I), and a 422-bp fragment specific only for *Os07g26930* cDNA was PCR amplified with primer *Os07g26930*-RNAi-F(5′-ggggtaccactagtCCAAAAACCGCTCA-CACTCG-3′, *Kpn*I, *Spe*I), and *Os07g26930*-RNAi-R (5′-ggggatccgagctcGAATTCGATCCCAGGTGGCT -3′, *Bam*HI, *Sac*I), which harbors restriction sites (set in lowercase letters) for cloning. The resulting PCR products were first digested by *Spe*I and *Sac*I and ligated into vector pTCK303 [28] to get the transitional vector. Then, the PCR products were digested by *Kpn*I and *Bam*HI and ligated into the transitional vector. The resulting products were sent for sequencing to confirm sense and antisense orientations of the inserts in pTCK303. The constructs were transformed into rice plants (Nipponbare) using an *Agrobacterium*-mediated method.

Pollen staining and microscopy

Mature pollens were obtained from anthers of 2–3 flowers. Anthers were disrupted on microscope slides using forceps and gently squashed in 4′, 6-diamidino-2-phenylindole (DAPI, Sigma-Aldrich) under a coverslip, and visualized with Olypus BX51 microscope.

Analysis of sphingolipid long-chain bases

LCBs were liberated from 10mg of lyophilized leaves or panicle by alkaline hydrolysis, extraction method [36]. Briefly, this is preformed using 10% BaOH and dioxane 1:1 v/v. The tubes are then capped and placed in a dry block at 110°C for 20 hours. TheLCBs are extracted in chloroform/dioxane/water (8/3/8, v/v/v). The LCBsare converted to dinitrophenyl derivatives with 0.2 ml of 0.5% (v/v) methanolic 1-fluoro-2, 4-dinitrobenzene and 0.8 ml of 2 M boric acid/KOH, at 60°C for 30 mins. LCBs are then extracted by phase partitioning with CHCl3/methanol/H2O, 2:1:1 (v/v/v). The organic phase is removed and washed with an equal volume of 0.1 M KOH and 0.5 M KCl. The organic phase is then blown down and resuspended with methanol, and analyzed by reversed-phase HPLC. Separation was performed using a C18 RP 250×4-mm column with a flow rate of 1 ml/min and a concave gradient from 80 to 100% methanol/acetonitrile/2-propanol, 10:3:1 (v/v/v), against water in 45 min. The elution was monitored with an Agilent 1200 DAD measuring at 350 nm and by ESI-MS/MS MRM on an ABSciex 4000 QTRAP. Three replicate extractions and LCB analysis were performed of each extraction method. Data presented as average (+/− standard error) in ng LCB per mg of dry weight.

Analysis of ceramide

Samples extracted were ground in liquid N2 to a fine powder. 10 mg of each sample was used for each replicate analysis. Analysis of the ceramide fraction was performed using LC-MS/MS as previously described [37]. Briefly, the ceramides were extracted in the lower phase of isopropanol/hexane/water (55:20:25 v/v/v) at 60 degrees C. Samples were dried down under a stream of nitrogen. The crude extract was de-esterified in 2 ml of 33% methylamine solution in ethanol/water (7:3 v/v) at 50 degrees for 1 hour. After hydrolysis the samples were dried down under a stream of nitrogen. The ceramides were resuspended in 1 mL of tetrahydrofuran (THF)/methanol/water (2:1:2 v/v/v) and 0.1% formic acid and 50 μL was analyzed on a 4000 QTRAP LC-MS/MS system (ABSciex) after HPLC using an Agilent 1200 fitted with a 100- μL sample loop. Separation was achieved on a SUPELCOSIL ABZ+Plus column 150×3 mm, 5-μm particle size, The sample was eluted at 1 mL min−1 with a binary gradient system consisting of solvent A, THF/methanol/5 mM ammonium formate (3:2:5v/v/v), and 0.1% formic acid, and solvent B, THF/methanol/5 mM ammonium formate (7:2:1v/v/v), and 0.1% formic acid. held at 40°C. The sample was eluted at 1 mL min^{-1} with a binary gradient system consisting of solvent A, THF/methanol/5 mM ammonium formate (3:2:5v/v/v), and 0.1% formic acid, and solvent B, THF/methanol/5 mM ammonium formate (7:2:1v/v/v), and 0.1% formic acid. The gradient started at 40% B rising to 75% B over 10 min. At the end of the gradient, the %B was increased to 100% over 1 min and held at 100% B for an additional 1 min to ensure complete elution of all compounds from the column. The probe was vertically positioned 11 mm from the orifice and charged with 5000 V. The temperature was held at 650°C, GS1 was set at 90 p.s.i., GS2 at 50 p.s.i., curtain gas at 20 p.s.i., and the interface heater was engaged. A minimum of 3 replicate extractions and analysis were performed. Data presented as average (+/− standard deviation). A t-test was performed to determine statistical significance, and in all cases data presented has a p-value of 0.05.

Supporting Information

Figure S1 Sequence alignment of the three transcripts of Os07g26940. Sequences of rice ORMDL were obtained from GRAMENE rice database.

Figure S2 Sequence alignment of the two transcripts of Os04g47970. The sequences were obtained from GRAMENE rice database.

Figure S3 Sequence alignment of the representative three transcripts of *ORMDL* genes. Highlight in yellow indicates sequences specific for Os07g26940. Highlight in gray indicates sequences used to make RNAi to knockdown all ORMDL genes. Green and red indicates primer sequences used for Os07g26940R-NAi and ORMDL RNAi, respectively. The sequences were obtained from GRAMENE rice database.

Table S1 qRT-PCR primer sequence.

Acknowledgments

We thank Professor Kang Chong from Research Center for Molecular and Developmental Biology; Key Laboratory of Photosynthesis and Environmental Molecular Physiology, Institute of Botany, the Chinese Academy of Sciences, Beijing 100093, China for providing the backbone vector of the

RNAi cassette. We acknowledge Dr. Tharathorn Teerakathiti and Dr. Phanramphoei Namprachan Frantz, for helpful discussion and Dr. Wananit Wimuttisuk for critical reading of the manuscript and her help with the quantitative RT-PCR. We also thank Dr. Channarong Seepiban and Ms. Amornpan Klanchui for their technical assistances. This work was supported by a grant from Cluster and Program Management Office (CPMO; P-00-20101), National Science and Technology Development Agency (NSTDA).

Author Contributions

Conceived and designed the experiments: AM. Performed the experiments: CC KP KC SC NS KS MS LM JN. Analyzed the data: AM LM JN. Wrote the paper: AM JN.

References

1. Galanter J, Choudhry S, Eng C, Nazario S, Rodríguez-Santana JR, et al. (2008) ORMDL3 gene is associated with asthma in three ethnically diverse populations. Am J Respir Crit Care Med 177: 1194–200.

2. Wu H, Romieu I, Sienra-Monge JJ, Li H, del Rio-Navarro BE, et al. (2009) Genetic variation in ORM1-like 3 (ORMDL3) and gasdermin-like (GSDML) and childhood asthma. Allergy 64: 629–35.

3. Fang Q, Zhao H, Wang A, Gong Y, Liu Q (2011) Association of genetic variants in chromosome 17q21 and adult-onset asthma in a Chinese Han population. BMC Med Genet 12(1): 133.

4. Moffatt MF, Kabesch M, Liang L, Dixon AL, Strachan D, et al. (2007) Genetic variants regulating ORMDL3 expression contribute to the risk of childhood asthma. Nature 448: 470–473.

5. Verlaan DJ, Berlivet S, Hunninghake GM, Madore AM, Larivière M, et al. (2009) Allele-specific chromatin remodeling in the ZPBP2/GSDMB/ORMDL3 locus associated with the risk of asthma and autoimmune disease. Am J Hum Genet 85: 377–93.

6. Cantero-Recasens G, Fandos C, Rubio-Moscardo F, Valverde MA, Vicente R. (2010) The asthma-associated ORMDL3 gene product regulates endoplasmic reticulum-mediated calcium signaling and cellular stress. Hum Mol Genet 19: 111–121.

7. Hjelmqvist L, Tuson M, Marfany G, Herrero E, Balcells S, et al. (2002) ORMDL proteins are a conserved new family of endoplasmic reticulum membrane proteins. Genome Biol 3: RESEARCH0027

8. Han S, Lone MA, Schneiter R, Chang A. (2010) Orm1 and Orm2 are conserved endoplasmic reticulum membrane proteins regulating lipid homeostasis and protein quality control. Proc Natl Acad Sci U S A. 107: 5851–6.

9. Breslow DK, Collins SR, Bodenmiller B, Aebersold R, Simons K, et al. (2010) Orm family proteins mediate sphingolipid homeostasis. Nature. 463: 1048–53.

10. Liu M, Huang C, Polu SR, Schneiter R, Chang A. (2012) Regulation of sphingolipid synthesis via Orm1 and Orm2 in yeast. J Cell Sci 125: 2428–2435.

11. Spassieva SD, Hille J (2003) Plant sphingolipids today—Are they still enigmatic? Plant Biol 5: 125–136.

12. Sperling P, Heinz E (2003) Plant sphingolipids: structural diversity, biosynthesis, first genes and functions. Biochim Biophys Acta 1632: 1–15.

13. Lynch DV, Dunn TM (2004) An introduction to plant sphingolipids and a review of recent advances in understanding their metabolism and function. New Phytol 161: 677–702.

14. Shi L, Bielawski J, Mu J, Dong H, Teng C, et al. (2007). Involvement of sphingoid bases in mediating reactive oxygen intermediate production and programmed cell death in Arabidopsis. Cell Res 17: 1030–1040.

15. Aubert A, Marion J, Boulogne C, Bourge M, Abreu S, et al. (2011) Sphingolipids involvement in plant endomembrane differentiation: the BY2 case. Plant J 65: 958–71.

16. Markham JE, Molino D, Gissot L, Bellec Y, Hématy K, et al. (2011) Sphingolipids containing very-long-chain Fatty acids define a secretory pathway for specific polar plasma membrane protein targeting in Arabidopsis. Plant Cell 23: 2362–78.

17. Chen M, Markham JE, Cahoon EB (2012) Sphingolipid $\Delta 8$ unsaturation is important for glucosylceramide biosynthesis and low-temperature performance in Arabidopsis. Plant J 69: 769–81.

18. Guo L, Mishra G, Markham JE, Li M, Tawfall A, et al. (2012) Connections between sphingosine kinase and phospholipase D in the abscisic acid signaling pathway in Arabidopsis. J Biol Chem 287: 8286–96.

19. Coursol S, Fan LM, Le Stunff H, Spiegel S, Gilroy S, et al. (2003) Sphingolipid signalling in Arabidopsis guard cells involves heterotrimeric G proteins. Nature 423(6940): 651–4.

20. Chen M, Markham JE, Dietrich CR, Jaworski JG, Cahoon EB (2008) Sphingolipid long-chain base hydroxylation is important for growth and

21. Chao DY, Gable K, Chen M, Baxter I, Dietrich CR, et al. (2011) Sphingolipids in the root play an important role in regulating the leaf ionome in Arabidopsis thaliana. Plant Cell 23: 1061–81.

22. Dietrich C, Han G, Chen M, Berg HR, Dunn TM, et al. (2008) Loss-of-function mutations and inducible RNAi suppression of Arabidopsis LCB2 genes reveal the critical role of sphingolipids in gametophytic and sporophytic cell viability. Plant J 54: 284–298.

23. Teng C, Dong H, Shi L, Deng Y, Mu J, et al. (2008) Serine palmitoyltransferase, a key enzyme for de novo synthesis of sphingolipids, is essential for male gametophyte development in Arabidopsis. Plant Physiol 146: 1322–1332.

24. Wang XL, Li XB (2009) The GhACS1 gene encodes an acyl-CoA synthetase which is essential for normal microsporogenesis in early anther development of cotton. Plant J 57: 473–86.

25. Imamura T, Kusano H, Kajigaya Y, Ichikawa M, Shimada H (2007) A rice dihydrosphingosine c4 hydroxylase (dsh1) gene, which is abundantly expressed in the stigmas, vascular cells and apical meristem may be involved in fertility. Plant Cell Physiol 48: 1108–1120.

26. Virmani SS, Kumar I (2004) Development and use of hybrid rice technology to increase rice productivity in the tropics. International Rice Research Notes 29: 10–19.

27. Pitnjam K, Chakhonkaen S, Toojinda T, Muangprom A (2008) Identification of a deletion in tms2 and development of gene-based markers for selection. Planta 228: 813–822.

28. Wang Z, Xu Y, Jiang R, Xu Z, Chong K (2004) A practical vector for efficient knockdown of gene expression in rice (Oryza sativa L.). Plant Molecular Biology Reports 22: 409–417.

29. Zauner S, Ternes P, Warnecke D (2010) Biosynthesis of sphingolipids in plants (and some of their functions). Adv Exp Med Biol 688: 249–263.

30. Chen R, Zhao X, Shao Z, Wei Z, Wang Y, et al. (2007) Rice UDP-glucose pyrophosphorylase1 is essential for pollen callose deposition and its cosuppression results in a new type of thermosensitive genic male sterility. Plant Cell 19: 847–61.

31. Yamaguchi Y, Ikeda R, Hirasawa H, Minami M, Ujihara A (1997) Linkage analysis of thermosensitive genic male sterility gene, tms-2 in rice(Oryza sativa L.). Breeding Sci 47: 371–373.

32. Lopez MT, Toojinda T, Vanavichit A, Tragoonrung S (2003) Microsatellite markers flanking the tms2 gene facilitated tropical TGMS rice line development. Crop Sci 43: 2267–2271.

33. Caldana C, Scheible WR, Mueller-Roeber B, Ruzicic S (2007) A quantitative RT-PCR platform for high-throughput expression profiling of 2500 rice transcription factors. Plant Methods 3(1): 7 doi: 10.1186/1746-4811-3-7.

34. Livak KJ, Schmittgen TD (2001) Analysis of relative gene expression data using real-time quantitative PCR and the 2(-Delta Delta C(T)) Method. Methods 25: 402–408. doi: 10.1006/meth.2001.1262.

35. Tamura K, Peterson D, Peterson N, Stecher G, Nei M, et al. (2011) MEGA5: molecular evolutionary genetics analysis using maximum likelihood, evolutionary distance, and maximum parsimony methods. Mol Biol Evol 28: 2731–2739. doi: 10.1093/molbev/msr121.

36. Ternes P, Sperling P, Albrecht S, Franke S, Cregg JM, et al. (2006) Identification of fungal sphingolipid C9-methyltransferases by phylogenetic profiling. J Biol Chem 281 5582–5592.

37. Markham JE, Jaworski JG (2007) Rapid measurement of sphingolipids from Arabidopsis thaliana by reversed-phase high-performance liquid chromatography coupled to electrospray ionization tandem mass spectrometry. Rapid Commun. Mass Spectrom. 21: 1304–1314.

7

Genetically Engineered Alginate Lyase-PEG Conjugates Exhibit Enhanced Catalytic Function and Reduced Immunoreactivity

John W. Lamppa[1], Margaret E. Ackerman[1], Jennifer I. Lai[2], Thomas C. Scanlon[1], Karl E. Griswold[1,3,4]*

1 Thayer School of Engineering, Dartmouth College, Hanover, New Hampshire, United States of America, 2 Department of Biological Engineering, Massachusetts Institute of Technology, Boston, Massachusetts, United States of America, 3 Department of Biological Sciences, Dartmouth College, Hanover, New Hampshire, United States of America, 4 Program in Molecular and Cellular Biology, Dartmouth College, Hanover, New Hampshire, United States of America

Abstract

Alginate lyase enzymes represent prospective biotherapeutic agents for treating bacterial infections, particularly in the cystic fibrosis airway. To effectively deimmunize one therapeutic candidate while maintaining high level catalytic proficiency, a combined genetic engineering-PEGylation strategy was implemented. Rationally designed, site-specific PEGylation variants were constructed by orthogonal maleimide-thiol coupling chemistry. In contrast to random PEGylation of the enzyme by NHS-ester mediated chemistry, controlled mono-PEGylation of A1-III alginate lyase produced a conjugate that maintained wild type levels of activity towards a model substrate. Significantly, the PEGylated variant exhibited enhanced solution phase kinetics with bacterial alginate, the ultimate therapeutic target. The immunoreactivity of the PEGylated enzyme was compared to a wild type control using *in vitro* binding studies with both enzyme-specific antibodies, from immunized New Zealand white rabbits, and a single chain antibody library, derived from a human volunteer. In both cases, the PEGylated enzyme was found to be substantially less immunoreactive. Underscoring the enzyme's potential for practical utility, >90% of adherent, mucoid, *Pseudomonas aeruginosa* biofilms were removed from abiotic surfaces following a one hour treatment with the PEGylated variant, whereas the wild type enzyme removed only 75% of biofilms in parallel studies. In aggregate, these results demonstrate that site-specific mono-PEGylation of genetically engineered A1-III alginate lyase yielded an enzyme with enhanced performance relative to therapeutically relevant metrics.

Editor: Roy Roop II, East Carolina University School of Medicine, United States of America

Funding: This work was supported by a Pilot and Feasibility Grant from the Cystic Fibrosis Foundation's Research Development Program at Dartmouth Medical School and by P20RR018787-06 from the National Center for Research Resources (NCRR), a component of the National Institutes of Health (NIH). The funders had no role in study design, data collection and analysis, decision to publish, or preparation of the manuscript.

Competing Interests: The authors have declared that no competing interests exist.

* E-mail: karl.e.griswold@dartmouth.edu

Introduction

The major contributor to mortality in cystic fibrosis (CF) patients is pulmonary infection by the Gram-negative bacterium *P. aeruginosa*. The majority (~75%) of CF-associated *P. aeruginosa* isolates exhibit a mucoid phenotype characterized by overproduction of alginate, an exopolysaccharide component of the biofilm matrix [1]. Bacterial alginate is one of the most studied *P. aeruginosa* virulence factors [2], with confirmed roles in protection of bacteria from host immune defenses [3,4], exacerbation of inflammatory tissue damage [5], and contribution to bacterial resistance towards conventional antibiotic therapies [6,7,8]. In addition, alginate has been shown to increase the viscosity of mucosal secretions contributing to respiratory tract obstructions [9]. Considering its important role in the pathology of *P. aeruginosa* infections of the CF lung, alginate represents an attractive target for developing novel therapeutic agents for CF patients.

Alginate lyase enzymes (EC 4.2.2.3) efficiently degrade alginate *via* β–elimination cleavage of glycosidic bonds in the polymer backbone. Numerous observations support the hypothesis that alginate lyases could be powerful therapeutic agents for treating mucoid *P. aeruginosa* infections. For example, alginate lyases have been shown to enhance phagocytosis of *P. aeruginosa* by human macrophages [10], increase the susceptibility of *P. aeruginosa* to a variety of antibiotic treatments [6,11,12,13], and decrease the viscosity of CF sputum [14]. The latter activity suggests a therapeutic application analogous to that of recombinant human DNase (Pulmozyme®), an inhaled enzyme therapy that degrades extracellular DNA, aids in clearance of viscous airway obstructions, and temporarily improves pulmonary function in CF patients [15]. Unfortunately, alginate lyases are invariably derived from non-human sources, and their exogenous origins may predispose them towards excessive immunogenicity in human patients. An immune response against biotherapeutic agents can manifest a spectrum of complications including increased rates of drug clearance, direct inhibition of therapeutic activity, and varying degrees of allergic reaction with the potential for life-threatening anaphylactic shock [16]. There is an increasing awareness of the risks associated with immune responses against biotherapeutic agents [17], and this knowledge is prompting the restructuring of biotherapeutic development strategies so as to address potential safety concerns earlier in the process [18]. Considering the tremendous potential of alginate lyase therapeutic agents, strategies to mitigate putative anti-enzyme immune reactions merit examination.

Chemical modification of therapeutic proteins with polyethylene glycol (PEG) is a common approach for modulating immunogenicity and stability [19]. Indeed, PEGylation of *Sphingomonas* sp. A1-III alginate lyase (A1-III), one therapeutic candidate, has been shown to reduce antibody binding *in vitro* [20]. Unfortunately, the random attachment of amine-reactive PEG molecules to solvent exposed lysines of A1-III resulted in a significant proportion of inactivated enzyme (>50% inactivation with 10 of 12 formulations). Thus, while PEGylation can successfully reduce the enzyme's immunoreactivity, maintaining a homogenous enzyme composition with high catalytic activity necessitates a more controlled PEG-conjugation strategy.

To facilitate precise control over both the site of PEG attachment and the extent of PEGylation, cysteine residues were engineered into the A1-III enzyme at five different surface accessible locations. These rationally substituted cysteine residues provided an orthogonal chemical handle for site-specific PEGylation reactions using maleimide activated PEG. It was anticipated that selective and controlled PEGylation would result in modified variants simultaneously demonstrating high catalytic proficiency and reduced immunoreactivity. In this study, solution phase reaction kinetics, biofilm disrupting activity, and *in vitro* antibody binding of genetically engineered PEG variants have been assessed and compared to the non-PEGylated wild type enzyme control. The results suggest that at least one modified enzyme meets or exceeds the experimental objectives, and thereby possesses enhanced potential as an antibacterial therapy.

Results

Construction of A1-III Mutants

To facilitate site-specific PEGylation, mutant A1-III genes encoding single cysteine substitutions were constructed by total gene synthesis [21]. The synthetic genes were codon optimized for expression in *E. coli*, and each gene encoded a single site-specific cysteine substitution: S32C, A41C, A53C, A270C or A328C, where residue numbering is per Yoon *et al.* [22] The five sites for mutation were selected based on an analysis of PDB structure 1HV6 [22]. Priority was placed on small amino acids that, when substituted with a cysteine, would result in a solvent exposed thiol group (Fig. 1 and Movie S1). Particular emphasis was given to residues with spatial proximity to the S32-C49 peptide segment, a motif that has previously been reported as constituting an immunodominant region of the enzyme [23]. A *C*-terminal hexahistidine tag (his-tag) was appended to each mutant to facilitate purification by immobilized metal-ion affinity chromatography (IMAC). A construct encoding the corresponding his-tagged version of the wild type enzyme (WT-his) was generated as a control.

Expression levels of the recombinant enzymes from a T7 driven pET vector system varied moderately. Following overnight shake flask induction, cell lysis, IMAC purification and dialysis, the WT-his enzyme and high yielding variants such as A53C-his produced upwards of 20 mg per liter of cell culture. In contrast, variant A41C-his yielded 3-fold less protein under the same expression conditions. Importantly, non-reducing SDS-PAGE gels showed that the cysteine variants were isolated predominantly as monomers. Only after extended storage were the genetically engineered proteins found to dimerize *via* intermolecular disulfide bond formation. Interestingly, the non-reducing SDS-PAGE analysis also indicated that one or both of the protein's two native disulfide bonds (C49–C112 and/or C188–C189) were not fully formed upon cell lysis and IMAC separation, but that oxidation to the fully disulfide bonded state occurred during the first 24 to

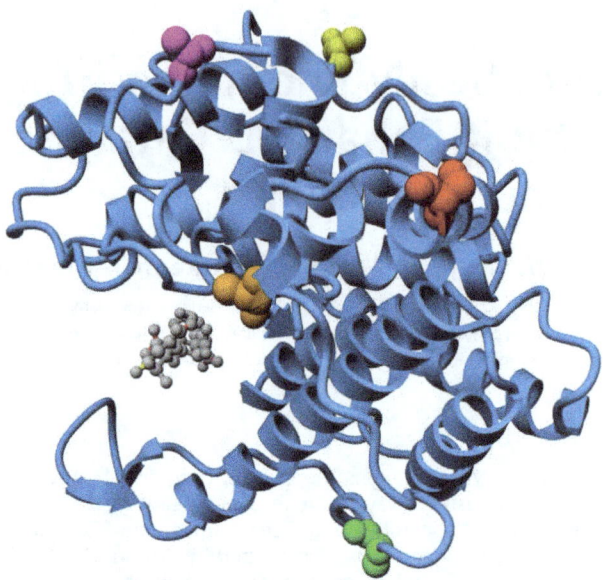

Figure 1. Sites of cysteine substitution. Ribbon diagram of alginate Lyase A1-III (PDB file 1HV6). A trisaccharide reaction product is bound in the active cleft and shown as a grey ball and stick model. Amino acid residues targeted for cysteine substitution are shown in space filling mode, and are color coded as follows: S32C = Red, A41C = Orange, A53C = Green, A270C = Yellow, and A328C = Purple.

48 hours after purification (data not shown). This observation is consistent with expression of the enzymes in the reducing environment of the *E. coli* cytoplasm and subsequent oxidation by molecular oxygen following lysis and storage.

PEG Conjugation and Purification

Exposed thiol groups of the engineered cysteine residues were conjugated to a 20 kDa methoxy-maleimide PEG. Reaction time, temperature and stoichiometry were the subject of detailed optimization studies, and it was ultimately determined that 1 hour reactions at 25°C with a 5:1 molar ratio of PEG:enzyme typically yielded maximal mono-PEGylated product, i.e., protein molecules each bearing a single PEG polymer chain. Mono-PEGylated reaction products were readily separated from unconjugated protein by FPLC size exclusion chromatography (Fig. 2), and optimized reactions typically produced 40 to 50% yields of >95% pure material.

Enzyme Kinetics

Alginate depolymerization kinetics were assessed for the various PEGylated enzymes and the WT-his control using brown seaweed alginate (BSWA) as a model substrate. Michaelis constants (K_m) and maximum reaction velocities (V_{max}) were determined by nonlinear regression of initial velocities vs. substrate concentration. All five of the PEGylated variants were found to possess catalytic efficiencies (V_{max}/K_m) exceeding that of the corresponding WT-his construct, although most exhibited a decrease of 2-fold or less in V_{max} (Table 1). Variant A41C-his-PEG was found to possess particularly low maximum reaction velocities, and it was therefore eliminated from further studies.

One modified enzyme, A53C-his-PEG, maintained V_{max} values similar to the WT-his control, and the activity of this variant was examined in greater detail. To separate the effects of the point mutation from the effects of PEGylation, the kinetics of A53C-his

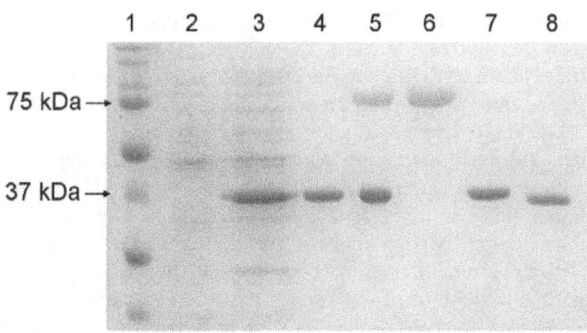

Figure 2. SDS-PAGE analysis of PEGylated variant production. Samples run on a reducing 12.5% gel, and stained for total protein with Coomassie brilliant blue. Lane 1: Bio-Rad Precision Plus Protein Ladder; Lane 2: Whole cell lysate of non-expressing cells; Lane 3: Whole cell lysate of induced cells; Lane 4: IMAC purified A53C-his; Lane 5: Crude A53C-his PEGylation reaction product; Lane 6: Size exclusion FPLC purified A53C-his-PEG; Lane 7: IMAC purified WT-his; Lane 8: FPLC purified native WT.

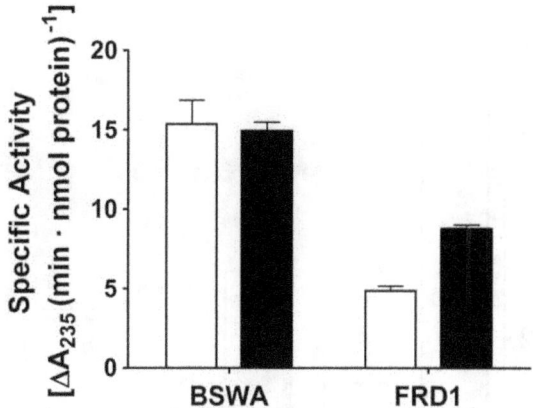

Figure 3. Comparison of reaction kinetics with BSWA and bacterial alginate. The specific activities of WT-his (white bars) and A53C-his-PEG (black bars) were determined with a model alginate substrate (BSWA) as well as with purified bacterial alginate (FRD1). The two enzymes are equally active with BSWA at saturating concentrations, but the PEGylated variant exhibits 80% faster kinetics with the bacterial substrate (p<0.01), which is the ultimate therapeutic target. Error bars represent standard deviation.

were measured both before and after PEG conjugation (Table 1). The A53C amino acid substitution drove a reduction in both V_{max} and K_m, but subsequent PEGylation produced a 60% increase in V_{max} restoring the variant's maximum reaction velocity to wild type levels while not altering the reduced K_m value. The result was an enzyme-PEG conjugate with a >2-fold improved catalytic efficiency compared to WT-his.

Alginate biopolymer produced by mucoid *P. aeruginosa* pathogens differs from that produced by brown seaweed in that bacterial alginate is partially acetylated at the C2 and C3 hydroxyls of mannuronate residues [24]. To evaluate activity on the bacterial substrate, alginate was purified from the mucoid *P. aeruginosa* clinical isolate FRD1. The specific activities of WT-his and A53C-his-PEG were determined using 0.1% (wt/vol) bacterial alginate. Unexpectedly, the PEGylated variant exhibited a 1.8-fold increased specific activity relative to the corresponding WT-his construct (Fig. 3).

It is possible that the altered catalytic activities of the PEGylated variants resulted from subtle structural perturbations to the enzyme's 3-dimensional fold. Such deviations from the native structure might have undesired consequences that compromise the potential for practical utility, e.g., decreasing enzyme stability during long term storage. To assess the impact of PEGylation on

storage stability, the activity of A53C-his-PEG was followed during more than two months of storage at 4°C. No loss of activity was observed during the course of the 70 day experiment (data not shown).

Antibody Binding and Immunogenicity

Polyclonal anti-A1-III IgG was purified by antigen affinity chromatography of pooled serum from two New Zealand white rabbits, which had both been immunized with the non-tagged, native enzyme (WT). The EC_{50} of the polyclonal antibody was determined for various PEGylated enzymes using standard ELISA techniques, and these values were compared to that for the non-PEGylated WT-his control. Genetically engineered variants S32C-his-PEG, A53C-his-PEG, A270C-his-PEG and A328C-his-PEG exhibited a 40–90% decrease in immunoreactivity relative to the WT-his enzyme counterpart (Fig. 4). Together, the high catalytic activity and decreased antibody binding of A53C-his-PEG distinguished this enzyme as a particularly promising candidate.

To evaluate immunogenicity by a metric with greater relevance to human patients, binding of a naïve human antibody repertoire to both WT-his and A53C-his-PEG was assessed. Each protein was biotinylated and immobilized at saturating mM surface densities on streptavidin-coated magnetic beads. These alginate lyase coated beads represent one of two key elements in the immunogenicity assays. The second element is a yeast library displaying 10^9 human scFv antibody fragments [25]. Yeast surface display produces a high degree of scFv multivalency, and when mixed with the alginate lyase coated beads, the resulting avid interactions facilitate capture of low affinity binders likely present in the human immune repertoire prior to affinity maturation. As a result, yeast cells expressing an scFv that recognizes an epitope on the candidate proteins are bound to the surface of the magnetic beads. Following a pre-screen to remove non-specific binders, the yeast library was incubated separately with either WT-his coated beads or A53C-his-PEG coated beads. After binding of the library, the beads were magnetically separated, unbound yeast in the supernatant were removed by aspiration, and the beads were resuspended in fresh buffer. An aliquot of this bead slurry was serial diluted, plated on yeast growth media, and outgrown to

Table 1. Kinetic parameters for alginate degradation.

Enzyme	V_{max} [ΔA_{235} (min · mg)$^{-1}$]	K_m (μg/ml)	V_{max}/K_m
WT-his	440±30	80±20	6±1
A53C-his	280±30	30±6	9±2
S32C-his-PEG	330±20	26±6	13±3
A41C-his-PEG	134±3	7.1±0.8	18±2
A53C-his-PEG	460±50	40±10	13±4
A270C-his-PEG	300±20	15±5	20±7
A328C-his-PEG	180±10	13±4	14±4

Depolymerization of BSWA was followed by monitoring absorbance at 235 nm. Kinetic parameters were determined by nonlinear regression of initial rate vs. substrate concentration data (Prism version 4.0).

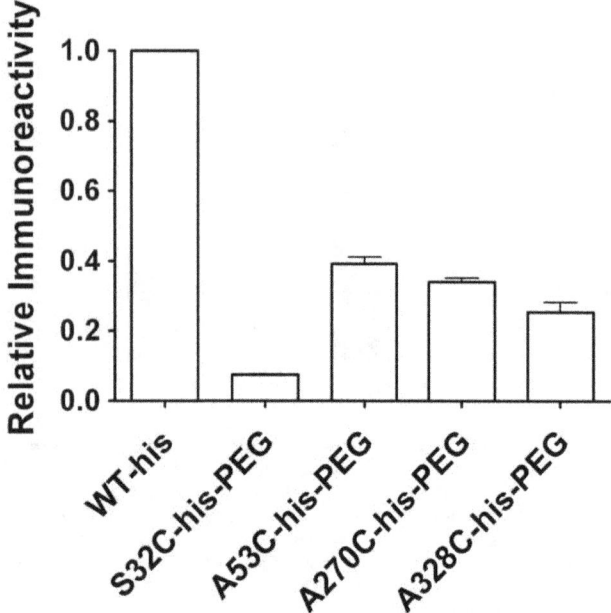

Figure 4. Immunoreactivity by ELISA. The antibody concentration required to achieve 50% maximum ELISA signal (EC_{50}) was determined for each enzyme using polyclonal anti-A1-III antibody purified from rabbit immune serum. The results are reported as fractional immunoreactivity based on normalization with the WT-his enzyme, which was included as an internal control in all experiments (see Experimental Procedures). All of the PEGylated enzymes were found to exhibit significantly reduced antibody binding relative to the WT-his control ($p < 0.01$ for each). Error bars represent standard deviation.

determine the number of colony forming units (cfu's) that remained bound to the beads. The washing and plating procedure was repeated two additional times, and the number of bead-bound yeast was determined for each wash step. The resulting cfu counts provide a means to assess the relative reactivity of a human antibody repertoire towards the two target proteins. Note that each yeast colony represents a single human scFv antibody fragment that specifically bound the target protein on the cognate

magnetic bead surface. Beads coated with the PEGylated enzyme target were found to bind up to 13-fold fewer yeast cells than those coated with the WT-his enzyme (Fig. 5a, wash 2). Because the A53C-his and WT-his proteins are nearly identical in amino acid sequence, the difference in binding counts indicates that the PEG moiety effectively blocks interactions between human scFvs and their corresponding immunogenic epitopes on the A1-III enzyme.

Bound yeast cells isolated during these initial experiments represent enriched populations displaying scFvs that specifically recognize epitopes of either WT-his or A53C-his-PEG. To assess the cross-reactivity of the scFvs, yeast selected as binders to the WT-his beads were propagated and employed in a second round of binding experiments against both proteins in parallel. Likewise, yeast that initially bound the A53C-his-PEG beads were similarly tested for cross-reactivity. Importantly, yeast originally isolated as binders to the WT-his enzyme had a reduced capacity to recognize the PEGylated variant. Furthermore, yeast originally isolated as binders to A53C-his-PEG more readily recognized the WT-his protein then their original PEGylated target (Fig. 5b). Collectively, this data set implies that, although the A1-III enzyme's human antibody epitopes have not been completely occluded, PEGylation effectively reduces access to these sites. In particular, site specific PEGylation of A1-III alginate lyase (i) substantially reduced binding of naïve human antibody repertoires (Fig. 5a), and (ii) blocked >50% of specific, human, anti-A1-III scFv antibody fragments (Fig. 5b).

Biofilm Disruption Studies

There exists considerable evidence that *P. aeruginosa* grows in biofilm communities during CF lung infection [26], and it is likely that disrupting alginate biofilms represents a key challenge in the fight to eradicate CF-associated *P. aeruginosa* infections. To assess this therapeutically relevant aspect of enzyme function, biofilms of the alginate-producing *P. aeruginosa* strain Xen5, a derivative of clinical isolate ATCC 19660, were first established by growth in 96-well plates. Subsequently, adherent biofilms were treated for one hour with 1 mg/ml of WT-his or A53C-his-PEG and then washed to remove degraded biofilm. The remaining adherent alginate biofilm matrix was quantified using a ConA lectin-HRP conjugate that binds to mannuronate residues of alginate [27]. The percentage of biofilm removed by each enzyme was

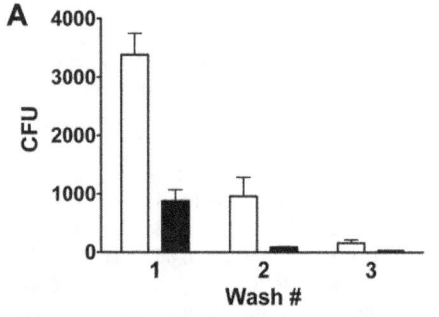
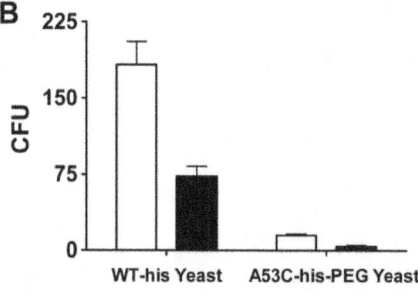

Figure 5. Human antibody binding. The WT-his (white bars) and A53C-his-PEG (black bars) protein targets were biotinylated and captured on the surface of streptavidin coated magnetic beads. A) The two bead preparations were independently incubated with a yeast surface displayed scFv antibody library derived from human immune cells. Following binding, the beads were magnetically separated and washed three times. The number of yeast that remained bound after each wash step was determined by plating serial dilutions of the resuspended beads and enumerating cfu's. The resulting yeast colonies represent human scFvs that specifically bound to the A1-III enzymes on the corresponding magnetic beads. A53C-his-PEG coated magnetic beads bound up to 13-fold fewer human antibodies than did the WT-his coated beads ($p < 0.01$ for each of the three washes). B) Characterization of first round binders from both protein targets. Yeast isolated as binders to either WT-his or A53C-his-PEG were propagated and subsequently incubated with magnetic beads bearing each protein target. For both yeast populations, the A53C-his-PEG coated beads (black bars) bound at least 60% fewer cells than did the WT-his beads (white bars), a result that demonstrates PEGylation effectively blocked key immunogenic epitopes ($p < 0.01$ for all differences). Error bars represent standard deviation.

determined by comparison to wells receiving a buffer control treatment. Both the wild type and PEGylated enzymes were found to effectively remove the majority of established biofilm from the wells (Fig. 6). Consistent with its enhanced solution phase activity towards bacterial alginate, A53C-his-PEG exhibited a significant (p = 0.025) increase in mucoid biofilm disruption relative to the WT-his protein (94% vs. 75% biofilm removal, respectively). These results suggest that the enhanced catalytic performance of the genetically engineered A53C-his-PEG enzyme may have relevance to clinical applications.

Discussion

Biofilms are thought to play a key role in refractory *P. aeruginosa* infections of the CF airway [26]. In particular, the transition of *P. aeruginosa* to a mucoid phenotype is associated with alginate overproduction, altered biofilm architecture, high level antibiotic-resistance, and accelerated deterioration of lung function [1,8]. As a consequence, inhaled alginate lyase enzymes could represent powerful new therapies for treating CF lung infections. To realize their therapeutic benefit in humans, however, the risks associated with the putative immunogenicity of these heterologous enzymes should be appropriately mitigated. It has been shown previously that random PEGylation of free amines on the surface of A1-III alginate lyase effectively blocked enzyme-specific antibody binding [20]. Unfortunately, this strategy resulted in a significant loss of catalytic proficiency, as most preparations exhibited >50% reduction in alginate degrading activity. For a given PEG chain length, the degree of inactivation was directly proportional to the degree of de-immunization, a result that likely derived from the stochastic nature of NHS-ester conjugation to protein surfaces.

To better leverage the de-immunizing properties of PEG, a site-specific PEGylation strategy has been developed for the A1-III enzyme. Several surface accessible residues of the native enzyme were substituted with cysteines, and site-specific mono-PEGylation of the genetically engineered variants was achieved using a maleimide-activated 20 kDa PEG chain. Importantly, all of the

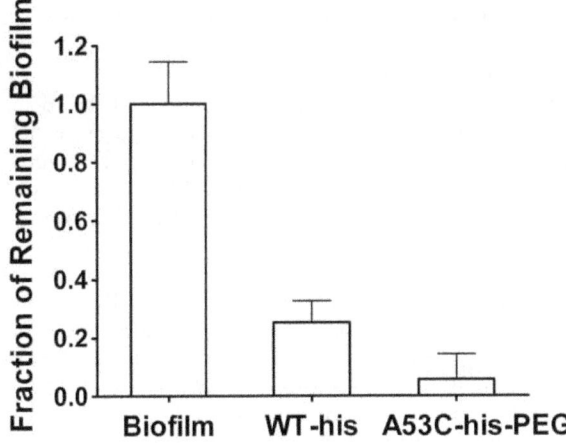

Figure 6. Disruption of mucoid *P. aeruginosa* biofilms. Adherent biofilms of a mucoid clinical isolate were established in 96-well plates and subsequently treated with 1 mg/ml enzyme for 1 hour. Remaining biofilm was then quantified using an alginate-sensitive lectin-HRP conjugate and ABTS substrate. Signals were normalized to a buffer only treatment. Both enzymes removed a significant proportion of biofilm relative to the buffer control (p<0.01). Importantly, theA53C-his-PEG enzyme removed >15% more biofilm than the WT-his enzyme (p<0.025). Error bars represent standard deviation.

site-specific, mono-PEGylated variants examined here were found to possess improved catalytic efficiency with a model substrate, BSWA. Maximum catalytic activity, however, was found to be critically dependent on the exact site of PEG attachment, as only variant A53C-his-PEG was found to maintain a V_{max} comparable to that of the wild type enzyme control. This outcome underscores a fundamental advantage of orthogonal conjugation chemistry: the site and extent of protein modification can be precisely controlled so as to yield a homogeneous enzyme preparation having uniformly high functionality.

Of particular relevance to treatment of bacterial infections, A53C-his-PEG was 80% more active than the WT-his control when assayed against solutions of bacterial alginate. The mechanistic origins of this enhanced activity on bacterial but not BSWA are not entirely clear. It is possible that PEGylation simply distorts the native enzyme structure so as to better accommodate the bulkier, acetylated, bacterial alginate. An alternative explanation, however, could relate to the acetylated alginate's greater hydrophobicity and increased extent of intermolecular interaction [28]. A high degree of substrate-substrate interaction could reduce enzyme accessibility to individual alginate chains and slow substrate degradation relative to non-acetylated BSWA. The amphipathic nature of PEG allows it to interact closely with both hydrophilic and hydrophobic molecules [19], and this property could facilitate insertion into amorphous higher-order structures of acetylated bacterial alginate. We speculate that the PEG moiety of A53C-his-PEG may disrupt enhanced substrate-substrate interactions in the enzyme's local environment, and thereby free individual alginate chains for more efficient enzymatic degradation. Loose parallels might be drawn to cellulases and chitinases, which efficiently degrade highly ordered, macromolecular, carbohydrate substrates. To do so, these enzymes employ non-catalytic binding domains that disrupt intermolecular polymer packing and enhance access to individual substrate chains [29,30]. Certainly, this analogy should be approached with caution, as alginate solutions are hydrogels as opposed to crystalline or semi-crystalline substrates. None-the-less, site-specific PEGylation of A1-III alginate lyase has yielded a functionally enhanced enzyme that degrades bacterial alginate with greater efficiency.

In addition to maintaining high level catalytic activity, de-immunization of the A1-III protein was a second critical design objective of these experiments. The PEGylated constructs showed a 60–90% reduction in immunoreactivity with rabbit anti-A1-III IgG antibodies. Of greater relevance to human use, A53C-his-PEG bound a substantially smaller fraction of a naïve human scFv antibody library, relative to the non-PEGylated WT-his control. Furthermore, human scFvs that specifically bound the WT-his enzyme were 2.5-fold less likely to bind A53C-his-PEG. These data suggest that site-specific PEGylation has yielded a general reduction in the enzyme's antibody reactivity, and the studies with human antibody fragments lead us to hypothesize that the reduced immunoreactivity could translate to the clinic.

In clinical applications of alginate lyases, high level solution-phase activity may not be sufficient to affect a therapeutic benefit in CF patients. Instead, the practical utility of alginate lyase therapies will likely be defined by their capacity to disrupt mucoid *P. aeruginosa* biofilms. During a one hour treatment, the modified A53C-his-PEG enzyme removed more than 90% of established biofilms, a >15% improvement over the non-PEGylated WT-his protein. This enhanced ability to disrupt biofilms is consistent with the improved solution-phase kinetics of the engineered enzyme, and may stem from a similar mechanistic origin. While biofilms in the human airway have properties distinct from those grown on abiotic surfaces [26], the fact that A53C-his-PEG virtually cleared adherent mucoid biofilms suggests that it or similar enzymes could yield therapeutic benefits in the treatment of CF associated, *P. aeruginosa* infections.

Materials and Methods

Oligonucleotides were purchased from IDT (Coralville, IA), and were purified by standard desalting. The gene sequences for the A1-III alginate lyase enzymes were derived from the wild type A1-III enzyme encoded in the genome of *Sphingomonas* sp. A1 (GenBank: AB011415). Restriction enzymes, Phusion polymerase, and T4 ligase were from New England Biolabs (Ipswich, MA), and were used as directed by the manufacturer. Expression vector pET28b was from Novagen (San Diego, CA). Plasmid purification kits, Ni-NTA agarose and corresponding columns were from QIAGEN (Valencia, CA). Gel extraction/DNA clean up kits were from Zymo Research (Orange, CA). 20 kDa methoxy-maleimide polyethylene glycol (PEG) was from JenKem Technology (Allen, TX). ÄKTA FPLC system and Superdex75 SEC resin were from GE Healthcare Life Sciences (Piscataway, NJ). Concanavalin A-horseradish peroxidase conjugate (ConA-HRP), medium viscosity brown seaweed alginate (BSWA) (cat #A2033), and 2,2′-azino-bis(3-ethylbenzthiazoline-6-sulphonic acid) (ABTS) were from Sigma-Aldrich (St. Louis, MO). BCA assay and AminoLink Plus Immobilization Kits were from Pierce Biotechnology (Rockford, IL). Polyclonal goat anti-rabbit HRP conjugate antibody was from Millipore (Billerica, MA). All other reagents were from Fisher Scientific (Pittsburgh, PA), unless specifically noted.

Data Analysis

Experiments were conducted in triplicate unless otherwise noted, and statistical significance was determined using two-tailed t-tests.

Ethics Statement

This study was carried out in strict accordance with the recommendations in the Guide for the Care and Use of Laboratory Animals of the National Institutes of Health. The protocol was approved by the Institutional Animal Care & Use Committee of Dartmouth College (Protocol Number: 07-07-11CL), and all efforts were made to minimize suffering.

Construction and Cloning of A1-III Encoding Genes

Following the procedure of Hoover and Lubkowski [21], synthetic genes, codon optimized for expression in *E. coli*, were assembled for both the wild type [31] and cysteine point mutant A1-III enzymes. The genes were appended with a 5′-methionine codon, a 5′-FatI restriction site spanning the ATG start, and a 3′-XhoI restriction site immediately following the terminal serine codon (appends a non-native, *C*-terminal LeuGlu sequence). Each point mutant gene encoded a single cysteine substitution at serine 32, alanine 41, alanine 53, alanine 270, or alanine 328 (numbering as per Yoon *et al.* [22]). The 1,089 base pair synthetic A1-III genes were digested with FatI and XhoI, and ligated into NcoI and XhoI digested pET-28b expression vector resulting in an in frame fusion with the hexahistidine tag encoded by the plasmid. Ligations were transformed into electrocompetent DH5alpha [F⁻Φ80*lacZ*ΔM15 Δ(*lacZYA-arg*F)U169*rec*A1*end*A1*hsd*R17(r$_K^-$ m$_K^+$) *phoA sup*E44 *thi*-1*gyr*A96*rel*A1 λ⁻], and the identities of the cloned genes were verified by sequencing plasmid isolated from individual clones. These plasmid constructs encoded his-tagged wild type (WT-his), or point mutant A1-III enzymes (S32C-his, A41C-his, A53C-his, A270C-his, and A328C-his). A gene encoding an untagged version of the wild type (WT) enzyme was constructed in a similar manner, but insertion of a dual stop codon (TGATAG) before the 3′ restriction site terminated translation prior to the hexahistidine coding sequence. Sequence verified plasmids were subsequently transformed into electrocompetent HMS174(DE3) expression hosts [F⁻*recA1hsdR*(r$_{K12}^-$ m$_{K12}^+$) (DE3) (Rif^R)]. The expression host also bore the pLysS plasmid to repress basal expression.

Protein Expression and Purification

Overnight cultures of expression hosts were grown in LB supplemented with 30 μg/ml kanamycin and 34 μg/ml chloramphenicol at 37°C, and then sub-cultured 1:100 into 500 ml of fresh media. Cultures were grown at 37°C to mid-log, equilibrated to 25°C, and induced with 0.5 mM IPTG for 20 hours. Following induction, cell cultures were centrifuged at 5000*g*, 4°C for 10 minutes, pellets were resuspended in 5 ml of native lysis buffer (50 mM NaH$_2$PO$_4$, 300 mM NaCl, 10 mM Imidazole, pH 8.0), transferred to a 10 ml Pyrex beaker, and equilibrated on ice for 20 minutes. Cells were disrupted by sonication (Fisher 550 Sonic Dismembrator). Whole cell lysate was dispensed into 2 ml eppendorf tubes and centrifuged at 17,000*g*, 4°C, for 20 minutes. The supernatant was removed, syringe filtered through a 0.22 μm PES membrane, and gently mixed with a 0.4 ml bed volume of Ni-NTA agarose, which had been equilibrated with native lysis buffer. After binding at 4°C for 1 hour, the column was drained and washed with 10 bed volumes of wash buffer (50 mM NaH$_2$PO$_4$, 300 mM NaCl, and 20 mM imidazole pH 8.0). Purified A1-III was eluted in a native elution buffer (50 mM NaH$_2$PO$_4$, 300 mM NaCl, 250 mM imidazole, pH 8.0), dialyzed into storage buffer (20mM NaH$_2$PO$_4$ pH 6.5), and kept at 4°C. The purity of enzyme preparations was typically >95% as assessed by Coomassie-stained SDS-PAGE gels. Enzyme concentrations were routinely determined by A$_{280}$ (NanoDrop 1000, Thermo Scientific, Waltham, Ma) using a standard curve that had been independently validated by BCA assay.

For the purpose of immunizing rabbits for antibody production, the non-tagged native enzyme (WT) was purified as described previously [32]. Enzyme purity was >99% as assessed by Coomassie-stained SDS-PAGE gels, and enzyme solutions were stored in phosphate buffered saline (PBS) at 4°C prior to use.

Covalent Conjugation to PEG

Preliminary optimization studies with the A53C-his point mutant examined the effects of time (5 minutes to overnight), temperature (25°C to 37°C), and stoichiometry (1:1 to 20:1 PEG:protein molar ratio) as reaction variables, and it was ultimately determined that 1 hour reactions at 25°C with a 5:1 molar ratio of PEG:enzyme typically yielded maximal mono-PEGylated product, i.e. protein molecules each bearing a single PEG polymer chain. Subsequently, purified A1-III cysteine mutants were covalently coupled with a 5 molar excess of 20 kDa methoxy-maleimide PEG. PEG was initially solubilized in DMSO at a concentration of 100 mg/ml, and 12 μl were added to 500 μl of a 1 mg/ml enzyme solution in 20 mM NaH$_2$PO$_4$ pH 6.5. Reactions were incubated at room temperature for 1 hour, and then loaded onto a 120 ml bed volume Superdex 75 size exclusion column. The column was eluted with 150 mM NaCl, 50 mM NaH$_2$PO$_4$ pH 7.0 at a flow rate of 0.6 ml/min. Mono-PEGylated A1-III product eluted at ~53 ml. Enzyme purity was typically >95% as assessed by Coomassie-stained SDS-PAGE gels, and enzyme solutions were stored at 4°C for later use. The concentrations of PEGylated enzymes were determined by A$_{280}$, as independent experiments demonstrated that conjugation to the PEG moiety did not alter enzyme molar absorptivity.

Enzyme Kinetic Analysis

Enzymatic activities were assessed in a 96-well plate format. Briefly, 5 µl of purified enzyme was added to each of 12 contiguous wells in a UV transparent, 96-well plate (Costar, Fisher #3635). Using a 12-channel pipette, 195 µl of alginate in reaction buffer (150 mM NaCl, 50 mM NaH$_2$PO$_4$ pH 7.0) was simultaneously added to each of the wells. BSWA concentrations were varied from 0.001% to 0.05% (wt/vol), and each concentration was assayed in triplicate. The 96-well plates were immediately transferred to a UV/Vis plate reader (SpectraMax 190, Molecular Devices, Sunnyvale, CA), and product formation was monitored by measuring absorbance at 235 nm every 15 seconds for 10 minutes. Initial velocities were taken from the linear portions of the absorbance verses time curves, and V$_{max}$ and K$_m$ values were determined by non-linear regression of initial reaction rates verses substrate concentration. Specific enzyme activities towards bacterial alginate, purified from *P. aeruginosa* FRD1 as described previously [33], were determined in triplicate at 0.1% (wt/vol) substrate concentration. Assays were carried out essentially as described above.

IgG Antibody Immunoreactivity

A1-III alginate lyase antiserum was obtained from Covance Research (Denver, PA). Two New Zealand white rabbits were initially immunized by subcutaneous injection of 250 µg of purified WT A1-III mixed with Freund's complete adjuvant (FCA). At twenty day intervals, the rabbits were boosted with subcutaneous injections of 125 µg of purified WT A1-III mixed with Freund's incomplete adjuvant (FIA). Ten days after the first and fourth boost, serum was collected and antibody titers were evaluated by determining the serum dilution required to produce a 50% ELISA signal against the WT immunogen. Polyclonal A1-III specific IgG antibodies were purified from immune serum using an AminoLink Plus A1-III affinity column prepared from purified WT-his enzyme as per the manufacturer's instructions. The purified primary antibody was aliquoted and stored at 700 µg/ml, −20°C in PBS. ELISAs were performed in high binding 96-well plates using purified alginate lyase enzymes, polyclonal rabbit IgG antibody, secondary goat anti-rabbit HRP conjugate and ABTS for detection. Dose response curves were fit to the data to obtain EC$_{50}$ values (half the maximal effective concentration of IgG). All ELISAs were performed in triplicate. The immunoreactivity of the PEGylated variants was defined as the ratio of the WT-his EC$_{50}$ to the EC$_{50}$ of the corresponding PEGylated enzyme. Equivalent binding of the WT-his and PEGylated variants to the 96-well ELISA plates was verified by activity assays of enzyme solutions pre- and post-binding. No statistically significant difference in the fraction of bound enzyme was observed for PEGylated or non-PEGylated enzymes (data not shown).

Human scFv Antibody Binding Studies

The immunogenicity of WT-his and A53C-his-PEG was further assessed using an *in vitro* assay that scores the relative reactivity of a protein of interest towards a human antibody fragment library displayed on the surface of yeast[1] [34]. Briefly, WT-his and A53C-his-PEG were biotinylated as per the manufacturer's instructions (Pierce Biotinylation Kit). Magnetic streptavidin beads (Invitrogen, Carlsbad, CA) were coated separately with each biotinylated enzyme overnight at 4°C. WT-his coated beads were combined with A53C-his-PEG coated beads, and the mixture was incubated with yeast expressing the human scFv library [25]. Following this binding step, the beads were magnetically separated and unbound yeast were removed by aspiration. The remaining bead:yeast mixture was placed in selective yeast growth media, and the selected yeast cells were regrown, induced, and selected against pooled beads a second time. This affinity-selected yeast population was regrown, induced and then independently incubated for one hour at 4°C with either WT-his or A53C-his-PEG coated beads. The beads were magnetically separated, and unbound yeast were aspirated and discarded. The yeast:bead mixtures were then resuspended in 1 ml of PBS, and a 50 µl aliquot was removed for serial dilution and plating on selective media to determine the number of yeast initially bound to each set of beads ("wash 1" population). The remainder of the resuspended yeast:bead mixture was then gently agitated at 4°C for 15 minutes, and the wash process was repeated twice more to generate "wash 2" and "wash 3" cell populations. The number of yeast binders to each protein target was quantified by plating serial dilutions of each wash population on selective growth media. Following a 2-day outgrowth, the number of cfu on each plate were determined and used to back calculate the total number of bead-bound yeast after each wash step. These values provide a relative metric for comparing the immunoreactivity of WT-his and A53C-his-PEG proteins towards a human scFv antibody fragment repertoire.

Following these initial studies, which yielded a relative count of antibody fragments capable of recognizing each individual enzyme, the cross-reactivity of yeast isolated against each protein target was evaluated. This analysis involved regrowth and induction of the wash 3 yeast populations, and subsequent magnetic bead selection against both protein targets in parallel.

A more detailed description of the methods for the human scFv antibody fragment studies is provided as Text S1.

Biofilm Disruption Assays

The capacity of the alginate lyase enzymes to disrupt bacterial biofilms was assessed *in vitro*. Briefly, mucoid Xen5 *P. aeruginosa* (Caliper Life Sciences, Hopkinton, MA) cultures were grown for 19 hours in 3 ml of TSB media at 37°C. Bacteria were then subcultured at a 1:5 ratio into fresh TSB media, and 100 µl aliquots were added in replicate to 96-well plates. Plates were covered with a gas permeable adhesive strip, and incubated without shaking for 20 hours at 37°C. Following biofilm growth, culture media and planktonic bacteria were shaken from the wells, and the remaining biofilms were rinsed with double distilled water. The adherent biofilms were treated with 200 µl aliquots of 1 mg/ml alginate lyase in 20 mM NaH$_2$PO$_4$ pH 6.5. Buffer only was used as a no treatment control. Each treatment was done in triplicate. Reactions proceeded at room temperature for 1 hour, after which enzyme solutions were shaken from the plate, and wells were again rinsed with double distilled water. Subsequently, 100 µl aliquots of 0.1 µg/ml ConA-HRP were added to all wells. Blank wells containing no biofilm were used as a background control. The ConA-HRP lectin was allowed to bind for 1 hour at room temperature, and the solution was then shaken from the plate followed by rinsing with double distilled water. Finally, 100 µl of ABTS substrate was added to all wells, and reactions were incubated for 15 minutes at room temperature before being quenched with 100 µl of 1% SDS. The absorbance of each well was measured at 405 nm, background signal from the blank wells was subtracted from experimental wells, and the percent decrease in biofilm was calculated by normalizing the signal from enzyme treated wells to that of wells receiving no enzyme treatment. Additional experiments directly monitored degraded alginate reaction products in treated biofilm supernatants, and the resulting data supported the conclusions drawn from the ConA-HRP lectin studies.

Supporting Information

Movie S1   Sites of cysteine substitution. Ribbon diagram of alginate Lyase A1-III (PDB file 1HV6). A trisaccharide reaction product is bound in the active cleft and shown as a grey ball and stick model. Amino acid residues targeted for cysteine substitution are shown in space filling mode, and are color coded as follows: S32C = Red, A41C = Orange, A53C = Green, A270C = Yellow, and A328C = Purple.

Acknowledgments

The authors would like to thank K. Dane Wittrup for sharing laboratory space and resources, Grant Henderson for helpful discussions regarding biofilm disruption assays, and George O'Toole for the kind gift of strain FRD1 and critical comments on the manuscript.   Ackerman et. al., manuscript in preparation.

Author Contributions

Conceived and designed the experiments: KEG JWL MEA. Performed the experiments: JWL MEA JL TCS. Analyzed the data: KEG JWL MEA. Contributed reagents/materials/analysis tools: JWL MEA TCS. Wrote the manuscript: KEG JWL MEA.

References

1. Elkin S, Geddes D (2003) Pseudomonal infection in cystic fibrosis: the battle continues. Expert Review of Anti-Infective Therapy 1: 609–618.
2. May TB, Shinabarger D, Maharaj R, Kato J, Chu L, et al. (1991) Alginate synthesis by Pseudomonas aeruginosa: a key pathogenic factor in chronic pulmonary infections of cystic fibrosis patients. Clinical Microbiology Reviews 4: 191–206.
3. Mai GT, Seow WK, Pier GB, McCormack JG, Thong YH (1993) Suppression of Lymphocyte and neutrophil Functions by Pseudomonas aeruginosa mucoid exopolysaccharide (alginate): Reversal by physicochemical, alginase, and specific monoclonal antibody treatments. Infection and Immunity 61: 559–564.
4. Simpson JA, Smith SE, Dean RT (1993) Alginate may accumulate in cystic fibrosis lung because the enzymatic and free radical capacities of phagocytic cells are inadequate for its degradation. Biochemistry and Molecular Biology International 6: 1021–1034.
5. Hoiby N, Krogh Johansen H, Moser C, Song Z, Ciofu O, et al. (2001) Pseudomonas aeruginosa and the in vitro and in vivo biofilm mode of growth. Microbes and Infection 3: 23–35.
6. Bayer AS, Park S, Ramos MC, Nast CC, Eftekhar F, et al. (1992) Effects of alginase on the natural history and antibiotic therapy of experimental endocarditis caused by mucoid Pseudomonas aeruginosa. Infection and Immunity 10: 3979–3985.
7. Bayer AS, Speert DP, Park S, Tu J, Witt M, et al. (1991) Functional role of mucoid exopolysaccharide (alginate) in antibiotic-induced and polymorphonu-clear leukocyte-mediated killing of Pseudomonas aeruginosa. Infect Immun 59: 302–308.
8. Hentzer M, Teitzel GM, Balzer GJ, Heydorn A, Molin S, et al. (2001) Alginate overproduction affects Pseudomonas aeruginosa biofilm structure and function. Journal of Bacteriology 183: 5395–5401.
9. Smedley YM, Marriott C, Hodges N, James SL (1986) Rheological interactions of cystic fibrosis tracheal mucin and Pseudomonas aeruginosa extracellular alginate. Journal of Pharmacy and Pharmacology 38: 54.
10. Eftekhar F, Speert DP (1988) Alginase treatment of mucoid Pseudomonas aeruginosa enhances phagocytosis by human monocyte-derived macrophages. Infection and Immunity 56: 2788–2793.
11. Alkawash MA, Soothill JS, Schiller NL (2006) Alginate lyase enhances antibiotic killing of mucoid Pseudomonas aeruginosa in biofilms. APMIS 114: 131–138.
12. Hatch RA, Schiller NL (1998) Alginate Lyase Promotes Diffusion of Aminoglycosides through the Extracellular Polysaccharide of Mucoid Pseudo-monas aeruginosa. Antimicrobial Agents and Chemotherapy 42: 974–977.
13. Alipour M, Suntres ZE, Omri A (2009) Importance of DNase and alginate lyase for enhancing free and liposome encapsulated aminoglycoside activity against Pseudomonas aeruginosa. J Antimicrob Chemother 64: 317–325.
14. Mrsny RJ, Lazazzera BA, Daugherty AL, Schiller NL, Patapoff TW (1994) Addition of a Bacterial Alginate Lyase to Purulent CF Sputum In Vitro Can Result in the Disruption of Alginate and Modification of Sputum Viscoelasticity. Pulmonary Pharmacology 7: 357–366.
15. Shire SJ (1996) Stability Characterization and Formulation Development of Recombinant Human Deoxyribonuclease I [Pulmozyme® (Dornase Alpha)]. In: Pearlman R, Wang YJ, eds. Pharmaceutical Biotechnology. New York: New York Kluwer Academic Publishers. pp 393–426.
16. Schellekens H (2002) Immunogenicity of therapeutic proteins: Clinical implications and future prospects. Clinical Therapeutics 24: 1720–1740.
17. Giezen TJ, Mantel-Teeuwisse AK, Straus SMJM, Schellekens H, Leufkens HGM, et al. (2008) Safety-Related Regulatory Actions for Biologicals Approved in the United States and the European Union. JAMA 300: 1887–1896.

18. Shankar G, Pendley C, Stein KE (2007) A risk-based bioanalytical strategy for the assessment of antibody immune responses against biological drugs. Nat Biotech 25: 555–561.
19. Kodera Y, Matsushima A, Hiroto M, Nishimura H, Ishii A, et al. (1998) Pegylation of proteins and bioactive substances for medical and technical applications. Progress in Polymer Science 23: 1233–1271.
20. Sakakibara H, Tamura T, Suzuki T, Hisano T, Abe S, et al. (2002) Preparation and Properties of Alginate Lyase Modified with Poly(ethylene Glycol). Journal of Pharmaceutical Sciences 91: 1191–1199.
21. Hoover DM, Lubkowski J (2002) DNAWorks: an automated method for designing oligonucleotides for PCR-based gene synthesis. Nucl Acids Res 30: e43-.
22. Yoon H-J, Hashimoto W, Miyake O, Murata K, Mikami B (2001) Crystal structure of alginate lyase A1-III complexed with trisaccharide product at 2.0 A resolution. Journal of Molecular Biology 307: 9–16.
23. Hashimoto W, Momma K, Miki H, Mishima Y, Kobayashi E, et al. (1999) Enzymatic and genetic bases on assimilation, depolymerization, and transport of heteropolysaccharides in bacteria. Journal of Bioscience and Bioengineering 87: 123–136.
24. Ramsey DM, Wozniak DJ (2005) Understanding the control of Pseudomonas aeruginosa alginate synthesis and the prospects for management of chronic infections in cystic fibrosis. Molecular Microbiology 56: 309–322.
25. Feldhaus MJ, Siegel RW, Opresko LK, Coleman JR, Feldhaus JMW, et al. (2003) Flow-cytometric isolation of human antibodies from a nonimmune Saccharomyces cerevisiae surface display library. Nat Biotech 21: 163–170.
26. Moreau-Marquis S, Stanton BA, O'Toole GA (2008) Pseudomonas aeruginosa biofilm formation in the cystic fibrosis airway. Pulmonary Pharmacology & Therapeutics 21: 595–599.
27. Strathmann M, Wingender J, Flemming H-C (2002) Application of fluorescently labelled lectins for the visualization and biochemical characterization of polysaccharides in biofilms of Pseudomonas aeruginosa. Journal of Microbio-logical Methods 50: 237–248.
28. Skjåk-Bræk G, Zanetti F, Paoletti S (1989) Effect of acetylation on some solution and gelling properties of alginates. Carbohydrate Research 185: 131–138.
29. Himmel ME, Ding S-Y, Johnson DK, Adney WS, Nimlos MR, et al. (2007) Biomass Recalcitrance: Engineering Plants and Enzymes for Biofuels Produc-tion. Science 315: 804–807.
30. Vaaje-Kolstad G, Horn SJ, van Aalten DMF, Synstad Br, Eijsink VGH (2005) The Non-catalytic Chitin-binding Protein CBP21 from Serratia marcescens Is Essential for Chitin Degradation. Journal of Biological Chemistry 280: 28492–28497.
31. Murata K, Inose T, Hisano T, Abe S, Yonemoto Y, et al. (1993) Bacterial alginate lyase: Enzymology, genetics and application. Journal of Fermentation and Bioengineering 76: 427–437.
32. Yoon H-J, Hashimoto W, Miyake O, Okamoto M, Mikami B, et al. (2000) Overexpression in Escherichia coli, purification, and characterization of Sphingomonas sp. A1 alginate lyases. Protein Expression and Purification 19: 84–90.
33. Wingender J, Strathmann M, Rode A, Leis A, Flemming HC (2001) Isolation and biochemical characterization of extracellular polymeric substances from Pseudomonas aeruginosa. Microbial Growth in Biofilms, Pt A. San Diego: Academic Press Inc. pp 302–314.
34. Ackerman M, Levary D, Tobon G, Hackel B, Orcutt KD, et al. (2009) Highly avid magnetic bead capture: An efficient selection method for de novo protein engineering utilizing yeast surface display. Biotechnology Progress 25: 774–783.

8

Protozoacidal Trojan-Horse: Use of a Ligand-Lytic Peptide for Selective Destruction of Symbiotic Protozoa within Termite Guts

Amit Sethi*¤, Jennifer Delatte, Lane Foil, Claudia Husseneder*

Department of Entomology, Louisiana State University Agricultural Center, Baton Rouge, Louisiana, United States of America

Abstract

For novel biotechnology-based termite control, we developed a cellulose bait containing freeze-dried genetically engineered yeast which expresses a protozoacidal lytic peptide attached to a protozoa-recognizing ligand. The yeast acts as a 'Trojan-Horse' that kills the cellulose-digesting protozoa in the termite gut, which leads to the death of termites, presumably due to inefficient cellulose digestion. The ligand targets the lytic peptide specifically to protozoa, thereby increasing its protozoacidal efficiency while protecting non-target organisms. After ingestion of the bait, the yeast propagates in the termite's gut and is spread throughout the termite colony via social interactions. This novel paratransgenesis-based strategy could be a good supplement for current termite control using fortified biological control agents in addition to chemical insecticides. Moreover, this ligand-lytic peptide system could be used for drug development to selectively target disease-causing protozoa in humans or other vertebrates.

Editor: Kostas Bourtzis, International Atomic Energy Agency, Austria

Funding: Funding was provided by the Biotechnology Agricultural Center Interdisciplinary Team Program at Louisiana State University Agricultural Center, and the state of Louisiana. The funders had no role in study design, data collection and analysis, decision to publish, or preparation of the manuscript.

Competing Interests: The authors have declared that no competing interests exist.

* Email: amit.sethi@pioneer.com (AS); chusseneder@agcenter.lsu.edu (CH)

¤ Current address: DuPont Pioneer Agricultural Biotechnology, Johnston, Iowa, United States of America

Introduction

One of the most important scientific achievements of the twentieth century has been the development of rapid and effective methods to control insect pests, principally through the use of chemical insecticides. However, the demand for new strategies has been growing due to an increasing recognition of the limitations associated with the use of chemical insecticides, such as insecticide resistance, concerns over environmental and human health impacts, and economic burdens. Therefore, biological control strategies that exploit insect-microbial relationships have been proposed as an alternative to chemical insecticides. The role of microbes in insects as well as the potential use of these microbes and their metabolic capabilities as biological control agents is well documented [1]. However, use of microbes as biological control agents has not been successful for some social insect systems mainly due to the presence of a suite of highly efficient synergistic defense mechanisms against entomopathogens, including behavioral responses (avoidance of pathogen and grooming), antimicrobial compounds, immunity, and competitive endogenous microbial fauna [2]. Thus, precise genetic manipulation to enable microbes that are not recognized as pathogens to interfere with host fitness has been identified as a novel tool to design more efficient biological control agents [2].

The use of genetically altered microorganisms to deliver gene products into a host organism is termed paratransgenesis. Specifically, in insects, genetically engineered microbes capable of colonizing the insect gut could be utilized as "Trojan-Horses" to produce effector molecules that kill the insect pest or eliminate the capacity of insects to act as vectors to transmit pathogenic agents [3,4]. Paratransgenesis (using genetically engineered bacteria, viruses or fungi) has been predominantly applied to prevent insects from transmitting pathogenic diseases [4–12]; only a few studies have used this biotechnology to actually kill the host, i.e. for insect control [13,14].

One of the major challenges in developing an efficient paratransgenesis system for insect control is the identification of mechanisms that allow microbes to spread efficiently among individuals. This challenge is easily overcome in social insects, such as termites, because they naturally exchange microbes among colony mates via social interactions, including trophallaxis (food exchange), coprophagy, and grooming [14,15]. Therefore, termites are ideal candidates for the development and application of a paratransgenesis model system for insect pest control [14–16].

Design of a control strategy using paratransgenesis requires identification of specific targets, and peptides with toxic effects against the identified target [17,18]. Subterranean termites are one of the most destructive urban and agricultural pests worldwide. The worker termites, which are responsible for foraging and feeding the colony, harbor cellulose-digesting protozoan symbionts in their hindguts [19]. Disruption of this obligate relationship has dramatic effects on the lifespan of individual termites and the entire colony, as termites deprived of their protozoa die presumably due to inefficient lignocellulose digestion. Thus, the protozoa are suitable targets for designing a paratransgenic system for termite control. Lytic peptides are a ubiquitous part of the non-specific eukaryotic immune system that destroys the integrity of protozoa membranes by disruption or pore formation by wedge-shaped insertion of monomers of the lytic peptide [20–22]. Lytic peptides have been shown to kill protozoan parasites in vertebrates [22,23] but have not been reported to harm the cell membranes of higher eukaryotes [22–24].

Recently, Husseneder and Collier [14] used lytic peptides to design a prototype of paratransgenesis for termite control using the Formosan subterranean termite (*Coptotermes formosanus*) as a model. First, they showed that lytic peptides (*Hecate*, *Cecropin*, and *Mellitin*) efficiently killed the three species of protozoa, *Pseudotrichonympha grassii*, *Holomastigotoides hartmanni*, and *Spirotrichonympha leidyi*, associated with the hindgut of *C. formosanus* workers. Furthermore, Husseneder and Collier [14] genetically engineered yeast (*Kluyveromyces lactis*) to express *Hecate*. After the yeast was ingested by termite workers, the lytic peptides expressed by the yeast killed the gut protozoa within 4 weeks, followed by the death of the termites within 6 weeks.

The top challenge in developing a paratransgenesis system is to enhance the efficiency of the technology while at the same time preventing from non-target effects of treatments. Lytic peptides have previously been shown to destroy specific cells (e.g., breast, testicular and prostate cancer cells) when they are conjugated with membrane receptor-recognizing molecules [20,25,26]. We followed the same concept and identified protozoa–recognition peptides to construct ligands that bind not only to symbiotic protozoa of *C. formosanus* but also symbiotic protozoa of another termite species, *Reticulitermes flavipes*, and free-living protozoa. Next, we genetically engineered the yeast *K. lactis* to express a fusion peptide (*Ligand-Hecate*) that specifically killed protozoa. Finally, we developed a target-specific bait containing genetically engineered yeast to kill termites.

Results

Identification of ligands that attach to protozoa

We used a phage library expressing variants of linear random heptapeptides to identify termite protozoa-recognizing peptides (Fig. S1A–G). Nineteen unique heptapeptide sequences that bound to protozoa were identified (Table S1). Two ligands, ALNLTLH (*Ligand-1*) and LPSLPAN (*Ligand-2*) showed homology to epitopes present on the variant surface glycoprotein (VSG) of *Trypanosoma brucei* and a single-pass type II membrane protein of *Thermosynechococcus elongatus*, respectively, when searched in the Database of Interacting Proteins (DIP, http://dip.doe-mbi.ucla.edu/). As it was not feasible to test all 19 selected candidate ligand peptide in our study, we selected *Ligand-1* and *Ligand-2* for synthesis based upon their predicted interactions described above. Next, the ligands were attached to the fluorophore *EDANS* (5-((2-Aminoethyl) amino) naphthalene-1-sulfonic acid) (Fig. S1H,I) to confirm their specific binding to termite protozoa under *in vitro* (protozoa culture) and *in vivo* (termite enema) conditions. Both of the ligands bound to all three species of protozoa of *C. formosanus* and not to the termite hindgut wall (Fig. 1). The ligands bound to the entire surface of protozoa, but were mostly concentrated in the anterior region of *P. grassii* clearly showing the axostyle (a sheet of microtubules) (Fig. 1A). For untreated protozoa, we only observed some patchy autofluorescence of wood particles ingested by the protozoa. However, the autofluorescence was easy to distinguish from specific binding of the ligands, since there is no autofluorescence of the surface and the axostyle region (Fig. S2A–C) These binding sites are likely to be present in all species of protozoa, as both the ligands also bound to all eleven species of protozoa [27] found in the hindgut of another termite

species *Reticulitermes flavipes* (Fig. 2A–H) and the four free-living aerobic protozoa species tested (*Tetrahymena pyriformis*, *Amoeba* sp., *Euglena* sp., and *Paramecium* sp.) (Fig. 2I–P). The ligands are most likely protozoa-specific as they did not bind to non-target microorganisms, such as gram negative *Escherichia coli*, gram positive *Pilibacter termitis* (a lactic acid bacterium exclusively found in the gut of *C. formosanus*) [28] and the yeast *K. lactis*.

Addition of ligand increases target specificity and efficiency of lytic peptides

Since both fluorescent ligand complexes showed similar binding characteristics, only *Ligand-1* was conjugated to *Hecate* (*Ligand-Hecate*, named hereafter) (Fig. S1J) to confirm its protozoacidal specificity and efficiency. One micromolar solution of *Ligand-Hecate* fusion peptide killed all three species of protozoa of *C. formosanus in vitro* in less than 10 min (Fig. 3A–E). However, the same concentration of *Hecate* alone (without the ligand) required more than 30 min to kill the protozoa (Fig. 3F). Increased efficiency of *Ligand-Hecate* compared to Hecate alone was also confirmed for the four species of free-living aerobic protozoa. Twenty-four hours after injection of *Ligand-Hecate* into the hindgut of *C. formosanus* workers via enemas, all three species of protozoa in the hindguts were dead. Treated termites died within two weeks after the loss of their protozoa. Target specificity was further confirmed by incubating non-targets *E. coli*, *P. termitis*, and *K. lactis* with *Ligand-Hecate* fusion peptide and *Hecate* alone. Median lethal dose (LD$_{50}$) of *Ligand-Hecate* was 8.3, 4.6 and 5.6-fold significantly higher than *Hecate* when tested against *E. coli*, *P. termitis*, and *K. lactis*, respectively (Fig. 4A). Thus, the addition of *Ligand-1* to *Hecate* increases not only the protozoacidal efficiency but also prevents immediate lysis of non-target species.

Termite bait containing protozoa-killing yeast strain

After confirming the target specific toxicity of the *Ligand-Hecate* fusion peptide, the commercially available *K. lactis* yeast was genetically engineered to express *Ligand-Hecate*. Simultaneously, another *K. lactis* strain expressing a red fluorescent protein *mPlum* was prepared to monitor ingestion and survival of yeast in the termite's guts, as well as spread of genetically engineered yeast among colony mates. Forty-eight hours old cultures of both the yeast strains secreted *mPlum* and *Ligand-Hecate*, respectively. Both the culture supernatant and pelleted *mPlum* yeast cells showed red fluorescence (Fig. 5 A,B). The culture supernatant from the *Ligand-Hecate* yeast strain caused 82% mortality compared to control in a free-living aerobic protozoa *T. pyriformis* after 24 h of treatment (Fig. 4B). We did not quantify the expression of *Ligand-Hecate* in the supernatant. Thus, the incubation experiment using culture supernatant was used as indirect evidence to suggest that the observed mortality of *T. pyriformis* could possibly be attributed to the *Ligand-Hecate* produced by the yeast.

The freeze-dried yeast strains (expressing *Ligand-Hecate*, *mPlum*, and a control containing only the vector plasmid with no inserted gene) were individually mixed with α-cellulose bait and control α-cellulose bait matrix without any yeast strain and were fed to termite workers (Fig. 4C,D). Addition of yeast in the bait matrix (α-cellulose) did not deter termite feeding and termites consumed similar amounts of bait among treatments (Fig. 4E).

After two weeks of bait consumption(Fig. 4D), we were able to confirm the ingestion of yeast strains by the termites via plating gut contents on *Kluyveromyces* differential medium and observing the growth of yeast colonies with the characteristic blue color. At the same time we also confirmed gene expression of the *mPlum* yeast strain by observing red fluorescence of yeast cells in the termite gut

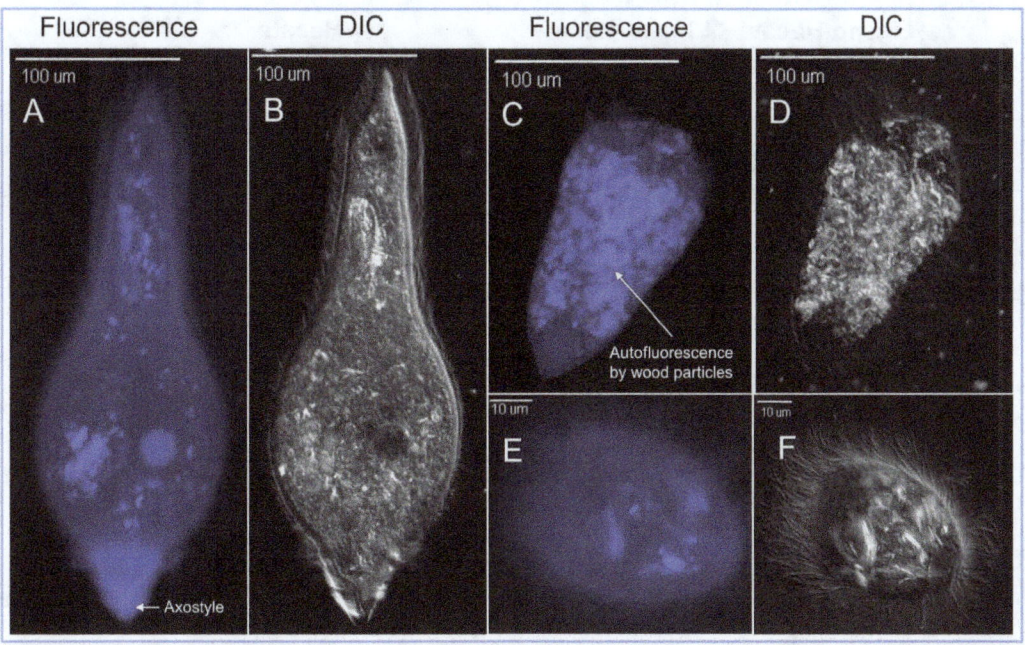

Figure 1. Visualization of binding of fluorescent *Ligand-1* **to gut protozoa of the Formosan subterranean termite,** *Coptotermes formosanus.* Blue fluorescence (excitation = 341 nm, emission = 471 nm) confirms that *Ligand-1* binds to all the three species of the termite protozoa. Phagocytosed wood particles within the protozoa cytoplasm show some patchy autofluorescence. (*A, B*) Fluorescent and differential interference contrast (DIC) exposures of *Pseudotrichonympha grassii*, respectively. Binding of ligands was concentrated in the anterior region of *P. grassii* clearly showing the axostyle (a sheet of microtubules). (*C, D*) Fluorescent and DIC exposures of *Holomastigotoides hartmanni*, respectively. (*E, F*) Fluorescent and DIC exposures of *Spirotrichonympha leidyi*, respectively. Binding of *Ligand-2* also showed a similar fluorescence pattern.

and yeast colonies cultured from gut contents (Fig. 5C–F). The number of yeast cells (counted as colony forming units after culture of gut contents on *Kluyveromyces* differential media) per worker gut significantly increased from the second week to third week after the termites began ingesting the baits containing yeast (Fig. 4F, Table S2). After three weeks of feeding on the bait

containing *Ligand-Hecate* expressing yeast, all three species of protozoa were dead and cellular debris of protozoa was found in the rectum of the workers (Fig. 6); all workers died within five weeks of continuous yeast ingestion.

Termites transfer genetically engineered yeast to nestmates via social interaction

Following visual detection of *mPlum* expressing yeast in termite guts after 2 weeks of ingesting the bait (see above), the remaining workers (donors, i.e. previously fed on mPlum yeast) were combined with an equal number of workers from the same colony that were fed on cellulose without yeast (recipients); the recipients were marked red by fat body stain (Sudan Red 7B) to distinguish them from donors (Fig. 4G). Both the donors and the recipients were fed on plain α-cellulose bait matrix without any yeast strain, and *mPlum* yeast was detected in the recipients at two weeks after both groups were combined. The number of *mPlum* yeast cells (CFU) significantly increased in the donors from the second week to fourth week, even though ingestion of yeast from bait was discontinued when donors and recipients were combined (Fig. 4H, Table S3). The number of yeast cells also increased in the recipients, but the increase was not significant within the measured time span.

Discussion

Paratransgenesis has been used primarily to control insect vector-borne diseases of humans and agricultural crops, where symbiotic microbes were genetically engineered to deliver molecules that block pathogen transmission [3–7,9,11,12,29]. Here, we provide the first example of a target-specific para-transgenesis system that has the potential to eliminate insect pests.

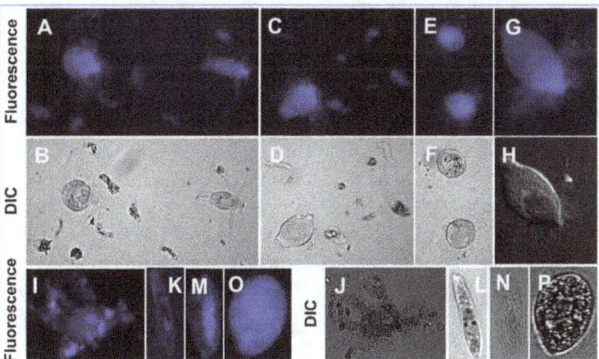

Figure 2. Visualization of binding of fluorescent *Ligand-1* **to other groups of protozoa:** (*A–H*) **twelve species of gut protozoa from the eastern subterranean termite,** *Reticulitermes flavipes,* **and** (*I–P*) **four species of free-living aerobic protozoa.** (*I, J*) Fluorescent and differential interference contrast (DIC) exposures of *Amoeba* sp., respectively. (*K, L*) Fluorescent and DIC exposures of *Euglena* sp., respectively. (*M,N*) Fluorescent and DIC exposures of *Paramecium* sp., respectively. (*O, P*) Fluorescent and DIC exposures of *Tetrahymena pyriformis*, respectively. Blue fluorescence confirms that *Ligand-1* binds to all the protozoa tested. Binding of *Ligand-2* also showed a similar fluorescence pattern.

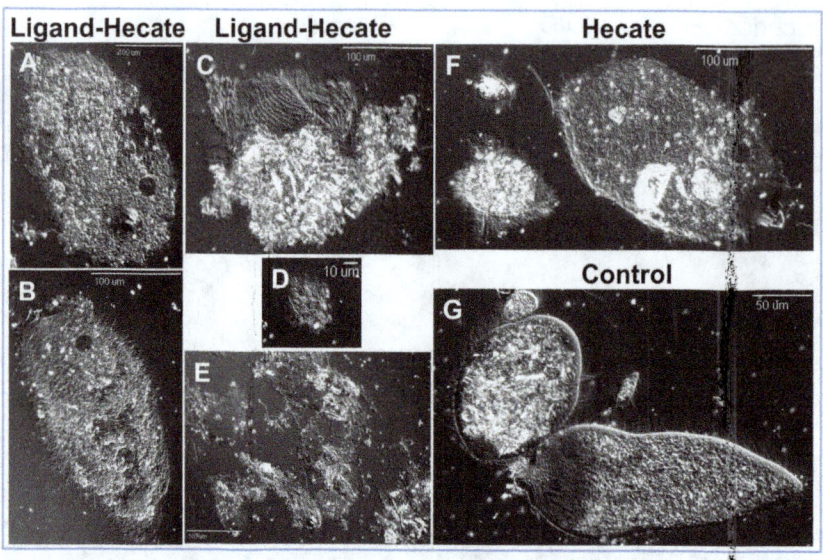

Figure 3. Enhanced toxicity of *Ligand-Hecate* **fusion peptide compared to** *Hecate* **alone.** Membranes of the termite protozoa lose their integrity five-fold faster when incubated with *Ligand-Hecate* fusion peptide as compared to incubation with *Hecate* alone at 1 µM concentration. (*A, B*) Differential interference contrast (DIC) images of *P. grassii* after 5 min of incubation with *Ligand-Hecate*. (*C*) DIC image of *H. hartmanni* after 5 min of incubation with *Ligand-Hecate*. (*D*) DIC image of *S. leidyi* after 5 min of incubation with *Ligand-Hecate*. (*E*) DIC images of all the three species of termite protozoa after 10 min of incubation with *Ligand-Hecate*. (*F*) DIC image of all the three species of termite protozoa after 10 min of incubation with *Hecate* alone. (*G*) DIC image of all the three species of termite protozoa after 10 min of incubation with the buffer without any peptide (control).

This paratransgenesis system uses a conjugate of recognition and lytic molecules (*Ligand-Hecate*) to kill the cellulose-digesting protozoa in the termite gut. The current findings demonstrate delivery, retention and biological activity of genetically engineered *K. lactis* yeast in the gut of Formosan subterranean termites.

The proof of concept of paratransgenesis in termites was first achieved by Husseneder and Grace [15], who genetically engineered *Enterobacter cloacae* isolated from the gut of *C. formosanus* to express ampicillin resistance markers and green fluorescent protein. The engineered bacteria were rapidly ingested by workers, efficiently transferred among nestmates and were detectable in termite guts for up to two months. Subsequently, Zhao et al. [13] genetically engineered *E. cloacae* to express insecticidal proteins from the bacterium *Photorhabdus luminescens* to kill termites. However, *E. cloacae* is not an ideal organism for paratransgenesis because it is ubiquitous in nature and causes a variety of infections and problems associated with humans. Moreover, insecticidal toxins from *P. luminescens* have considerable mammalian toxicity [30,31]. Thus, we choose *K. lactis* yeast and lytic peptides to develop a prototype of paratransgenesis to control termites [14]. The yeast is non-pathogenic for vertebrates and lytic peptides are not known to harm higher eukaryotes [21,23]. Moreover, our approach is to kill termites indirectly via targeting obligate gut protozoa linked to cellulose digestion and other processes.

To further enhance environmental safety of a termite paratransgenesis system, we designed effector molecules to specifically target the protozoa [3,4]. We conjugated lytic peptides to protozoa-specific ligands. Based on the database search, the two ligands showed homology to epitopes present on the membrane proteins (peripheral and transmembrane, respectively) that are involved in several trafficking pathways. *Ligand-1* showed homology with VSG of *T. brucei*, which is a glycosylphosphatidylinositol-anchored glycoprotein expressed on the external surface of the protozoan at extreme density (∼ five million) [32,33]. *Ligand-2* showed homology with the biopolymer

transport protein (single-pass type II ExbD family), which is involved in the transport of vitamin B_{12}, iron siderophores, sucrose, nickel and sulfates [34–36]. Since *Ligand-1* showed homology with the epitopes of VSG, it might be possible that *Ligand-1* binds to a VSG-like protein that is endocytosed in a similar way as described for trypanosomes [37–43]. Thus, the binding of *Ligand-Hecate* on a VSG-like protein could have facilitated rapid membrane internalization and could lead to increased protozoacidal activity of the fusion peptide by five-fold over *Hecate* alone. At the same time, the conjugation of *Ligand-1* to *Hecate* increased the LD_{50} of the fusion peptide for non-target microbes by four to eight-fold over *Hecate* alone. Based on the therapeutic research on ligand-lytic peptide conjugates [20,25,26], it is possible that fusion of *Ligand-1* to *Hecate* interferes with *Hecate*'s insertion into the cell membrane and thereby decreases its affinity against non-target species. On the other hand in case of termite protozoa, it is possible that fusion of *Ligand-1* to *Hecate* provides more stability to *Hecate* and exposes *Hecate* molecules in a close proximity of the cell membrane after *Ligand-1* binds to the membrane receptors. Thus, the conjugation of the lytic peptide to a ligand increases not only the activity but also enhances the selectivity. Similar results have been found in cancer treatment studies using hormone ligand-lytic peptide conjugates [20,25,26]. Lytic peptides are required in less than one micromolar range to effectively kill protozoa [44]; the linking of ligand to lytic peptide even further reduces the minimal activity range. Development of resistance to lytic peptides has not yet been observed, possibly due to the pore-forming mode of action and the rapid environmental degradation that reduces selection pressure [22,45]. Hence, it appears that ligand–lytic peptide combinations are an ideal effector molecule to specifically kill termite protozoa with low risk to non-target organisms.

Another important feature of a successful paratransgenesis system is uncompromised fitness of the Trojan-Horse in the insect gut [3,4]. The Trojan-Horse should be able to survive and multiply in the insect gut and further propagate in the insect

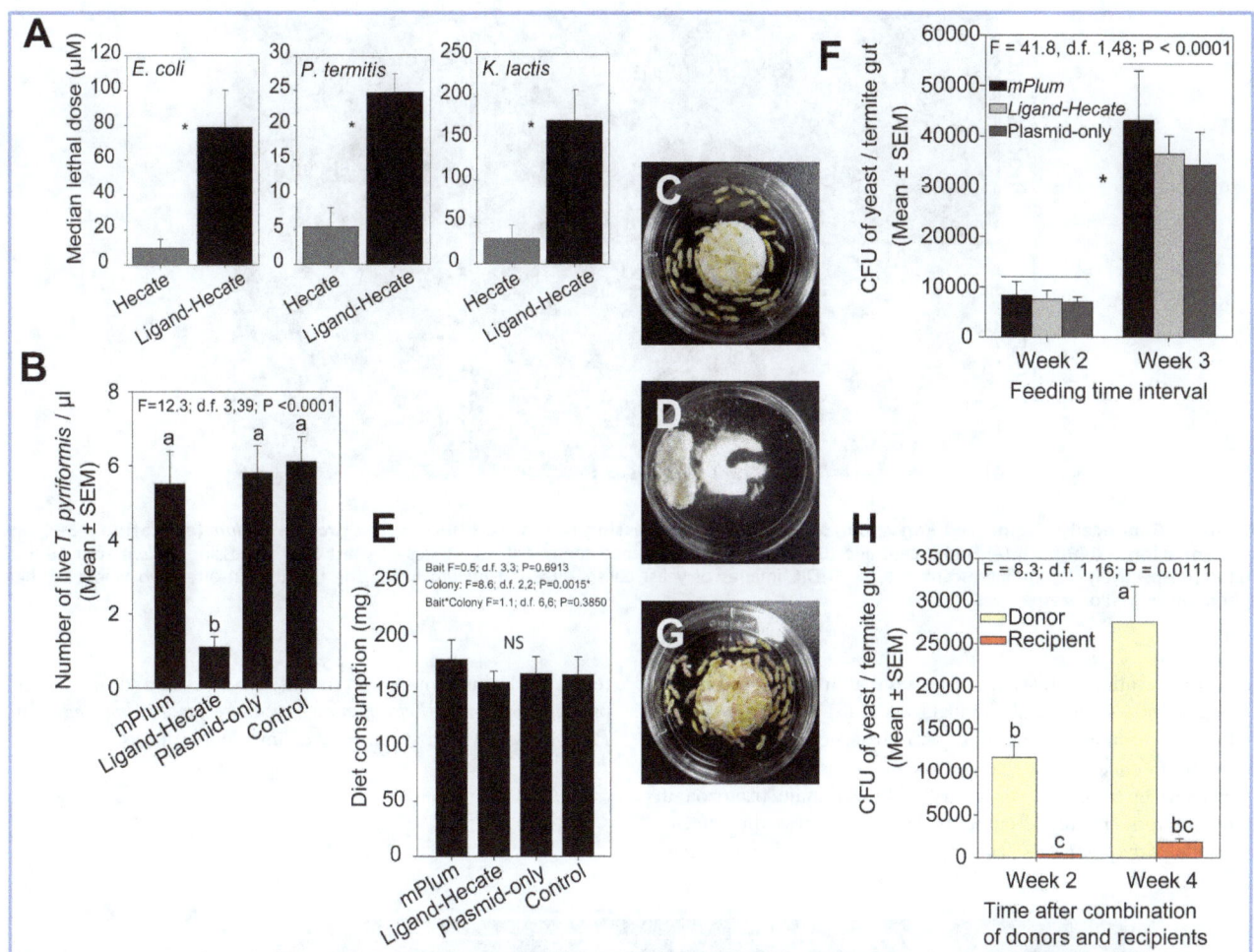

Figure 4. Assays using *Ligand-Hecate* and *Hecate* peptides, and genetically engineered yeast strains. (*A*) Mean lethal doses (LD$_{50}$) of *Ligand-Hecate* and *Hecate* peptides against non-target microorganisms *Escherichia coli*, *Pilibacter termitis*, and *Kluyveromyces lactis*. The linking of *Ligand-1* with *Hecate* significantly enhanced the mean lethal dose for each non-target microorganism. * indicates significant difference between treatments. (*B*) Toxicity of culture supernatants of different yeast strains against aerobic protozoa *T. pyriformis*. (*C*) Termite workers feeding on α-cellulose bait disk containing genetically engineered yeast cells in a bioassay setup. (*D, E*) Bait consumed by termite workers after five weeks. Addition of yeast into α-cellulose matrix did not deter termites from feeding and no significant difference was found in the diet consumption among different treatments. (*F*) Increasing number of yeast cells in the termite gut at two and three weeks of ingesting α-cellulose bait containing genetically engineered yeast strains. Control bait containing only α-cellulose did not show any CFU of *K. lactis*. (*G*) Bioassay setup to test transfer of the genetically engineered yeast cells to other nestmates. Termites fed on α-cellulose bait containing *mPlum* expressing yeast strain for two weeks (donors) were mixed with an equal number of workers from the same colony that were not fed on yeast bait (recipients, stained red with 1% Sudan Red 7B) and the mixed termites were fed on plain α-cellulose bait without any yeast in a Petri dish. (*H*) Number of *mPlum* expressing yeast cells (CFU) recovered from the donor and recipient termite guts two and four weeks after combining donors and recipients.

population. In our studies, we found that the genetically engineered yeast was retained in termite guts and multiplied without continuous feeding on the yeast bait. Further, infected termites transferred the yeast to other nestmates via social interaction and it propagated in recipients. Since two weeks was the first observation time in the transfer experiment, transfer of yeast from infected termites to recipients is likely to occur more rapidly, as Hussenecer and Grace [15] previously reported that transfer of bacteria from infected to recipients occurs within hours and even ratios as low as 1 infected termite: 25 recipients were sufficient to spread the bacteria throughout laboratory colonies. Thus, our termite paratransgenesis system using the yeast fulfills the requirement of Trojan-Horse colonization.

In summary, we present evidence for a novel, functional, target-specific and potentially environmentally-friendly termite baiting

system with a living agent that expresses a continuous source of effector molecules in the termite colony. Such paratransgenesis-based termite control is attractive due to easy mass production of yeast in bioreactors [46] and relatively easy delivery of the Trojan Horse in the form of baits containing a lyophilized delivery system [47]. Amalgamation of paratransgenic yeast into current termite baiting systems or in conjunction with soil treatments would also likely contribute to enhancing the efficacy of chemical insecticides against termites. Uptake and horizontal transfer of the bait containing the yeast can be further enhanced as demonstrated in chemical insecticide baits by adding known feeding stimulants, such as sugars, amino acids and lipids [48]. Similar to Bt transgenic crops, an additional environmentally-friendly feature can be added to the paratransgenesis system by expressing the effector molecule in inactive form (pro-peptide) [49,50] that

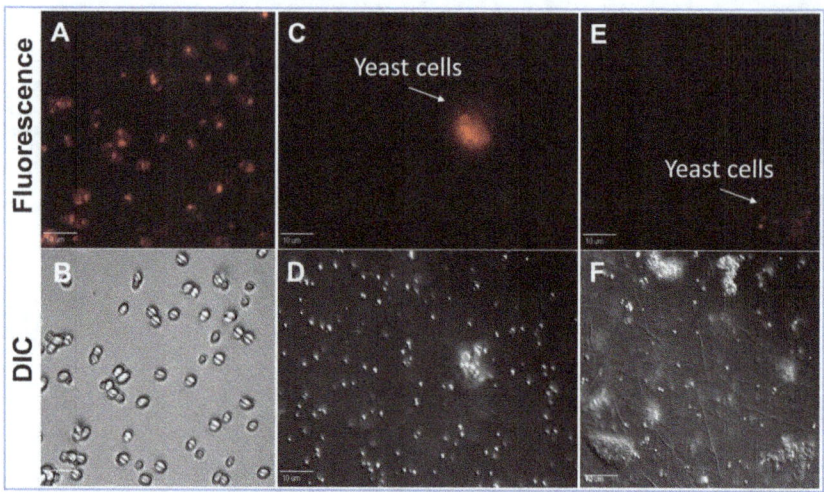

Figure 5. Genetically engineered *Kluyveromyces lactis* yeast expressing the far red fluorescent protein *mPlum* (excitation – 590 nm and emission – 649 nm). (*A*) Fluorescent and (*B*) Differential interference contrast (DIC) images of yeast cells expressing *mPlum* after 48 h of culture, respectively. (*C, E*) Fluorescent and (*D, F*) DIC images of yeast cells expressing *mPlum* inside the termite gut after two weeks of their continuous ingestion, respectively.

requires activation by digestive proteases that are produced by the protozoa and/or the termite hindgut tissue [51].

Besides termite control, this paratransgenesis biotechnology could be modified for use to control other insect pests that are dependent on symbiotic microbes or to eliminate protozoa in insect vectors. Finally, from a wider perspective, the effector molecule (*Ligand-Hecate*) efficiently killed all protozoa species

tested and thus could also be used to develop drugs against parasitic protozoa (*Leishmania, Trypanosoma, Trichomonas*, and *Plasmodium*) within vertebrates or invertebrate hosts.

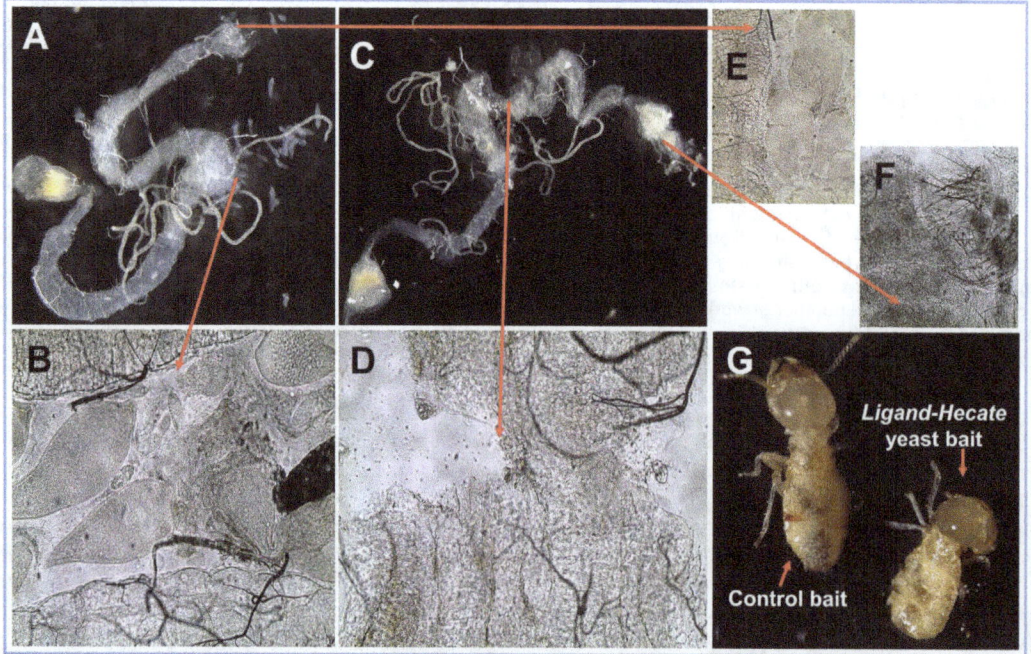

Figure 6. α-cellulose bait containing *Ligand-Hecate* expressing yeast strain kills all the three species of protozoa found in *C. formosanus* workers within three weeks and the workers die within five weeks of bait ingestion. (*A*) The gut of a worker with live protozoa at three weeks of ingesting plain α-cellulose bait. (*B*) Healthy protozoa exude out of worker gut when the gut is cut open. (*C, D*) The empty paunch of a worker possessing no protozoa at three weeks of ingesting the bait containing *Ligand-Hecate* yeast. (*E*) Healthy and (*F*) cellular debris of protozoa in the rectum of a worker at three weeks of ingesting the bait containing no yeast and *Ligand-Hecate* yeast, respectively. (*G*) Workers at five weeks of ingesting α-cellulose bait containing no yeast (left) and *Ligand-Hecate* expressing yeast (right). The worker fed on *Ligand-Hecate* bait is dead.

Materials and Methods

Termite collection and protozoa isolation

Three colonies of *C. formosanus* and one colony of *R. flavipes* were collected from New Orleans, Louisiana. The termite species collected herein are not endangered or protected. Thus, no specific permissions were required for the collection of termites. Claudia Riegel, Kenneth Brown and Edward Freytag from New Orleans Mosquito, Termite & Rodent Control Board helped in collecting the termites. After collection, the termites were maintained on damp cardboard in plastic buckets at $26\pm2°C$ and 85% R.H. Three groups of 50 worker guts were extirpated from each colony and placed in 100 µl Trager U media (pH 7.0) sparged with gas mixture of nitrogen (92.5%), carbon dioxide (5%) and hydrogen (2.5%) on a glass slide under anaerobic conditions in a glove box (Coy Laboratories Inc., MI, USA) [52]. The hindguts were pierced with a pair of sterile fine dissecting probes to release the protozoa. The gut contents were transferred into a 1 ml microcentrifuge tube containing 900 µl Trager U media. After allowing for sedimentation of gut wall fragments (~5 sec), the supernatant (900 µl) was transferred into a fresh tube. Then, the protozoa (Fig. S1A–C) were centrifuged at $30\times g$ for 10 min at $4°C$. The pellet was collected after rinsing it twice with Trager U.

Identification of termite protozoa recognition peptides using phage display

We used phage display libraries (Ph.D. 7 Phage Display Peptide Library Kit, New England Biolabs Inc., MA, USA) to identify protozoa recognition peptides by an *in vitro* selection process called panning (Fig. S1G). The pellet (protozoa) was suspended in sparged ice-cold 10 mM Tris-HCl buffer at pH 7.0 that contained 2 mM phenylmethyl sulphonyl fluoride and 2 mM $MgCl_2$ [53]. The cells were allowed to swell in the hypotonic buffer for 1 h in an anaerobic chamber. The cells were homogenized and cell breakage was monitored by phase contrast microscopy. The homogenate was layered over a two step gradient consisting of 8 ml of 0.5 M mannitol over 4 ml of 0.58 M sucrose, both in Tris buffer and was centrifuged at $250\times g$ for 30 min. The pellet was resuspended in 3 ml Tris buffer and homogenized again. The second homogenate was layered on a single step gradient that consisted of 20% sucrose in Tris buffer and centrifuged at $250\times g$ for 30 min. The supernatant was collected and centrifuged at $40,000\times g$ for 1 h. The obtained pellet containing plasma membrane was resuspended directly in Tris buffer and stored at $-20°C$ for future use. The purity of the plasma membrane layer was assessed by electron microscopy [54] (Fig. S1D–F).

Isolated plasma membranes were coated on plates and incubated with the phage library as per manufacturer's instruction. After washing of unbound phages, the specifically bound phages were eluted and amplified in *E. coli*. Additional 3 rounds of panning were performed to achieve positive selection (Fig. S1G). After positive selection, a pool of ninety phages (10 phages per replication per termite colony) were purified and sequenced to identify the displayed heptapeptide sequences.

Sequence analysis

The obtained heptapeptide sequences of the phages were compared to those in Genbank using Swissprot: BLAST (http://www.expasy.ch/tools/blast/) for identification of potential protozoa recognition peptides (ligands). A minimum 1000 E-value was used for the search. Next, ligand identity was used in database of interacting proteins (DIP, http://dip.doe-mbi.ucla.edu/) to determine potential binding partners. Ligand sequences were deposited

in the NCBI Probe database under Probe Unique Identifiers (PUIDs) 16719496–16719514.

Conjugation of ligands to fluorophore and lytic peptide

Two heptapeptides were selected out of 19 identified unique sequences to synthesize two ligands *Ligand-1* and *Ligand-2*. Each ligand was coupled to a fluorophore *EDANS* (5-((2-Aminoethyl) amino) naphthalene-1-sulfonic acid) via solid state peptide synthesis using NovaTag resin (EMD Biosciences) at the Louisiana State University peptide facility (Fig. S1H,I) to prepare two Ligand-EDANS complexes (*Ligand-1-EDANS* and *Ligand-2-EDANS*). EDANS can be directly visualized in fluorescence microscopy by the use of an UV light source and a DAPI filter [55]. Since both fluorescent ligand complexes showed similar binding characteristics (see results), only *Ligand-1* out of the two ligands was conjugated to Hecate to prepare ligand-lytic peptide fusion peptide (*Ligand-Hecate*) (Fig. S1J). *Ligand-Hecate* and *Hecate* (without ligand) were synthesized at the Interdisciplinary Center for Biotechnology Research, University of the Florida, USA.

Testing binding of *Ligand-EDANS* to protozoa, bacteria and yeast

Termite protozoa were isolated as described above and control cultures of the aerobic protozoa *Tetrahymena pyriformis*, *Amoeba* sp., *Euglena* sp., and *Paramecium* sp. (Carolina Biological Supply Company, NC, USA) as well as cultures of *E. coli* and *K. lactis* (New England Biolabs Inc., MA, USA) were prepared according to the supplier instructions. Cultures of *Pilibacter termitis* (American Type Culture Collection, KS, USA) were prepared according to the methods given in Higashiguchi et al. [28]. All microorganisms were fixed in 10% formaldehyde at $4°C$ for 12 h [52]. Fixing is necessary to prevent movement of the microorganisms for detailed observation, picture and documentation of fluorescence. In addition, termite protozoa are strictly anaerobic. Without fixing, the fluorescent signal cannot be properly detected as the protozoa cells disrupt due to slight exposure to oxygen during slide preparation for fluorescence microscopy.

For *in vitro* testing, all microorganisms were incubated for 1 h with two *Ligand-EDANS* solutions (*Ligand-1-EDANS* and *Ligand-2-EDANS*), separately at 1 µM final concentration and observed under a fluorescent microscope (excitation = 341 nm, emission = 471 nm; Model: DMRxA2, Leica Microsystems Inc.) at 400 × magnification. For *in vivo* testing, each worker was injected into the rectum with 0.3 µl of 1 µM *Ligand-EDANS* solutions using micromanipulators (Leitz micromanipulators, Vermont Optechs Inc., VT, USA) and a pedal-driven high-speed electronic injection system [52]. Control termites were injected with the buffer only. The experiment had three replications with 20 workers in each replication. After injections, the workers were placed into separate Petri dishes with damp filter paper and kept at $26\pm2°C$ with 85% R.H. Guts from the injected workers were extirpated after 24 h and the protozoa were collected, fixed and observed.

Testing toxicity of *Ligand-Hecate* against protozoa, bacteria and yeast

Cultures of all microorganisms were prepared as described above. For *in vitro* testing, termite protozoa were incubated for 1 h with *Ligand-Hecate* solution (end concentration 1 µM). Controls included: (a) protozoa incubated with *Hecate* solution (end concentration 1 µM), and (b) protozoa incubated with the buffer without any peptides. Survival of protozoa was observed after 5, 10, 30 and 60 min of incubation. For *in vivo* testing, each worker was injected into the rectum with 0.3 µl of either (a) 1 µM *Ligand-Hecate* solution, (b) 1 µM *Hecate* solution, or (c) the buffer without any peptide using micromanipulators and a pedal-driven high-speed electronic injection system [52]. The experiment consisted of three replications with 20 workers in each replication. After injections, the workers were placed into separate Petri dishes with damp filter paper and kept at 26±2°C with 85% R.H. Guts from five injected workers were extirpated after 24 h and the protozoa were collected, fixed (as explained above) and observed for mortality. Once the death of protozoa in the termite gut was confirmed, the mortality of the remaining termites was assessed daily.

Cultures of *E. coli*, *P. termitis* and *K. lactis* were incubated for 1 h with six end-concentrations (1, 10, 25, 50, 75 and 100 µM) of *Ligand-Hecate* solution or *Hecate* solution. For controls, the cultures were incubated with the corresponding volume of the buffer without any peptide. The experiment was replicated three times. After 1 h, three ten-fold serial dilutions of the cultures were plated in triplicates on BHI media and incubated at 37°C overnight. The number of colony forming units on each plate was then recorded. Median lethal dose (LD_{50}) was calculated for both *Ligand-Hecate* and *Hecate* using probit analysis (dose-response curve) for each microorganism.

Genetic engineering of *K. lactis* to express recombinant proteins

The commercially available yeast-based protein expression system (*K. lactis*, New England Biolabs Inc., MA, USA) was genetically engineered to produce two strains to express and secrete two types of proteins: (a) a far red fluorescent protein, *mPlum* (Clontech Laboratories Inc., CA, USA), and (b) *Ligand-Hecate* fusion peptide. DNA sequences of *mPlum* and *Ligand-Hecate* were codon optimized for expression in *K. lactis* by GenScript Ltd., NJ, USA. The *mPlum* gene was amplified using primers (forward - 5'TTATGCTTCCGGCTCGTATG 3'and reverse - 5'AGGCCTATTATTTTTGACACCAGA3'). The *Ligand-Hecate* gene was amplified using primers (forward - 5'GTAAAACGACGGCCAGT3'and reverse -5'CAGGAAA-CAGCTATGAC3'). The amplified *mPlum* fragment was cloned into the *BamHI – EcoRI* site of pKLAC2 downstream of the *K. lactis* α-mating factor domain (α-MF) according to cloning strategy given in the instruction manual. Similarly, the amplified *Ligand-Hecate* fragment was cloned into *XhoI – NotI* site of pKLAC2 (Fig. S3A,B). For control, pKLAC2 without any foreign gene (*plasmid-only*) was included. All three constructs were cloned into competent *E. coli* cells (NEB # C2992, New England Biolabs Inc., MA, USA). Each vector was isolated and digested with a pair of respective restriction endonucleases to determine the presence of the insert.

All three pKLAC2 vectors (*mPlum*, *Ligand-Hecate* and *plasmid-only*) were linearized with *SacII* to generate the expression cassettes. The lineralized expression cassettes were introduced into competent *K. lactis* cells at the LAC4 locus according to the manufactures instructions. Cells of three yeast strains (*mPlum*, *Ligand-Hecate* and *plasmid-only*) were grown separately on yeast carbon base (YCB) agar medium containing 5 mM acetamide at 30°C for 2 days. Colonies of each strain were picked and resuspended in 2 ml YPGal medium and then incubated with shaking at 250 rpm for 2 days at 30°C. Cells of all the three yeast strains were harvested by centrifugation at 7000×g for 30 sec and the culture supernatants were transferred to fresh tubes.

Yeast cells with correct integration of the expression cassettes into the *K. lactis* genome were identified by PCR using the primers (Primer 1 -5'ACACACGTAAACGCGCTCGGT3' and Primer 2 - 5'ATCATCCTTGTCAGCGAAAGC 3') supplied with the *K. lactis* kit. Fresh colonies of each yeast strain were picked and resuspended in 25 µl of 1 M sorbitol containing 2 mg/ml lyticase. Cells were mixed by vortexing and incubated at 30°C. After 1 h, the lyticase-treated cells were lysed at 98°C for 10 min in a thermocyler. PCR was performed according to the *K. lactis* instruction manual. In case of each yeast strain, integration of the expression cassette at the LAC4 locus in the *K. lactis* genome resulted in amplification of a 2.4 kb product (the promoter region of the LAC4 locus) (Fig. S3C,D).

All harvested yeast strains (pelleted cells as well as culture supernatants) were tested for fluorescence under a fluorescent microscope (excitation – 590 nm and emission – 649 nm; Model: DMRxA2, Leica Microsystems Inc.) at 400 × magnification to confirm that red fluorescence was only produced by the *mPlum* yeast strain. Biological activity of the culture supernatants of yeast strains was determined against aerobic protozoa of the species *T. pyriformis*. Fifty microliter of the culture supernatants was incubated with 50 µl of *T. pyriformis* culture. After 24 h, live protozoa were counted using a Sedgewick-Rafter cell (Pyser-SGI Limited, Kent, UK) under a microscope (Model: DMLB, Leica Microsystems Inc.) at 200 × magnification.

Termite feeding bioassays using genetically engineered yeast strains

All freshly harvested yeast strains were freeze-dried overnight using a lyophilizer. Freeze-drying does not affect the viability of yeast strains [56,57]. Freeze-dried yeast strains were mixed separately at the rate of 7.5 mg with 1500 mg of α-cellulose powder and 3 ml water. Three disks of 0.5-cm thickness were punched out the mixture using a 1.5-cm-diameter cork borer. Each feeding experiment was set up in a Petri dish using a bait disk and 75 worker and 5 soldier termites. The Petri dishes were placed in a tray with moist paper towels and kept in an incubator at 26±2°C and 85% R.H. Each bait disk was hydrated with 300 µl autoclaved deionized water every 48 h. Termites were treated with four different bait disks containing: (a) *mPlum* yeast strain, (b) *Ligand-Hecate* yeast strain, (c) *plasmid-only* yeast strain (control), or (d) plain α-cellulose without any yeast (control). Three replicates were set up for each treatment. The whole experiment was repeated three times using three different termite colonies. The termite colonies were collected as explained above.

Fifteen worker guts from each replication were extirpated using sterile forceps at two- and three-week intervals of feeding and then divided into three groups of five guts for their use in three assays: (a) plating on *Kluyveromyces* differential medium, (b) testing for fluorescence, and (c) observing the status of gut protozoa.

(a) Plating on *Kluyveromyces* differential medium: A group of five guts (per replication) was homogenized in a microcentrifuge tube containing 500 µl of autoclaved deionized water. Three ten-fold serial dilutions of homogenized gut contents were prepared and plated in triplicates on *Kluyveromyces* differential medium [58]. Plates were incubated at 30°C for 48 h. Cells of *K. lactis* produced blue colonies on the medium due to the presence of X-Gal/IPTG. All remaining yeast species produced white, cream or pink color colonies [58]. Blue colonies were counted to assess uptake, survival and multiplication of yeast cells inside the termite guts.

(b) Testing for fluorescence: The gut contents were prepared from a group of five workers (per replication) as explained above and viewed under a fluorescent microscope (excitation – 590 nm and emission – 649 nm; Model: DMRxA2, Leica Microsystems

Inc.) at 400 × magnification for the presence of *mPlum* fluorescence in yeast strains.

(c) Observing status of gut protozoa: Five worker hindguts (per replication) were cut open in sparged Trager U media on a glass slide using fine probes in an anaerobic glovebox and the status of the gut protozoa was checked under both stereo (Model: MZ16, Leica Microsystems) and (Model: DMLB, Leica Microsystems) compound microscopes at 50 and 200 × magnification, respectively.

After five weeks of termite feeding, dry weight of bait consumed was calculated for comparison among the treatments. To determine bait dry weight, an additional 10 bait disks from each treatment (*mPlum, Ligand-Hecate, plasmid-only* and control) were weighed individually (disk fresh weight) before they were put into an oven at 50±5°C. After 48 h, these bait disks were reweighed individually (disk dry weight). A dry/fresh weight ratio was calculated for each bait disk and averaged over the 10 disks. The bait fresh disk from each treatment was weighed prior to the start of the feeding experiment, and dry weight was computed by multiplying with the corresponding average dry/fresh weight ratio. After five weeks of exposure to termite feeding, the bait disks were dried in the oven for 48 h at the same temperature. The dry weight of bait consumed was calculated as the difference between initial and final dry weights [59].

Testing transfer of the genetically engineered yeast to other colony members

Following confirmation of *mplum* yeast strain in the termite gut (after 2 weeks of feeding on the bait containing *mplum* yeast strain), the remaining termites (donors) were mixed with an equal number of workers (recipients) from the same colony that were fed on a bait of α-cellulose without yeast. The recipient termites were marked red by feeding them with filter paper containing 1% (w/w, 6.0 mg stain per paper) Sudan Red 7B to distinguish between donors and recipients termites [15]. The mixed termites were fed on plain α-cellulose bait without any yeast in a Petri dish. The whole experiment was carried out with three replicates and was repeated twice using two different termite colonies. Termites were collected as explained above.

The guts of five donors and five recipients were extirpated at two- and four-week intervals of feeding and homogenized in water. The homogenate was spread on *Kluyveromyces* Differential Medium as described above to quantify the presence of yeast in donors and recipients and confirm the transfer of yeast to the recipients by counting the number of blue colonies. Since the yeast strain contained the *mPlum* gene, we also confirmed transfer and gene expression of the yeast by viewing the gut homogenate under a fluorescent microscope as described above.

Statistical analyses

The dose-response data on *Ligand-Hecate* and *Hecate* against *E. coli, P. termitis,* and *K. lactis* was subjected to probit analysis and the values obtained for mean lethal dose (LD$_{50}$) were compared within each microorganism using t-test (JMP software, SAS Institute). The data on: (1) the number alive protozoa *T. pyriformis* in biological activity assay using the culture supernatants of yeast strains, (2) the number of yeast cells (CFU) per termite gut in yeast feeding assays, and (3) diet consumption in yeast feeding assay were analyzed using analysis of variance. Then, Tukey's honestly significant difference (HSD) test with a significance level of α = 0.05 was used for post hoc means separation (JMP software, SAS Institute).

Supporting Information

Figure S1 Identification and construction of ligands that bind to protozoa living in the hindgut of the Formosan subterranean termite, *Coptotermes formosanus.* SEM images of the three species of protozoa: (*A*) *Pseudotrichonympha grassii,* (*B*) *Holomastigotoides hartmanni,* and (*C*) *Spirotrichonympha leidyi.* (*D*) Cross section of the three species of protozoa. (*E*) SEM and (*F*) TEM images of isolated plasma membrane from the protozoa. (*G*) Scheme explaining panning of isolated plasma membrane with a phage library consists of linear heptapeptides (Ph.D. 7). (*H, I*) Two selected ligands (*Ligand-1* and *Ligand-2*) attached to a fluorophore *EDANS* (5-((2-Aminoethyl) amino) naphthalene-1-sulfonic acid). (*J*) Fusion peptide consisting of *Ligand-1* and *Hecate.*

Figure S2 Visualization of untreated gut protozoa of the Formosan subterranean termite, *Coptotermes formosanus* under fluorescence microscope. (*A, B, C*) Superimposed fluorescent (excitation = 341 nm, emission = 471 nm) and differential interference contrast (DIC) exposures of *Pseudotrichonympha grassii, Holomastigotoides hartmanni* and *Spirotrichonympha leidyi,* respectively. Phagocytosed wood particles within the protozoa cytoplasm show some patchy autofluorescence.

Figure S3 Genetic engineering of *Kluyveromyces lactis* yeast to produce two strains *mPlum* and *Ligand-Hecate*. (*A*) The pKLAC2 expression vector. (*B*) Cloning strategy for *mPlum* and *Ligand-Hecate* into pKLAC2. (*C*) Genomic integration of two expression cassettes, *mPlum* and *Ligand-Hecate* in the *K. lactis* genome. Vector pKLAC2 containing either *mPlum* or *Ligand-Hecate* was digested with SacII and introduced into *K. lactis* cells. The 5′ PLAC4 and 3′ PLAC4 sequences directed insertion of the cassette into the promoter region of the LAC4 locus in the *K. lactis* genome. (*D*) Genetically engineered *K. lactis* cells in which the expression cassette had correctly integrated into the *K. lactis* genome were identified by PCR using supplied Integration Primers 1 and 2 to amplify a 2.4 kb product (the promoter region of the LAC4 locus).

Table S1 Protozoa recognition peptides identified using phage display libraries. Two heptapeptide sequences (shown in red) were selected to synthesized two ligands, *Ligand-1* and *Ligand-2,* respectively.

Table S2 ANOVA of the number of yeast CFU per termite gut at two and three weeks of ingesting α-cellulose diets.

Table S3 ANOVA of the number of *mPlum* yeast CFU per termite gut at two and four weeks after combining the donors and recipients.

Acknowledgments

We thank Claudia Riegel, Kenneth Brown and Edward Freytag (New Orleans Mosquito, Termite & Rodent Control Board) for providing the termites. We also thank Allison Richard (Louisiana State University peptide facility) for advice on ligand development and synthesizing the fluorescent ligand, Savita Shanker and Sixue Chen (Interdisciplinary Center for Biotechnology Research, University of Florida) for sequencing and peptide synthesis, and Mathew Brown and Ying Xio (Socolovsky

Microscopy Center, Louisiana State University) for providing access to fluorescence and electron microscopes. We thank Mike Scharf, Ramandeep Kaur, Rhitoban Raychoudhury and Ameya Gondhalaker for critical reviews of the manuscript.

References

1. Hajek AE, Tobin PC (2010) Micro-managing arthropod invasions: eradication and control of invasive arthropods with microbes. Biol Invasions 12:2895–2912.
2. Chouvenc T, Grace JK, Su NY (2011) Fifty years of attempted biological control of termites – analysis of a failure. Biol Control 59:69–82.
3. Coutinho-Abreu IV, Zhu KY, Ramalho-Ortigao M (2010) Transgenesis and paratransgenesis to control insect-borne diseases: current status and future challenges. Parasitol Int 59:1–8.
4. Hurwitz I, Fieck A, Read A, Hillesland H, Klein N, et al. (2011) Paratransgenic control of vector borne diseases. Int J Biol Sci 7:1334–1344.
5. Durvasula RV, Gumbs A, Panackal A, Kruglov O, Aksoy S, et al. (1997) Prevention of insect-borne disease: an approach using transgenic symbiotic bacteria. Proc Natl Acad Sci USA 94:3274–3278.
6. Riehle MA, Moreira CK, Lampe D, Lauzon C, Jacobs-Lorena M (2007) Using bacteria to express and display anti-Plasmodium molecules in the mosquito midgut. Int J Parasitol 37:595–603.
7. Ramirez JL, Perring TM, Miller TA (2008) Fate of a genetically modified bacterium in foregut of glassy-winged sharpshooter (Hemiptera: Cicadellidae). J Econ Entomol 101:1519–1525.
8. Ren X, Hoiczyk E, Rasgon JL (2008) Viral paratransgenesis in the malaria vector Anopheles gambiae. PLoS Pathog 4:e1000135.
9. Cirimotich CM, Ramirez JL, Dimopoulos G (2011) Native microbiota shape insect vector competence for human pathogens. Cell Host Microbe 10:307–310.
10. Fang W, Vega-Rodriguez J, Ghosh AK, Jacobs-Lorena M, Kang A, et al. (2011) Development of transgenic fungi that kill human malaria parasites in mosquitoes. Science 331:1074–1077.
11. Dandekar AM, Gouran H, Ibáñez AM, Uratsu SL, Agüero CB, et al. (2012) An engineered innate immune defense protects grapevines from Pierce disease. Proc Natl Acad Sci USA 109:3721–3725.
12. Vooght LD, Caljon G, Stijlemans B, De Baetselier P, Coosemans M, et al. (2012) Expression and extracellular release of a functional anti-trypanosome Nanobody in Sodalis glossinidius, a bacterial symbiont of the tsetse fly. Microb Cell Fact 11:23.
13. Zhao R, Han R, Qiu X, Yan X, Cao L, et al. (2008) Cloning and heterologous expression of insecticidal genes from Photorhabdus luminescens TT01 in Enterobacter cloacae for termite control. Appl Environ Microbiol 74:7219–7226.
14. Husseneder C, Collier RE (2009) Paratransgenesis for termite control. In Insect Symbiosis, vol. 3., eds Bourtzis K, Miller TA (CRC Press, Boca Raton), pp. 361–376.
15. Husseneder C, Grace JK (2005) Genetically engineered termite gut bacteria deliver and spread foreign genes in termite colonies. Appl Microbiol Biotechnol 68:360–367.
16. Husseneder C, Grace JK, Oishi DE (2005) Use of genetically engineered bacteria (Escherichia coli) to monitor ingestion, loss and transfer of bacteria in termites. Curr Microbiol 50:119–123.
17. Miller TA (2011) Paratransgenesis as a potential tool for pest control: review of applied arthropod symbiosis. J App Entomol 135:474–478.
18. Crotti E, Balloi A, Hamdi C, Sansonno L, Marzorati M, et al. (2012) Microbial symbionts: a resource for the management of insect-related problems. Microb Biotechnol 5:307–317.
19. Cleveland LR (1923) Symbiosis between termites and their intestinal protozoa. Proc Natl Acad Sci USA 9:424–428.
20. Leuschner C, Hansel W (2004) Membrane disrupting lytic peptides for cancer treatments. Curr Pharm Des 10:2299–2310.
21. Guaní-Guerra E, Santos-Mendoza T, Lugo-Reyes SO, Terán LM (2010) Antimicrobial peptides: general overview and clinical implications in human health and disease. Clin Immunol 135:1–11.
22. Bell A (2011) Antimalarial peptides: the long and the short of it. Curr Pharm Design 17: 2719–2731.
23. Mutwiri GK, Henk WG, Enright FM, Corbell LB (2000) Effect of the antimicrobial peptide, D-Hecate, on trichomonads. J Parasitol 86:1355–1359.
24. Javadpour MM, Juba MM, Lo WC, Bishop SM, Alberty JB, et al. (1996) De novo antimicrobial peptides with low mammalian cell toxicity. J Med Chem 39:3107–3113.
25. Hansel W, Enright FM, Leuschner C (2007) Destruction of breast cancers and their metastases by lytic peptide conjugates in vitro and in vivo. Mol Cell Endocrinol 260-262:183–189.
26. Yates C, Sharp S, Jones J, Topps D, Coleman M, et al. (2011) LHRH-conjugated lytic peptides directly target prostate cancer cells. Biochem Pharma 81:104–110.
27. Sethi A, Kovaleva ES, Slack JM, Brown S, Buchman GW, et al. (2013) A GHF7 cellulase from the protist symbiont community of Reticulitermes flavipes enables more efficient lignocellulose processing by host enzymes. Arch Insect Biochem Physiol 84:175–193.
28. Higashiguchi DT, Husseneder C, Grace JK, Berestecky JM (2006) Pilibacter termitis gen. nov. sp. nov., a novel lactic acid bacterium from the hindgut of the Formosan subterranean termite (Coptotermes formosanus). Int J Syst Evol Microbiol 56:15–20.
29. Rasgon JL (2011) Using infections to fight infections: paratransgenic fungi can block malaria transmission in mosquitoes. Future Microbiol 6:851–853.
30. Hares MC, Hincliffe SJ, Strong PC, Eleftherianos I, Dowling AJ, et al. (2008) The Yersinia pseudotuberculosis and Yersinia pestis toxin complex is active against cultured mammalian cells. Microbiol 154:3503–3517.
31. Lang AE, Schmidt G, Schlosser A, Hey TD, Larrianua IM, et al. (2010) Photorhabdus luminescens toxins ADP-ribosylate actin and RhoA to force actin clustering. Science 327:1139–1142.
32. Mehlert A, Zitzmann N, Richardson JM, Treumann A, Ferguson MAJ (1998) The glycosylation of the variant surface glycoproteins and procyclic acidic repetitive proteins of Trypanosoma brucei. Mol Biochem Parasitol 91:145–152.
33. Stijlemans B, Conrath K, Cortez-Retamozo V, Van Xong H, Wyns L, et al. (2004) Efficient targeting of conserved cryptic epitopes of infectious agents by single domain antibodies. J Biol Chem 279:1256–1261.
34. Sennett C, Rosenberg LE, Mellman IS (1981) Transmembrane transport of cobalamin in prokaryotic and eukaryotic cells. Annu Rev Biochem 50:1053–1086.
35. Watanabe F, Ito T, Tabuchi T, Nakano Y, Kitaoka S (1988) Isolation of pellicular cobalamin binding proteins of the cobalamin uptake system of Euglena gvacilis. J Gen Microbiol 134:67–74.
36. Braun V (1995) Energy-coupled transport and signal transduction through the gram-negative outer membrane via TonB-ExbB-ExbD-dependent receptor proteins. FEMS Microbiol Rev 16:295–307.
37. De Baetselier P, Beschin A, Lucas R, Bilej M, Stijlemans B, et al. (2002) The functional relevance of the lectin-like activities of cytokines: TNF as an illustrative example. In: Cooper EL, Beschin A, Bilej M, editors. A new model for analyzing antimicrobial peptides with biomedical applications. Amsterdam: IOS Press. pp. 157–166.
38. Engstler M, Thilo L, Weise F, Grünfelder CG, Schwarz H, et al. (2004) Kinetics of endocytosis and recycling of the GPI-anchored variant surface glycoprotein in Trypanosoma brucei. J Cell Sci 117: 1105–1115.
39. Grunfelder CG, Engstler M, Weise F, Schwarz H, Stierhof YD, et al. (2003) Endocytosis of a glycosylphosphatidylinositol-anchored protein via clathrin-coated vesicles, sorting by default in endosomes, and exocytosis via RAB11-positive carriers. Mol Biol Cell 14: 2029–2040.
40. Koumandou VL, Boehm C, Horder KA, Field MC (2013) Evidence for recycling of invariant surface transmembrane domain proteins in African trypanosomes. Eukaryot. Cell 12:330–342.
41. Manna PT, Boehm C, Leung KF, Natesan SK, Field MC (2014) Life and times: synthesis, trafficking, and evolution of VSG. Trends Parasitol. 30:251–258.
42. Pal A, Hall BS, Jeffries TR, Field MC (2003) Rab5 and Rab11 mediate transferrin and anti-variant surface glycoprotein antibody recycling in Trypanosoma brucei. Biochem J 374: 443–451.
43. Vansterkenburg EL, Coppens I, Wilting J, Bos OJ, Fischer MJ, et al. (1993) The uptake of the trypanocidal drug suramin in combination with low-density lipoproteins by Trypanosoma brucei and its possible mode of action. Acta Trop 54: 237–250.
44. Wade D, Boman A, Wåhlin B, Drain CM, Andreu D, et al. (1990) All-D amino acid-containing channel-forming antibiotic peptides. Proc Natl Acad Sci USA 87:4761–4765.
45. Bechinger B (2004). Structure and function of membrane-lytic peptides. Crit Rev Plant Sci 23:271–292.
46. Gomes-solecki MJC, Brisson DR, Dattwyler RJ (2006) Oral vaccine that breaks the transmission cycle of the Lyme disease spirochete can be delivered via bait. Vaccine 24:4440–4449.
47. Böer E, Piontek M, Kunze G (2009) Xplor 2—an optimized transformation/expression system for recombinant protein production in the yeast Arxula adeninivorans. Appl Microbiol Biotechnol 84:583–594.
48. Saran RK, Rust MK (2008) Phagostimulatory sugars enhance uptake and horizontal transfer of hexaflumuron in the western subterranean termite (Isoptera: Rhinotermitidae). J Econ Entomol 101:873–879.
49. Kreil G, Mollay C, Kaschnitz R, Haiml L, Vilas U (1980) Prepromelittin: specific cleavage of the pre- and the propeptide in vitro. Ann N Y Acad Sci 343:338–346.
50. Boman HG, Boman IA, Andreu D, Li ZQ, Merrifield RB, et al. (1989) Chemical synthesis and enzymatic processing of precursor forms of cecropin A and B. J Biol Chem 264:5852–5860.
51. Sethi A, Xue QG, La Peyre JF, Delatte J, Husseneder C (2011) Dual origin of gut proteases in Formosan subterranean termites (Coptotermes formosanus Shiraki). Comp Biochem Physiol A 159:261–267.
52. Husseneder C, Sethi A, Delatte J, Foil L (2010) Testing protozoacidal activity of ligand-lytic peptides against termite gut protozoa in vitro (protozoa culture) and in vivo (microinjection into termite hindgut). J Vis Exp 46:e2190.

Author Contributions

Conceived and designed the experiments: AS LF CH. Performed the experiments: AS JD. Analyzed the data: AS. Contributed reagents/materials/analysis tools: AS LF CH. Wrote the paper: AS LF CH.

53. Aley SB, Scott WA, Cohn ZA (1980) Plasma membrane of *Entamoeba histolytica*. J Exp Med 152:391–404.

54. Clarke BJ, Hohmn TC, Bowers B (1988) Purification of plasma membrane from *Acanthamoeba castellanii*. J Protozool 35:408–413.

55. Manzoni C, Colombo L, Bigini P, Diana V, Cagnotto A, et al. (2011) The molecular assembly of amyloid abeta controls its neurotoxicity and binding to cellular proteins. PLoS One 6: e24909.

56. Sakane T, Kuroshima K (1997) Viabilities of dried cultures of various bacteria after preservation for 20 years and their production by the accelerated storage test. Microbiol Cult Collect 13:1–7.

57. Abadias M, Benabarre A, Teixidó N, Usall J, Viñas I (2001) Effect of freeze drying and protectants on viability of the biocontrol yeast *Candida sake*. Int J Food Microbiol 65:173–182.

58. Valderrama MJ, De Siloniz MI, Gonzalo P, Peinado JMA (1999) Differential medium for the isolation of *Kluyveromyces marxianus* and *Kluyveromyces lactis* from dairy products. J Food Prot 62:189–193.

59. Sethi A, McAuslane HJ, Alborn HT, Nagata RT, Nuessly GS (2008) Romaine lettuce latex deters feeding of banded cucumber beetle: a vehicle for deployment of biochemical defenses. Entomol Exp Appl 128:410–420.

9

Ectopic Expression of the RING Domain of the Arabidopsis PEROXIN2 Protein Partially Suppresses the Phenotype of the Photomorphogenic Mutant *De-Etiolated1*

Mintu Desai[1¤], Navneet Kaur[1], Jianping Hu[1,2]*

1 Michigan State University-Department of Energy Plant Research Laboratory, Michigan State University, East Lansing, Michigan, United States of America, 2 Plant Biology Department, Michigan State University, East Lansing, Michigan, United States of America

Abstract

The Arabidopsis CONSTITUTIVE PHOTOMORPHOGENIC/DE-ETIOLATED 1/FUSCA (COP/DET1/FUS) proteins repress photo-morphogenesis by degrading positive regulators of photomorphogenesis, such as the transcription factor LONG HYPOCOTYL5 (HY5). The gain-of-function mutant *ted3*, which partially suppresses the *det1* mutant, contains a missense mutation of a Val-to-Met substitution before the C-terminal RING finger domain of the peroxisomal membrane protein PEROXIN2 (PEX2). We hypothesized that a truncated PEX2 protein, which only contains the C-terminal RING domain, is initiated by the *ted3* mutation and by-passes the function of DET1 in the nucleus. Although we have not been able to detect this hypothetic peptide *in vivo*, we show in this study that, when fused with a fluorescent protein and overexpressed, the PEX2 RING domain can localize to the nucleus, where it is able to interact with HY5, and PEX2 RING domain overexpression in *det1* also partially suppresses the *det1* phenotype. Compared with *det1*, *ted3 det1* plants have significantly decreased levels of the HY5 protein and the expression of most of the analyzed HY5 target genes is altered to levels comparable to those in *hy5*. We conclude that compromised activity of HY5 may have been mainly responsible for the partial reversal of the *det1* phenotype in *ted3 det1*. Our data support the notion that, when appropriately localized, some RING finger domains may be able to achieve neomorphic effects in the cell.

Editor: Vladimir N. Uversky, University of South Florida College of Medicine, United States of America

Funding: This work was supported by the National Science Foundation (http://www.nsf.gov), MCB 0618335 and MCB 1330441, to JH and the Chemical Sciences, Geosciences and Biosciences Division, Office of Basic Energy Sciences, Office of Science, U.S. Department of Energy (http://science.energy.gov/) (DE-FG02-91ER20021) to JH. The funders had no role in study design, data collection and analysis, decision to publish, or preparation of the manuscript.

Competing Interests: The authors have declared that no competing interests exist.

* Email: huji@msu.edu

¤ Current address: Tyton BioEnergy Systems, LLC, Danville, VA, United States of America

Introduction

In response to the changing light regime, seedlings of higher plants undergo two drastically different programs. Skotomorpho-genesis (etiolation) takes place in the dark, during which seedlings develop a long hypocotyl and hooked/undeveloped cotyledons. When exposed to light, seedlings go through photomorphogenesis (de-etiolation), where hypocotyl growth is inhibited, cotyledons open, chloroplasts develop, and genes involved in photosynthesis and light-regulated development are expressed. Light signals are transduced from photoreceptors, early signaling factors, central integrators, to downstream effectors, resulting in changed expression of hundreds of genes [1,2,3].

As central regulators of photomorphogenesis, CONSTITU-TIVE PHOTOMORPHOGENIC/DE-ETIOLATED 1/FUSCA (COP/DET1/FUS) proteins comprise three distinct protein complexes in a ubiquitin (Ub)-proteasome system. This system targets key positive regulators of light response, such as the photoreceptor phytochrome A (phyA) and transcription factors

Long Hypocotyl5 (HY5)/HY5 Homolog (HYH), Long Hypocotyl in Far-Red1 (HFR1), and Long After Far-Red Light1 (LAF1), for degradation. COP1 is a RING finger-containing E3 ligase that acts as a central component of a CULLIN4 (CUL4)-based E3 complex. Besides being a chromatin regulator to repress gene expression, DET1 is part of another CUL4-based E3 complex that functions to enhance the activity of the COP1 complex with the help of the COP9 signalosome (CSN) [4,5]. The bZIP transcription factor HY5 is a master regulator of photomorphogenesis that controls the expression of a repertoire of light-response genes [6]. In the dark, COP1 transits from cytoplasm to the nucleus, where it interacts with HY5 and mediates its ubiquitination by the concerted activity of COP/DET1/FUS protein complexes, resulting in significant reduction of HY5 abundance due to protein degradation by the 26S proteasome [7]. As such, dark-grown loss-of-function mutants of most *COP/DET1/FUS* genes show developmental patterns akin to that in light-grown wild-type seedlings (i.e. de-etiolated), whereas seedlings of photomorphogenesis-promoting factors such as HY5 often have long hypocotyls in the light [4].

Light regulates the development and function of subcellular organelles as well. In addition to its well-known impact on chloroplasts, light has also been linked to peroxisomes, essential eukaryotic organelles that mediate a variety of metabolic processes, such as photorespiration, fatty acid β–oxidation, and biosynthesis and metabolism of hormones in plants [8,9]. Light up-regulates the expression of genes encoding enzymes involved in photorespiration – a process that accompanies photosynthesis, while it represses genes involved in fatty acid β–oxidation and the glyoxylate cycle – processes that provide energy to seedling establishment before photosynthesis begins [10]. Light also promotes the proliferation of peroxisomes in Arabidopsis seedlings through phyA and the bZIP transcription factor HYH, the latter of which directly binds to the promoter and presumably activates the expression of the peroxisome proliferation factor gene *PEX11b* [11,12]. This is consistent with the idea that during photomorphogenesis, an increase in peroxisomal population takes place besides the activation of the expression of photorespiratory genes and the import of their products into the peroxisome.

Before the discovery of DET1 as part of the protein complexes that degrade positive regulators of photomorphogenesis, the *det1-1* allele was used as the background to isolate extragenic suppressors to investigate the function of the DET1 protein [13]. One partial suppressor, *ted3* (for reversal of *det*), turned out to carry a gain-of-function mutation in the peroxisome biogenesis factor *PEROXIN2* (*PEX2*) [14]. PEX2 is a conserved RING finger domain-containing peroxisomal membrane protein involved in peroxisomal protein import in diverse species. PEX2 or its RING domain possesses E3 ubiquitin ligase activity in yeast *Saccharomyces cerevisiae* [15], mammals [16], and Arabidopsis [17].

Multiple models have been proposed to explain the partial suppression of *det1* by *ted3* [10]. One model postulated that DET1 is a key positive regulator of peroxisomal functions and that *ted3* possesses enhanced peroxisomal activities to suppress *det1-1*. Some of the phenotypes in *det1-1* are similar to those in peroxisomal β-oxidation mutants, such as sugar-dependent seedling establishment and partial resistance to indole-3-butyric acid (IBA), a protoauxin that is converted to the bioactive auxin indole-3-acetic acid (IAA) by β–oxidation [14]. However, viable loss-of-function peroxisomal mutants do not have opened cotyledons like *det1* despite having shorter hypocotyls on media without sucrose, arguing that peroxisomes do not play a major role in photomorphogenic development but rather represent one of the many downstream branches in DET1's regulatory network in growth and development. In addition, DET1 represses photomorphogenesis yet light activates photorespiration and peroxisomal proliferation, suggesting that DET1 is not a primary regulator of general peroxisomal function.

A second hypothesis favored the scenario that *ted3* encodes a gain-of-function product, which bypasses the function of DET1 in photomorphogenesis. The *ted3* mutation contains a G-to-A transition that leads to a Val-to-Met substitution one amino acid upstream from the first Cys of the C-terminal RING finger domain [14]. It is conceivable that in *ted3 det1*, this new Met may initiate the translation of a cryptic peptide that comprises the RING finger domain. Alternatively, changing from Val to Met may increase the accessibility of the protein to cytoplasmic proteases, which cleave off the cytosolically exposed RING domain of PEX2. This RING domain from PEX2, which has been shown to contain E3 ubiquitin ligase activity *in vitro* [17], may be mobilized to the nucleus because of its small size (~6 kDa)

and substitute for the function of the COP1-DET1 E3 ligase complexes in degrading some of the positive regulators of photomorphogenesis. We have not been able to detect this small peptide *in vivo*. However, in this study we have provided evidence that the RING domain of PEX2 when overexpressed is able to partially rescue the *det1* phenotype. PEX2's RING domain can enter the nucleus, where it interacts with the transcription factor HY5 and presumably reduces its function. We postulate that this alteration of HY5 activity may largely account for the partial reversal of the *det1* phenotypes in the *ted3 det1* dominant mutant during photomorphogenesis.

Materials and Methods

Plant growth, light conditions and genetic crosses

The wild-type Arabidopsis plants used in this study were from the Columbia-0 (Col-0) ecotype. *hy5-1*, *cop1-4*, *det1-1* and *ted3 det1* were in the Col-0 background. These mutants were confirmed by their respective dark-grown phenotypes, and genotyped by PCR analysis to ensure their homozygosity. Seeds were surface sterilized with 20% Clorox and 0.025% Triton X-100, washed 5 times with sterile water. To measure hypocotyl length, sterilized seeds were plated on 0.5X MS medium supplemented with 0.5% sucrose and solidified with 0.6% phytagar, stratified at 4°C for 3d, exposed to white light (100 μm m^{-2}s^{-1}) for 1 h to induce synchronous germination, and returned to the darkness for 4d at 22°C. Hypocotyl lengths of >30 seedlings from each genotype were measured using ImageJ software (http://imagej.nih.gov/ij/). Three biological replicates were undertaken. After having acquired their first true leaves, the seedlings were transferred to soil and grown in growth chambers with 100 μm m^{-2}s^{-1} white light, 16/8 h photoperiod, and at 22°C.

Confocal laser scanning microscopy (CLSM) and epifluorecence microscopy

Plant tissues (as indicated in the text) were incubated with DAPI (Invitrogen, Carlsbad, CA) at 300 nM concentration in 1X PBS at room temperature, covered with aluminum foil for 15 min followed by 3–4 washes to remove excess stain, and directly mounted in distilled water to be analyzed by CLSM (Zeiss LSM 510 META). A 488-nm, 514-nm argon ion laser and 401-nm diode were used for excitation; emission filters of 505–530 nm, 520–555 nm band-pass and 433-nm long-pass were used for GFP, YFP and DAPI respectively. Images were acquired at 63X with oil. Epifluorescence microscopy was performed with an Axio Imager M1 microscope (Carl Zeiss) for visualization of the BiFC between HY5-YFPct and YFPnt-PEX2RF proteins (excitation 500±12 nm; emission 542±13.5 nm).

RT-PCR analyses

Total RNA was isolated from 4d dark-grown seedlings using SV total RNA isolation system kit (Promega, Madison, WI). For RT-PCR analysis, 2 μg total RNA was reverse-transcribed with the Omniscript RT kit (Qiagen, Valencia, CA). PEX2 RF-specific primers FW (5'-GTGACTTGCCCTATTTGC-3') and RE (5'-TCATTTGCCACTTGAAAC-3') were used to amplify a 0.1-kb product that covered the entire C-terminal end containing the RF domain from *PEX2* cDNA. *UBQ10*-FW (5'-TCAATTCTCTC-TACCGTGATCAAGATGCA-3') and *UBQ10-RE* (5'-GGTGTCAGAACTCTCCACCTCAAGAGTA-3') from the *UBQ10* gene (At4g05320) were used to amplify a product of ~320 bp that served as an internal control. For *PEX2* RF domain and *UBQ10* amplification, PCR was performed with the following

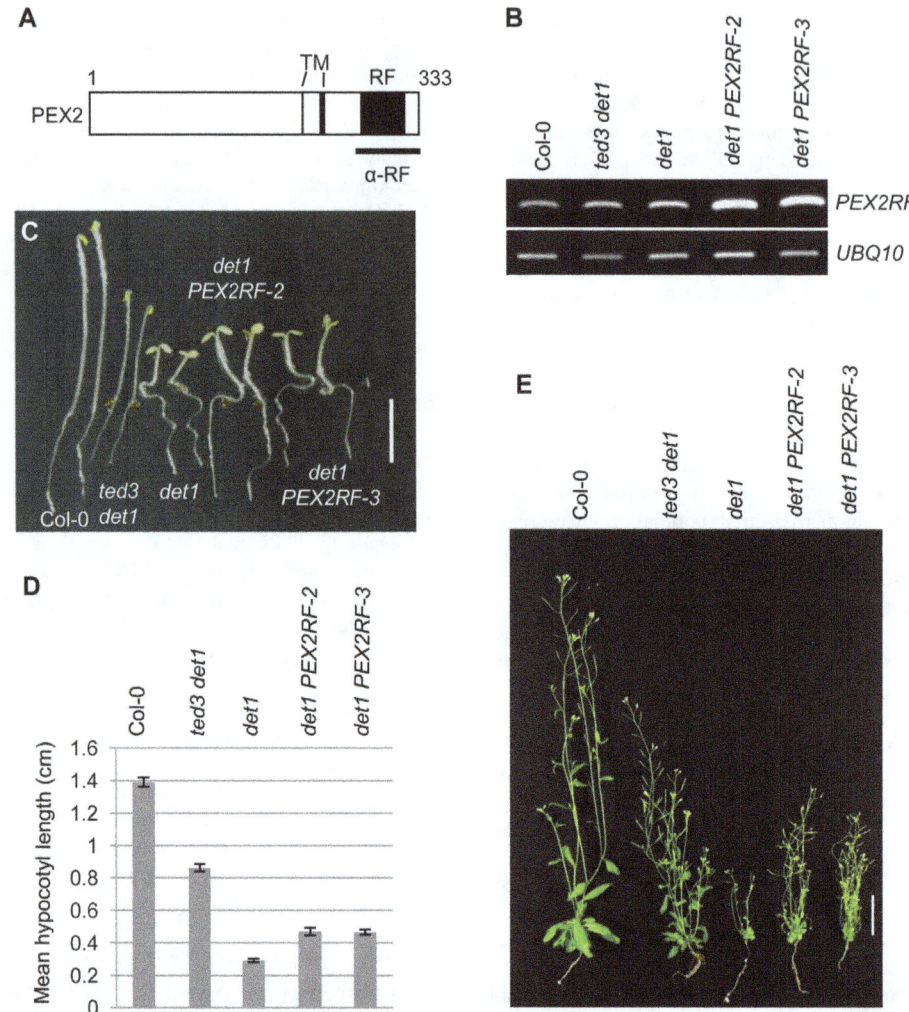

Figure 1. Overexpression of a peptide that contains the PEX2 RING domain suppresses det1. (A) Schematic of the Arabidopsis PEX2 protein, showing positions of the transmembrane (TM) and RING finger (RF) domains, and the region (indicated by horizontal bar) used as PEX2RF and as antigen for antibody generation. (B) RT-PCR analyses of the *PEX2RF* transcript in two transgenic lines (lines2 & 3) overexpressing PEX2RF in the *det1* background. *UBQ10* is the internal control. (C) Phenotype of 4d dark-grown seedlings grown on 0.5X MS supplemented with 0.5% sucrose. Scale bar = 0.5 cm. Two seedlings are shown for each genotype. (D) Hypocotyl length measurements of 4d dark-grown seedlings shown in (C). n>30 for each genotype. Student *t*-test, P<0.0001 for all lines vs. *det1*. Error bars indicate s.e.m. (E) Four-week plants. Scale bar = 3 cm.

conditions: 94°C for 2 min, 30 cycles of 94°C for 30 s, 57°C for 30 s, 72°C for 30 s, and a final extension at 72°C for 4 min.

Quantitative real-time PCR

For transcript analysis, whole Arabidopsis seedlings grown under constant light at 22°C on 0.5X MS media plates were used. Harvested seedling samples were frozen in liquid N_2 and total RNA was extracted using RNeasy plant mini kits (Qiagen, Valencia, CA) followed by treatment with DNase I (Qiagen, Valencia, CA) according to manufacturer's instructions. Synthesis of cDNA was performed with the Omniscript Reverse Transcription system (Qiagen, Valencia, CA) using random primers with 0.1 μg of total RNA in a 20 μl volume RT reaction, and incubated for 1 hr at 42°C. The RT reaction mixture was diluted 10-fold and 1 μl was used as a template in 10-μl PCR reaction, using the Applied Biosystems FAST7500 Real-Time PCR systems in fast mode and FAST SYBR GREEN PCR Master Mix (Applied Biosystems, Foster City, CA), following the manufacturer's protocol. Cycling conditions were as follows: 8 min at 95°C, 40

cycles of 10 s at 95°C, 30 s at 58°C, and 30 s at 72°C, followed by a 60 to 95°C dissociation protocol. The primers for transcript analysis were designed by the primer express software (Applied Biosystems, Foster City, CA) and are listed in Table S1. All reactions were performed in triplicate and the products were checked by melting curve analysis. Sequence of the PCR products had been confirmed. The transcript level was measured by normalizing the level with that of the *UBQ10 as* reference transcript. Each experiment was repeated at least 2 times. The values are average of three biological replicates which yielded consistent results.

Plasmid construction

For all the plasmid construction, PFU turbo (Invitrogen, Carlsbad, CA) was used. PEX2RF was amplified from pCHF3-PEX2 [14] by PCR with primers introducing a *Kpn*-I site at the 5′ of FW and *Sac*-I at the 3′ end of RE. The amplified PCR product was confirmed by sequencing and cloned into pCHF3:GFP [14] to generate pHU006 and into a pCAMBIA vector (Cambia,

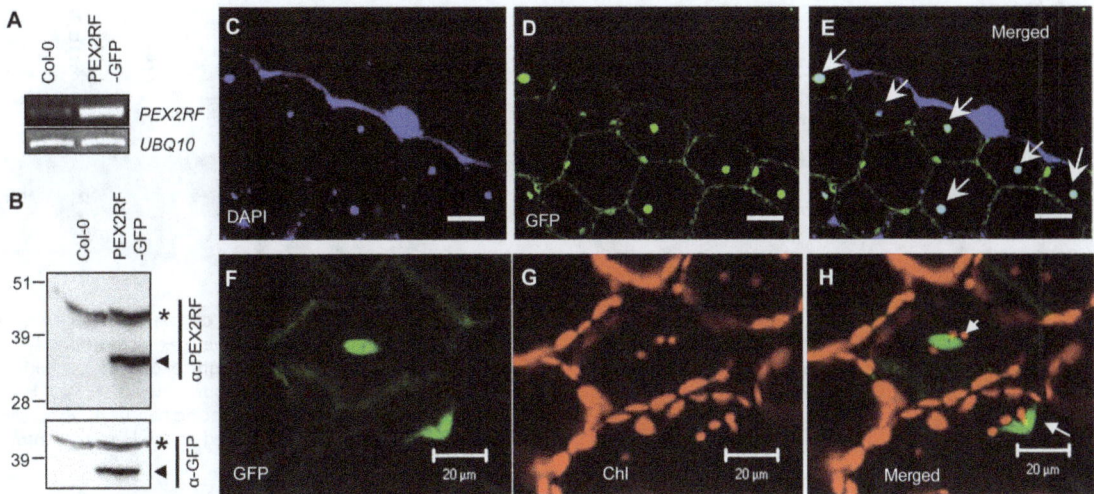

Figure 2. Nuclear localization of PEX2RF-GFP in transgenic plants. (A) RT-PCR analysis of *PEX2RF* mRNA and the *UBQ10* control in Col-0 and 35S::PEX2RF-GFP lines. (B) Immunoblot analyses of proteins from Col-0 and PEX2RF-GFP-expressing plants, using α–PEX2RF and α–GFP antibodies respectively. Asterisks indicate cross-reacting bands, and arrowheads point to the PEX2RF-GFP fusion protein. Numbers on the left indicate molecular weight markers in kDa. (C–H) Confocal images of transgenic plants from hypocotyl cells of 10d seedlings (C–E) and leaf mesophyll cells of two-week plants (F–H). DAPI stains the nucleus, green signals are from PEX2RF-GFP, and red signals are from chlorophyll autofluorescence. Arrows in the merged images indicate the nucleus. Scale bars = 10 μm in (C–E) and = 20 μm in (F–H).

Canberra) to generate pHU007. By floral dipping [18], pHU006 was transformed into Col-0 for PEX2RF-GFP the localization study, using Hygromycin for selection, and pHU007 was transformed into *det1-1* for the complementation study, using Kanamycin for selection. T2 transgenic plants were used for further analyses.

To express the PEX2RF protein for antibody generation, specific oligonucleotides were synthesized and cloned at *Nco* I and *Xho* I sites. For amplification of the PEX2 RING finger domain, 5′-CATGCCATGGGGCATGACTTGCCCTATTTGC-3′ and 5′-CCGCTCGAGTCATTTGCCACTTGAAAC-3′ were used to PCR-amplify PEX2RF with the *pfu* turbo enzyme (Stratagene,

La Jolla, CA) from Arabidopsis total cDNA from light-grown seedlings. The product was cloned into *Nco*I and *Xho*I sites of the bacterial pET28a+ expression vector (Novagen, Madison, WI) to generate pHU010. Insert was confirmed by sequencing. Recombinant PEX2RF fused to 6xHis in the pET28a+ vector was expressed in bacteria and purified with nickel nitrilotriacetic acid agarose (Qiagen, Valencia, CA) according to the manufacturer's protocol.

Constructs pHU011 and pHU012 were made by cloning the coding region of PEX2 RING finger and HY5, which had been amplified using the following PCR primer sets: 5′-GCGCAG-GAGCTCATGACGCCGTCTACGCCTGC-3′ and 5′-GAC-

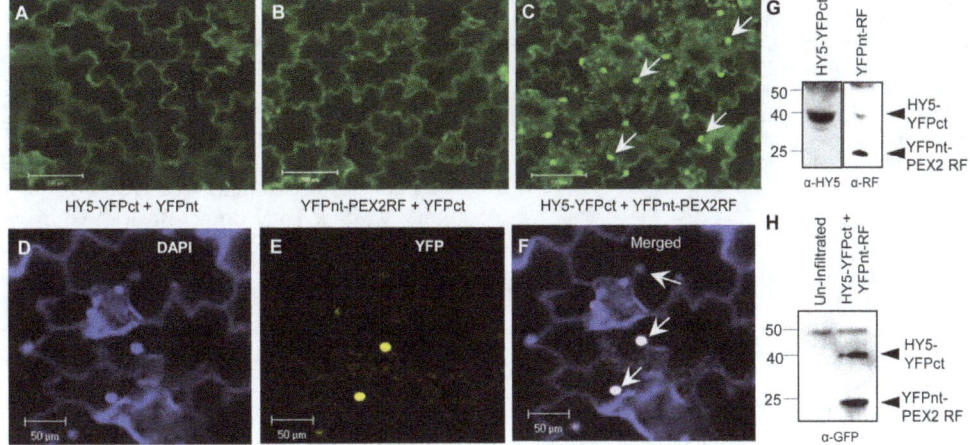

Figure 3. PEX2RF and HY5 interact in the nucleus. (A–C) Epifluorescence micrographs of tobacco leaf epidermal cells infiltrated with the indicated gene constructs. Strong YFP signals (BiFC) in the nucleus, as indicated by arrows in (C), were observed only when HY5-YFPct and YFPnt-PEX2RF were co-expressed. Scale bars = 100 μm. (D–F) Confocal micrographs of tobacco leaf epidermal cells co-infiltrated with HY5-YFPct and YFPnt-PEX2RF constructs. DAPI stains the nucleus (D), and BiFC signals are indicated by YFP fluorescence (E). Arrows in the merged image (F) indicate the overlaps of DAPI and BiFC. Scale bars = 50 μm. (G–H) Immunoblot analyses showing expression of HY5-YFPct and YFPnt-PEX2RF proteins in tobacco tissue. In (G), tissues were from plants shown in (A) and (B) respectively and α-HY5 (left) and α-RF (right) antibodies were used. In (H), tissue was from plant shown in (C), and α-GFP was used. Molecular weight markers in kDa are shown to the left of the blots.

A

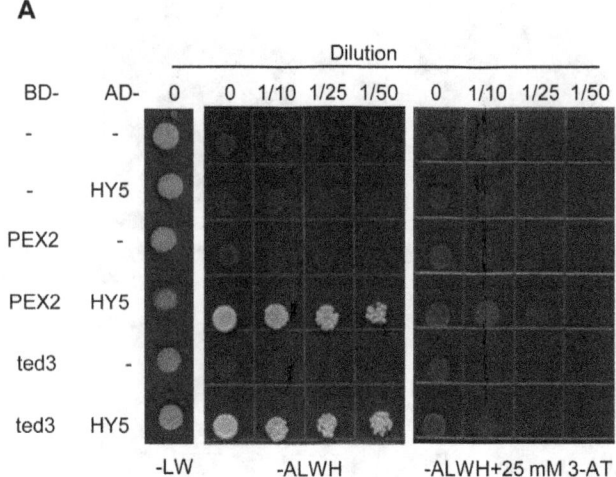

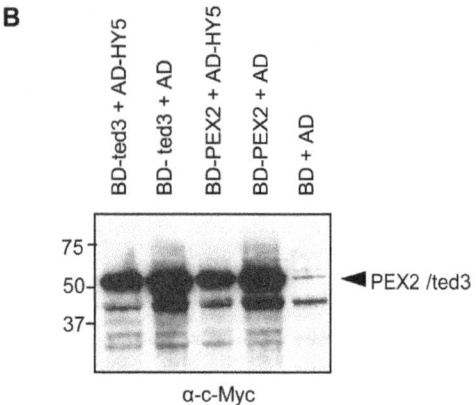

Figure 4. PEX2 and HY5 interact in yeast two-hybrid assays. (A) Yeast two-hybrid assays to show interaction between PEX2 and HY5. Yeast transformants containing the indicated GAL4 DNA binding domain (BD) and GAL4 activation domain (AD) fusion constructs were grown overnight in liquid culture and spotted on selection media plates (lacking leucine and tryptophan; –LW) and interaction media plates (lacking adenine, leucine, tryptophan and histidine; –ALWH, or –ALWH supplemented with 25 mM 3-amino-1,2,4-triazole; –ALWH+25 mM 3-AT). Growth on –ALWH and –ALWH+25 mM 3-AT media indicates protein interaction. (B) Immunoblot analysis of BD fusion constructs. Proteins extracted from transformed yeast cells shown in (A) were subjected to immunoblotting using α-c-Myc antibody. Numbers on the left of the blot indicate protein molecular weight markers in kDa.

TAGTTCATTTGCCACTTGAAACACCTTC-3' for PEX2RF with *Sac* I and *Spe* I sites (restriction sites are underlined); and 5'-GCGCAGGAGCTCATGCAGGAACAAGCGAC-TAGCTCTTTAGC-3' and 5'-CATGACCGTCGA-CAAAAGGCTTGCATCAGCATTAGAAC-3' for HY5 (At5g11260) with *Sac* I and *Sal* I sites (restriction sites underlined). Restriction enzyme-digested PCR product was cloned at the *Sal* I and *Sac* I sites of pSY735 to generate pHU011 and *Sac* I and *Spe* I sites to generate pHU012. Both these constructs were verified by sequencing and subsequently digested with *Hind* III and subcloned into binary vector pZP221 for generating BiFC constructs pHU014 and pHU015. All constructs were confirmed by sequencing.

Vectors used in this study are described in Table S2.

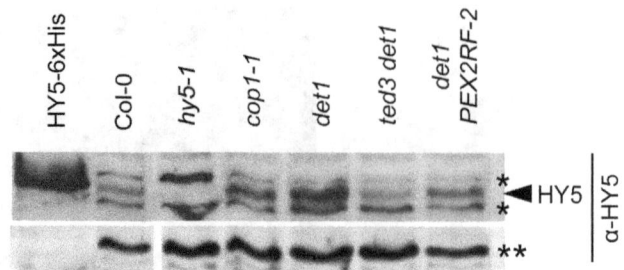

Figure 5. Immunoblot analysis of the HY5 protein in various genetic backgrounds. Proteins were extracted from 4d dark-grown seedlings exposed to 1 hr white light and detected with the α–HY5 antibody. Purified HY5–6xHis from our previous study [11] was used as a control. Asterisks indicate non-specific bands. A cross-reacting band indicated by a double asterisk served as the loading control.

Antibody production

Polyclonal antibody was raised in rabbit against the PEX2 RING finger domain (aa 275–333) that had been purified to homogeneity from *E. coli* cells expressing the PEX2RF. ImmunoPure (Protein A) IgG Purification Kit (Thermo Fisher Scientific, Rockford, IL) was used to isolate IgG from the rabbit sera according to the manufacturer's instructions. Purified IgG was desalted using Zeba desalt column using phosphate buffer (pH 7.2). A 1:500 dilution of the desalted IgG fraction was used for all subsequent immunoblot assays.

Yeast two-hybrid analysis

Full-length HY5, PEX2, ted3, and PEX2RF were restriction cloned into pGBKT7 (PEX2/ted3/PEX2RF) and pGADT7 (HY5) plasmids of the GAL4 Y2H system (Clontech, Mountain-view, CA), using the method as previously described [17]. The yeast strain Y190 was transformed with the respective constructs and transformants selected on minimal media lacking leucine and tryptophan (–LW). Interactions were assessed by growing transformants in liquid culture at 30°C and spotting serial dilutions on –LW, –ALWH and –ALWH+25 mM 3-AT media. Plates were imaged after 2d of growth at 30°C. Immunoblotting of yeast extracts was carried out as previously described [17].

Immunoblot analysis

Plant tissues (as indicated in the text) were ground to fine powder with liquid N_2 and resuspended in 200 μl of buffer (400 mM sucrose, 50 mM Tris-HCl pH 7.5, 2.5 mM EDTA, 10 mM PMSF). Total extract was cleared by centrifugation and supernatant was mixed with 5x Lamelli buffer and resolved in a PAGE (polyacrylamide gel electrophoresis). Resolved protein was then transferred to PVDF membrane and blocked with 5% milk and 0.5% Tween-20 for 2 hr at room temperature and subsequently incubated with 1:500 dilution of α-PEX2RF (Covance, Princeton, NJ), 1:200 dilution of α-HY5 (Xing Wang Deng lab), or 1:20,000 dilution of α-GFP (Abcam) overnight at 4°C. 1:20,000 goat anti-rabbit IgG (Thermo Fisher Scientific, Rockford, IL) was used as the secondary antibody. The PVDF membrane was washed four times with 1X TBST for 10 min each time before the signals were visualized with SuperSignal West Dura Extended duration substrate (Thermo Fisher Scientific, Rockford, IL).

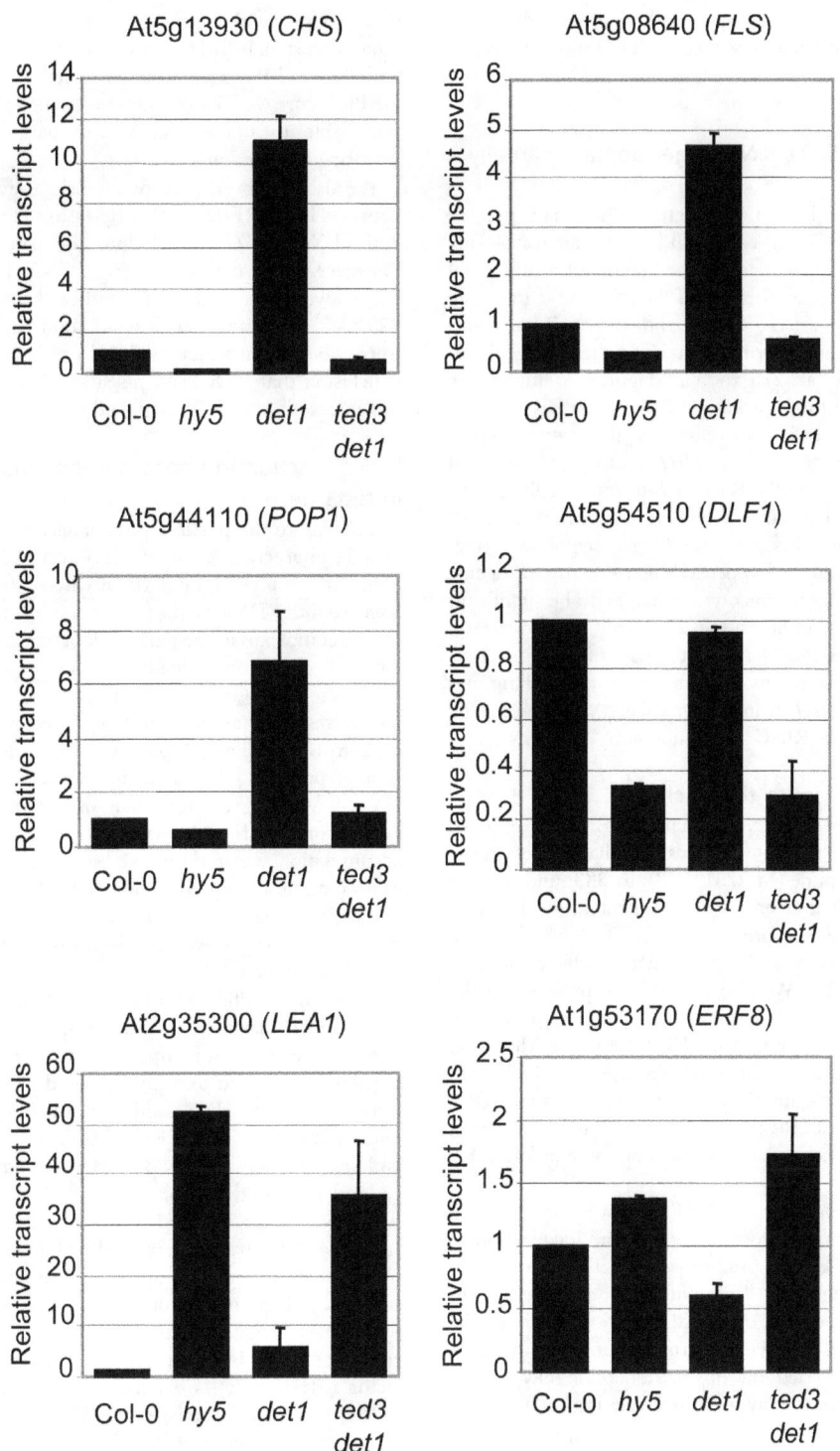

Figure 6. qRT-PCR analysis of the expression of some of HY5's target genes. RNA was extracted from 4d dark-grown seedlings in different genetic backgrounds. Three biological replicates of qRT reaction were performed for individual primer sets. The transcript level of each gene in the mutant is represented as arbitrary unit relative to the transcript level of the same gene in the wild-type plant, which was set to 1.0. The transcript level (relative expression) is the ratio between the transcript abundance of the studied gene and the transcript abundance of *UBQ10*. Values correspond to the mean and s.d. of three biological replicates. The experiments were repeated twice with consistent results.

Transient protein expression assays

Agrobacterium tumefaciens (strain GV3101) were transformed with the BiFC constructs and transformants were selected with 50 μg/ml kanamycin and 30 μg/ml gentamycin. Overnight bacterial cultures (28°C) of GV3101 containing the plasmid of interest was harvested by centrifugation, washed in water and resuspended in induction medium. Leaf infiltration was done as previously described [19]. Infiltrated plants were grown for 2 to

3 d in growth chambers before the leaf epidermal cells were examined for BiFC with epifluorescence or confocal microscopy.

Results

Overexpression of PEX2's RING finger domain partially rescues det1

To test the hypothesis that *ted3* creates a small peptide containing PEX2's RING finger domain, which can translocate to the nucleus to partially compensate for the loss of a functional DET1, we first tested whether this RING domain is able to rescue the mutant phenotypes of *det1*. To this end, the *det1-1* mutants were transformed with a construct containing the RING finger (RF) domain of PEX2 (aa 275 to 333, Figure 1A) under the control of the 35S constitutive promoter. After RT-PCR analysis, two transgenic lines showing increases in the expression of *PEX2RF* mRNA compared with the *det1* control were selected for further analysis (Figure 1B). Dark-grown *det1* seedlings had short hypocotyls and opened cotyledons, whereas transgenic seedlings overexpressing PEX2RF had longer hypocotyls (Figure 1C). Quantification of the hypocotyl lengths of the transgenic seedlings proved this longer-hypocotyl phenotype to be significant ($P<0.0001$; Figure 1D). Further, adult transgenic plants were on average two times taller than *det1-1* although smaller than *ted3 det1-1* (Figure 1E). These results suggested that the seedling and adult phenotypes of *det1* can be partially suppressed by overexpression of PEX2's RING finger domain.

PEX2 RING-GFP localizes to the nucleus

To determine whether the RING domain of PEX2 is capable of entering the nucleus, we generated a construct that expressed the PEX2RF-containing peptide (aa $275^{Val->Met}$ to 333) and fused it in-fame with a C-terminal green fluorescent protein (GFP). After generating transgenic lines expressing 35S::PEX2RF-GFP, semiquantitative RT-PCR analysis was performed to check for gene overexpression (Figure 2A). We also checked the presence of the PEX2RF-GFP protein with immunoblots, using a polyclonal antibody generated against PEX2's RING domain (see Methods). This antibody detected the presence of overexpressed PEX2RF-GFP protein in plants (Figure 2B) and the overexpressed MBP-PEX2RF protein in yeast cells (Figure S1), but it failed to detect the hypothetical endogenous small peptide that contains PEX2RF in *ted3 det1*.

Transgenic plants expressing PEX2RF-GFP were subjected to confocal laser-scanning microscopy. Besides some localization in the cytosol, PEX2RF-GFP was primarily found in the nucleus in seedling hypocotyl (Figure 2C–2E) and leaf mesophyll cells (Figure 2F–2H). The presence of PEX2 RING domain in the nucleus and PEX2RF's ability to partially suppress the *det1* phenotypes together suggested that this small peptide may be able to function in the nucleus to play a positive role in skotomorphogenesis, i.e. etiolation in the dark.

PEX2RF interacts with HY5 in the nucleus

Since HY5 is a key nuclear regulator of photomorphogenesis, we hypothesized that the nuclear localized PEX2RF may have an effect on HY5's function. For example, it may physically interact with HY5 and allosterically modify its activity or stability. To determine whether PEX2's RING finger domain and HY5 physically interact, we performed a Bimolecular Fluorescence Complementation (BiFC) assay [20] using tobacco (*Nicotiana tabacum*) plants. HY5 and PEX2RF were fused to the C- and N-terminal halves of YFP respectively to generate HY5-YFPct and YFPnt-PEX2RF. Epifluorescence and confocal microscopy anal-

yses of infiltrated tobacco leaves revealed strong YFP complementation signals (BiFC) only when both proteins were expressed, and these YFP signals were enriched in the nucleus labeled by DAPI (Figure 3). These results confirmed that HY5 and PEX2RF were able to interact in the nucleus, where HY5 normally performs its function.

We also employed yeast two-hybrid assays to test the interaction between PEX2RF and HY5 by fusing HY5 into the prey vector and PEX2/ted3/PEX2RF into the bait vector (see Methods). However, constructs containing PEX2RF autoactivated (Figure S2), so we focused on PEX2 and ted3 (i.e. PEX2 containing the $275^{Val->Met}$ substitution) instead. Both PEX2 and ted3 proteins were able to interact with HY5 (Figure 4), supporting the conclusion that PEX2 can physically interact with HY5 and that this interaction is likely mediated by the RING domain of PEX2.

HY5's function in photomorphogenesis is compromised in ted3 det1

To explore the possible physiological relevance of this protein-protein interaction between HY5 and PEX2RF, we checked the abundance of HY5 in dark-grown seedlings in various genetic backgrounds. HY5 is the target for degradation by the COP-DET1 complexes in the dark; lack of or significant reduction of the level of this protein leads to long hypocotyls in light-grown seedlings [6]. Conversely, in mutants of the COP-DET1 complexes such as *det1* and *cop1*, HY5 is stabilized and thus dark-grown seedlings display a de-etiolated phenotype by having short hypocotyls [7]. Similar to what had been shown previously, HY5 showed higher accumulation in *cop1-1* and *det1-1* mutants when compared with wild-type Col-0, whereas this higher accumulation was reduced in *ted3 det1-1* and to lesser degrees, in *det1* mutant overexpressing PEX2RF (Figure 5). These results led us to speculate that PEX2's RING finger domain in the nucleus may be involved in inactivation and/or turnover of HY5 directly or indirectly.

Given the significant reduction of the level of HY5 in *ted3 det1*, we reasoned that the downstream events regulated by HY5 may also be reversed in this suppressor to revert *det1*'s phenotype. To test this, we selected six light-regulated genes that are known to be direct targets of HY5, and performed expression profiling by quantitative real-time PCR of these genes in Col-0, *hy5-1*, *det1-1*, and *ted3 det1-1*. Those positively regulated by HY5 included genes that encode chalcone synthase (CHS) and flavonol synthase (FLS), which are involved in anthocyanin/flavonoid biosynthesis [21], the ABC transporter POP1 (P-loop containing nucleoside triphosphate hydrolases superfamily protein) [22], and the auxin signal transduction component Dwarf in Light1 (*DFL1*) [23]. The two genes negatively regulated by HY5 encoded the late embryogenesis abundant protein LEA1 and ethylene response factor ERF8 [22]. As expected, transcript levels of *CHS, FLS, POP1, and DLF1* decreased in *hy5-1* but increased in *det1.*, For the genes negatively regulated by HY5, ERF8 was up-regulated in *hy5-1* and down-regulated in *det1*, whereas *LEA1* was up-regulated in both *hy5-1* and *det1* (Figure 6). In *ted3 det1*, the altered expression pattern shown in *det1* was reversed for five of the six genes to levels similar to those in the *hy5* mutant (Figure 6), supporting the notion that HY5 activity in *ted3 det1* is compromised, which may be a major cause for the partial reversal of the *det1* phenotype.

Discussion

COP/DET1/FUS are global repressors of light-regulated development, functioning in the proteolysis of positive regulators of photomorphogenesis in the nucleus [4,5]. The identification of a dominant peroxisomal mutation that suppressed the de-etiolated phenotype of *det1* was intriguing [14]. One plausible explanation was that the Met created at position 275 in *ted3* initiated the translation of a small peptide that contains the C-terminal RING finger domain of PEX2. Alternatively, this Val-to-Met change renders the protein more susceptible to proteases, which cleave off the RING domain of PEX2 in the cytosol. The RING domain-containing peptide then translocates to the nucleus to substitute the function of DET1 in photomorphogenesis. We have not been able to unequivocally prove the above hypothesis in this study, as we could not detect the hypothetical small peptide derived from the Val-to-Met substitution in *ted3 det1* or *ted3* overexpressors. This is possibly due to insufficient avidity of the PEX2 antibody we generated and/or the low abundance/instability of this peptide. However, we have shown in this study that overexpression of a small peptide containing PEX2 RING domain in *det1* can indeed partially suppress the *det1* phenotype. Majority of the PEX2RF-GFP protein was seen in the nucleus, and only a small portion of the fusion protein was visible in the cytoplasm. We speculate that PEX2RF passively enters the nucleus due to its small size (~6 kDa), although we do not rule out the possibility that it goes to the nucleus through active targeting or other mechanisms.

The RF domains of PEX2 and COP1 both belong to the C_3HC_4 type. When overexpressed in wild-type Arabidopsis plants, an N-terminal fragment of COP1 that contained both the RING finger and coiled-coil domains was found in the nucleus and conferred a dominant negative effect that mimicked the phenotype of *cop1*. The phenotype was believed to be caused by the interaction between this peptide and the endogenous COP1, which resulted in the interference with COP1's normal function [24]. In our study, PEX2 RF alone was overexpressed in the mutant *det1* background and conferred phenotype opposite to that of the COP1 study. Our BiFC and yeast two-hybrid assays demonstrated the interaction between PEX2 and HY5. In addition, the accumulation of the HY5 protein in *det1* was reduced in *det1 PEX2RF* and *ted3 det1*. Furthermore, the altered expression of five out of the six analyzed HY5 target genes in *det1* was reversed in *ted3 det1*, prompting us to speculate that this reduced activity of HY5 was at least in part responsible for the suppression of *det1*. This is also consistent with a previous report, which showed that HY5 inactivation in *cop1* and *det1* mutants resulted in reversal of their dark-grown phenotypes [25]. How does ted3 reduce the level of HY5? Given that PEX2RF contains E3 ubiquitin ligase activity [17], ted3 may be directly involved in the degradation of HY5. *ted3* also partially suppressed *cop1* but not *det2*, a de-etiolated mutant deficient in an enzyme in brassinosteroid biosynthesis [14]. Therefore, *ted3* seems to have some specificity toward the COP/DET1-associated photomorphogenic pathway, which makes sense given that HY5 is a major target of the COP/DET1 proteolytic complexes. Finally, since HY5 is not the only target of DET1's function, reducing HY5 activity may not be sufficient to completely rescue the *det1* mutant phenotypes.

Although we have not been able to prove this hypothesis, we predict that the partial suppression of *det1* by *ted3* is primarily due to the creation of a RING finger-containing peptide that replaces the function of DET1 in the nucleus, and not due to changes in peroxisomal function. Replacement of a Val, which is nonpolar, by the partially charged Met was shown to affect the function of proteins related in human diseases [26,27,28]. Similarly, substitution of Val by Met may distort the overall configuration of the cytoplasmic end of the PEX2 protein thus affecting the activity of the RING finger domain, resulting in a PEX2 protein with mildly reduced activity in peroxisome biogenesis.

RING-type E3 ligases mediate ubiquitination and are implicated in diverse developmental processes across kingdoms [29]. Our work supports the possibility that a gain-of-function mutation in a peroxisomal gene can have a marked effect on the function of a nuclear protein. Given the conservation of the RING finger domain among proteins in various genomes, it is interesting to speculate that some other RING domains when appropriately localized may also cause neomorphic phenotypes.

Supporting Information

Figure S1 Specificity of the PEX2RF antibody. (A) SDS-PAGE gel showing induction of the expression of the fusion of maltose binding protein (MBP) and PEX2RF in bacterial protein lysates. U, Is and Ip stand for uninduced, soluble and pellet fractions, respectively. Protein expression constructs have been previously described in Kaur *et al.*, 2013 [17]. Arrow and arrowhead point to MBP alone and MBP-PEX2RF respectively. (B) Immunoblot analysis of MBP-RF expression in induced bacterial protein lysates, as detected by the PEX2RF antibody. Numbers on the left of the blots indicate molecular weight markers in kDa.

Figure S2 PEX2RF auto-activates in yeast two-hybrid assays. (A) Yeast cells transformed with BD and AD constructs were spotted on selection (−LW) and interaction media (−ALWH and −ALWH+25 mM 3-AT). Strains containing BD-PEX2RF grow on interaction media even in the absence of HY5, indicating that the RF autoactivates. (B) Immunoblot analysis to detect the expression of BD-PEX2RF fusion proteins in yeast cells, using anti-c-Myc and anti-PEX2RF antibodies respectively.

Acknowledgments

We would like to express our thanks to Dr. Vandana Yadav for help with the qRT-PCR analysis and comments on the manuscript, Jilian Fan for assistance with the construction of pHU006, the Arabidopsis Biological Resources Center for providing all the mutant seeds, and Dr. Xing-Wang Deng for sharing the HY5 antibody.

Author Contributions

Conceived and designed the experiments: MD NK JH. Performed the experiments: MD NK. Analyzed the data: MD NK JH. Contributed reagents/materials/analysis tools: MD NK JH. Contributed to the writing of the manuscript: MD NK JH.

References

1. Kami C, Lorrain S, Hornitschek P, Fankhauser C (2010) Light-regulated plant growth and development. Curr Top Dev Biol 91: 29–66.
2. Chen M, Chory J, Fankhauser C (2004) Light signal transduction in higher plants. Annu Rev Genet 38: 87–117.
3. Jiao Y, Lau OS, Deng XW (2007) Light-regulated transcriptional networks in higher plants. Nat Rev Genet 8: 217–230.
4. Lau OS, Deng XW (2012) The photomorphogenic repressors COP1 and DET1: 20 years later. Trends Plant Sci 17: 584–593.

5. Nezames CD, Deng XW (2012) The COP9 signalosome: its regulation of cullin-based E3 ubiquitin ligases and role in photomorphogenesis. Plant Physiol 160: 38–46.

6. Oyama T, Shimura Y, Okada K (1997) The Arabidopsis HY5 gene encodes a bZIP protein that regulates stimulus-induced development of root and hypocotyl. Genes Dev 11: 2983–2995.

7. Osterlund MT, Hardtke CS, Wei N, Deng XW (2000) Targeted destabilization of HY5 during light-regulated development of Arabidopsis. Nature 405: 462–466.

8. Hu J, Baker A, Bartel B, Linka N, Mullen RT, et al. (2012) Plant peroxisomes: biogenesis and function. Plant Cell 24: 2279–2303.

9. Beevers H (1979) Microbodies in higher plants. Ann Rev Plant Physiol 30: 159–193.

10. Kaur N, Li J, Hu J (2013) Peroxisomes and photomorphogenesis. Subcell Biochem 69: 195–211.

11. Desai M, Hu J (2008) Light induces peroxisome proliferation in Arabidopsis seedlings through the photoreceptor phytochrome A, the transcription factor HY5 HOMOLOG, and the peroxisomal protein PEROXIN11b. Plant Physiol 146: 1117–1127.

12. Hu J, Desai M (2008) Light control of peroxisome proliferation during Arabidopsis photomorphogenesis. Plant Signal Behav 3: 801–803.

13. Pepper AE, Chory J (1997) Extragenic suppressors of the Arabidopsis det1 mutant identify elements of flowering-time and light-response regulatory pathways. Genetics 145: 1125–1137.

14. Hu J, Aguirre M, Peto C, Alonso J, Ecker J, et al. (2002) A role for peroxisomes in photomorphogenesis and development of Arabidopsis. Science 297: 405–409.

15. Platta HW, El Magraoui F, Baumer BE, Schlee D, Girzalsky W, et al. (2009) Pex2 and Pex12 function as protein-ubiquitin ligases in peroxisomal protein import. Mol Cell Biol 29: 5505–5516.

16. Okumoto K, Noda H, Fujiki Y (2014) Distinct modes of ubiquitination of peroxisome-targeting signal type 1 (PTS1)-receptor Pex5p regulate PTS1 protein import. J Biol Chem 289: 14089–14108.

17. Kaur N, Zhao Q, Xie Q, Hu J (2013) Arabidopsis RING Peroxins are E3 Ubiquitin Ligases that Interact with Two Homologous Ubiquitin Receptor Proteins(F). J Integr Plant Biol 55: 108–120.

18. Clough SJ, Bent AF (1998) Floral dip: a simplified method for Agrobacterium-mediated transformation of Arabidopsis thaliana. Plant J 16: 735–743.

19. Sparkes IA, Runions J, Kearns A, Hawes C (2006) Rapid, transient expression of fluorescent fusion proteins in tobacco plants and generation of stably transformed plants. Nat Protoc 1: 2019–2025.

20. Bracha-Drori K, Shichrur K, Katz A, Oliva M, Angelovici R, et al. (2004) Detection of protein-protein interactions in plants using bimolecular fluorescence complementation. Plant J 40: 419–427.

21. Song YH, Yoo CM, Hong AP, Kim SH, Jeong HJ, et al. (2008) DNA-binding study identifies C-box and hybrid C/G-box or C/A-box motifs as high-affinity binding sites for STF1 and LONG HYPOCOTYL5 proteins. Plant Physiol 146: 1862–1877.

22. Lee J, He K, Stolc V, Lee H, Figueroa P, et al. (2007) Analysis of transcription factor HY5 genomic binding sites revealed its hierarchical role in light regulation of development. Plant Cell 19: 731–749.

23. Nakazawa M, Yabe N, Ichikawa T, Yamamoto YY, Yoshizumi T, et al. (2001) DFL1, an auxin-responsive GH3 gene homologue, negatively regulates shoot cell elongation and lateral root formation, and positively regulates the light response of hypocotyl length. Plant J 25: 213–221.

24. McNellis TW, Torii KU, Deng XW (1996) Expression of an N-terminal fragment of COP1 confers a dominant-negative effect on light-regulated seedling development in Arabidopsis. Plant Cell 8: 1491–1503.

25. Ang LH, Deng XW (1994) Regulatory hierarchy of photomorphogenic loci: allele-specific and light-dependent interaction between the HY5 and COP1 loci. Plant Cell 6: 613–628.

26. Kazemi-Esfarjani P, Beitel LK, Trifiro M, Kaufman M, Rennie P, et al. (1993) Substitution of valine-865 by methionine or leucine in the human androgen receptor causes complete or partial androgen insensitivity, respectively with distinct androgen receptor phenotypes. Mol Endocrinol 7: 37–46.

27. Murray EW, Giles AR, Lillicrap D (1992) Germ-line mosaicism for a valine-to-methionine substitution at residue 553 in the glycoprotein Ib-binding domain of von Willebrand factor, causing type IIB von Willebrand disease. Am J Hum Genet 50: 199–207.

28. Orth U, Fairweather N, Exler MC, Schwinger E, Gal A (1994) X-linked dominant Charcot-Marie-Tooth neuropathy: valine-38-methionine substitution of connexin32. Hum Mol Genet 3: 1699–1700.

29. Metzger MB, Pruneda JN, Klevit RE, Weissman AM (2014) RING-type E3 ligases: master manipulators of E2 ubiquitin-conjugating enzymes and ubiquitination. Biochim Biophys Acta 1843: 47–60.

10

The Meganuclease I-SceI Containing Nuclear Localization Signal (NLS-I-SceI) Efficiently Mediated Mammalian Germline Transgenesis via Embryo Cytoplasmic Microinjection

Yong Wang[1]*◐, **Xiao-Yang Zhou**[1]◐, **Peng-Ying Xiang**[1], **Lu-Lu Wang**[1], **Huan Tang**[2], **Fei Xie**[1], **Liang Li**[1], **Hong Wei**[1]*

1 Department of Laboratory Animal Science, College of Basic Medical Sciences, Third Military Medical University, Chongqing, China, **2** China Three Gorges Museum, Chongqing, China

Abstract

The meganuclease I-SceI has been effectively used to facilitate transgenesis in fish eggs for nearly a decade. I-SceI-mediated transgenesis is simply via embryo cytoplasmic microinjection and only involves plasmid vectors containing I-SceI recognition sequences, therefore regarding the transgenesis process and application of resulted transgenic organisms, I-SceI-mediated transgenesis is of minimal bio-safety concerns. However, currently no transgenic mammals derived from I-SceI-mediated transgenesis have been reported. In this work, we found that the native I-SceI molecule was not capable of facilitating transgenesis in mammalian embryos via cytoplasmic microinjection as it did in fish eggs. In contrast, the I-SceI molecule containing mammalian nuclear localization signal (NLS-I-SceI) was shown to be capable of transferring DNA fragments from cytoplasm into nuclear in porcine embryos, and cytoplasmic microinjection with NLS-I-SceI mRNA and circular I-SceI recognition sequence-containing transgene plasmids resulted in transgene expression in both mouse and porcine embryos. Besides, transfer of the cytoplasmically microinjected mouse and porcine embryos into synchronized recipient females both efficiently resulted in transgenic founders with germline transmission competence. These results provided a novel method to facilitate mammalian transgenesis using I-SceI, and using the NLS-I-SceI molecule, a simple, efficient and species-neutral transgenesis technology based on embryo cytoplasmic microinjection with minimal bio-safety concerns can be established for mammalian species. As far as we know, this is the first report for transgenic mammals derived from I-SceI-mediated transgenesis via embryo cytoplasmic microinjection.

Editor: Atsushi Asakura, University of Minnesota Medical School, United States of America

Funding: YW was supported by grants from Natural Science Fund of China [31171280, 31271330], National 973 Project of China [2011CB944102], Chongqing Natural Science Fund [cstc2011jjA10049]. HW was supported by grants from Natural Science Fund of China [81173126], National 973 Project of China [2011CBA01006] and National Science and Technology Support Program [2011BAI15BO2]. The funders had no role in study design, data collection and analysis, decision to publish, or preparation of the manuscript.

Competing Interests: The authors have declared that no competing interests exist.

* Email: yongw7528@gmail.com (YW); weihong63528@163.com (HW)

◐ These authors contributed equally to this work.

Introduction

Genetic modification of mammalian genomes is of great importance for bio-medical researches such as deciphering gene functions, investigating disease mechanisms and searching and validating therapeutic targets, and also a potential method to generate farm animals with improved economic traits for agricultural purposes.

Mammalian genetic modification includes transgenesis, gene disruption and random mutation of genomes. Gene disruption was once a sophisticated and labor-intensive process which was based on DNA homologous recombination (HR) in embryonic stem cells (ESCs). However, this DNA HR-based technology achieved very limited success in mammalian species other than mice due to the lack of ESCs derived from these species. Recently, with the development of powerful site-specific engineered endonucleases(EENs), especially Zinc Finger Nucleases(ZFNs) [1–4], Transcription Activator-like Effector Nucleases (TALENs)[5–10] and Clustered Regularly Interspaced Short Palindromic Repeats/CRISPR-associated system 9 (CRISPR/Cas9) [11–15], which are capable of disrupting genes efficiently by making double strand breaks (DSBs) at target sites, gene disruption has become a much more efficient and convenient process which is independent on ESCs and achieved significant success in mammalian species other than mice. Random mutation of mammalian genomes is regularly efficient using powerful chemical mutagens such as ENU or insertional viral vectors. In contrast, mammalian transgenesis, especially for species other than mice, remains to be further optimized.

Transgenesis is a process of adding exogenous and (or) artificially constructed genes to animal genomes, which is indispensable for generating mammalian models with gain of functions for bio-medical researches or genetically modified farm animals with additional economic traits. Currently, the available technologies for mammalian transgenesis include embryo pronuclear microinjection, somatic cell nuclear transfer (SCNT) using transgenic cells as nuclear donors, sperm-mediated gene transfer (SMGT), lentiviral transgenesis using retro-viral vectors derived from lentiviruses as vehicles to deliver transgenes into animal genomes and transposon-mediated gene transfer. Embryo pronuclear microinjection is a reliable and traditional method to produce transgenic mammals, but the inaccessibility to pronuclear of many mammalian species other than mice and the low efficiency of transgene integration largely limits its effectiveness and utility [16,17]. SCNT is a reproducible method to produce transgenic mammals, but SCNT is a sophisticated and complex procedure with a rather low efficiency [18,19] and a large number of oocytes are needed. Practically, many mammalian species of biological or biomedical importance, such as non-human primates or other none-economic animals, are not able to be cloned due to the lack of regular ovary sources. Besides, the unpredictable abnormalities related to cloned individuals limit the usage of resulted transgenic animals to model human diseases, and the antibiotic resistant genes, the necessary selection markers for transgenic nuclei donor cell culture which are finally added into the genomes of resulted transgenic individuals by SCNT process, brings additional uncertainties for the application of derived transgenic animals. SMGT is reported to be a simple and inexpensive method for transgenic animal production, however extremely variant data has been reported from different labs and the highly unstable outcome of this technology limits its application. Lentiviral transgenesis has been recognized as an extremely efficient method to generate transgenic animals of different species [20–22]. However, the preparation of high titre lentiviral particle suspensions is a complicated procedure and the viral vectors integrated into animal genomes are of bio-safety concerns. Transposon systems have been used for animal transgenesis [23–27], but transposons are mobile genetic elements, and the derived transgenic animals are of similar bio-safety concerns as those derived from lentiviral transgenesis.

On the basis of these points mentioned above, it is valuable to develop an efficient, simple, and species-neutral transgenesis technology for mammals, which is of minimal bio-safety concerns being without the involvement of viral or mobile vectors and independent on SCNT process or the accessibility to embryo pronuclear. The meganuclease I-SceI, which is derived from the mitochondria of *Saccharomyces cerevisiae* and has a long (>18 bp) recognition sequence that does not exist in animal genomes naturally, has been effectively used to facilitate trangenesis in fishes via embryo cytoplasmic microinjection [28–32]. However, in this study we found that the native I-SceI molecule failed to efficiently facilitate transgenesis in mammalian embryos as it did in fish eggs after cytoplasmic microinjection along with the plasmids of transgene vector containing two inversely flanking I-SceI recognition sequences, suggesting that in mammalian embryos, the native I-SceI molecule did not exhibit the efficacy on transgenesis in the same way as that in fish eggs. By adding a mammalian nuclear localization (NLS) signal to the N-terminus of I-SceI molecule, the I-SceI molecule containing NLS (NLS-I-SceI) was found to be capable of translocating DNA fragments from mammalian embryo cytoplasm into nuclear, and the I-SceI recognition sequence-containing transgene vector plasmids, which was injected into cytoplasm along with NLS-I-SceI mRNA,

exhibited expression in both mouse and porcine embryos. By transferring the embryos cytoplasmically co-injected with NLS-I-SceI mRNA and the transgene plasmids into synchronized female recipients, transgenic founder animals were efficiently generated and transgenes were found to be capable of germline transmission. These data suggested that using the NLS-I-SceI molecule, a simple, efficient and species-neutral transgenesis technology, which was based on embryo cytoplasmic microinjection and without the involvement of viral or transposon vectors, can be established for mammals.

Materials and Methods

Animals

Mice of FVBN inbred strain and Bama minipigs, which are of one local minipig strain in China, were used in this study. The mice were purchased from SLAC Laboratory Animal Co., Ltd (Shanghai, China) and maintained under specific pathogen-free conditions in Laboratory Animal Centre of our university. The minipigs used in this study were derived from the closed colony regularly maintained in Laboratory Animal Centre. All the protocols involving the use of animals were approved by the Institutional Animal Care and Use Committee of Third Military Medical University (Approval ID: SYXK-PLA-2007036).

Construction of NLS-I-SceI molecule and transgene vector

The NLS-I-SceI molecule was constructed by adding a modified version of 3×SV40 NLS sequence containing a HA epitope to the N-terminal of the native I-SceI molecule. The coding sequence for NLS-I-SceI, of which the initiation codon was surrounded by a kozak sequence for optimal translation initiation and the codons were optimized for both pigs and mice, was artificially synthesized and subcloned into the mammalian expression vector PCI (Promega) downstream T7 promoter, and the resulted vector was designated as PCI-T7-NLS-I-SceI in this article. For the convenience of transgene vector construction, an intermediate vector designated as p2IS was constructed by subcloning a synthesized DNA fragment containing a long multi cloning sites (MCS) inversely flanked by two I-SceI recognition sequences into pUC18 vector at the two restriction sites BsmBI and SapI to substitute the original MCS region. To construct the transgene vector used in this study, a DNA fragment containing human Ubiquitine C (UBC) promoter, eGFP CDS and a poly (A) signal sequence was cut off from FUGW plasmid (Addgene, #14883) using the two endonucleases PacI and PmeI and then subcloned into p2IS vector at the same two restriction sites, and the resulted transgene vector was designated as p2IS-UBC-eGFP.

Preparation of mRNA

NLS-I-SceI mRNA was prepared by *in vitro* transcription using linearized PCI-T7-NLS-I-SceI plasmid as templates. The plasmid was linearized by restrictive digestion at the ClaI site which was located downstream NLS-I-SceI CDS. After complete digestion, the reaction system were treated with proteinase K (100 µg/mL) and SDS (0.5% (v/v)), and then further treated with one equal volume of phenol:chloroform mixture. After centrifuge at 12000 g, 4°C for 10 min, the supernatant was carefully collected and the DNA was precipitated by adding 2.5 volumes of ice-cold absolute alcohol and one tenth volume of RNase-free 5 M NaAc solution. After washing in 75% alcohol, the DNA precipitate was finally dissolved into RNase-free deionized water after drying. Using the purified linearized plasmids as templates, NLS-I-SceI mRNA was produced by *in vitro* transcription using the mMESSAGE mMACHINE@T7 Ultra Kit (Life Technologies, AM1345) as described in the manual. After transcription was terminated, 1 μL of transcription products was saved prior to poly(A) tailing as a control to assess the tailing quality after poly(A) tailing procedure was completed. To prepare purified mRNA for embryo microinjection, the poly(A)-tailed mRNA products were recovered from reaction system using RNeasy Mini Kit (Qiagen, 74104) and eluted with RNase-free deionized water. The quality of mRNA samples was assessed by agarose gel electrophoresis.

Embryo microinjection, observation and transfer

The circular or linearized transgene vector plasmids p2IS-UBC-eGFP used for embryo microinjection were treated and purified in the same way as that for *in vitro* transcription templates. For microinjection, the purified p2IS-UBC-eGFP plasmids were mixed with different concentrations of NLS-I-SceI mRNA or included in the digestive reaction system of I-SceI endonuclease (NEB) as the substrate as previously described for fish transgenesis [31]. The I-SceI nuclease was stored at −80°C in 2 μL aliquots and added into the reaction system prior to microinjection as described [31], and its activity was confirmed by digestion of the plasmid p2IS-UBC-eGFP. To observe the localization of the injected DNA, two completely complementary 130 bp-long Cy3-labeled single strand DNA fragments containing two inversely flanking I-SceI recognition sequences at both ends were synthesized, denatured and annealed to be double-stranded, and then used for embryo cytoplasmic microinjection with NLS-I-SceI mRNA in the same way as transgene vector plasmids.

Microinjection was performed as described [33], except that the materials were injected into cytoplasm instead of pronuclear in this study. The mouse or porcine embryos subjected to microinjection were collected from mated female individuals and cultured as described [33,21]. The porcine oocytes were collected from ovaries and subjected to *in vitro* maturation (IVM) as described [34]. The matured oocytes at metaphase of meiosis II (MII phase) with extruded first polar body were selected and subjected to microinjection post parthenogenetic activation by direct current electrical pulses (1.2 KV/cm, 30 μs, two times, 1 sec interval) as described [34]. The parthenogentically activated porcine oocytes (parthenogenetic embryos) were cultured as that for the collected porcine embryos.

The cultured embryos were observed under fluorescence microscopy or laser scanning confocal microscopy (LSCM, Zeiss LSM 780) to examine transgene expression or the localization of injected Cy3-labeled DNA fragments. To stain chromosomal DNAs, embryos were incubated in culture media containing 15 μg/mL Hoechst 33342(Sigma) for 30 min prior to microinjection and washed thoroughly in fresh media. To obtain transgenic founders, injected embryos were surgically transferred into oviducts of synchronized recipient female mice or sows as described [33,21].

Analysis of the presence of uncut I-Scel recognition site in embryos by polymerase chain reaction (PCR)

Total DNA samples were extracted from individual embryos by incubating each embryo in 10 μL of lysis buffer (KCl: 50 mM; MgCl2:1.5 mM; Tris-Cl (pH8.0): 10 mM; Nonidet P-40:0.5% (w/v); Tween-20:0.5% (v/v); proteinase K: 100 μg/mL) at 65°C for 1 h. After heated at 95°C for 10 min to inactivate proteinase K, the lysate was used as template for PCR. A set of primer pair IS-site-F1/R1 (IS-site-F1:5′-CCACTGACCTTTGGATGGTG-3′; IS-site-R1:5′-TACCGCCTTTGAGTGAGCTG-3′; product size: 518 bp), of which the PCR product covered the I-SceI recognition sequence 3′ to the transgene cassette, was designed to detect the

presence of uncut I-SceI site. Another primer pair set eGFP-F1/R1 (eGFP-F1:5′-ACTGGAGAACTCGGTTTGTCGT-3′; eGFP-R1:5′-ACGGCCAGAATTTAGCGGAC -3′; product size: 453 bp) was used to detect the presence of eGFP CDS. The total DNA samples were further subjected to quantitative PCR (qPCR) analysis in a system based on SybrGreen qPCR Master Mix(2×) (ABI). The primer pair set for qPCR analysis of uncut I-SceI sites was IS-site-F2/R2 (IS-site-F2:5′-AACTAGGGAACC-CACTGCTT-3′; IS-site-R2:5′-AACTAGGGAACC-CACTGCTT-3′; product size: 171 bp), and that for qPCR of the eGFP CDS (the internal control) was eGFP-F2/R2 (eGFP-F2:5′-CAGAAGAACGGCATCAAGGT-3′; eGFP-R2:5′-TCTCGTTGGGGTCTTTGCT-3′; product size: 172 bp). Using the p2IS-UBC-eGFP plasmids diluted to different concentrations as standard samples, the qPCR analysis was performed in an absolute quantitation manner.

Transgenic animal screen

Transgenic animals were screened by PCR and Southern blot assay. The primer pair set used for transgenic mouse screen by PCR was eGFP-F3/R3, of which the sequences were 5′-ATGGTGAGCAAGGGCGAGGA-3′ (eGFP-F3) and 5′-TGCCGTCCTCGATGTTGTGG-3′ (eGFP-R3), and the product size was 526 bp. The primer pair used for transgenic pig screen was eGFP-F1/R1 as described above. The probe for Southern blot assay was prepared by PCR using PCR DIG Probe Synthesis Kit (Roche) as described in the kit manual. The primer pair set used for probe preparation was Probe-DIG-F/R, of which the sequences were 5′-GCAGAAGAACGGCATCAAGGT-3′ (Probe-DIG-F) and 5′-TAGGGAGGGGGAAAGCGAA-3′ (Probe-DIG-R), which covered the junction region between eGFP CDS and the poly(A) signal sequence. Southern blot was performed using DIG-High Prime DNA Labeling and Detection Starter Kit II (Roche) as described in manual using genomic DNAs (>10 μg) completely digested by PstI. The *in vivo* green fluorescence in transgenic animals was detected using a GFP Macroscopy system (BLS, Hungarian) by exposure to blue excitation light with wave length of 460–495 nm and observed through a filter.

Results

The NLS-I-Scel molecule and transgene construct

The NLS-I-SceI molecule consists of 3×SV40 NLS, an HA tag epitope and the native I-SceI molecule as shown in Fig. 1 B. 3×SV40 NLS is highly potent for nuclear localization, and HA tag epitope can be used to detect NLS-I-SceI molecule distribution in cells once it was expressed in cytoplasm. The NLS-I-SceI CDS was optimized for both porcine and murine codon usage preferences (the NLS-I-SceI CDS and amino acid sequence were shown in Fig. S1 and S2). High quality NLS-I-SceI mRNA was produced by T7 promoter-driven *in vitro* transcription using the linearized PCI-T7-NLS-I-SceI vector as templates (Fig. 1 C). The complete sequence of transgene vector p2IS-UBC-eGFP was shown in Fig. S3. In the transgene vector, a UBC promoter-driven eGFP expression cassette was flanked by two inversed I-SceI recognition sequences at both ends (Fig. 1 A). After cut by NLS-I-SceI molecule, the transgene vector plasmid was linearized, and the NLS-I-SceI protein was expected to be bound to the fragment containing transgene expression cassette, and thereby protect the transgene fragments from degradation and transfer the fragments from cytoplasm into nuclear (Fig. 1 D), for I-SceI protein exhibited high affinity in binding to the downstream cleavage product [36].

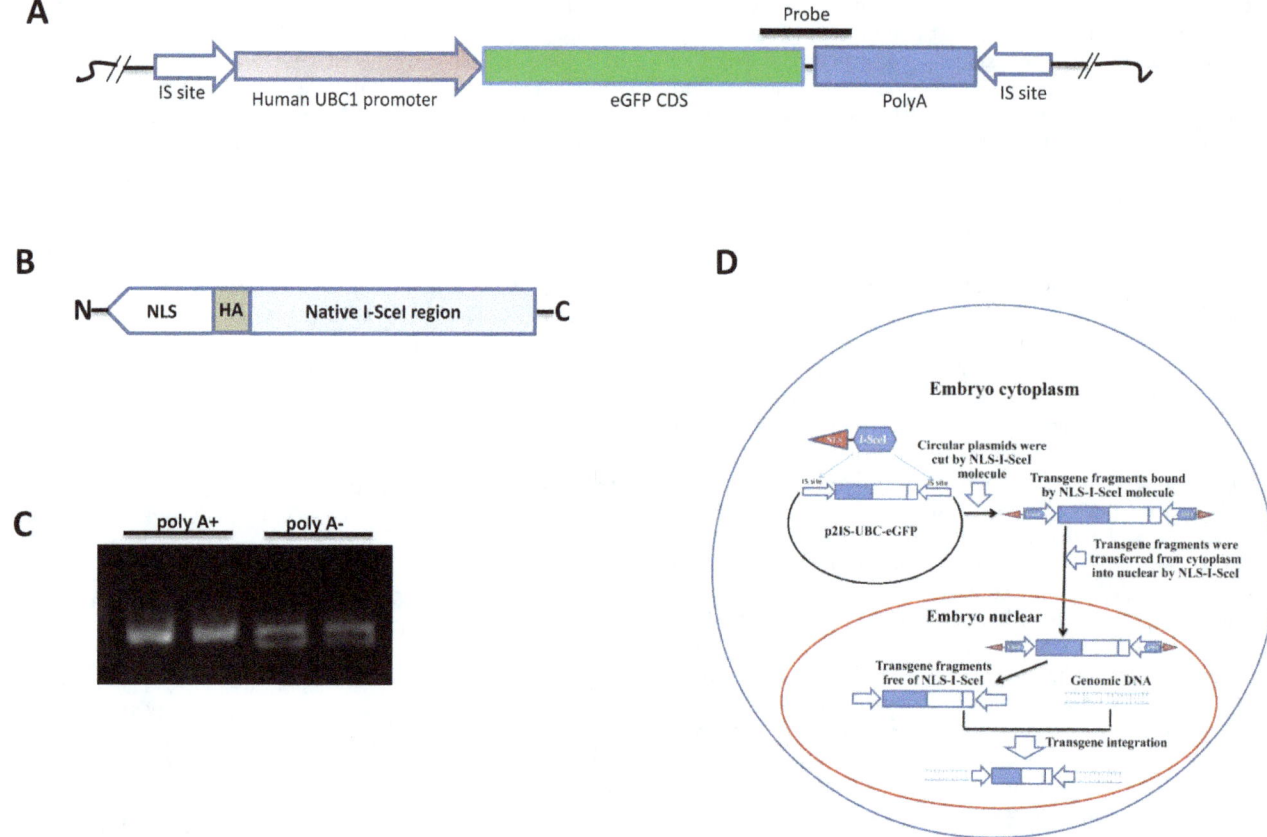

Figure 1. Transgene construct and the NLS-I-SceI molecule. A: The schematic structure of p2IS-UBC-eGFP vector. IS site: the inversely flanking I-SceI recognition sequence; the black bar indicates the position of the probe used for Southern blot assay. B: The schematic structure of NLS-I-SceI molecule. C: The *in vitro* transcribed NLS-I-SceI mRNA. polyA+: the mRNA with polyA tail; polyA-: the mRNA without polyA tail. D: The expected working principle of NLS-I-SceI-mediated transgenesis.

The NLS-I-SceI molecule was capable of cutting circular transgene plasmids and transferring DNA fragments from cytoplasm into nuclear in mammalian embryos

To investigate whether the NLS-I-SceI molecule was capable of cutting the I-SceI recognition sequence-containing circular plasmids in mammalian embryos, total DNAs were extracted from single porcine parthenogenetic blastocysts developed from oocytes co-injected with NLS-I-SceI mRNAs and the circular p2IS-UBC-eGFP plasmids (30 ng/μL each), and subjected to PCR analysis to assess the extent to which the circular plasmids were digested. The uncut I-SceI site was quantitatively detected by qPCR using a primer pair covering the I-SceI recognition sequence 3′ to the transgene expression cassette, and the eGFP CDS detected as internal control. Prior to qPCR, a qualitative PCR was performed to confirm the existence of plasmids in embryos. As shown in Fig. 2 A, in the embryos co-injected with NLS-I-SceI mRNA and the circular transgene plasmids, the band intensities of PCR products covering I-SceI site were remarkably lower than those of eGFP CDS. In contrast, in embryos injected only with circular plasmids, the uncut I-SceI site and eGFP CDS were simultaneously detected or not in these samples (Fig. 2 A), and the band intensities of PCR products covering I-SceI site were comparable to those of eGFP CDS (Fig. 2 A), indicating that in these embryos the levels of uncut I-SceI site and eGFP CDS were comparable and varied proportionally. The samples with expected PCR products were subjected to qPCR analysis, of which the

Amplification Plots, Melt Curves and Standard Curves were shown in Fig. S4. The qPCR data further showed that the levels of uncut I-SceI site relative to eGFP CDS in embryos co-injected with NLS-I-SceI mRNA and circular plasmids were largely lower than those in embryos injected only with circular plasmids (P< 0.001, Fig. 2 B), indicating that the NLS-I-SceI molecule produced from mRNAs in mammalian embryos was bio-active and capable of cutting circular plasmids. Moreover, these data further demonstrated that in the embryos co-injected with NLS-I-SceI mRNA and circular plasmids, although the relative levels of uncut I-SceI site were much lower, the eGFP CDS copy numbers were remarkably higher than those in embryos injected only with circular plasmids (P<0.001, Fig. 2 C), suggesting that the linearized transgene DNA fragments were protected from degradation by NLS-I-SceI molecule after plasmids were cut at I-SceI sites.

To display the localization of injected DNAs in living embryos, Cy3-labeled double-stranded DNA (Cy3-DNA) fragments containing two inversely flanking I-SceI recognition sequences at both ends, of which the schematic structure was shown in Fig. 3 A and sequence Fig. S5, were co-injected with NLS-I-SceI mRNA into the cytoplasm of activated porcine MII oocytes (parthenogenetic embryos) of which the chromosomal DNAs were stained with Hoechst 33342 prior to microinjecction. The injected embryos were cultured and observed under LSCM at 16 and 24 h post activation. In control groups, the embryos were injected with only Cy3-DNA fragments or Cy3-DNA fragments included in the

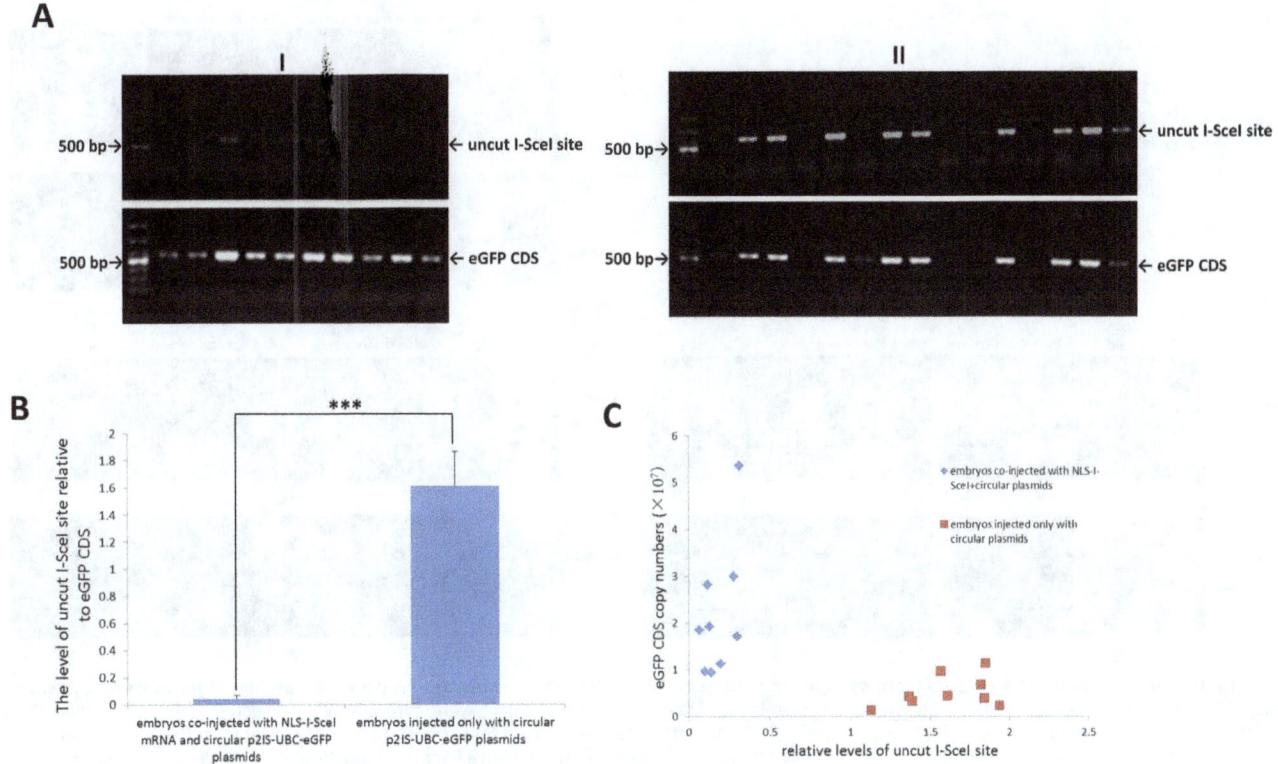

Figure 2. The NLS-I-SceI molecule was capable of cutting circular p2IS-UBC-eGFP plasmids in porcine parthernogenetic embryos. A: Detection of uncut I-SceI site and eGFP CDS by PCR in the embryos cytoplasmically injected with circular p2IS-UBC-eGFP plasmids plus NLS-I-SceI mRNA and only with circular p2IS-UBC-eGFP plasmids. I: embryos cytoplasmically injected with circular p2IS-UBC-eGFP plasmids plus NLS-I-SceI mRNA; II: embryos cytoplasmically injected only with circular p2IS-UBC-eGFP plasmids. B: The levels of uncut I-SceI site relative eGFP CDS detected by qPCR in the injected embryos. C: The eGFP CDS copy numbers in the injected embryos. *: statistical significance.

native I-SceI endonuclease digestive reaction system as that for I-SceI-mediated transgenesis in fish. At 16 h post activation, in the embryos co-injected with NLS-I-SceI mRNAs and Cy3-DNAs, of which the chromosomes were in a relaxed state and loosely assembled suggesting that meiosis was proceeding to the telophase and the nuclear was under construction (Fig. 3 B), the Cy3-DNA fragments (red fluorescence) were found to be clustered and located near to the chromosomes (blue fluorescence) (Fig. 3 B). At 24 h post activation, in the embryos co-injected with NLS-I-SceI mRNAs and Cy3-DNAs, the blue fluorescence was concentrated suggesting that chromosomes were compactly aggregated, meiosis completed and nuclear was constructed, and the clustered Cy3-DNA fragments were observed to be completely co-localized with chromosomes (Fig. 3 B), indicating that the Cy3-DNA fragments were transferred into nuclear. In contrast, in the embryos of control groups, the red fluorescence was scattered and extremely weak, and no Cy3-DNAs were observed to be clustered, located closely to or co-localized with the chromosomes at 16 h or 24 h post activation (Fig. 3 C, D), suggesting that the Cy3-DNA fragments were diffusely distributed in cytoplasm or degraded. These data provided a direct demonstration that the NLS-I-SceI molecule was capable of transferring DNA fragments from cytoplasm into nuclear in mammalian embryos, and this transfer process was co-incident with the process of nuclear formation during meiosis (or mitosis) of embryos, while the native I-SceI molecule was not, being consistent with previously reported data for the native I-SceI-mediated transgenesis in fish embryos [30].

NLS-I-SceI molecule was capable of facilitating transgenesis in mammalian embryos via cytoplasmic microinjection

To investigate whether NLS-I-SceI molecule was capable of mediating transgenesis and resulting in transgene expression in early mammalian embryos, mouse eggs were subjected to cytoplasmic microinjection with the mixture of NLS-I-SceI mRNA and circular transgene plasmid p2IS-UBC-eGFP, for mouse eggs have visible pronuclear and the materials can be confirmed to be injected into cytoplasm. With a given plasmid concentration (30 ng/μL), NLS-I-SceI mRNAs at different concentrations (10, 20 and 30 ng/μL) were co-injected with circular transgene plasmids into cytoplasm of 1-cell mouse eggs, and green fluorescence in the blastocysts developed from injected eggs were observed and counted at 5 d post injection. To avoid cellular lysis after injection, a very small volume (about 5 pL) of solution, which was less than that for pronuclear microinjection, was injected into embryo cytoplasm. Results showed that the NLS-I-SceI molecule mediated transgenesis in a dose-dependent manner (Fig. 4 A). In the group injected with 30 ng/μL of NLS-I-SceI mRNA, the fluorescence intensity was significantly higher than those in groups injected with 10 and 20 ng/μL of NLS-I-SceI mRNA (Fig. 4 A). In contrast, in the group injected with 30 ng/μL of circular p2IS-UBC-eGFP plasmid included in the native I-SceI endonuclease digestive reaction system, no fluorescent blastocysts were observed (Fig. 4 A), indicating that without the added NLS signal, the native I-SceI molecule was not capable of facilitating transgenesis and further resulting in transgene expres-

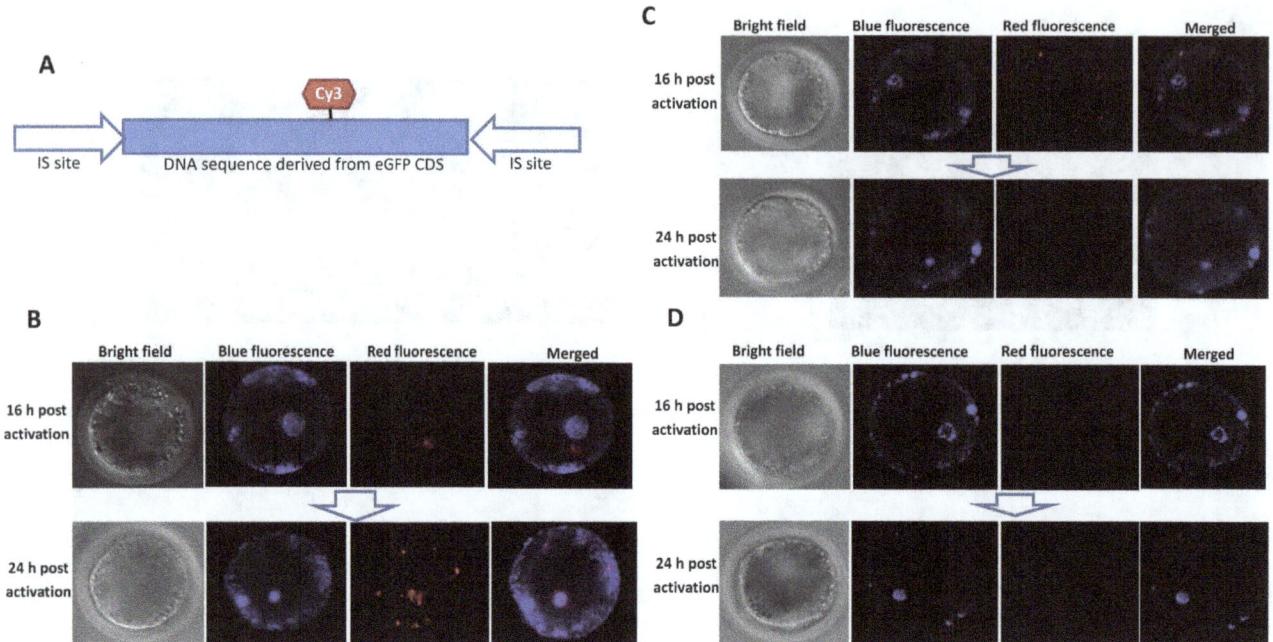

Figure 3. Transfer of DNA fragments from cytoplasm into nuclear by NLS-I-SceI molecule in porcine parthernogenetic embryos. The activated porcine MII oocytes (1-cell parthernogenetic embryos) stained with Hoest33342 were cytoplasmically injected with Cy3-labelled DNA fragments plus NLS-I-SceI mRNA, and the localization of DNA fragments were observed under LSCM at 16 and 24 h post microinjection respectively. In control groups, the embryos were injected with Cy3-DNA fragments included into the native I-SceI endonuclease digestive reaction system or only with Cy3-DNA fragments. A: The structure of Cy3-labeled DNA fragments. B: The localization of Cy3-DNA fragments co-injected with NLS-I-SceI mRNA. C: The localization of Cy3-DNA fragments co-injected with the native I-SceI nuclease. D: The localization of Cy3-DNA fragments injected alone. Red fluorescence: the Cy3-DNA fragments; Blue fluorescence: the chromosomal DNAs.

sion in mouse embryos. In the groups cytoplasmically injected only with circular or linearized plasmids at the same concentration, no fluorescence was observed in the derived blastocysts either, although fluorescence was observed in a few developmentally arrested embryos in the circular plasmid injection group (Fig. 4 A). In all the groups, the blastocyst development rates (blastocysts/ cleaved eggs) were comparable to the untreated group (data not shown), suggesting that the injected materials did not interfere with *in vitro* development once embryos survived the microinjection process. The dynamics of transgene expression in the embryos cytoplasmically co-injected with NLS-I-SceI mRNA and circular transgene plasmid was similar to that in embryos subjected to pronuclear microinjection only with circular transgene plasmid, although the fluorescence intensity in the cytoplasmic injection group was lower (Fig. 4 B), suggesting that the transgene fragments delivered into cytoplasm were transferred into pronuclear by NLS-I-SceI molecule as early as embryo cleavage started, which was consistent with the results of LSCM observation. The lower fluorescence intensity may be due to the less copies of transgene fragment in pronuclear transferred from cytoplasm by NLS-I-SceI molecule compared to those of transgene fragment directly delivered into pronuclear by microinjection.

To address whether NLS-I-SceI molecule was capable of mediating transgenesis in mammalian embryos of species other than mice, 1- or 2-cell porcine eggs surgically collected from mated sows were subjected to cytoplasmic co-injection with NLS-I-SceI mRNAs and circular transgene plasmids (30 ng/μL each), for pig is a typical mammalian species of which the pronuclear is usually invisible and refractory to pronuclear microinjection. Porcine eggs have a relatively larger size and are much more tolerant to cytoplasmic microinjection compared to mouse eggs, and a much

larger volume (40–60 pL) of solution, which contained 1.2–1.8 pg of transgene plasmids and NLS-I-SceI mRNAs respectively, was injected into cytoplasm. As shown in Fig. 5 A, in the porcine embryos derived from eggs co-injected with NLS-I-SceI mRNA and circular p2IS-UBC-eGFP plasmids, strong fluorescence was observed on 3 d post injection, and the majority of derived blastocysts exhibited strong fluorescence on 6 d post injection. In contrast, in the embryos injected with circular p2IS-UBC-eGFP plasmids (30 ng/μL) included into the native I-SceI endonuclease digestive reaction system, only weak fluorescence was observed in a few embryos on 3 d post injection, and on 6 d, no fluorescence was observed in the derived blastocysts, although fluorescence was observed in a few developmentally arrested embryos (Fig. 5 A). This different fluorescence was not because of the difference in eGFP CDS copy numbers, for the eGFP CDS was readily detected in all the injected embryos (Fig. 5 B), and the eGFP CDS copy numbers in the embryos injected with circular plasmids plus NLS-I-SceI mRNA were comparable to those in embryos injected with circular plasmids at the same concentration included into the native I-SceI endonuclease digestive reaction system ($P>0.1$, Fig. 6 A), which were much higher than those in embryos injected only with circular plasmids ($P<0.001$, Fig. 6 A). The uncut I-SceI site was detected in the injected embryos (Fig. 5 B), and its levels relative to eGFP CDS were also comparable between the two groups injected with circular plasmids plus NLS-I-SceI mRNA and native I-SceI nuclease ($P>0.1$, Fig. 6 B), but were significantly lower than those of the group injected only with circular plasmids ($P<0.001$, Fig. 6 B), suggesting that the NLS-I-SceI molecule derived from mRNA cut circular plasmids to a similar degree to the native I-SceI nuclease in porcine embryos. The presence of uncut I-SceI site indicated that there existed residual circular

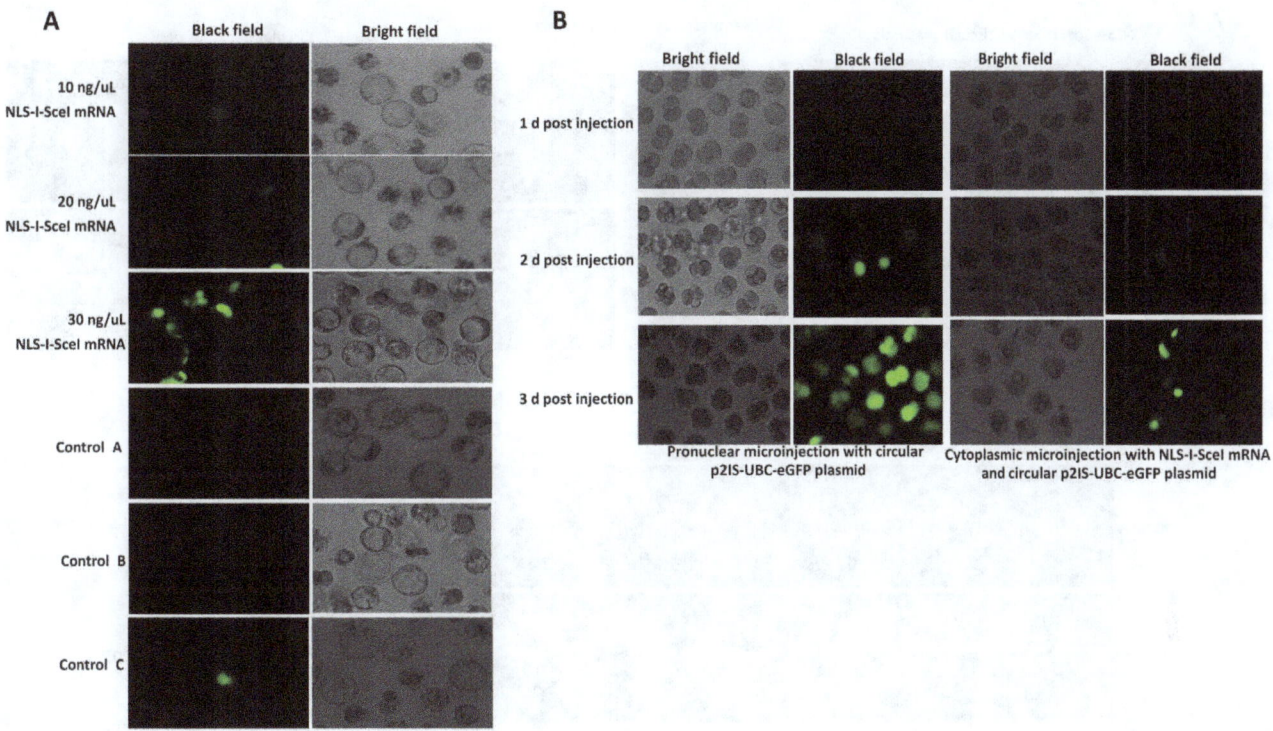

Figure 4. Transgene expression in cytoplasmically injected mouse embryos. A: Mouse eggs cytoplasmically injected with 30 ng/µL of circular p2IS-UBC-eGFP plasmids plus NLS-I-SceI mRNAs at different concentratiions. Controls A-C were the control groups injected with 30 ng/µL circular plasmids included into the native I-SceI endonuclease digestive reaction system (control A), linearized plasmids (control B) or circular plasmids (control C). B: The dynamics of transgene expression in embryos subjected to cytoplasmic microinjection with circular p2IS-UBC-eGFP plasmids plus NLS-I-SceI mRNA or pronuclear microinjection only with circular p2IS-UBC-eGFP plasmids.

plasmids in the injected embryos, which may be a reason for the fluorescence in the few porcine embryos injected with the circular plasmids included into the native I-SceI endonuclease digestive reaction system. The circular plasmids were resistant to endogenous nuclease and could be passively diffused into the nuclear during embryo cleavage as indicated by a previous report [36]. Consistently, in this work, the porcine embryos injected only with circular plasmids also exhibited fluorescence (Fig. 5 A), while those injected with linearized plasmids did not (data not shown). However, the circular plasmid rarely results in transgene integration in mammalian embryos even introduced directly into pronuclear in large amounts [27,36]. Totally, these data indicated that the NLS-I-SceI molecule was capable of efficiently facilitating transgenesis in porcine embryos, while the native I-SceI molecule was not.

The transgenesis mediated by NLS-I-SceI molecule in mammalian embryos efficiently resulted in transgenic animals

To answer whether the NLS-I-SceI-mediated transgenesis in mammalian embryos via cytoplasmic microinjection was able to result in transgenic animals, 411 fertilized mouse eggs were collected from nine super-ovulated and mated female mice, and 330 eggs with visible pronuclear were selected and randomly and equally divided into two groups. One group was subjected to cytoplasmic microinjection with the mixture of NLS-I-SceI mRNA and circular transgene plasmids (30 ng/µL each), and the other group (control) injected with circular transgene plasmid (30 ng/µL) included into the native I-SceI endonuclease digestive reaction system as described above. 116 eggs which survived the

microinjection process and cleaved the next day were transferred into 4 surrogate mice. Totally, 23 founder pups were born, of which 10 pups were derived from eggs of control group, and 13 pups from eggs co-injected with NLS-I-SceI mRNA and circular plasmids. As shown in Fig. 7 A, in the founders derived from eggs co-injected with NLS-I-SceI mRNA and circular plasmids, 6 pups were detected to be transgenic by PCR, while in the control group no transgenic pub was detected. The transgenic rate in founders of NLS-I-SceI-mediated transgenesis group was 46.2% (6/13), and the transgenesis efficiency (transgenic founders/transferred eggs) was 10.7% (6/56), which were both higher than the data for pronuclear microinjection in our lab (unpublished). However, the survival rate of cytoplasmically microinjected mouse eggs (35.2% (116/330)) was remarkably lower than that of eggs subjected to pronuclear microinjection in our lab (usually 50%), indicating that mouse eggs were more vulnerable to cytoplasmic microinjection than to pronuclear microinjection. To test the germline transmission competence of transgene, the transgenic founder mouse with the strongest PCR product band were mated with wild-type mice, and transgenic individuals were detected from the resulted offspring (Fig. 7 B). *In vivo* fluorescence was not observed in the transgenic founder mice or the transgenic individuals of F1 offspring, and transgene integration was detected by Southern blot only in one founder mouse (Fig. 7 C). However, the *in vivo* fluorescence was observed after the transgenes were enriched by mating between transgenic individuals consecutively over at least three generations (Fig. 7 D), indicating that the NLS-I-SceI-mediated transgenesis did resulted in transgene integration in mouse genome although not detected by Southern blot assay in most founders. These results indicated that the NLS-I-SceI-

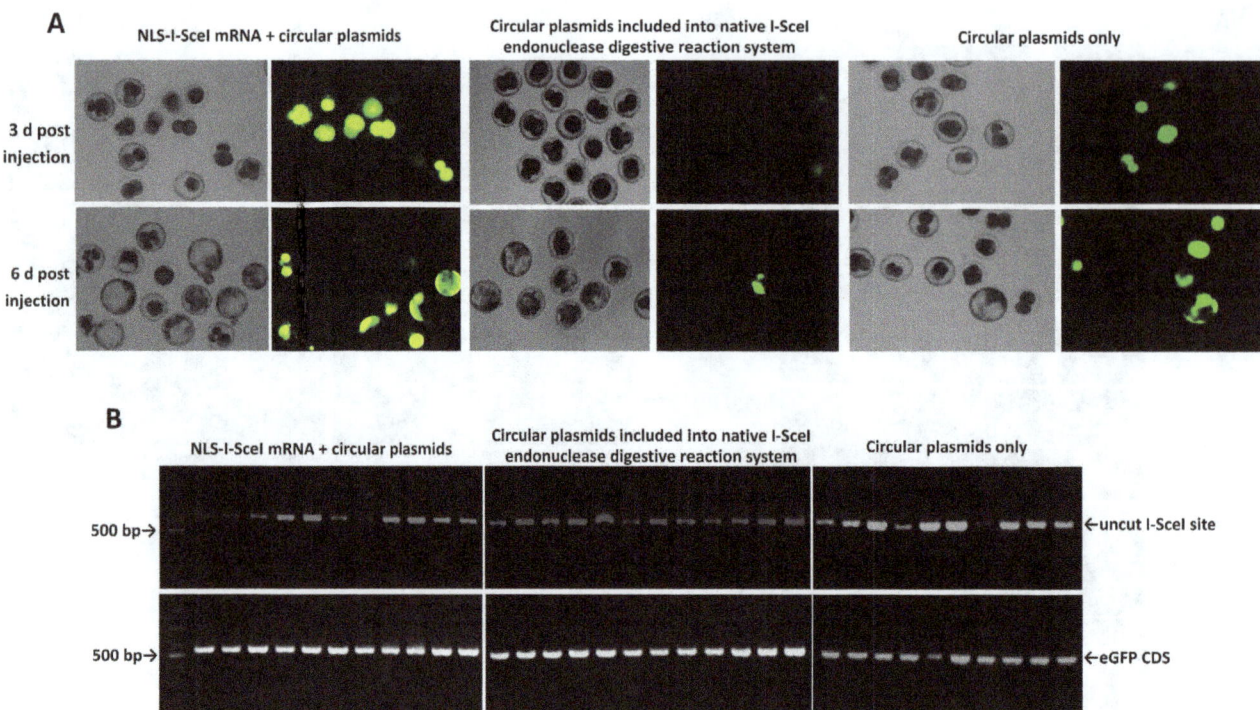

Figure 5. Transgene expression and detection of uncut I-SceI site and eGFP CDS by PCR in cytoplasmically injected porcine embryos. A: Transgene expression in the porcine embryos cytoplasmically injected with circular plasmids (p2IS-UBC-eGFP) plus NLS-I-SceI mRNA, circular plasmids included into the native I-SceI nuclease digestive reaction system and circular plasmids only. B: Detection of uncut I-SceI site and eGFP CDS by PCR in the cytoplasmically injected porcine embryos as described in A.

mediated transgenesis was capable of resulting in transgenic mice, while the native I-SceI nuclease was not.

To further test whether the NLS-I-SceI-mediated transgenesis would result in transgenic animals of species other than mice, 36 porcine eggs at 1- or 2-cell stage surgically collected from mated sows were subjected to cytoplasmic co-injection with NLS-I-SceI mRNA and the circular p2IS-UBC-eGFP plasmids, and then transferred into two synchronized surrogate sows. One recipient was pregnant and four piglets were born. The *in vivo* fluorescence was observed in three of the four founder pigs (Fig. 8 A). However, all of the founder pigs were detected to be transgenic by PCR screen and transgene integration was confirmed by Southern blot assay (Fig. 8 B, C). The lack of *in vivo* fluorescence in one transgenic founder pig (4#) may be due to the low copy number of integrated transgenes as indicated by the Southern blot data (Fig. 8 C). The founder pig with the strongest fluorescence (1#)

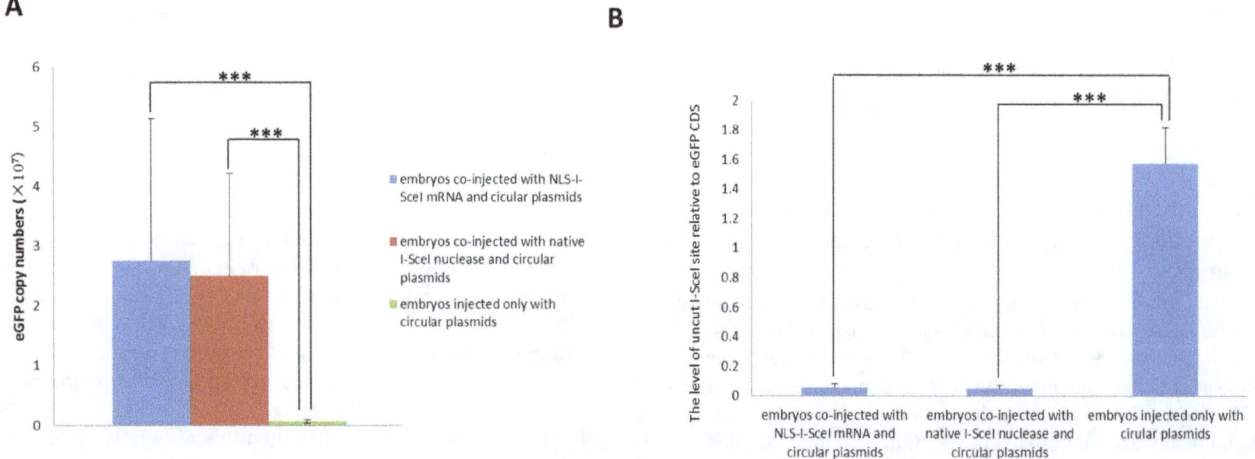

Figure 6. Quantitative analysis of uncut I-SceI site and eGFP CDS by qPCR in cytoplasmically injected porcine embryos. A: The eGFP CDS copy numbers in the cytoplasmically injected porcine embryos as described in Fig. 5. B: The uncut I-SceI site levels relative to eGFP CDS in the cytoplasmically injected porcine embryos as described in Fig. 5. *: statistical significance.

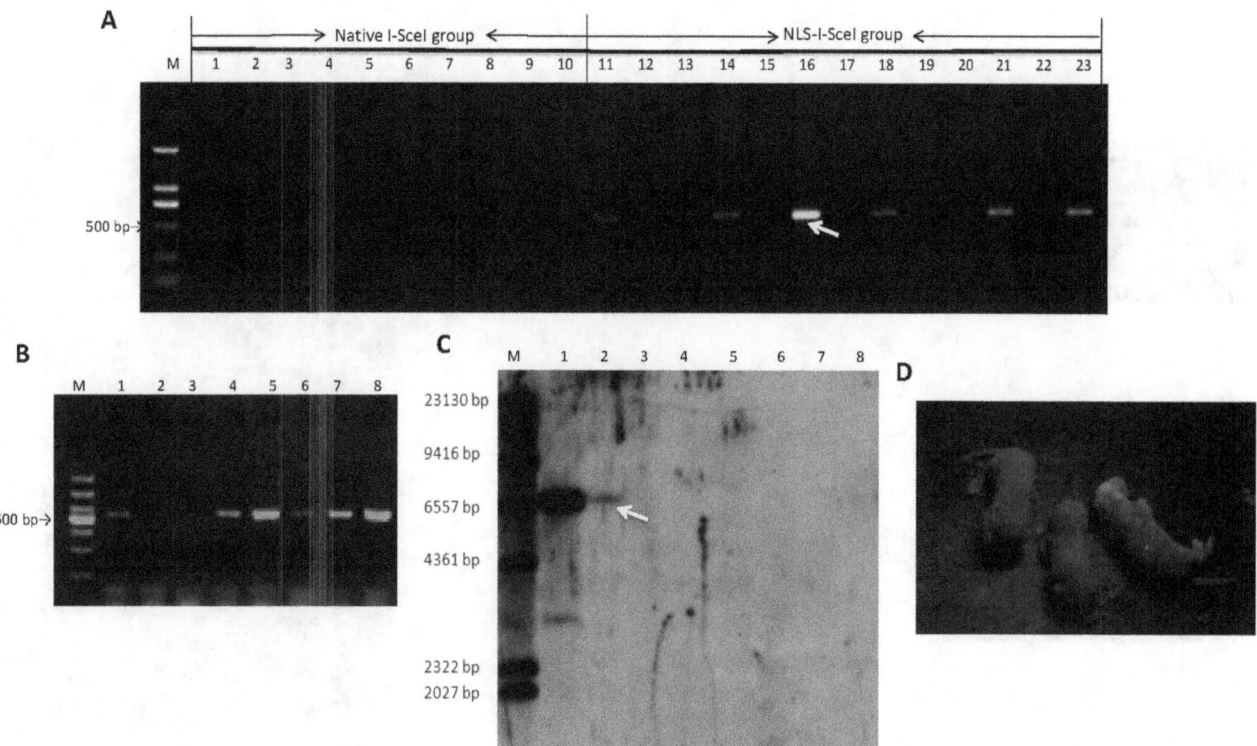

Figure 7. Genetic screen of transgenic mice derived from cytoplasmically microinjected eggs. A: Screen of transgenic founder mice by PCR. M: DL2000 DNA marker; 1–10: the founder mice derived from cytoplasmic microinjection with circular p2IS-UBC-eGFP plasmids (30 ng/μL) included into the native I-SceI nuclease digestive reaction system; 11–23: The founder mice derived from cytoplasmic microinjection with circular p2IS-UBC-eGFP plasmids plus NLS-I-SceI mRNA (30 ng/μL each). B: Screen of transgenic individuals of F1 offspring derived from transgenic founder mice by PCR. M: DNA marker; 1–8: Genomic DNA samples of F1 individuals. C: Genetic screen of transgenic founder mice by Southern blot assay. M: DNA molecular weight marker II; 1: plasmids; 2–7: genomic DNA samples of founder mice; 8: negative control (wild-type mouse genomic DNA). D: The transgenic mice exhibiting *in vivo* fluorescence derived from breeding between transgenic individuals over three consecutive generations. The arrow indicates the founder mouse detected to be transgenic by both Southern blot and PCR screen.

was mated with wild-type pig to test the germline transmission competence of transgenes. As shown in Fig. 8 D, in the seven individuals of F1 offspring, four were detected to be transgenic by Southern blot, indicating that the transgenes were capable of germline transmission. After gemline transmission was confirmed, the founder pig (1#) was sacrificed due to disease related to respiratory system infection, and genomic DNA samples of different organs were subjected to Southern blot assay. As shown in Fig. 8 E, transgene was detected in all the organs except skin and lung in a similar band distribution pattern. However, the failure to detect transgene in these two organs was due to the experimental procedure but not to the lack of transgene integration, for the genomic DNAs of the two organs were not thoroughly digested and separated in gel electrophoresis as a result before DNA was transferred to membrane (Fig. S6 A), and transgene was finally detected in these two organs with a similar band distribution pattern by a repeated Southern blot assay after the genomic DNAs were completely digested (Fig. S6 B, C), suggesting that this founder pig was not transgenically mosaic and transgene integration occurred at a very early stage of embryo development. The death of the founder pig was not due to transgenesis, for some wild-type pigs in the farm also died of the same disease at that time. The rest transgenic pigs, including the offspring of the dead founder pig, kept healthy. These results demonstrated that the NLS-I-SceI-mediated transgenesis in mammalian embryos was capable of efficiently resulting in transgenic animals with germline transmission competence,

especially in species other than mice which was refractory to embryo pronuclear microinjection but exhibited higher tolerance to embryo cytoplasmic microinjection.

Discussion

Embryo microinjection is a simple and reproducible method for mammalian transgenesis, however the dependence on visible pronuclear largely limits its application to mammalian species other than mice, especially those large animal species of which the pronuclear is usually invisible. Currently, transgenisis via embryo cytoplasmic microinjection has achieved limited success in mammalian species. Page et al (2005) produced transgenic mice using Polylysine/DNA mixture by cytoplasmic microinjection of eggs, however the transgenic rate (born transgenic pups/transferred embryos) was much lower than that of pronuclear microinjection (12.8% vs 21.7%) [37]. Garrels et al (2011) efficiently produced transgenic pigs by Sleeping Beauty (SB) transposon-mediated transgenesis via embryo cytoplasmic microinjection with circular plasmids of SB transposon-based transgene vector and SB tranposase expression vector, and the transgenic rate of founder pigs was as high as 47.3% [23]. Nonetheless, transposons are mobile genetic elements and transgenic organisms derived from transposon-mediated transgenesis would be of biosafety concerns. Recently, Wilson et al (2013) has described a sophisticated system termed intracellular electroporetic nanoinjection (IEN) to propel transgene fragments from cytoplasm into

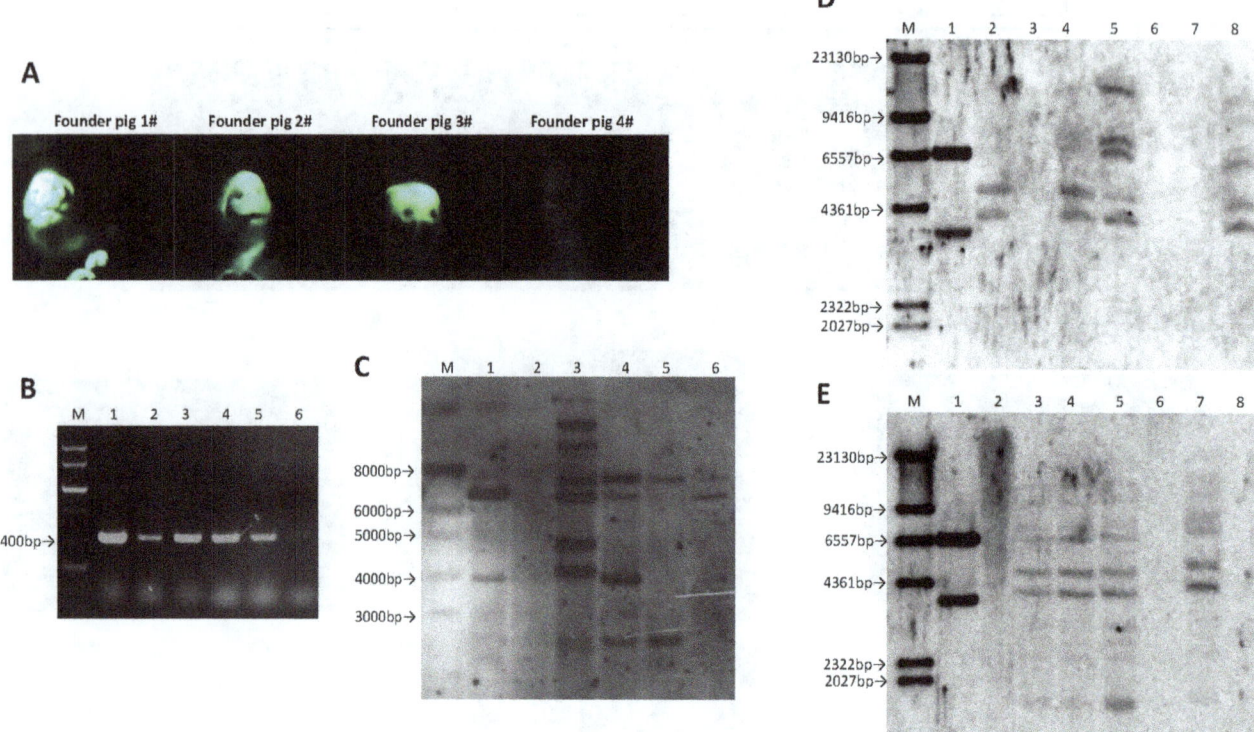

Figure 8. Genetic screen of transgenic pigs derived from embryos cytoplasmically microinjected with circular p2IS-UBC-eGFP plasmids plus NLS-I-SceI mRNA. A: *In vivo* fluorescence in founder pigs. B: Screen of transgenic founder pigs by PCR. M: DNA marker; 1–4: genomic DNA samples of 1–4# founder pigs; 5: positive controls (wild-type pig genomic DNAs containing p2IS-UBC-eGFP plasmids); 6: negative control (wild-type pig genomic DNA). C: Southern blot assay of transgenic founder pigs. M: DNA marker (1KB DNA Ladder); 1: positive control (plasmids); 2: wild-type pig genomic DNA as negative control; 3–6:1–4# founder pigs. D: Southern blot analysis of F1 offspring individuals derived from founder pig 1#. M: DNA molecular weight marker II; 1: plasmid as positive control; 2–8: the F1 offspring individuals. E: Southern blot analysis of genomic DNAs extracted from different organs of founder pig 1#. M: DNA molecular weight marker II; 1: positive control (plasmids); 2: skin; 3: heart; 4: liver; 5: spleen; 6: lung; 7: kidney; 8: wild-type pig genomic DNA as negative control.

pronuclear, however this method required additional complicated equipment and experimental skills besides conventional microinjection, and more importantly, the transgenesis efficiency of IEN system was not higher (actually slightly lower) than that of pronuclear microinjection [38].

I-SceI has been effectively used to facilitate transgenesis in fish eggs for several years. Because only plasmid vectors containing I-SceI recognition sequences are involved in the I-SceI-mediated transgenesis, regarding the transgenesis process and the application of the resulted transgenic organisms, the I-SceI-mediated transgenesis is of minimal bio-safety concerns. In this work, we efficiently generated transgenic mammals (pigs and mice) simply by co-injecting circular transgene vector plasmids containing I-SceI recognition sequences and the mRNAs coding NLS-I-SceI molecule into embryo cytoplasm. As far as we know, this is the first report for efficient generation of transgenic mammals via embryo cytoplasmic microinjection using the I-SceI molecule.

Our work demonstrated that the native I-SceI molecule was not capable of efficiently facilitating transgenesis in mammalian embryos as it did in fish eggs, which may be due to the much smaller size of mammalian embryos compared to that of fish eggs and much less plasmid copies that can be delivered into mammalian embryos as a result. In contrast, the NLS-I-SceI molecule, which contains mammalian NLS sequence at its N-terminal, was shown to be capable of cutting transgene fragments off from circular plasmids, protecting transgene fragments from degradation and efficiently facilitating transgenesis in both mouse

and porcine embryos, indicating that the artificially added mammalian NLS signal largely promoted the efficacy of I-SceI-mediated transgenesis. The ability of NLS-I-SceI molecule to facilitate transgenesis in mammalian embryos was directly demonstrated by the localization of Cy3-labeled DNA fragments containing inversely flanking I-SceI cutting sites at both ends which were co-injected with NLS-I-SceI mRNA into the cytoplasm of porcine pathenogenically activated oocytes at MII stage (parthenogenetic embryos). The reason for the use of porcine MII oocytes was that the nuclear was breakdown at this stage and to be constructed upon activation, providing a time window to observe the localization of DNA fragments during the process of mammalian pronuclear construction, and that in addition, the lack of nuclear excluded the probability that materials happened to be injected into nuclear by chance due to the invisibility of pronuclear. Data showed that only the DNA fragments co-injected with NLS-I-SceI molecule were clustered and co-localized with chromosomes in parthenogenetic porcine embryos, while those co-injected with the native I-SceI molecule were diffusely distributed in the cytoplasm and not clustered or co-localized with chromosomes, indicating that the NLS-I-SceI was capable of transferring DNA fragments from cytoplasm into nuclear, while the native I-SceI molecule was not, and the transferring process was co-incident with the procedure of nuclear formation. These results were consistent with the observation that the porcine blastocysts developed from eggs co-injected with NLS-I-SceI mRNA and circular transgene plasmids p2IS-UBC-eGFP exhibited strong fluorescence, while those co-injected with the native I-SceI nuclease and circular transgene plasmids at the same concentration did not, although the eGFP CDS was detected at similar levels in these embryos, suggesting that although the transgene fragments were efficiently cut off from circular plasmids and protected from degradation by the native I-SceI nuclease in porcine embryos, the transgene fragments were not efficiently translocated from cytoplasm into nuclear by this molecule to result in expression. The efficient production of transgenic mice and pigs and the germline transmission competence of the resulted transgenic animals further confirmed that the NLS-I-SceI molecule can be used as a potent tool to facilitate mammalian transgenesis.

The NLS-I-SceI-mediated transgenesis resulted in random integration in mammalian genome. With the advent of powerful EENs such as ZFN, TALEN and CRIPR/Cas9 system, the NLS-I-SceI molecule can be used in combination with EENs to facilitate targeted transgene integration into mammalian embryo genomes. Recently, it has been reported that *in vivo* cleavage of circular plasmids by EENs effectively facilitated targeted integration of transgenes into the DSBs created by the same or another EEN molecule through none-homologous end joining (NHEJ) mechanism in the genomes of mammalian somatic cells [39,40]. However, the ability of the EENs to bind to the cleaved DNAs and further transfer DNA fragments from cytoplasm into nuclear of mammalian embryos remains to be investigated, although these molecules have NLS signal. More recently, Cas9-sgRNA complex was shown to be capable of binding cleaved DNA with high affinity, however the both ends of cleaved DNA were tightly bound to Cas9-sgRNA complex and the cleaved circular plasmids were still in circular form [41], which would hinder transgene integration. On this basis, considering the confirmed ability of NLS-I-SceI molecule to cut transgene fragments off from circular transgene plasmids, protect transgene fragments from degradation and transfer transgene fragments from cytoplasm into nuclear in mammalian embryos, NLS-I-SceI molecule can be used in combination with EENs to facilitate targeted transgene integration into mammalian embryo genomes, and thereby a simple, efficient and species-neutral technology for targeted transgenesis in mammalian animals can be established, especially for large mammalian species such as pig, cattle and none-human primates.

In this work, we found that the native I-SceI molecule without mammalian NLS signal did not efficiently facilitate transgenesis in mouse or porcine embryos. However, Bevacqua et al (2013) recently reported that the native I-SceI-mediated transgenesis resulted in transgenic eGFP expression in bovine blastocysts derived from *in vitro* fertilization [42]. This inconsistency may be partly due to the much higher concentration (50 ng/μL) of circular transgene plasmids used in this study compared to that in our work (30 ng/μL). Such a high concentration may result in the presence of more uncut circular plasmids in embryos, which were resistant to degradation in cells and can be passively diffused into nuclear during embryo cleavage as suggested by a previous report [36] and our data in this work. However, circular plasmids rarely integrated into genome even directly delivered into pronuclear in a large amount, and no transgenic cattle was produced in this report either. Besides, the fluorescence in bovine blastocysts resulting from the native I-SceI-mediated transgenesis with plasmids of a natural promoter (Pax6)-driven eGFP expression vector was rather weak, which was comparable to that in the few fluorescent porcine embryos co-injected with the native I-SceI nuclease and circular p2IS-UBC-eGFP plasmids in this study. The moderately stronger fluorescence, of which the intensity was remarkably lower compared to that in the porcine embryos co-injected with NLS-

I-SceI mRNAs and p2IS-UBC-eGFP plasmids in our work, resulted from transgenesis with another artificially synthesized strong promoter(CAG)-driven eGFP expression vector plasmids, suggesting that the relatively stronger fluorescence was due to the much higher activity of the CAG promoter, but not to the more transgene copies in nuclear, and the native I-SceI molecule did not actively or efficiently transfer transgene fragments from cytoplasm into nuclear in the *in vitro* fertilized bovine embryos either.

Because the circular DNA plasmids can be passively diffused into nuclear during embryo cleavage and the NLS-I-SceI molecule is nuclear-localized, we can't exclude the possibility that the efficient NLS-I-SceI-mediated transgenesis in mammalian embryos was partly derived from *in situ* cleavage of circular plasmids by NLS-I-SceI molecule in nuclear. *In situ* cleavage of circular transgene plasmids in cells was shown to protect transgene fragments from degradation and facilitate transgene integration as a result [39,40]. On this basis, considering that cytoplasmic microinjection with circular bacterial artificial vector (BAC) plasmids also resulted in transgene expression in mammalian embryos, suggesting that circular BAC plasmids can be passively diffused into nuclear once introduced into cytoplasm of embryos [35], the NLS-I-SceI molecule can be used to facilitate BAC transgenesis in mammalian embryos only if the I-SceI recognition sequences were included in BAC vectors.

In summary, this work demonstrated that the NLS-I-SceI molecule was capable of efficiently facilitating mammalian transgenesis, and using this molecule, a simple and efficient general transgenesis technology with minimal bio-safety concerns can be established for mammals. For fully validating this method, a transgenic animal model with exclusive characteristics can be generated via NLS-I-SceI-mediated transgenesis as a quality control, such as the transgenic pig model for human Huntington's disease exhibiting apoptosis in brain neurons similar to human that is not observed in murine models harboring the same transgene [43]. In addition, to fully characterize this technology, the variation of transgene integration sites can be investigated in the future when more transgenic individuals were derived from NLS-I-SceI-mediated transgenesis.

Supporting Information

Figure S1   The coding sequence of NLS-I-SceI molecule. The codon usage was optimized for both mice and pigs on the basis that possible splice sites were excluded.

Figure S2   The amino acid sequence of NLS-I-SceI molecule.

Figure S3   The sequence of p2IS-UBC-eGFP vector. The bold and underlined sequences are inversely flanking I-SceI recognition sequences, and the bold sequence in green is the eGFP CDS.

Figure S4   The Amplification Plots, Melt Curves and Standard Curves for qPCR of the uncut I-SceI site and eGFP CDS.

Figure S5   The sequence of the Cy3-labeled DNA fragment. The underlined sequences are the inversely flanking I-SceI recognition sequences, and the bold base in red is the one where the Cy3 fluorophore is linked.

Figure S6 Repeated Southern blot analysis of transgene integration in the skin and lung of transgenic founder pig 1#. The genomic DNA samples of the skin and lung of founder pig1# were not thoroughly digested with PstI endonuclease in the first Southern blot assay (A). In the repeated Southern blot analysis, the same genomic DNA samples were completely digested as indicated by gel electrophoresis (B), and then transgene integration was detected by Southern blot in the two organs (C). M: DNA marker (1 Kb ladder in gel electrophoresis, and DNA molecular weight marker II in Southern blot assay).

Acknowledgments

We'd like to thank Dr. Baltimore of California Institute of Technology (USA) for his kind offer of the vector FUGW.

Author Contributions

Conceived and designed the experiments: YW HW. Performed the experiments: XYZ PYX LLW HT FX LL. Analyzed the data: YW XYZ. Contributed reagents/materials/analysis tools: LLW. Contributed to the writing of the manuscript: YW.

References

1. Yang D, Yang H, Li W, Zhao B, Ouyang Z, et al. (2011) Generation of PPARγ mono-allelic knockout pigs via zinc-finger nucleases and nuclear transfer cloning. Cell Research, 21: 979–982.
2. Orlando SJ, Santiago Y, DeKelver RC, Freyvert Y, Boydston EA, et al. (2010) Zinc-finger nuclease-driven targeted integration into mammalian genomes using donors with limited chromosomal homology. Nuclear Acids Research, 38 : e152.
3. Mashimo T, Takizawa A, Voigt B, Yoshimi K, Hiai H, et al. (2010) Generation of knockout rats with X-linked severe combined immunodeficiency (X-SCID) using Zinc-Finger Nucleases. PLoS ONE, 2010, 5: e8870.
4. Geurts AM, Cost GJ, Freyvert Y, Zeitler B, Miller JC, et al. (2009) Knockout Rats via Embryo Microinjection of Zinc-Finger Nucleases. Science, 325: 433–434.
5. Boch J, Scholze H, Schornack S, Landgraf A, Hahn S, et al. (2009) Breaking the Code of DNA Binding Specificity of TAL-Type III Effectors. Science, 326: 1509–1512.
6. Liu H, Chen Y, Niu Y, Zhang K, Kang Y, et al. (2014) TALEN-mediated gene mutagenesis in Rhesus and Cynomolgus monkeys. Cell Stem Cell, 14: 323–328.
7. Carlson DF, Tan W, Lillico SG, Stverakova D, Proudfoot C, et al. (2012) Efficient TALEN-mediated gene knockout in livestock. PNAS, 43: 17382–17387.
8. Mussolino C, Morbitzer R, Lutge F, Dannemann N, Lahaye T, et al. (2011) A novel TALE nuclease scaffold enables high genome editing activity in combination with low toxicity. Nuclear Acids Research, 39: 9283–9293.
9. Miller CJ, Tan S, Qiao G, Barlow KA, Wang J, et al. (2011) A TALE nuclease architecture for efficient genome editing. Nature Biotechnology, 29: 143–149.
10. Cermak T, Doyle EL, Christian M, Wang L, Zhang Y, et al. (2011) Efficient design and assembly of custom TALEN and other TAL effector-based constructs for DNA targeting. Nuclear Acids Research, 39: e82.
11. Niu Y, Shen B, Cui Y, Chen Y, Wang J, et al. (2014) Generation of gene-modified Cynomolgus monkey via Cas9/RNA-mediated gene targeting in one-cell embryos. Cell, 156: 1–8.
12. Hai T, Teng F, Guo R, Li W and Zhou Q (2014) One-step generation of knockout pigs by zygote injection of CRISPR/Cas9 system. Cell Research, doi:10.1038/cr.2014.11.
13. Wang H, Yang H, Shivalila CS, Dawlaty MM, Cheng AW, et al. (2013) One-step generation of mice carrying mutation in multiple genes by CRISPR/Cas9-mediated genome engineering. Cell, 153: 1–9.
14. Cong L, Ran FA, Cox D, Lin S, Barretto R, et al. (2013) Multiplex genome engineering using CRISPR/Cas9 system. Science, 339: 819–823.
15. Mali P, Yang L, Esvelt KM, Aach J, Guell M, et al. (2013) RNA-guided genome engineering via Cas9, Science, 339: 823–826.
16. Wall RJ (1996) Transgenic livestock: Progress and prospects for the future. Therionology, 45: 57–68.
17. Hammer RE, Pursel VG, Rexroad CE Jr, Wall RJ, Bolt DJ, et al. (1985) Production of transgenic rabbits, sheep and pigs by microinjection. Nature, 315: 680–683.
18. Zhao J, Ross JW, Hao Y, Spate LD, Walters EM, et al. (2009) Significant Improvement in Cloning Efficiency of an Inbred Miniature Pig by Histone Deacetylase Inhibitor Treatment after Somatic Cell Nuclear Transfer. Biology of Reproduction, 81: 525–530.
19. Huang Y, Tang X, Xie W, Zhou Y, Li D, et al. (2011) Histone Deacetylase Inhibitor Significantly Improved the Cloning Efficiency of Porcine Somatic Cell Nuclear Transfer Embryos. Cellular Reprogramming, 13: 513–520.
20. Lois C, Hong EJ, Pease S, Brown EJ, Baltimore D (2002) Germline Transmission and Tissue-Specific Expression of Transgenes Delivered by Lentiviral Vectors. Science, 295: 868–872.
21. Whitelaw CB, Radcliffe PA, Ritchie WA, Carlisle A, Ellard FM, et al. (2004) Efficient generation of transgenic pigs using equine infectious anaemia virus (EIAV) derived vector. FEBS Letter, 571: 233–236.
22. Brem G, Wolf E, Pfeifer A (2003) Efficient transgenesis in farm animals using lentiviral vectors. EMBO Reports, 4: 1054–1060.
23. Garrels W, Mátés L, Holler S, Dalda A, Taylor U, et al. (2011) Germline transgenic pigs by Sleeping beauty transposition in porcine zygote and targeted integration in pig genome. PLoS ONE, 6: e23573.
24. Ivics Z, Garrels W, Mátés L, Yau TY, Bashir S, et al. (2014) Germline transgenesis in pigs by cytoplasmic microinjection of Sleeping Beauty transposons. Nature Protocol, 2014, 9: 810–827.
25. Ivics Z, Hiripi L, Hoffmann OI, Mátés L, Yau TY, et al. (2014) Germline transgenesis in rabbits by pronuclear microinjection of Sleeping Beauty transposons. Nature Protocol, 2014, 9: 794–809.
26. Rostovskaya M, Naumann R, Fu J, Obst M, Mueller D, et al. (2013) Transposon mediated BAC transgenesis via pronuclear injection of mouse zygotes. Genesis, 51: 135–141.
27. Ding S, Wu X, Li G, Han M, Zhuang Y, et al. (2005) Efficient Transposition of the piggyback (PB) Transposon in Mammalian Cells and Mice. Cell, 122: 473–483.
28. Ogino H, McConnell WB, Grainger RM (2006) Highly efficient transgenesis in Xenopus tropicalis using I-SceI endonuclease. Mechanism of Development, 123: 103–113.
29. Thermes V, Grabher C, Ristoratore F, Bourrat F, Choulika A, et al. (2002) I-SceI mediated highly efficient transgenesis in fish. Mechanism of Development, 118: 91–98.
30. Pan FC, Chen Y, Loeber J, Henningfeld K, Pieler T (2006) I-SceI meganuclease-mediated transgenesis in Xenopus. Developmental Dynamics, 235: 247–252.
31. Rembold M, Lahiri K, Foulkes NS, Wittbrodt J (2006) Transgenesis in fish: efficient selection of transgenic fish by co-injection with a fluorescent report construct. Nature Protocols, 1: 1133–1139.
32. Grabher C, Wittbrodt J (2007) Meganuclease and transposon mediated transgenesis in medaka. Genome Biology, 8: S10.
33. Brigid H, Rosa B, Frank C, Elizabeth L (1996) Mouse embryo manipulation (2nd edition). Cold Spring Habour Press.
34. Betthauser J, Forsberg E, Augenstein M, Childs L, Eilertsen K, et al. (2000) Production of cloned pigs from in vitro system. Nature Biotechnology, 18: 1055–1059.
35. Perrin A, Buckle M, Dujon B (1993) Asymmetrical recognition and activity of the I-SceI endonuclease on its site and on intron-exon junctions. The EMBO Journal, 12: 2939–2947.
36. Iqbal K, Barg-Kues B, Broll S, Bode J, Niemann H, et al. (2009) Cytoplasmic injection of circular plasmids allows targeted expression in mammalian embryos. Biotechniques, 47: 959–68.
37. Page RL, Butler SP, Subramanian A, Gwazdauskas FC, Johson JL, et al. (1995) Transgenesis in mice by cytoplasmic injection of plylysine/DNA mixtures. Transgenic Research, 4: 353–360.
38. Wilson AM, Aten QT, Toone NC, Black JL, Jensen BD, et al. (2013) Transgene delivery via intracellular electrophoretic nanoinjection. Transgenic Research, 22: 993–1002.
39. Maresca M, Lin VG, Guo N, Yang Y (2013) Obligate ligation-gated recombination (ObLigaRe): custom-designed nuclease-mediated targeted integration through nonhomologous end joining. Genome Research, 23: 539–546.
40. Cristea S, Freyvert Y, Santiago Y, Holmes MC, Urnov FD, et al. (2013) In vivo Cleavage of Transgene Donors Promotes Nuclease-Mediated Targeted Integration. Biotechnology and Bioengineering, 110: 871–880.
41. Sternberg SH, Redding S, Jinek M, Greene EC, Doudna JA (2014) DNA interrogation by the CRISPR RNA-guided endonuclease Cas9. Nature, 507: 62–67.
42. Bevacqua RJ, Canel NG, Hiriart MI, Sipowicz P, Rozenblum GT, et al. (2013) Simple gene transfer technique based on I-SceI meganuclease and cytoplasmic injection in IVF bovine embryos. Theriogenology, 80: 104–113.
43. Yang DS, Wang CE, Zhao BT, Li W, Ouyang Z, et al. (2010) Expression of Huntington's disease protein results in apoptotic neurons in the brains of cloned transgenic pigs. Human Molecular Genetics, doi:10.1093/hmg/ddq313.

11

Transgenic Mice Expressing Yeast CUP1 Exhibit Increased Copper Utilization from Feeds

Xiaoxian Xie[❂], Yufang Ma[❂], Zhenliang Chen, Rongrong Liao, Xiangzhe Zhang, Qishan Wang, Yuchun Pan*

School of Agriculture and Biology, Department of Animal Sciences, Shanghai Jiao Tong University, Shanghai, PR China, Shanghai Key Laboratory of Veterinary Biotechnology, Shanghai, PR China

Abstract

Copper is required for structural and catalytic properties of a variety of enzymes participating in many vital biological processes for growth and development. Feeds provide most of the copper as an essential micronutrient consumed by animals, but inorganic copper could not be utilized effectively. In the present study, we aimed to develop transgenic mouse models to test if copper utilization will be increased by providing the animals with an exogenous gene for generation of copper chelatin in saliva. Considering that the *S. cerevisiae CUP1* gene encodes a Cys-rich protein that can bind copper as specifically as copper chelatin in yeast, we therefore constructed a transgene plasmid containing the *CUP1* gene regulated for specific expression in the salivary glands by a promoter of gene coding pig parotid secretory protein. Transgenic CUP1 was highly expressed in the parotid and submandibular salivary glands and secreted in saliva as a 9-kDa copper-chelating protein. Expression of salivary copper-chelating proteins reduced fecal copper contents by 21.61% and increased body-weight by 12.97%, suggesting that chelating proteins improve the utilization and absorbed efficacy of copper. No negative effects on the health of the transgenic mice were found by blood biochemistry and histology analysis. These results demonstrate that the introduction of the salivary *CUP1* transgene into animals offers a possible approach to increase the utilization efficiency of copper and decrease the fecal copper contents.

Editor: Vladimir V. Kalinichenko, Cincinnati Children's Hospital Medical Center, United States of America

Funding: This work was supported by the National Transgenic Breeding Program (grant no.: 2014ZX08006-004; 2014ZX08009-003-006). The funders had no role in study design, data collection and analysis, decision to publish, or preparation of the manuscript.

Competing Interests: The authors have declared that no competing interests exist.

* Email: panyuchun1963@aliyun.com

❂ These authors contributed equally to this work.

Introduction

Copper is an essential trace element and required for survival by a wide range of species, from yeast to mammals [1]. It functions as a cofactor and is required for structural and catalytic properties of a variety of enzymes because of its capacity to act as an intermediary in the transfer of electrons that makes it central to the catalytic activity of the enzymes [2,3], which are involved in a number of vital biological processes, such as cellular respiration and iron transport, required for growth and development [4]. Thus, Cu supplements were used to treat anemia in animals in the 1920s, and later in chicks [5], pigs [6], infants [7], and adult humans with good success [8].

Feeds provide most of the copper as an essential micronutrient consumed by animals, and drinking water contributes about 6–13% of average daily intake of copper [9,10,11,12]. Most of ingested copper is absorbed in the small intestine, and very small amounts in the stomach [10]. The absorption of copper in the body depends on a variety of factors including its chemical form [10]. Chelated copper has been proven to improve the utilization of copper, which is absorbed more efficiently through an amino acid transport system [13] by increasing intestinal absorption and renal tubular reabsorption of copper, and the chelated form displays increased retention in the body compared with its inorganic form [14,15,16], as has been demonstrated in many compounds, such as copper-lysine [17], organic copper chelates [18], copper carbonate [19] and copper-metallothionein (copper-MT) complex [15].

If inorganic coppers are transformed to copper-MT through binding by the MT produced endogenously by animals, such organic copper could be utilized effectively, and then lower doses of inorganic copper could be added in feed, which would, in turn, lead to reduced fecal copper contents. To investigate the feasibility of this hypothesis, we developed transgenic mice that secrete MT proteins in their saliva. The transgenes used in this study contain the *S. cerevisiae CUP1* gene. It encodes a Cys-rich protein with a low molecular weight that can bind copper as specifically as copper chelatin in yeast. The protein is characterized principally by its high copper-binding capacity and unusual amino acid composition, and it contains 20% cysteine residues [20,21]. Mammalian MT is a metal-binding protein that is present in most tissues. The protein was first found to bind cadmium and zinc [22], but it also binds copper and is a major copper-binding protein in the liver, in addition to binding other metals [16]. Yeast CUP1 has highly divergent primary sequences compared to mammalian MT by reconstructing the phylogenetic tree [23]. However, these proteins all possess identical functional sequence motifs, Cys-X-Cys or Cys-X-X-Cys, and binding of copper to these motifs occurs through the Cys residues [21,24]. In this study, we took advantage of the yeast *CUP1* gene to establish a transgenic mouse model to determine whether endogenous expression of CUP1 can increase copper utilization by mice.

Materials and Methods

Ethics Statement

The FVB and ICR mice varieties were used in this research. All animal procedures received approval from the Institutional Animal Care and Use Committee (IACUC) of Shanghai city, China. The mice were housed in the Animal Care Facility at Shanghai Jiao Tong University (IACUC permit numbers: SYXK (Shanghai) 2013-0052).

Construction of the recombinant plasmids expressing the *CUP1* gene

A fragment, which contained the complete open reading frame (ORF) of the *CUP1* gene, was synthesized based on the published sequence in GenBank (NM_001179185). The vector pPSP (pig parotid secretory protein) was a gift from Ning Li (College of Biological Sciences, China Agricultural University, Beijing, China). The recombinant plasmid pPSP-CUP1 was constructed by insertion of the fragment containing the ORF of the *CUP1* gene, which was digested with *Asc* I (TaKaRa, Japan), into the same endonuclease-digested pPSP vector. The recombinant plasmid was confirmed by restriction analysis and DNA sequencing.

Transgene purification, quantification and pronuclear microinjection

The linear DNA fragment containing the pPSP promoter, signal peptide and *CUP1* gene was obtained by digestion with *Xho* I and *Not* I and subsequently purified by agarose gel electrophoresis as described by Yin et al. [25]. DNA was resuspended in microinjection buffer, which consisted of 0.1 mmol/L ethylenedi-aminetetraacetic acid and 10 mmol/L Tris Cl (pH 7.4) at a concentration of 20 ng/μL, and stored at −20°C.

The injection of transgene DNA was performed according to Hogan et al. [26].

Transgenic examination by PCR and southern blot

The presence of the *CUP1* transgene in the transgenic founders and offspring was confirmed by PCR analysis of genomic DNA derived from tail biopsies and DNA sequencing. PCR was performed with specific primers (forward primer 5′TGTGTAAGCGTGGTAGGTGCTCATC 3′, reverse primer 5′GACACCTACTCAGACAATGCGATGC 3′), and the transgene length was 337 bp. The transgenic founders (G0) were confirmed by Southern blot analysis. Genomic DNA was isolated from the mouse tails using a ZR Genomic DNA-Tissue MiniPrep Kit (Zymo Research, USA). Twenty micrograms of DNA was restriction digested with *Bgl* II and *Ssp* I, fractionated in a 0.8% agarose gel electrophoresis, and transferred to a nylon membrane (Millipore, UK). The fragment containing the complete ORF of yeast *CUP1* was amplified by PCR using specific primers (forward primer 5′TGGGGAATCAGTAGGAAGTCTTGGC 3′, reverse primer 5′CCCCAGAATAGAATGACACCTACTC 3′), and the fragment length was 832 bp. The fragment was then purified with Qiagen PCR purification kits before its use as a probe. The membrane was hybridized with a DIG-labeled CUP1 DNA probe (20 ng). Pre-hybridization and hybridization were performed according to the procedures described by Van Rijs et al. [27].

Pre-hybridization was performed for 1 h at 45°C, hybridization for 6 h at 45°C, and then the membrane was washed twice in 2×standard saline citrate (SSC), 0.1% SDS at 65°C for 20 min, and 0.5×SSC with 0.1% SDS once. Hybridization signals were examined using a Roche DIG DNA Labeling and Detection Kit according to the manufacturer's instructions.

Western blot analysis

Polyclonal antibodies (Santa Cruz Biotechnology, USA) were raised against amino acids 1–61 taken from the CUP1 and represented the full-length CUP1 sequence of *S. cerevisiae*. The β-actin antibodies were purchased from Sigma-Aldrich (St Louis, MO). Western blot analysis was performed according to the methods of Spencer et al. [28].

Approximately 200 μL of saliva per mouse was collected from 6-wk-old mice as described by Hu et al. [29] and stored at -80°C. The proteins were extracted from the tissues and saliva by homogenization in lysis buffer, separated by SDS-PAGE electrophoresis, and transferred to nitrocellulose. Immunoblotting was performed with antibodies against CUP1 (1:400) and β-actin (1:3000), which served as loading controls. As the secondary antibody, a goat anti-rabbit horseradish peroxidase-conjugated antibody was diluted to 1:5000.

Detection of the copper content in mouse manure and changes of body weight

Prepared feed was purchased from Shanghai SLAC Laboratory Animal Co., Ltd., China. The feed contained 10 mg/kg content of copper, which is in concordance with the national nutrition standard GB 14924.3–2010 [30].

The G1 offspring were weaned at 4 wk of age, and the transgenic mice were confirmed as described above. Then, the transgenic and control mice were individually caged under controlled temperature (22±2°C), humidity (40–60%) and lighting (12 h light; 12 h darkness) and fed the prepared feed and water ad libitum. Body weight was recorded once daily for 2 wk. Fecal samples were collected once daily for 2 wk and then placed into 2 mL sterile tubes and dried immediately. The samples were dried (130°C) for 48 h and ashed at 600°C for 4 h in a muffle furnace. Next, they were cooled, weighed, and digested in nitric acid (Merck, Germany) at 95°C for at least 2 h. After filtration, the contents of copper in mouse manure were measured by ICP-MS (7500 Series ICP-MS system; USA). Each digested sample volume was standardized to 5 mL.

Blood biochemistry and histology analysis

Blood samples of approximately 1 mL per mouse were obtained from the retro-orbital venous plexus of the transgenic and control mice using heparinized capillary tubes. Five mice at 6 wk of age and ten mice at 1 yr of age in each group were used for the blood biochemistry analysis.

The blood samples were centrifuged at 3000 rpm for 10 min for the sera. The sera were stored at −80°C prior to blood biochemistry analysis. Nineteen blood biochemical parameters, including Ca (Calcium ion), Fe (ferrum ion), GLU (glucose), CRE (creatinine), CHO (cholesterol), BUN (blood urea nitrogen), AMY (amylase), ALT (alanine aminotransferase), AST (aspartate aminotrasferase), and ALP (alkaline phosphatase), were detected using an auto-analyzer (Hitachi 7180, Hitachi, Japan).

After blood drawing, the mice were sacrificed for histopathology analysis. Tissues (heart, liver, spleen, stomach, kidney, intestine, brain, parotid gland and submandibular gland) were collected and fixed in PBS buffered 10% formalin. The specimens, after paraffin embedding, were sectioned horizontally at 5 μm thickness, stained with hematoxylin and eosinaccording to standard protocol, and observed using a microscope (Nikon, Japan) at an excitation wavelength of 559 nm.

Statistical analysis

The phenotypic data (the fecal ash copper contents and the body-weight increases of transgenic and control mice) were analyzed separately based on a general linear model (SAS 9.3): $y_{ijk} = \mu + s_i + d_j + g_k + e_{ijk}$ Where

y_{ijk}: the phenotypic value

μ: an overall mean

s_i: a fixed paternal effect

d_j: a fixed maternal effect

g_k: a fixed CUP1 gene effect

e_{ijk}: a residual error effect with a normal distribution N $(0, \sigma^2)$

Results

Generation of transgenic mice

The 12.5-kb linear transgene pPSP-CUP1 (construction shown in Fig. S1) was generated by digestion with *Xho* I and *Not* I and introduced into fertilized mouse oocytes through pronuclear injection.

Four male (No. 5, 6: FVB mice; No. 20, 22: ICR mice) and two female (No. 15: FVB mouse; No. 26: ICR mouse) transgenic founder (G0) mice obtained from 29 mice were confirmed by PCR screening and DNA sequencing. Southern blotting was further used to identify the transgene integrated into the genome of the transgenic mice. The transgenes shares the same 832-bp sequences containing the complete ORF of yeast *CUP1* as the probe. The results indicated that the transgene was integrated into the genome (Fig. 1A). Six transgenic founders were mated twice with wild-type mice, of which 4 males transmitted the transgene to their offspring, and 2 females did not pass the transgene to their progeny. A total of 46 G1 transgenic mice were confirmed by PCR amplification from genomic DNA and sequencing among 77 offspring (Table S1).

Expression of yeast *CUP1* transgene in the salivary glands

The *CUP1* transgene mRNA expression in the salivary glands of transgenic founder was analyzed by reverse transcription PCR. The results revealed that *CUP1* was expressed in the parotid and submandibular glands and was barely expressed in the heart, liver, spleen, stomach, kidney, intestine, and brain tissue of the transgenic founders. However, the CUP1 gene was not expressed in all tissues of the control mice (Fig. S2).

The CUP1 protein was detected in the parotid and submandibular glands using anti-CUP1 antibodies to probe western blot analysis in transgenic founders. The level of β-actin in each sample was determined as the control for protein loading. The results indicated the presence of CUP1 in both detected tissues and indicated relatively high expression in the submandibular glands after normalization against β-actin. In contrast, relatively low expression was observed in the parotid glands (Fig. 1B). CUP1 protein was also detected in the salivary fluid of the transgenic mice by western blot analysis (Fig. 1B), and no CUP1 protein was detected in the saliva of the control mice. The molecular mass of the protein containing CUP1 was 9 kDa as identified in the salivary glands and the secreted saliva.

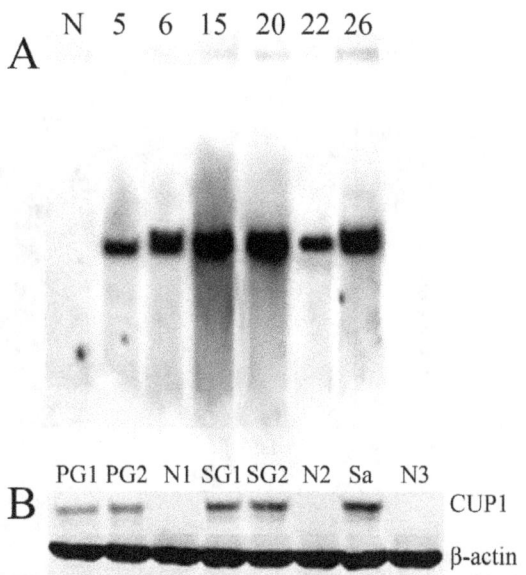

Figure 1. Identification of yeast *CUP1* transgene by southern blot and western blot analysis. (A) Southern blot analysis of the transgenes. Purified genomic DNA from each transgenic founder was digested with *Bgl* II and *Ssp* I, and analyzed by southern blotting using probes specific for *CUP1*. Transgenic founders numbered 5, 6, 15, 20, 22, and 26; N: genomic DNA of control mice as a negative control. (B) Western blot was performed to confirm expression of CUP1 in the parotid and submandibular glands and in the saliva of transgenic lines. The expected protein size is 9 kDa. The lower band is β-actin (42 kDa, used as an internal control). The results demonstrated high expression of CUP1 in the parotid and submandibular glands after normalization against β-actin, as well as that in the saliva of transgenic mice. PG1 and PG2: the parotid glands of transgenic mice; SG1 and SG2: the submandibular glands of transgenic mice; Sa: the saliva of transgenic mice; N1, N2, and N3: the parotid gland, the submandibular gland and the saliva of control mice as negative controls, respectively.

The contents of copper in mouse manure ash and changes of body weight

Forty-four G1 offspring were selected from the total G1 mice considering similar weights and used for further experiments, of which 28 were transgenic mice, and 16 were control mice. To determine the effect of the expressed CUP1 on the transgenic mice, the usage efficiency of copper in the prepared feed was investigated by detecting the contents of copper in mouse manure from the transgenic and control mice at a dietary level of 10 mg/kg copper. For the first week, the transgenic mice exhibited manure ash copper contents of 168.285±18.849 mg/kg, a reduction of 18.41% ($P = 0.0022 < 0.01$) compared with the control mice (206.263±42.307 mg/kg) raised under the same conditions. At the second week, the manure ash copper contents of transgenic mice (171.449±10.767 mg/kg) were significantly lower ($P = 0.0003 < 0.01$) compared with that (218.713±49.831 mg/kg) of the control mice. This represents a reduction of 21.61% in the ash copper contents under the same conditions (Fig. 2A). The effect of the expressed CUP1 on body weight of the transgenic mice was also analyzed. At day 0, the body weight of the control group was 20.531±1.099 g and that of the transgenic group was 20.835±1.214 g. After 1 wk, the body-weight increases of the transgenic mice (6.906±0.998 g) were significantly greater ($P = 0.025 < 0.05$) compared with those (5.063±1.214 g) of the control mice (Fig. 2B). On average, the transgenic mice were 7.2% heavier than the control mice raised under the same conditions.

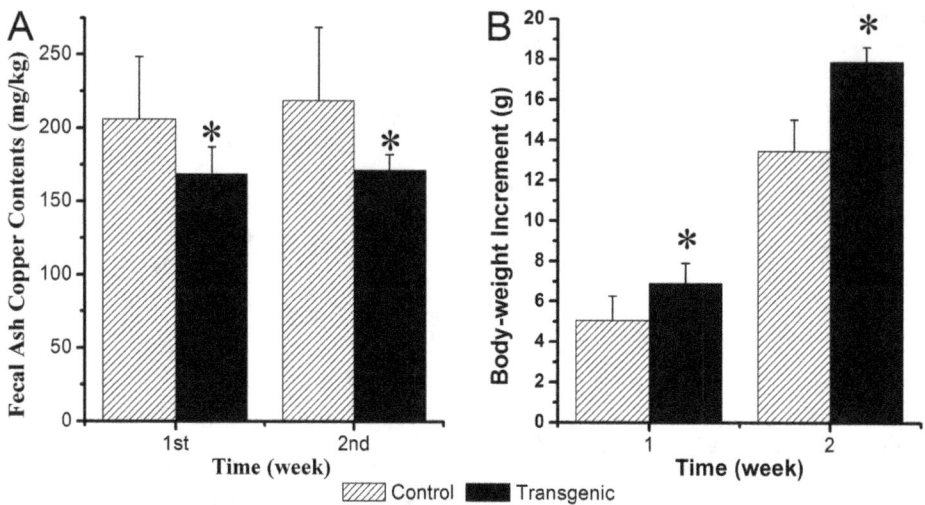

Figure 2. Fecal ash copper contents and changes of body weight of the transgenic and control mice. At 4 wk of age, all G1 offspring were weaned and fed with the prepared feed (copper content, 10 mg/kg). Subsequently, the fecal samples were collected and the body weights of the mice were recorded once daily for 2 wk. (A) The fecal copper contents of the mice were analyzed. The fecal ash copper contents of transgenic mice (n = 28) were significantly lower than the control mice (n = 16), presenting a reduction of 18.41% (*P = 0.0022<0.01) and 21.61% (*P = 0.0003< 0.01) for the first week and the second week, respectively. (B) The changes of body weight of the mice were analyzed. After 1 and 2 wk, the body-weight of transgenic mice (n = 28) increased significantly (*P = 0.025<0.05; *P = 0.019<0.05, respectively) compared with those of the control mice (n = 16) at 7.2% and 12.97%, respectively. 1st: the first; 2nd: the second.

After 2 wk, the body-weight increases of the transgenic mice (17.884±0.728 g) were significantly greater (P = 0.019<0.05) compared with those (13.475±1.556 g) of the control mice (Fig. 2B). On average, the transgenic mice were 12.97% heavier than the control mice raised under the same conditions.

Analysis of blood biochemistry and histology

The blood biochemistry results revealed that the levels of GLU, BUN, AMY, ALT, AST, and ALP in serum were slightly elevated in the transgenic mice at 6 wk of age compared with the control mice. In contrast, the blood concentration of CRE was slightly decreased, and all differences were not significant (P>0.05; Table 1). The similar results were observed in the mice at 1 yr of age, but the level of ALP was slightly decreased in the transgenic mice (Table S2). Differences between the transgenic and control

mice were hardly observed for other serum biochemical parameters.

At the ages of 6 wk and 1 yr, the transgenic mice were in good health and did not exhibit any gross pathological abnormalities or illness. Histological analysis was performed in the tissues (heart, liver, spleen, stomach, kidney, intestine, brain, parotid gland, and submandibular gland) of the mice, and no obvious changes were observed in the tissues of the transgenic mice compared with those of the control mice, and the results were shown in Fig. 3 and Fig. S3, respectively.

Discussion

Metallothionein is a highly conserved family of closely related proteins.

Table 1. Blood biochemistry results in the transgenic and control mice at 6 wk of age.

	ALB (g/L)	GLOB (g/L)	A/G	TP (g/L)	GLU (mmol/L)	CHO (mmol/L)	TG (mmol/L)
Transgenic	18.167±1.002	24.333±0.577	0.757±0.038	42.2±1.323	5.165±0.332	2.133±0.153	1.3±0.386
Control	18.400±1.365	24.5±2.082	0.7±0.026	42.967±3.362	4.335±0.458	1.717±0.097	1.347±0.375
	Ca (mmol/L)	**Fe (mmol/L)**	**HDL (mmol/L)**	**LDL (mmol/L)**	**UA (μmol/L)**	**BUN (mmol/L)**	**CRE (μmol/L)**
Transgenic	1.58±0.115	36.833±1.501	1.513±0.101	0.463±0.148	187.5±25.03	6.6±0.917	3.145±0.518
Control	1.433±0.076	33.767±3.465	1.533±0.163	0.427±0.154	176.867±15.689	5.567±0.586	6.76±0.679
	LDH (U/L)	**AMY (U/L)**	**ALT (U/L)**	**AST (U/L)**	**ALP (U/L)**		
Transgenic	1165.867±161.645	1801.43±195.493	35.433±3.047	130.633±15.314	64.5±8.839		
Control	1173.167±97.779	1627.097±94.849	29.6±0.566	105.467±10.919	51.7±2.83		

The differences between the transgenic (n = 5) and control mice (n = 5) were not significant (P>0.05) for all examined serum biochemical parameters.
ALB: albumin; GLOB: globulin; A/G: ALB/GLOB; TP: total protein; GLU: glucose; CHO: cholesterol; TG: triglyceride; Ca: calcium ion; Fe: ferrum ion; HDL: high density lipoprotein; LDL: low-lipid lipoprotein; UA: uric acid; BUN: blood urea nitrogen; CRE: creatinine; LDH: lactate dehydrogenase; AMY: amylase; ALT: alanine aminotransferase; AST: aspartate aminotrasferase; ALP: alkaline phosphatase.

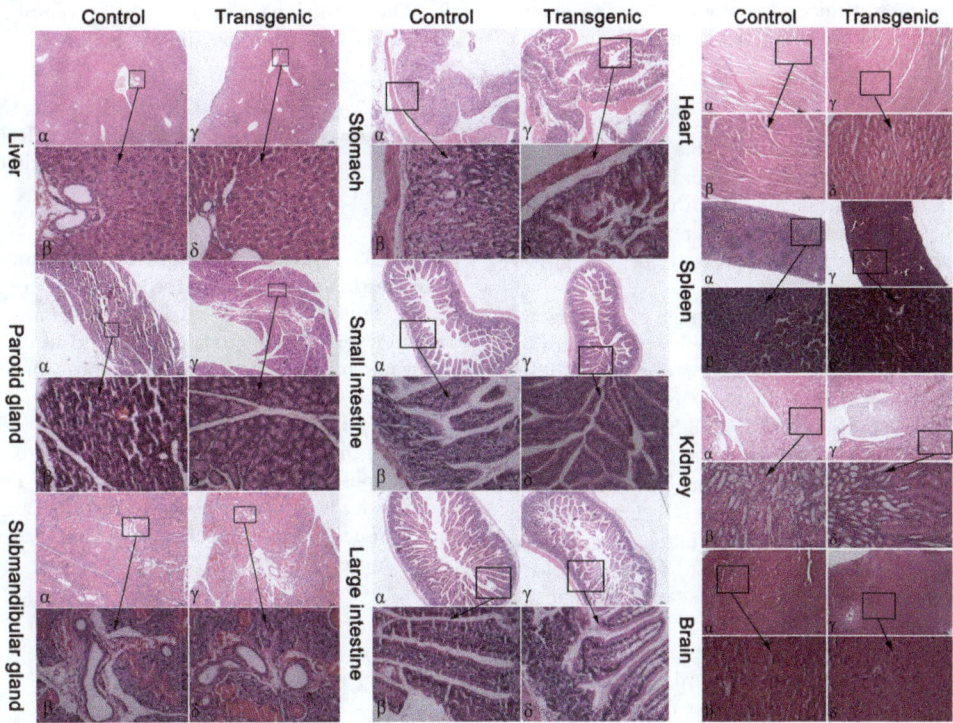

Figure 3. Histological analysis of the tissues of the transgenic and control mice. The heart, liver, spleen, stomach, kidney, small intestine, large intestine, brain, parotid gland, and submandibular gland tissue samples from the transgenic mice (transgenic; n = 5) and control mice (n = 5) at 6 wk of age were analyzed by histology observation. In the above pictures, α and γ are whole tissues, and β and δ are amplified regions of the tissues. The length of the scale bar is 100 μm in all micrographs. The profiles of the tissues of the transgenic and control mice were determined. No obvious changes were observed in the tissues of the transgenic mice compared with those of the control mice.

Yeast CUP1 is a member of the MT family and accounts for copper-binding in *S. cerevisiae* [31]. Mammals have the *MT* gene, and it possesses multiple isoforms [32]. In this study, we took advantage of yeast *CUP1* for transgene as it shares functional sequence identity to mammalian *MT*. Thus, yeast CUP1 binds to copper through Cys residues by the formation of metal-thiolate linkages, as well as the mammalian MT proteins [33], suggesting that this gene may be effective in copper-binding in mammals. This gene has been used for transgene in some organisms. Yeast *CUP1* was introduced to tobacco plants, and its expression contributed to copper content because of its role in copper-binding [32]. In *Drosophila*, this gene was selected as the transgene instead of the endogenous *Mtn* gene, which has a similar structure and function with yeast *CUP1*, to determine its role in binding copper [34].

Many researchers have used the promoter of the mouse salivary gland-specific *PSP* gene to express the exogenous genes such as *phytase* gene in the saliva of transgenic mice, which has been confirmed to be feasible [25,35,36]. A similar phenomenon was examined in our results, and constitutive expression in the PSP/CUP1 mice was notably specific for the parotid and submandibular glands. A three-fold higher expression was detected in submandibular glands compared with the parotid glands, and a similar phenomenon was detected by Mikkelsen et al. (1992) [37], and the opposite observation was reported by Golavan et al. (2001) [36].

Copper as a feed additive is effective in growth enhancement and disease prevention in weanling pigs, and it is widely used in pork production around the world, especially in China [38,39]. However, copper in pig diets heavily exceeds the minimum requirements for normal performance (5–25 mg/kg copper for different classes of pigs) [40], and most of the indigested copper, acting as promotants, by pigs is excreted in the manure (>90%) [41]. The concentrations of copper in pig manure are 5–12 times those in pig feeds with additives [42], which are higher than for other agricultural animals, such as cattle and sheep [43]. The application of pig manure directly onto agricultural land as fertilizer is common practice in China [39]. Because of copper's low mobility and non-degradation, copper can accumulate in soils [44], which leads to environmental consequences. When manure is repeatedly applied as fertilizer, copper can cause surface pollution with severe biological consequences, e.g., causing toxicity to plants, elevating bacterial resistance to toxic metals and increasing human exposure to copper via the food chain [39,40,45]. Still, the present inputs of copper are too high and reducing the contents of copper in the diet should reduce concentrations in the pig manure [46]. In the present study, we validated a method to increase copper utilization efficiency and to decrease the fecal copper content by providing the animals with an exogenous gene for generation of copper chelatin in the saliva. We determined that this approach can reduce mouse fecal copper content by 21.61%. Therefore, this might provide an important clue for preventing the pollution caused by the fecal copper in the pig production.

In addition, we reduced the dietary supply of copper with a 10 mg/kg concentration and demonstrated its feasibility for decreasing fecal copper content. Similar results were observed by Jondreville et al. (2003) [47]. The *CUP1* transgene mice, at a dietary level of 10 mg/kg copper, displayed a body weight-increase response, with the transgene mice 12.97% heavier compared with the control mice after 2 wk. These results are similar to those obtained when animals were fed 250 mg/kg of dietary copper [48]. The decreased fecal copper contents and increased body-weight increases caused by the transgene yeast CUP1 suggest that the *CUP1* transgene most likely enhanced the usage of copper in the diet. Copper is able to stimulate the secretion of several neuropeptides and growth hormones [49], in addition to being a component of the growth factor Iamin [48]. Therefore, copper could influence the growth regulatory system in many ways and might be the main reason for the growth stimulation.

No significant difference was found in a range of markers and the histology of tissues of the transgenic mice compared with the controls. These results suggest that the *CUP1* transgene did not affect the blood composition and histology of the mice. The transgenic mice were confirmed to be in good health and did not exhibit any gross pathological abnormalities or illness.

In summary, we have demonstrated that the repertoire of copper-chelating proteins produced by a model animal can be modified by introduction of *CUP1* transgene into its genome. The salivary copper-chelating proteins in these mice lead to a significant reduction of fecal copper levels and a significant increase of body weight, suggesting the enhancement of the utilization efficiency of the dietary copper by transgenic mice. Our findings provide the essential data toward elucidating the physiological functions of *MT* gene on copper metabolism.

Supporting Information

Figure S1   Construction of the recombinant plasmids expressing *CUP1* gene and confirmation by PCR and restriction. (A) The recombinant plasmid pPSP-CUP1 was constructed by insertion of the fragment containing the ORF of the *CUP1* gene into the same endonuclease-digested pPSP vector.

(B) The recombinant plasmid was confirmed by restriction analysis, DNA sequencing, and by PCR.

Figure S2   RT-PCR analysis of yeast *CUP1* transgene expression. (A, B) The *CUP1* transgene mRNA expression was analyzed by RT-PCR in the salivary glands of the transgenic founders and the control mice, respectively.

Figure S3   Histological analysis of the tissues of the transgenic and control mice at 1 yr of age. The heart, liver, spleen, stomach, kidney, small intestine, large intestine, brain, parotid gland, and submandibular gland tissue samples from the transgenic mice (transgenic; n = 10) and control mice (n = 10) at 1 yr of age were analyzed by histology observation. In the above pictures, α and γ are whole tissues, and β and δ are amplified regions of the tissues. The length of the scale bar is 100 μm in all micrographs. The profiles of the tissues of the transgenic and control mice were determined. No obvious changes were observed in the tissues of the transgenic mice compared with those of the control mice.

Author Contributions

Conceived and designed the experiments: YP XX. Performed the experiments: XX YM ZC RL. Analyzed the data: QW XX XZ. Contributed reagents/materials/analysis tools: QW XX. Wrote the paper: XX.

References

1. Pena MM, Lee J, Thiele DJ (1999) A delicate balance: homeostatic control of copper uptake and distribution. J Nutr 129: 1251–1260.
2. Shils ME, Shike M (2006) Modern nutrition in health and disease: Lippincott Williams & Wilkins.
3. Gambling L, Kennedy C, McArdle HJ (2011) Iron and copper in fetal development. Semin Cell Dev Biol 22: 637–644.
4. Gaetke LM, Chow CK (2003) Copper toxicity, oxidative stress, and antioxidant nutrients. Toxicology 189: 147–163.
5. Elvehjem C, Hart E (1929) The relation of iron and copper to hemoglobin synthesis in the chick. J Biol Chem 84: 131–141.
6. Elvehjem C, Hart E (1932) The necessity of copper as a supplement to iron for hemoglobin formation in the pig. J Biol Chem 95: 363–370.
7. Elvehjem C, Duckles D, Mendenhall DR (1937) Iron versus iron and copper in the treatment of anemia in infants. Am J Dis Child 53: 785–793.
8. Harris ED (2003) Basic and clinical aspects of copper. Crit Rev Clin Lab Sci 40: 547–586.
9. Sandstead HH (1995) Requirements and toxicity of essential trace elements, illustrated by zinc and copper. Am J Clin Nutr 61: 621S–624S.
10. Turnlund JR, Scott KC, Peiffer GL, Jang AM, Keyes WR, et al. (1997) Copper status of young men consuming a low-copper diet. Am J Clin Nutr 65: 72–78.
11. Fitzgerald DJ (1998) Safety guidelines for copper in water. Am J Clin Nutr 67: 1098S–1102S.
12. Potrykus J, Ballou ER, Childers DS, Brown AJ (2014) Conflicting Interests in the Pathogen–Host Tug of War: Fungal Micronutrient Scavenging Versus Mammalian Nutritional Immunity. PLoS Pathog 10: e1003910.
13. Jacob RA, Skala JH, Omaye ST, Turnlund JR (1987) Effect of varying ascorbic acid intakes on copper absorption and ceruloplasmin levels of young men. J Nutr 117: 2109–2115.
14. Coffey R, Cromwell G, Monegue H (1994) Efficacy of a copper-lysine complex as a growth promotant for weaning pigs. J Anim Sci 72: 2880–2886.
15. Bunch R, McCall J, Speer V, Hays V (1965) Copper supplementation for weanling pigs. J Anim Sci 24: 995–1000.
16. Bremner I (1987) Involvement of metallothionein in the hepatic metabolism of copper. J Anim Sci 117: 19–29.
17. Apgar G, Kornegay E, Lindemann M, Notter D (1995) Evaluation of copper sulfate and a copper lysine complex as growth promoters for weanling swine. J Anim Sci 73: 2640–2646.
18. Stansbury W, Tribble L, Orr D (1990) Effect of chelated copper sources on performance of nursery and growing pigs. J Anim Sci 68: 1318–1322.
19. Armstrong T, Cook D, Ward M, Williams C, Spears J (2004) Effect of dietary copper source (cupric citrate and cupric sulfate) and concentration on growth performance and fecal copper excretion in weanling pigs. J Anim Sci 82: 1234–1240.
20. Fogel S, Welch JW, Cathala G, Karin M (1983) Gene amplification in yeast: CUP1 copy number regulates copper resistance. Curr Genet 7: 347–355.
21. Karin M, Najarian R, Haslinger A, Valenzuela P, Welch J, et al. (1984) Primary structure and transcription of an amplified genetic locus: the CUP1 locus of yeast. Proc Natl Acad Sci U S A 81: 337–341.
22. Kägi JH, Vallee BL (1960) Metallothionein: a cadmium-and zinc-containing protein from equine renal cortex. J Biol Chem 235: 3460–3465.
23. Ecker DJ, Butt T, Sternberg E, Neeper M, Debouck C, et al. (1986) Yeast metallothionein function in metal ion detoxification. J Biol Chem 261: 16895–16900.
24. Jensen LT, Howard WR, Strain JJ, Winge DR, Culotta VC (1996) Enhanced effectiveness of copper ion buffering by CUP1 metallothionein compared with CRS5 metallothionein in Saccharomyces cerevisiae. J Biol Chem 271: 18514–18519.
25. Yin H, Fan B, Yang B, Liu Y, Luo J, et al. (2006) Cloning of pig parotid secretory protein gene upstream promoter and the establishment of a transgenic mouse model expressing bacterial phytase for agricultural phosphorus pollution control. J Anim Sci 84: 513–519.
26. Hogan B, Costantini F, Lacy E (1986) Manipulating the mouse embryo: a laboratory manual: Cold spring harbor laboratory Cold Spring Harbor, NY.
27. Van Rijs J, Giguère V, Hurst J, Van Agthoven T, van Kessel AG, et al. (1985) Chromosomal localization of the human Thy-1 gene. Proc Natl Acad Sci U S A 82: 5832–5835.
28. Spencer RL, Kalman BA, Cotter CS, Deak T (2000) Discrimination between changes in glucocorticoid receptor expression and activation in rat brain using western blot analysis. Brain Res 868: 275–286.

29. Hu Y, Nakagawa Y, Purushotham KR, Humphreys-Beher MG (1992) Functional changes in salivary glands of autoimmune disease-prone NOD mice. Am J Physiol- Endocrinol Metab 263: E607–E614.

30. Lv J, Nie Z-K, Zhang J-L, Liu F-Y, Wang Z-Z, et al. (2013) Corn Peptides Protect Against Thioacetamide-Induced Hepatic Fibrosis in Rats. J Med Food 16: 912–919.

31. Richards MP (1989) Recent developments in trace element metabolism and function: role of metallothionein in copper and zinc metabolism. J Nutr 119: 1062–1070.

32. Thomas JC, Davies EC, Malick FK, Endreszl C, Williams CR, et al. (2003) Yeast metallothionein in transgenic tobacco promotes copper uptake from contaminated soils. Biotechnol Prog 19: 273–280.

33. Kaegi JH, Schaeffer A (1988) Biochemistry of metallothionein. Biochemistry 27: 8509–8515.

34. Meyer JL, Hoy MA, Jeyaprakash A (2006) Insertion of a yeast metallothionein gene into the model insect Drosophila melanogaster (Diptera: Drosophilidae) to assess the potential for its use in genetic improvement programs with natural enemies. Biol Control 36: 129–138.

35. Madsen HO, Hjorth JP (1985) Molecular cloning of mouse PSP mRNA. Nucleic Acids Res 13: 1–13.

36. Golovan SP, Hayes MA, Phillips JP, Forsberg CW (2001) Transgenic mice expressing bacterial phytase as a model for phosphorus pollution control. Nat Biotechnol 19: 429–433.

37. Mikkelsen TR, Brandt J, Larsen HJ, Larsen BB, Poulsen K, et al. (1992) Tissue-specific expression in the salivary glands of transgenic mice. Nucleic Acids Res 20: 2249–2255.

38. Cromwell GL, Stahly TS, Monegue HJ (1989) Effects of source and level of copper on performance and liver copper stores in weanling pigs. J Anim Sci 67: 2996–3002.

39. Xiong X, Yanxia L, Wei L, Chunye L, Wei H, et al. (2010) Copper content in animal manures and potential risk of soil copper pollution with animal manure use in agriculture. Resour Conserv Recy 54: 985–990.

40. de Lange K, Nyachoti M, Birkett S (1999) Manipulation of diets to minimize the contribution to environmental pollution. Adv Pork Prod 10: 173–186.

41. Delahaye R, Fong P, Van Eerdt M, Van der Hoek K, Olsthoorn C (2003) Emissie van zeven zware metalen naar landbouwgrond. CBS, Voorburg/Heerlen.

42. Isobe H, Sekimoto H (1999) A survey of the contents of heavy metals in blended feeds, feces and composts of swine. Jpn J Soil Sci Plant Nutr 70: 39–44.

43. Ogiyama S, Sakamoto K, Suzuki H, Ushio S, Anzai T, et al. (2005) Accumulation of zinc and copper in an arable field after animal manure application. Soil Sci Plant Nutr 51: 801–808.

44. Graber I, Hansen JF, Olesen SE, Petersen J, Ostergaard H, et al. (2005) Accumulation of copper and zinc in Danish agricultural soils in intensive pig production areas. Geografisk Tidsskrift-Danish J 105: 15.

45. Poulsen HD (1998) Zinc and copper as feed additives, growth factors or unwanted environmental factors. J Anim Feed Sci 7: 135–142.

46. Aarnink A, Verstegen M (2007) Nutrition, key factor to reduce environmental load from pig production. Livest Sci 109: 194–203.

47. Jondreville C, Revy P, Dourmad J (2003) Dietary means to better control the environmental impact of copper and zinc by pigs from weaning to slaughter. Livest Prod Sci 84: 147–156.

48. Zhou W, Kornegay ET, Lindemann MD, Swinkels JW, Welten MK, et al. (1994) Stimulation of growth by intravenous injection of copper in weanling pigs. J Anim Sci 72: 2395–2403.

49. Tsou R, Dailey R, McLanahan C, Parent A, Tindall G, et al. (1977) Luteinizing hormone releasing hormone (LHRH) levels in pituitary stalk plasma during the preovulatory gonadotropin surge of rabbits. Endocrinology 101: 534–539.

12

Thy1-GCaMP6 Transgenic Mice for Neuronal Population Imaging *In Vivo*

Hod Dana, Tsai-Wen Chen, Amy Hu, Brenda C. Shields, Caiying Guo, Loren L. Looger, Douglas S. Kim, Karel Svoboda*

Janelia Farm Research Campus, Howard Hughes Medical Institute, Ashburn, Virginia, United States of America

Abstract

Genetically-encoded calcium indicators (GECIs) facilitate imaging activity of genetically defined neuronal populations *in vivo*. The high intracellular GECI concentrations required for *in vivo* imaging are usually achieved by viral gene transfer using adeno-associated viruses. Transgenic expression of GECIs promises important advantages, including homogeneous, repeatable, and stable expression without the need for invasive virus injections. Here we present the generation and characterization of transgenic mice expressing the GECIs GCaMP6s or GCaMP6f under the *Thy1* promoter. We quantified GCaMP6 expression across brain regions and neurons and compared to other transgenic mice and AAV-mediated expression. We tested three mouse lines for imaging in the visual cortex *in vivo* and compared their performance to mice injected with AAV expressing GCaMP6. Furthermore, we show that GCaMP6 *Thy1* transgenic mice are useful for long-term, high-sensitivity imaging in behaving mice.

Editor: Benjamin Arenkiel, Baylor College of Medicine, United States of America

Funding: Funding provided by Howard Hughes Medical Institute. The funders had no role in study design, data collection and analysis, decision to publish, or preparation of the manuscript.

Competing Interests: The authors have declared that no competing interests exist.

* Email: svobodak@janelia.hhmi.org

Introduction

Optical imaging of calcium dynamics is commonly used for monitoring activity in neuronal ensembles and micro-compartments. For example, using 2-photon microscopy the activity of hundreds of cells has been measured during behavior [1,2]. Continued development of genetically encoded calcium indicators (GECIs) has enabled a shift from synthetic indicators, such Fluo-4 and Oregon Green BAPTA-1 [3], to protein indicators [4,5,6,7,8,9]. GECIs can be introduced to the brain using relatively noninvasive gene transfer methods such as viral infection using adeno-associated viruses (AAVs) [4,10]. Neurons expressing GECIs can be monitored over weeks [1,4,9,11]. The recently developed GCaMP6 indicators allow sensitive detection of activity, under favorable circumstances down to single action potentials (APs) [7].

AAVs can produce the high intracellular GECI concentrations (~10–100 μM) required for *in vivo* imaging [1,12]. However, AAVs produce different expression levels in neighboring neurons and gradients in expression levels across the infection site [4,7]. In addition, GECI expression levels continue to rise over time until they can cause aberrant cell health [4,12]. The time window for GECI imaging is thus typically limited to a few weeks, depending on the promoter construct, viral titer, injection volume, and other factors. Finally, AAV-mediated gene transfer requires challenging surgeries. Best-practice procedures demand tiny injection volumes (approximately 50 nl) [1], which can result in variable numbers of infected cells with variable GECI expression levels.

Transgenic methods can produce stable expression of GECIs over longer time scales [12,13,14], potentially over the entire lifetime of the mouse, without invasive procedures for gene transfer. Expression patterns and levels are reproducible across different individual animals [12]. Several transgenic GECI mouse lines have been developed [12,13,14,15,16,17,18,19,20], which have demonstrated the advantages of transgenic control of protein expression. Here we present the development and characterization of transgenic mouse lines expressing GCaMP6s and GCaMP6f GECIs under the *Thy1* promoter [20,21,22]. We characterize the brain-wide expression patterns of each line, and the performance of selected lines for cellular *in vivo* imaging.

Materials and Methods

All surgical and experimental procedures were in accordance with protocols approved by the Janelia Farm Institutional Animal Care and Use Committee and Institutional Biosafety Committee.

Transgenic mice

Here we report on GENIE Project (GP) lines GP4.x (where 'x' refers to the founder number) expressing GCaMP6s, and GP5.x expressing GCaMP6f. *Thy1*-GCaMP6-WPRE transgenic mice were generated using standard techniques [23].We included the WPRE (Woodchuck hepatitis virus post-transcriptional regulatory element), which increases mRNA stability and protein expression [24,25]. Genotyping primers were 5′-CATCAGTGCAGCA-GAGCTTC-3′ (forward, anneals to calmodulin sequence in

GCaMP6) and 5′-CAGCGTATCCACATAGCGTA-3′ (reverse, anneals to WPRE sequence). Mouse lines GP4.3, 4.12, 5.5, 5.11 and 5.17 were deposited at The Jackson Laboratory (acquisition numbers provided at end).

Expression analysis

Adult mice (P42–P56) were deeply anesthetized with isofluorane and transcardially perfused with 10 ml 1× Dulbecco's phosphate-buffered saline (DPBS, Life Technologies), followed by 50 ml 4% paraformaldehyde in 0.1 M phosphate buffer. After perfusion, the brains were removed and post-fixed overnight at 4°C. The brains were embedded in 5% agarose in DPBS, and cut into 50 μm thick coronal sections with a vibratome (Leica VT 1200S). Since DPBS contains a saturating concentration of calcium (0.9 mM) GCaMP brightness will be maximal. Every other section was dehydrated with DPBS and coverslipped with Vectashield mounting medium (H-1400, Vector laboratories). The coverslipped sections were imaged using a slide scanner (Nanozoomer, Hamamatsu). Confocal images (LSM 710, Zeiss) were collected for selected brain regions (Fig. 1 and 2, Fig. S1 and S3) [26], using an 20× 0.8 NA objective and standard GFP imaging filters. Individual images were tiled and stitched using commercial software (Zeiss).

For a subset of mouse lines (GP4.3, GP4.12, GP5.5, GP5.11, and GP5.17) we visualized neurons using NeuN to measure the fraction of neurons expressing GCaMP. Staining was performed on sections that were not used for quantification of expression. Sections were blocked with 2% BSA and 0.4% Triton X-100 solution for 1 hour at room temperature to prevent nonspecific antibody binding, followed by incubation overnight at 4°C with mouse anti-NeuN primary antibody (1:500; Millipore, MAB 377) and incubation with Alexa594-conjugated goat-anti-mouse secondary antibody (1: 500; Life Technologies, A11032) for 4 hours at room temperature. Sections were mounted on microscope slides with Vectashield mounting medium (H-1400, Vector laboratories).

We analyzed primary motor cortex (M1), primary somatosensory cortex (S1), primary visual cortex (V1) and hippocampus (CA1, CA3, and Dentate Gyrus, DG) using confocal microscopy. For sample images in each area we identified all labeled cells, segmented their somata, and calculated the somatic GCaMP fluorescence brightness for each cell. For cortical regions, cells were grouped into layer 2/3 (L2/3) and layer 5 (L5) cells. We also counted the fraction of GCaMP labeled cells (green channel) as a fraction of the NeuN stained cells (red channel). To compensate for variations of imaging conditions across time (e.g. changes in the excitation light source intensity), images of a fluorescence

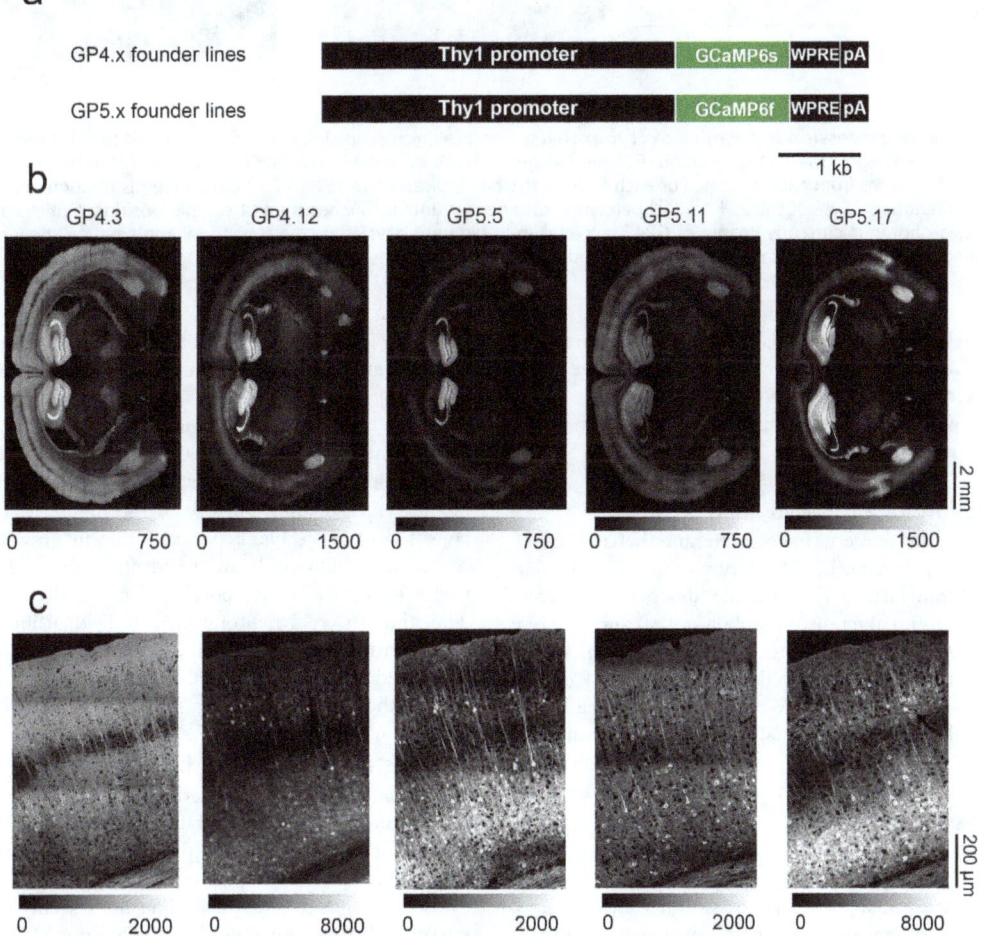

Figure 1. *Thy1* transgenic mice expressing GCaMP6s or GCaMP6f. a. Schematic of the transgene cassettes used to generate GP4.x (top) and GP5.x (bottom) lines. WPRE = Woodchuck hepatitis virus post-transcriptional regulatory element, pA = poly-adenylation tail. **b.** Wide-field images of coronal sections showing GCaMP fluorescence in various transgenic lines. **c.** Representative confocal images (tiled and stitched to show larger field of view) from the somatosensory cortex of the same lines as in **b.** All images show GCaMP6 fluorescence.

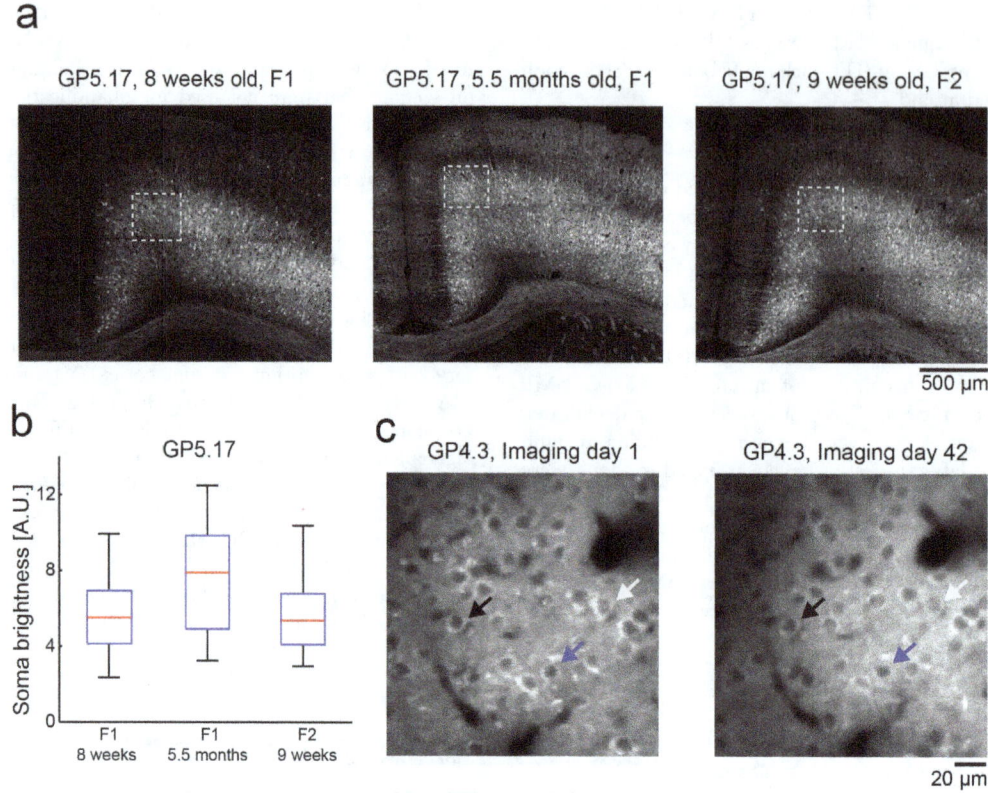

a

GP5.17, 8 weeks old, F1   GP5.17, 5.5 months old, F1   GP5.17, 9 weeks old, F2

500 μm

b   GP5.17

c   GP4.3, Imaging day 1   GP4.3, Imaging day 42

20 μm

Figure 2. Stable GCaMP expression in GP mice over months. a. Confocal microscope images of fixed coronal sections from the motor cortex of GP5.17 mice at different ages (F1- first generation, F2- second generation). **b**. Somatic GCaMP6f brightness for all neurons inside the white rectangles in **a** (51–65 neurons from each animal). For each mouse, the box indicates the 25th to 75th percentile distribution, red line indicates the median, and whisker length is 150% of the 25th to 75th percentile distance, or until it touches the last sample position. Outliers are marked in red crosses. **c**. *In vivo* two-photon microscopy images of GP4.3 mice taken at different days after cranial window implantation show similar expression pattern without filled nuclei. Arrowheads point to three individual cells in both images.

standard, 3.8 μm fluorescent beads (Ultra Rainbow Fluorescent Particles, Bangs Laboratories), were acquired. The average bead brightness was used to normalize the GCaMP signal.

In addition we performed a coarse analysis of expression levels across numerous brain regions (Table 1; Data S1).

Mouse preparation for V1 *in vivo* imaging

For cranial window surgery mice were anesthetized using isoflurane (2.5% for induction, 1.5–2% during surgery). A circular craniotomy (2–2.5 mm diameter) was made above V1 (centered 2.7 mm left, and 0.2 mm anterior to Lambda suture) and covered with 1% agarose. A 3 mm round glass coverslip (no. 1 thickness, Warner Instruments) was cemented to the brain using black dental cement (Contemporary Ortho-Jet). A custom titanium head post was cemented to the skull. The animal was then placed under a microscope on a warm blanket (37°C) and kept anesthetized using 0.5% isoflurane and sedated with chlorprothixene (20–40 μl at 0.33 mg/ml, i.m.) [27].

In vivo mouse imaging in V1

Imaging was performed with a custom-built two-photon microscope with a resonant scanner (designs available at http://research.janelia.org/svoboda/). The light source was a Mai Tai HP 100 femtosecond-pulse laser (Spectra-Physics) running at 940 nm. The objective was a 16× water immersion lens with 0.8 NA (Nikon). Images were acquired using ScanImage 4 (vidriotechnologies.com) [28]. Functional images (512×512 pixels,

250×250 μm^2) of L2/3 cells (100–250 μm under the pia) were collected at 15 Hz. Laser power was 145 mW at the front aperture of the objective.

Visual stimuli were moving gratings generated using the Psychophysics Toolbox [29,30] in MATLAB (Mathworks), presented using an LCD monitor (30×40 cm), placed 25 cm in front of the center of the right eye of the mouse. Each stimulus trial consisted of a 4 s blank period (uniform gray display at mean luminance) followed by a 4 s drifting sinusoidal grating (0.05 cycles/degree, 1 Hz temporal frequency, 8 different directions). The stimuli were synchronized to individual image frames using frame-start pulses provided by ScanImage 4. The monitor subtended an angle of ±38° horizontally and ±31° vertically around the eye of the mouse.

Analysis of V1 functional imaging

All analyses were performed in MATLAB. Regions of interest (ROIs) corresponding identifiable cell bodies were selected using a semi-automated algorithm [6,7]. Depending on the neuron's appearance, annular [4] or circular ROIs were placed over the cytosolic regions of each cell. The fluorescence time course was measured by averaging all pixels within the ROI, after correction for neuropil contamination [31]. The neuropil signal $F_{neuropil}(t)$ surrounding each cell was measured by averaging the signal of all pixels within a 20 μm circular region from the cell center (excluding all somata). The fluorescence signal of a cell body was estimated as

Table 1. GCaMP6 expression at the level of brain regions.

Line	Olfactory bulb	M1	Piriform area	Amygdala	S1	Hippo-campus	Thala-mus	Hypo-thalamus	V1	Cere-bellum	Mid-brain	Pons	Medulla
GP4.1		++	+	+	++	+++	++	+	++	+	+	+	+
GP4.2	+	-	-	-	+	++	+	-	+	+	+	+	+
GP4.3	+	++	+	+	++	+++	+	-	++	-	+	+	+
GP4.4	+	+	+	+	+	++	+	-	+	+	+	+	+
GP4.6	+	+	+	-	+	++	+	-	+	+	+	+	+
GP4.7	+	L6 ++	-	-	L6 ++	-	+	-	L6 ++	+	-	+	-
GP4.9	-	L5a ++	-	+	L5a ++	+++	-	+	L5a ++	-	-	+	-
GP4.12	++	+++	++	++	++	+++	+	+	++	-	+	+	-
GP4.14	-	++	-	-	++	+++	-	-	+	-	-	+	
GP4.15	+	-	-	-	+	+	+	+	+	+	+	+	+
GP4.17	+	++	++	-	++	+++	+	+	+	-	-	+	-
GP5.1	++	++	++	++	++	+++	+	-	++	+	+	+	-
GP5.3	++	++	++	+	++	+++	+	+	++	-	-	++	+
GP5.5	++	++	+	++	++	+++	+	+	++	-	-	+	-
GP5.9		++	+	++	++	+++	+	+	++	+	+	+	-
GP5.10	-	+	-	-	+	+	+	-	+		-	+	-
GP5.11	++	++	+	+	++	++	+	+	++	+	-	+	-
GP5.12	+	+	++	+	++	++	++	+	++	+	+	+	+
GP5.14	+	+	+	+	+	+	+	+	+	+	-	+	+
GP5.15	+	+	+	+	+	+	+	+	+	-	+	+	+
GP5.17	++	+++	+	++	+++	+++	+	+	+++	-	+	++	+
GP5.18	+	+	+	-	+	+	+	+	+	+	-	+	+
GP5.21	+	++	+	++	++	+++	+	+	++	+		+	+

GCaMP6 brightness was scored based on widefield microscope images of coronal sections. Note: no effort was made to separate axonal projections and somato-dendritic signal. Legend: − no signal, + weak signal, ++ moderate signal, +++ strong signal. See Data S1 for raw images.

$$F_{cell\_true}(t) = F_{cell\_measured}(t) - r \cdot F_{neuropil}(t),$$

with $r = 0.7$ [7]. Although we used one value for r across preparations, for optimal neuropil correction r may have to be adjusted for different mouse lines and experimental conditions. Neuropil correction was applied only to cells with baseline fluorescence (F_0) signal stronger than the surrounding neuropil signal by more than 3%; other cells (approximately 10%) were excluded from the analysis because F_0 could not be reliably estimated. After neuropil correction, the $\Delta F/F_0$ of each trial was calculated as $(F - F_0)/F_0$, where F_0 was averaged over a 2 s period for GCaMP6f experiments and 1 s for GCaMP6s experiments immediately before the start of grating stimulation. Visually responsive neurons were defined as cells with $\Delta F/F_0 > 0.05$ during at least one stimulus period, and using ANOVA across blank and eight direction periods ($p < 0.01$) [32].

For calculating the mean response to the preferred stimulus, traces for cells with large responses ($\Delta F/F_0 > 1$) were averaged. Because each cell responded at slightly different times, depending on its receptive field structure, each trace was shifted so that their maxima align. For visual display, traces were smoothed with a 3 sample moving average kernel.

We calculated the decay time of fluorescence after the end of the preferred stimulus. For each cell we averaged responses from five trials; baseline fluorescence and standard deviation were calculated from 1 s (GCaMP6s) or 2 s (GaMP6f) before the start of the stimulus. Only responsive cells with fluorescence response 5 times the standard deviation of the baseline during the last 1 s of the stimulus were analyzed. The time required for each trace to reach half of its peak value (baseline fluorescence subtracted) was calculated by linear interpolation.

AAV injection

Adult mice (P42–56) were anesthetized and injected with AAV-synapsin1-GCaMP6s (AAV-6s) or AAV-synapsin1-GCaMP6f (AAV-6f) into the primary visual cortex (2 injections, 25 nl each, centered 2.5 and 2.9 mm left, and 0.2 mm anterior to Lambda suture) [7]. Ai38 mice were injected with Cre-expressing AAV virus under the human synapsin1 promoter. 3–4 weeks post-injection, mice were implanted with a cranial window and imaged on the same day. After the imaging session, mice were perfused and their brains were fixed. Confocal microscopy images of V1 were collected for expression analysis (Fig. S1); cells with filled nuclei (~5% of the total) were excluded from analysis.

In vivo imaging in anterior lateral motor cortex during behavior

A circular craniotomy (2–3 mm diameter) was made above left anterior lateral motor cortex (ALM) cortex (centered at 2.5 mm anterior and 1.5 mm lateral to Bregma) [33]. The imaging window, constructed from two layers of microscope coverglass [1], was fixed to the skull using cyanoacrylate glue and dental acrylic. Behavioral training started ~2 weeks after window surgery [34]. Imaging was performed using a resonant scanning two-photon microscope controled by ScanImage 4 [28]. Images (512×512 pixels) covering a field of view of ~600×600 μm were acquired at 15 Hz. The laser wavelength was 940 nm and the power used was between 70–120 mW at the front aperture of the objective.

Results

We screened 13 lines of Thy1-GCaMP6s mice and 16 of Thy1-GCaMP6f mice. 11 Thy1-GCaMP6s lines expressed GCaMP6s

(lines 'GP4.x', where 'x' refers to the founder) and 12 Thy1-GCaMP6f expressed GCaMP6f (lines 'GP5.x') (Fig. 1a). Because of the strong dependence of expression level and labeling pattern on transgene cassette integration site in the founder mouse genome [20,21], significant differences were found between these lines. Lines GP4.3, GP4.12, GP5.5, GP5.11, and GP5.17 showed robust expression with some unique features and were analyzed in more depth (Fig. 1b–c; Table 1; Fig. S2, Data S1). Images of tissue sections and an analysis of expression levels across brain regions are available in the supplemental materials (Table 1; Data S1).

Characterization of the GP lines

Expression was similar across different individual mice from the same line (Fig. 2 a, b; Fig. S2). In long-term imaging experiments in adult mice GECI concentration was stable over time (Fig. 2c). With AAV infection, long-term expression can cause accumulation of GCaMP6 in the nucleus, a correlate of cytomorbidity [4,7,12]. In the GP lines GCaMP6 remained excluded from the nucleus (Fig. 2c), even in 11-month old mice in all brain regions examined (data not shown).

We quantified expression across brain regions (Table 1; Data S1) and individual neurons by quantifying somatic native fluorescence in fixed tissue (Fig. 3a). The highest expression was typically seen in the hippocampus. For cortical regions, expression in layers 5 and 6 was usually higher than for layer 2/3, whereas layer 4 cells did not express (Fig. 1b, c, Fig. 3a–d), consistent with other Thy1 transgenic mice [20]. Expression was detected in multiple other brain regions (Table 1; Data S1).

Expression patterns varied across lines. For instance, distinct cortical regions expressed GECIs and different fractions of cells were labeled in these brain regions (Fig. 3). For some lines, e.g. GP4.3 and GP5.11, the majority of pyramidal cells in a particular cortical region were labeled. For other lines, e.g. GP5.17 and GP5.5, only a minor population of the cells was labeled (Fig. 3e–h). Expression levels were generally lower than seen with AAV infection. However, in some brain regions GP4.12 and GP5.17 lines showed expression levels comparable to AAV infection (Fig. 3b, c).

Thy1 transgenics can exhibit transgene expression in specific subsets of cortical projection neurons [20]. In GP5.3 mice L2, but not L3 cells, were labeled over large parts of the cortex; in the neocortex of GP4.7 mice mainly L6 neurons were labeled; in GP4.9 mice mainly L5a cells were labeled; GP4.14 shows sparse labeling of cells in multiple cortical regions (Data S1).

Imaging activity in the visual cortex in vivo

We next performed in vivo functional imaging in L2/3 of the primary visual cortex of transgenic mice (Fig. 4a). Three transgenic lines were tested: GP4.3 with moderate expression levels and a majority of L2/3 pyramidal neurons labeled; GP4.12 with higher expression levels and ~45% of L2/3 pyramidal neurons labeled; GP5.17 with the highest expression levels and ~30% of L2/3 pyramidal neurons labeled (Fig. 3f). Anesthetized mice were presented with oriented gratings moving in eight different directions (Fig. 4a, Methods) [7].

Subsets of GCaMP6 positive cells showed tuned responses to the stimulus (Fig. 4b–d). A majority of the responsive neurons were modulated at the temporal frequency of the moving grating (1 Hz). GP5.17 showed stronger modulation at 1 Hz than GP4.3 and GP4.12, presumably due to the faster kinetics of GCaMP6f vs. GCaMP6s [7]. The transgenic lines showed stronger or similar modulation than AAV-infected mice with the same indicator (Fig. 4d, e).

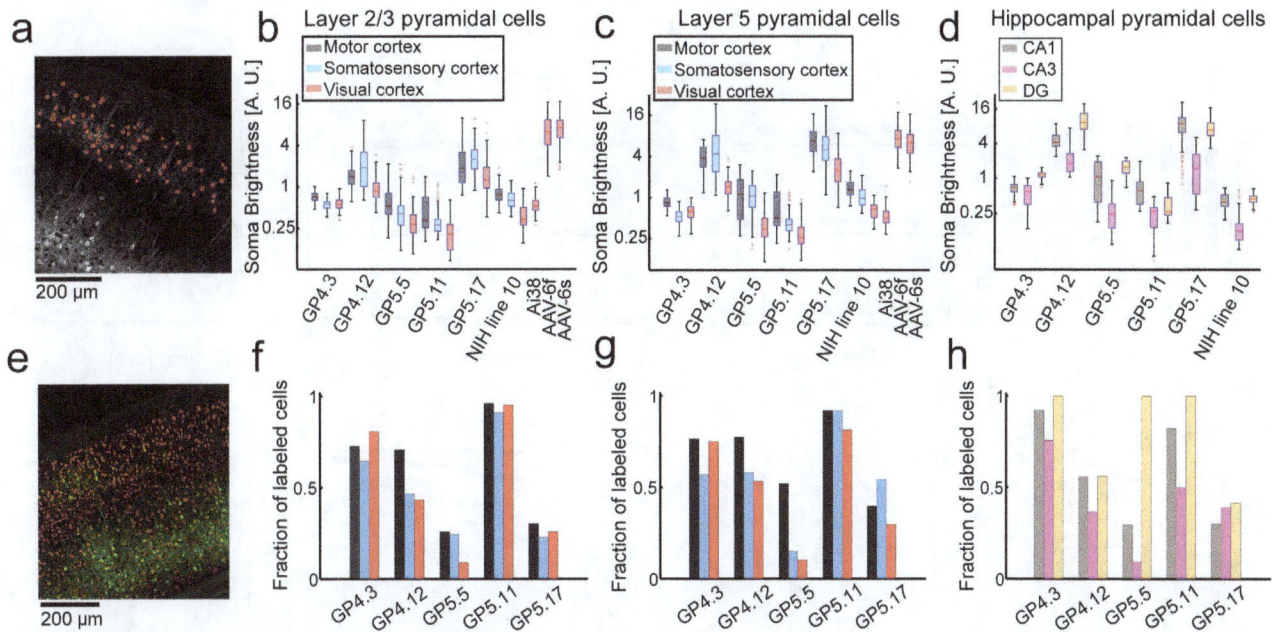

Figure 3. Quantification of GCaMP expression. a. Demonstration of the analysis method used for calculating single-neuron brightness distribution across brain regions. Confocal microscopy images of fixed brain slices were used for segmentation into cell bodies (red rings, nuclei were excluded from somata). Somatic brightness was calculated by averaging all pixels in each segmented cell. **b–d**. Neuronal (somatic) GCaMP6 brightness of labeled neurons in various transgenic and AAV infected mice. NIH line 10 is a *Thy1*-transgenic mouse line expressing GCaMP3 [38]. Ai38 is a Cre-dependent reporter mouse expressing GCaMP3 in the ROSA26 locus [12], here injected with *synapsin1*-Cre AAV. Each box indicates the 25th to 75th percentile distribution with different colors for each brain region, red line indicates the median, and whisker length is 150% of the 25th to 75th percentile distance, or until it touches the last sample position. Outliers are marked in red crosses. **b**, L2/3 pyramidal cells (86–289 cells per line; median, 181). **c**, L5 pyramidal cells (45–230 cells per line; median, 148). **d**, Hippocampal pyramidal cells (25–342 cells per line; median, 113). **e**. Confocal image of GP5.17 fixed tissue (green) counterstained with NeuN (red). **f–h**. Fraction of neurons that are GCaMP6-positive, estimated by counterstaining with NeuN, corresponding to **b–d**, respectively (54–186 cells per line; median, 102).

We analyzed the half-decay time of fluorescence traces after the last response peak during stimulus presentation (Fig. 4f, Methods). The averaged half-decay time was faster for the GP lines than for AAV infected mice (GP5.17, 140 ± 70 ms, $n=52$; versus AAV-GCaMP6f, 350 ± 300 ms, $n=136$; mean$\pm$s.d.) (GP4.3, 360 ± 300 ms, $n=75$; GP4.12, 510 ± 400 ms, $n=446$; versus AAV-GCaMP6s, 510 ± 460 ms, $n=235$). Line GP5.17 shows the fastest responses measured, with decay times in the 150 ms range, close to the decay time expected for cytoplasmic calcium after an action potential [35]. This faster observed kinetics may be associated with lower GECI concentration, consistent with previous experiments [35,36].

Response amplitudes of GP4.3 and GP4.12 were higher than for GP5.17 (Fig. 4b–d, g), as expected from the higher sensitivity of GCaMP6s vs. GCaMP6f [7]. The percentage of L2/3 cells detected as responding varied across the different lines. For the highly expressing GCaMP6s line the fraction was similar to AAV expression (GP4.12, $42.7\pm11.1\%$; 2 mice, 33 FOVs, 1325 cells vs. AAV-6s, $50.9\pm13.7\%$; 3 mice, 23 FOVs, 672 cells). Similarly, for the highly expressing GCaMP6f line the fraction was also comparable to AAV expression (GP5.17, $19.5\pm13.7\%$; 3 mice, 32 FOVs, 731 cells vs. AAV-6f, $27.5\pm17.7\%$; 3 mice, 29 FOVs, 871 cells). For the lower expressing GCaMP6s line the fractions were lower (GP4.3, $8.3\pm7.9\%$; 2 mice, 19 FOVs, 1130 cells), probably because of low fluorescence signal and reduced signal-to-noise ratio, SNR (Fig. 4h).

Imaging activity during behavior

We trained head-fixed GP4.3 mice in a whisker-based object location discrimination task [34,37]. In each trial, a vertical pole was presented in one of two positions (anterior or posterior) during a sample epoch (1.3 s) (Fig. 5a, b). Mice learned to discriminate the location of the pole using their whiskers. During a subsequent delay epoch (1.3 s) mice prepared for the upcoming response. An auditory "go" cue (0.1 s) signaled the beginning of the response epoch, when mice reported the perceived pole position by licking one of two lickports (posterior→"lick right", anterior→"lick left") (Fig. 5a). Mice achieved high levels of performance (mean percent correct>70%). Imaging was performed in the anterior lateral motor cortex (ALM), which is known to be involved in planning and execution of voluntary licking [33,34]. Consistent with histological data (Fig. 2, 3), we observed densely labeled GCaMP6s expressing neurons in L2/3 (Fig. 5b). Neurons were active during specific periods of the trial and for specific licking directions (Fig. 5c). The stable expression level allowed us to image the same cells over times of many weeks. Direction selective cells showed consistent responses (Fig. 5e). The half-decay time constant of calcium transients was 0.65 ± 0.35 s (mean$\pm$s.d., $n=127$, Fig. 5f), significantly faster than GCaMP6s expressed using AAV (1.8 ± 1.1s, $n=84$, $p<10^{-23}$, t-test). We conclude that GP mice are suitable for long-term mapping of behavior-related neuronal dynamics across large cortical areas.

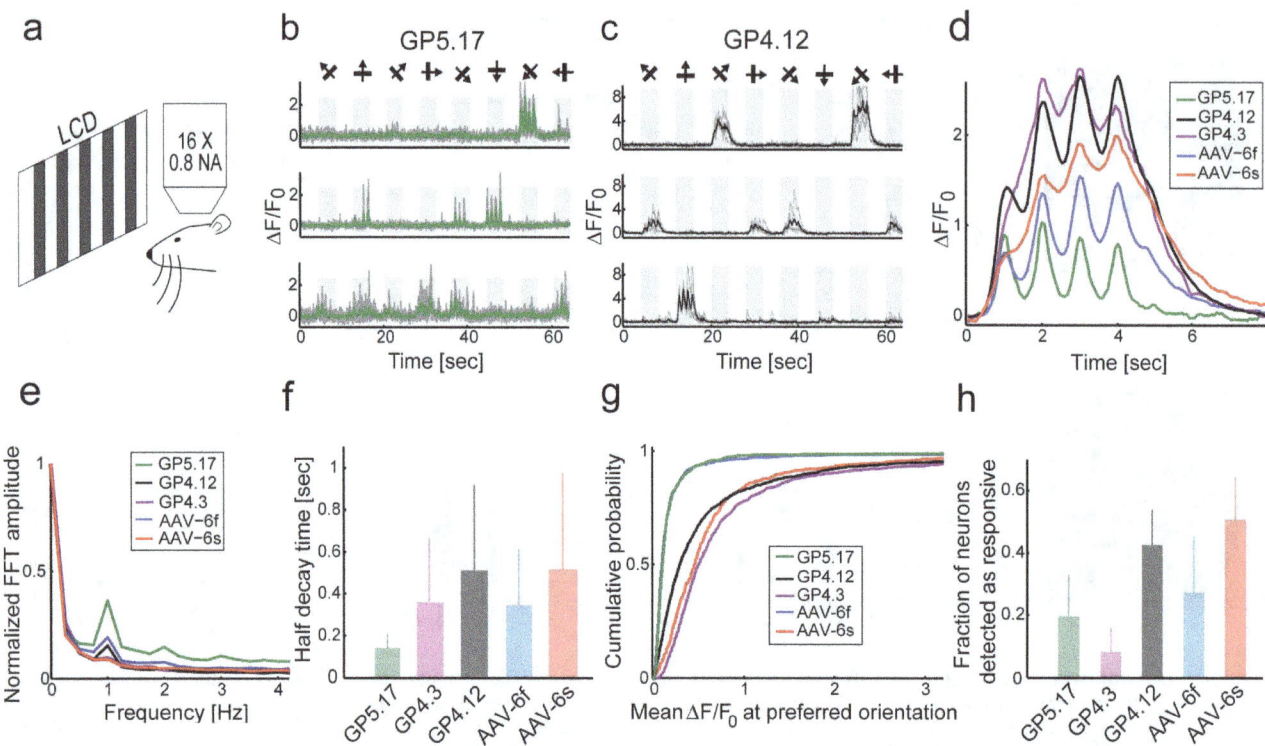

Figure 4. Functional imaging in the visual cortex (V1) of transgenic and AAV infected mice. a. Schematic of the experimental setup. **b.** Responses of three GP5.17 example cells to eight oriented grating stimuli. **c.** Responses of three GP4.12 example cells. **d.** Mean $\Delta F/F_0$ responses to the preferred stimulus for all cells with peak $\Delta F/F_0 > 1$. Cells were aligned according to their response maximum to one of four time points (1, 2, 3, or 4 s), and each stimulus lasts 4 s (average of 91 cells for GP5.17, 124 cells for GP4.3, 362 cells for GP4.12, 83 cells for AAV-6f, and 224 cells for AAV-6s). **e.** Fourier transform of the response to the preferred stimulus (median across cells). The 1 Hz peak corresponds to the frequency of the drifting grating. **f.** Half-decay time (mean±s.d.) after the last response peak during stimulus presentation (n = 52, 75, 446, 136, and 235 cells for GP5.17, GP4.3, GP4.12, AAV-6f, and AAV-6s respectively). **g.** Distribution of $\Delta F/F_0$ responses to the preferred stimulus. **h.** Fraction of statistically significant responsive cells (mean±s.d., n = 731,1130, 1325, 871, and 672 cells, for GP5.17, GP4.3, GP4.12, AAV-6f, and AAV-6s respectively).

Discussion

We generated multiple transgenic mouse lines with stable and reproducible expression of GCaMP6s and GCaMP6f under the control of the *Thy1* promoter ('*Thy1*-GCaMP6' lines). Each line has a unique expression pattern, a hallmark of *Thy1* transgenics [20,22]. Expression was distributed across numerous brain regions and cell types. Selected lines showed sufficient expression levels for cellular *in vivo* imaging with good signal-to-noise ratio, obviating the need for AAV injection and the associated surgery. GCaMP6 expression was stable across many months, without signs of cytotoxicity. The sensitivity and kinetics of GCaMP6s and GCaMP6f make the *Thy1*-GCaMP6 mice a preferred choice for long-term cellular imaging of neuronal populations in the intact brain.

We characterized GCaMP expression and functional signals in the *Thy1*-GCaMP6 mice and compared them to previously published transgenic mice and AAV infected mice (Table 1, Fig. S1 and S2, Data S1). Neocortical expression levels in *Thy1*-GCaMP6 mice are 2–10 fold lower than typical conditions of AAV infection (Fig. 3). Since AAV infection produces [GCaMP] of approximately 80 μM [1,12], we estimate that expression levels in the *Thy1* mice are on the order of 8–40 μM. The *Thy1*-GCaMP6 mice showed higher cortical expression levels than two GCaMP3 transgenic lines, Ai38 [12] and NIH line 10 [38]. Two lines, GP4.12 and GP5.17, showed comparable expression levels to AAV infected animals in selected cortical regions. No signs of

cytomorbidity (*e.g.* nuclear filling of cells) was observed in any of the GP transgenic lines, even after many months of expression.

We tested *Thy1*-GCaMP6 mice for *in vivo* imaging in anesthetized and awake, behaving mice. We observed robust signals with faster kinetics compared to AAV infected mice (Fig. 4d,f, Fig. 5f). This enhanced speed may be explained by the lower concentration of GCaMP6 in the *Thy1*-GCaMP6 mice (Fig. 3b, c) and weaker calcium buffering [35,36]. In *Thy1*-GCaMP6 mice expression was stable over time (Fig. 2). Maintaining stable expression level over time is challenging with AAV-mediated expression [1]. AP detection following visual stimulation was similar to AAV infected mice, *i.e.* superior to Oregon Green BAPTA-1 [7].

There are several drawbacks in using *Thy1*-GCaMP6 mice. First, the lower expression levels (Fig. 3) require higher laser power (~50%–100% higher). However, we did not observe laser-induced damage in brain tissue even after multiple imaging sessions involving continuous imaging over one hour. Photobleaching was negligible. Second, *Thy1*-GCaMP6 mice show mosaic expression, unevenly distributed across different brain regions and cortical layers. Experiments performed across multiple brain areas may require different transgenic mouse lines. Third, the *Thy1* promoter drives expression mostly in projection neurons. Other types of neurons, including GABAergic neurons, are not accessible using this strategy.

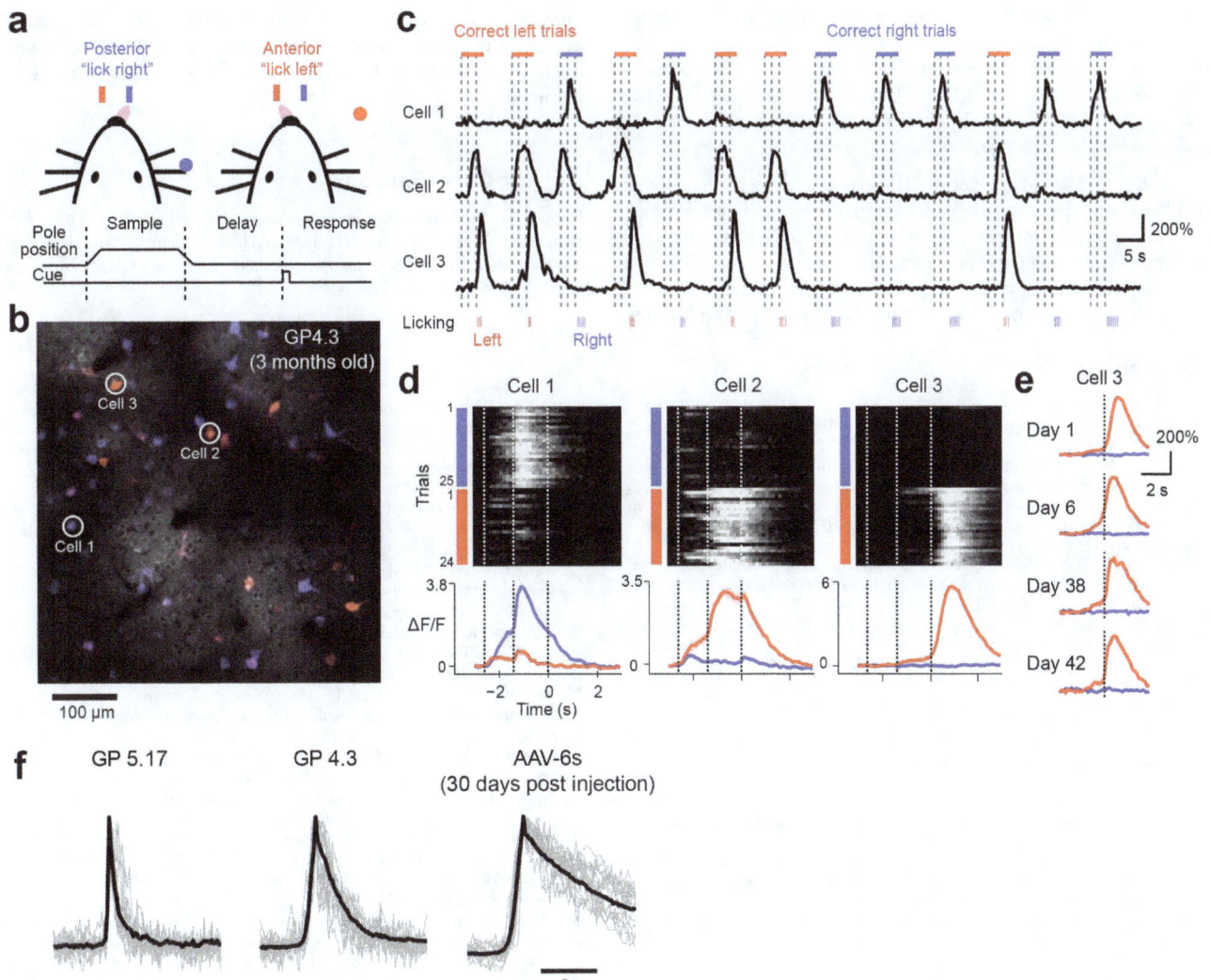

Figure 5. ALM functional imaging using GP4.3 during behavior. a. Schematic of the object localization behavior (see ref. [34] for details). **b.** An image of GCaMP6s labeled neurons in a GP4.3 mouse, 160 µm below the pia. Blue and red colors indicate cells that responded during lick-right and lick-left trials, respectively. Cue = auditory "go" signal. **c.** Fluorescent responses of three example cells indicated in **b**. **d.** Top, single-trial responses of the same neurons in **b** sorted according to trial type (blue: lick-right, red: lick-left). Bottom, trial-averaged response. **e.** The responses of cell 3 measured over 4 behavioral sessions spanning more than one month. **f.** Peak normalized fluorescent transients of GCaMP expressing neurons in GP5.17, GP4.3 and AAV-GCaMP6s injected mice (0.2±0.1 s, 0.65±0.35 s, and 1.8±1.1s mean±s.d; n = 207, 771, and 369 cells for GP5.17, GP4.3, and AAV-6s respectively).

Availability

Lines GP 4.3, 4.12, 5.5, 5.11, and 5.17 are available at Jackson Laboratories (http://jaxmice.jax.org), with stock numbers 024275, 025776, 024276, 024339, and 025393, respectively. For other lines please contact the GENIE project (kimd@janelia.hhmi.org). AAV viruses are available at the University of Pennsylvania Vector Core (http://www.med.upenn.edu/gtp/vectorcore/Catalogue.shtml).

Supporting Information

Figure S1 AAV-mediated expression *vs*. transgenic expression. Images of fixed tissue coronal sections of AAV mediated expression (GCaMP6s, left image) and transgenic expression (GP4.12, right image). For the AAV-injected mouse, two injections (25 nl each, *synapsin1*-AAV GCaMP6s) were made in adjacent locations (0.4 mm) in mouse V1, resulting in typical inhomogeneous expression with several nuclear-filled cells (imaged 4 weeks after the AAV injection). The transgenic GCaMP expression shows no filled cells (P56).

Figure S2 Quantification of GCaMP expression for multiple GP lines. Somatic GCaMP6 brightness of labeled neurons in various transgenic GP lines. For 5 lines (GP4.3, GP4.12, GP5.5, GP5.11, and GP5.17) more than one mouse was analyzed, and GCaMP brightness for each individual animal is presented (*i.e.* GP4.3A, GP4.3B, *etc.*). Somatic brightness distribution for GP4.x (upper row), GP5.x (lower row), layer 2/3 cells (left column) and layer 5 cells (right column) is shown. Each box indicates the 25th to 75th percentile distribution in different colors for each brain region, red line indicates the median, and whisker length is 150% of the 25th to 75th percentile distance, or until it touches the last sample position. Outliers are marked in red crosses.

Figure S3 GCaMP6 expression in the olfactory bulb. Confocal microscope images of fixed coronal sections show different expression patterns in the olfactory bulb. Mitral cells are labeled in GP4.3 and GP5.11 lines, whereas lines GP4.12, GP5.5, and GP5.17 show brighter signal in the granule layer.

Data S1 Widefield images of GP lines coronal sections. Widefield microscopy images were taken using a slide scanner (Nanozoomer, Hamamatsu) with a ×20 0.75 NA air objective (Olympus). Imaging conditions were kept constant across time, but note the different greyscale range used for presenting the different lines. Sections thickness was 50 μm; every second sections was

mounted and used for imaging (see Methods section for details). Sections were mounted from anterior to posterior. For several lines (such as GP4.2 and GP5.18) only a subset of sections were mounted.

Author Contributions

Conceived and designed the experiments: HD TWC LLL DSK KS. Performed the experiments: HD TWC AH BCS. Analyzed the data: HD TWC. Contributed reagents/materials/analysis tools: CG. Wrote the paper: HD TWC LLL DSK KS.

References

1. Huber D, Gutnisky D, Peron S, O'Connor D, Wiegert J, et al. (2012) Multiple dynamic representations in the motor cortex during sensorimotor learning. Nature 484: 473–478.
2. Ahrens MB, Li JM, Orger MB, Robson DN, Schier AF, et al. (2012) Brain-wide neuronal dynamics during motor adaptation in zebrafish. Nature 485: 471–477.
3. Stosiek C, Garaschuk O, Holthoff K, Konnerth A (2003) In vivo two-photon calcium imaging of neuronal networks. Proc Natl Acad Sci USA 100: 7319–7324.
4. Tian L, Hires SA, Mao T, Huber D, Chiappe ME, et al. (2009) Imaging neural activity in worms, flies and mice with improved GCaMP calcium indicators. Nature Methods 6: 875–881.
5. Horikawa K, Yamada Y, Matsuda T, Kobayashi K, Hashimoto M, et al. (2010) Spontaneous network activity visualized by ultrasensitive Ca(2+) indicators, yellow Cameleon-Nano. Nature methods 7: 729–732.
6. Akerboom J, Chen TW, Wardill TJ, Tian L, Marvin JS, et al. (2012) Optimization of a GCaMP Calcium Indicator for Neural Activity Imaging. The Journal of Neuroscience 32: 13819–13840.
7. Chen TW, Wardill TJ, Sun Y, Pulver SR, Renninger SL, et al. (2013) Ultrasensitive fluorescent proteins for imaging neuronal activity. Nature 499: 295–300.
8. Thestrup T, Litzlbauer J, Bartholomäus I, Mues M, Russo L, et al. (2014) Optimized ratiometric calcium sensors for functional in vivo imaging of neurons and T lymphocytes. Nature methods 11: 175–182.
9. Margolis DJ, Lütcke H, Schulz K, Haiss F, Weber B, et al. (2012) Reorganization of cortical population activity imaged throughout long-term sensory deprivation. Nature Neuroscience 15: 1539–1546.
10. Wallace DJ, Zum Alten Borgloh SM, Astori S, Yang Y, Bausen M, et al. (2008) Single-spike detection in vitro and in vivo with a genetic Ca(2+) sensor. Nature Methods 5: 797–804.
11. Peters AJ, Chen SX, Komiyama T (2014) Emergence of reproducible spatiotemporal activity during motor learning. Nature 510: 263–267.
12. Zariwala HA, Borghuis BG, Hoogland TM, Madisen L, Tian L, et al. (2012) A Cre-dependent GCaMP3 reporter mouse for neuronal imaging in vivo. The Journal of Neuroscience 32: 3131–3141.
13. Heim N, Garaschuk O, Friedrich MW, Mank M, Milos RI, et al. (2007) Improved calcium imaging in transgenic mice expressing a troponin C–based biosensor. Nature Methods 4: 127–129.
14. Direnberger S, Mues M, Micale V, Wotjak CT, Dietzel S, et al. (2012) Biocompatibility of a genetically encoded calcium indicator in a transgenic mouse model. Nature Communications 3: 1031.
15. Hasan MT, Friedrich RW, Euler T, Larkum ME, Giese G, et al. (2004) Functional fluorescent Ca2+ indicator proteins in transgenic mice under TET control. PLoS Biology 2: e163.
16. Díez-García J, Matsushita S, Mutoh H, Nakai J, Ohkura M, et al. (2005) Activation of cerebellar parallel fibers monitored in transgenic mice expressing a fluorescent Ca2+ indicator protein. European Journal of Neuroscience 22: 627–635.
17. Tallini YN, Ohkura M, Choi B-R, Ji G, Imoto K, et al. (2006) Imaging cellular signals in the heart in vivo: Cardiac expression of the high-signal Ca2+ indicator GCaMP2. Proceedings of the National Academy of Sciences of the USA 103: 4753–4758.
18. Tallini YN, Brekke JF, Shui B, Doran R, Hwang S-m, et al. (2007) Propagated Endothelial Ca2+ Waves and Arteriolar Dilation In Vivo Measurements in Cx40BAC–GCaMP2 Transgenic Mice. Circulation Research 101: 1300–1309.
19. Atkin SD, Patel S, Kocharyan A, Holtzclaw LA, Weerth SH, et al. (2009) Transgenic mice expressing a cameleon fluorescent Ca2+ indicator in astrocytes and Schwann cells allow study of glial cell Ca2+ signals in situ and in vivo. Journal of Neuroscience Methods 181: 212–226.
20. Chen Q, Cichon J, Wang W, Qiu L, Lee S-JR, et al. (2012) Imaging Neural Activity Using Thy1-GCaMP Transgenic Mice. Neuron 76: 297–308.
21. Caroni P (1997) Overexpression of growth-associated proteins in the neurons of adult transgenic mice. Journal of Neuroscience Methods 71: 3–9.
22. Feng G, Mellor RH, Bernstein M, Keller-Peck C, Nguyen QT, et al. (2000) Imaging neuronal subsets in transgenic mice expressing multiple spectral variants of GFP. Neuron 28: 41–51.
23. Behringer R, Gertsenstein M, Vintersten Nagy K, Nagy A (2013) Manipulating the Mouse Embryo: A Laboratory Manual. Cold Spring Harbor, NY, USA: Cold Spring Harbor Laboratory Press.
24. Donello JE, Loeb JE, Hope TJ (1998) Woodchuck hepatitis virus contains a tripartite posttranscriptional regulatory element. Journal of Virology 72: 5085–5092.
25. Loeb JE, Cordier WS, Harris ME, Weitzman MD, Hope TJ (1999) Enhanced expression of transgenes from adeno-associated virus vectors with the woodchuck hepatitis virus posttranscriptional regulatory element: implications for gene therapy. Human Gene Therapy 10: 2295–2305.
26. Dong HW (2008) The Allen reference atlas: A digital color brain atlas of the C57Bl/6J male mouse. Hoboken, NJ, USA: John Wiley & Sons Inc.
27. Niell CM, Stryker MP (2008) Highly selective receptive fields in mouse visual cortex. J Neurosci 28: 7520–7536.
28. Pologruto TA, Sabatini BL, Svoboda K (2003) ScanImage: flexible software for operating laser scanning microscopes. Biomed Eng Online 2: 13.
29. Brainard DH (1997) The Psychophysics Toolbox. Spatial Vision 10: 433–436.
30. Pelli DG (1997) The VideoToolbox software for visual psychophysics: transforming numbers into movies. Spatial Vision 10: 437–442.
31. Kerlin AM, Andermann ML, Berezovskii VK, Reid RC (2010) Broadly tuned response properties of diverse inhibitory neuron subtypes in mouse visual cortex. Neuron 67: 858–871.
32. Ohki K, Chung S, Ch'ng YH, Kara P, Reid RC (2005) Functional imaging with cellular resolution reveals precise micro-architecture in visual cortex. Nature 433: 597–603.
33. Komiyama T, Sato TR, O'Connor DH, Zhang YX, Huber D, et al. (2010) Learning-related fine-scale specificity imaged in motor cortex circuits of behaving mice. Nature 464: 1182–1186.
34. Guo ZV, Li N, Huber D, Ophir E, Gutnisky DA, et al. (2014) Flow of cortical activity underlying a tactile decision in mice. Neuron 81: 179–194.
35. Helmchen F, Imoto K, Sakmann B (1996) Ca2+ buffering and action potential-evoked Ca2+ signaling in dendrites of pyramidal neurons. Biophys J 70: 1069–1081.
36. Hires SA, Tian L, Looger LL (2008) Reporting neural activity with genetically encoded calcium indicators. Brain Cell Biol 36: 69–86.
37. Guo ZV, Hires SA, Li N, O'Connor DH, Komiyama T, et al. (2014) Procedures for behavioral experiments in head-fixed mice. PloS ONE 9: e88678.
38. Xu NL, Harnett MT, Williams SR, Huber D, O'Connor DH, et al. (2012) Nonlinear dendritic integration of sensory and motor input during an active sensing task. Nature 492: 247–251.

13

Evaluation of the Agronomic Performance of Atrazine-Tolerant Transgenic *japonica* Rice Parental Lines for Utilization in Hybrid Seed Production

Luhua Zhang[1☉]**, Haiwei Chen**[1☉¤]**, Yanlan Li**[1]**, Yanan Li**[1]**, Shengjun Wang**[2]**, Jinping Su**[2]**, Xuejun Liu**[2]**, Defu Chen**[1]*****, Xiwen Chen**[1]*****

1 Laboratory of Molecular Genetics, College of Life Sciences, Nankai University, Tianjin, China, **2** Tianjin Crop Research Institute, Tianjin, China

Abstract

Currently, the purity of hybrid seed is a crucial limiting factor when developing hybrid *japonica* rice (*Oryza sativa* L.). To chemically control hybrid seed purity, we transferred an improved atrazine chlorohydrolase gene (*atzA*) from *Pseudomonas* ADP into hybrid *japonica* parental lines (two maintainers, one restorer), and Nipponbare, by using *Agrobacterium*-mediated transformation. We subsequently selected several transgenic lines from each genotype by using PCR, RT-PCR, and germination analysis. In the presence of the investigated atrazine concentrations, particularly 150 µM atrazine, almost all of the transgenic lines produced significantly larger seedlings, with similar or higher germination percentages, than did the respective controls. Although the seedlings of transgenic lines were taller and gained more root biomass compared to the respective control plants, their growth was nevertheless inhibited by atrazine treatment compared to that without treatment. When grown in soil containing 2 mg/kg or 5 mg/kg atrazine, the transgenic lines were taller, and had higher total chlorophyll contents than did the respective controls; moreover, three of the strongest transgenic lines completely recovered after 45 days of growth. After treatment with 2 mg/kg or 5 mg/kg of atrazine, the atrazine residue remaining in the soil was 2.9–7.0% or 0.8–8.7% respectively, for transgenic lines, and 44.0–59.2% or 28.1–30.8%, respectively, for control plants. Spraying plants at the vegetative growth stage with 0.15% atrazine effectively killed control plants, but not transgenic lines. Our results indicate that transgenic *atzA* rice plants show tolerance to atrazine, and may be used as parental lines in future hybrid seed production.

Editor: Jin-Song Zhang, Institute of Genetics and Developmental Biology, Chinese Academy of Sciences, China

Funding: This work was supported by the Key Project of Tianjin Science and Technology Support Program (11ZCGYNC01000), the Key Program of the Natural Science Foundation of Tianjin (12YFJZJC01700, 14JCZDJC34100), the grant of Natural Science Foundation of China (No. 31070273, 31070717) and the 111 Project (No. B08011). The funders had no role in study design, data collection and analysis, decision to publish, or preparation of the manuscript.

Competing Interests: The authors have declared that no competing interests exist.

* Email: chendefu@nankai.edu.cn (DC); xiwenchen@nankai.edu.cn (XC)

☉ These authors contributed equally to this work.

¤ Current address: College of Life Sciences, Chifeng College, Chifeng, China

Introduction

Rice (*Oryza sativa*) is one of the most important staple food crops globally. According to the National Grain and Oil Information Center, the area of China planted with rice in 2012 was 3.0×10^7 hm², including 9.0×10^6 hm² of *japonica* rice [1]. *Japonica* rice is mainly planted in the northern region of the Qinling Mountains–Huai River, and its planted area has increased in recent years because of its high quality and good taste. *Japonica* rice production is currently dominated by conventional varieties, with hybrid rice accounting for only 3% of the cultivated area. On the other hand, *indica* hybrid rice represents 70–80% of the total planted area of *indica* rice [2]. Therefore, there is considerable potential for the development of *japonica* hybrid rice. An increase in the annual planted area of *japonica* hybrid rice from 3% to 50%, i.e., to reach 4.0×10^6 hm², is estimated to lead to the production of 3.5×10^9 kg of high-quality grain (www.cngrain.com/Publish/qita/200503/207290), thereby contributing considerably to meet consumer's demand for high-quality food both in China and globally.

The three-line system is a traditional and effective production method for hybrid *japonica* rice seed [3]. The most widely used male sterile line in the system is BT-type cytoplasmic male sterile (CMS). However, the panicle of this line is loosely enclosed when heading, and this appearance closely resembles that of the maintainer. This makes it difficult for farmers to distinguish the BT-CMS line when eliminating off-type plants [2]. Furthermore, the BT CMS line has good restorability, and may therefore be easily pollinated with exotic pollens that contaminated during mechanical harvesting and storage of seeds, and also with exotic pollens from other plants [4]. The use of contaminated CMS lines in seed production results in decreased hybrid seed purity. Therefore, off-type contamination must be eliminated as early as possible. This is largely a manual process and requires considerable labor input, particularly in Asia. On the one hand, the need for increased labor will increase the price of hybrid seeds, while on the other hand, the increase in manual procedures may lead to the production of false hybrids. Furthermore, as the Chinese economy develops, increasing numbers of young men are leaving their home towns to seek work in the cities, leaving the elderly and women to work on the farms. The transformation of heavy and complex farming to light and simple farming is therefore becoming increasingly important. Thus, ensuring hybrid seed purity and reducing labor costs are two key issues in hybrid *japonica* rice seed production.

Genetic engineering, especially herbicide resistance engineering, provides an efficient means of controlling purity in hybrid seed production. Yan first proposed a strategy of utilizing herbicide resistance genes to chemically control purity in hybrid seed production [5]. Since then, two-line hybrid rice production has been extensively investigated [6–12] and progress has recently been reviewed [13]. Additionally, some transgenic hybrid rice combinations have been used in field trials [7,8,10,12]. However, the research has mainly focused on the *bar* gene isolated from *Streptomyces hygroscopicus* [6,8,10,12]; other genes, such as the EPSPS (5-enolpyruvylshikimate-3-phosphate synthase) gene from *Agrobacterium* strain CP4, and the protoporphyrinogen oxidase gene from *Bacillus subtilis*, have rarely been investigated [11]. If the strategy is proven to be effective, chemical control of hybrid seed purity will be mainly dependent on an herbicide with a single mode of action, and this will hinder sustainable weed management.

Atrazine (6-chloro-N^2-ethyl-N^4-isopropyl-1,3,5-triazine-2,4-diamine) is a triazine herbicide, and is commonly used in maize, sorghum, and sugarcane fields [14]. By inhibiting electron transport to plastoquinone in the photosystem PSII, atrazine terminates photosynthesis and kills weeds [15]. Atrazine was once the most widely used herbicide worldwide, because of its low cost and high effectiveness [14]. We previously isolated an atrazine chlorohydrolase gene (*atzA*) from a soil bacterium *Pseudomonas* ADP [16], and modified this gene by using directed evolution, to improve the enzymatic activity [17]. In the present study, we transferred the improved *atzA* gene into breeding hybrid *japonica* parental lines. Our results indicate that the transgenic *atzA* rice lines show tolerance to atrazine, and may be used as parental lines to chemically improve seed purity in hybrid seed production.

Materials and Methods

Construction of plant expression vector

Ubiquitin promoter and an improved atrazine chlorohydrolase gene *atzA*-22-4 [17] were respectively amplified from pSTAR-LING, an RNAi intermediate vector for monocots (a kind gift from the Commonwealth Scientific and Industrial Research Organization, Australia), and *AtzA*-22-4 using primers SL-Ubi-F/SL-Ubi-R and atzA-TJ-F/atzA-TJ-R (Table 1). After respectively inserted into TaKaRa pMD19T-simple vector and confirmed by sequencing, the pAtzA-22-4-19T-simple was restricted with *Eco*RV to collect the 1.4 kb fragment, and then lignated with pUbi-19T-simple. The resulting plasmid was restricted with *Sac*I/*Spe*I to collect the 3.4 kb fragment, and then ligated with pCAMBIA1301. The recombinant plasmid p1301-ubi-22-14 was then introduced into *Agrobacterium tumefaciens* EHA105 by the freeze-thaw method [18].

Genetic transformation and plant regeneration

Oryza sativa L. Nipponbare and *japonica* hybrid rice parental lines in the three-line system, Jindao7 (maintainer), Jindao8 (maintainer) and Jinhui3 (restorer) were used for transformation. Mature seeds were dehulled, surface-sterilized and placed on NB medium (N6 macro elements, B5 micro elements and vitamins)

supplemented with 2 g/L proline, 3 mg/L 2, 4-D and 300 mg/L casein hydrolysate in dark at 28°C. After 2–3 weeks, the scutellum-derived calli were excised and subcultured every four weeks on the same medium but with 0.5 g/L proline, 2 mg/L 2, 4-D in dark at 28°C. The highly embryogenic compact calli (3–5 mm in diameter) that subcultured for less than five generations, were selected and co-cultivated with *A. tumefaciens* EHA105 harboring p1301C-ubi-22-14 on the co-cultivation medium (subculture medium but with 100 μM acetosyringone) for 3 days in dark at 28°C. Following that, the explants were then transferred into selection medium (subculture medium but with 50 mg/L hygromycin and 500 mg/L cefotaxime) in dark at 28°C for selection. After two cycles of selection, hygromycin-resistant calli were transferred onto pre-regeneration medium (NB medium with 0.5 g/L proline, 2 mg/L 6-BA, 1 mg/L NAA, 5 mg/L ABA, 300 mg/L casein hydrolysate and 50 mg/L hygromycin) for 14 to 21 days in dark at 28°C, then to regeneration medium (pre-regeneration medium but without 5 mg/L ABA) for 30 days under 54 μmol/m^2/s light at 28°C and finally to the rooting medium (MS medium with 1 mg/L IBA) for 15 days under light at 28°C.

Molecular analysis of transgenic lines

Genomic DNA was isolated from young leaves using a modified CTAB method [19]. PCR was performed to preliminarily select the transformed plants using primers atzA7/atzA8 [20] and SL-Ubi-C-F/SL-Ubi-C-R (Table 1). RT-PCR was performed to further confirm the expression of *atzA* in the PCR-positive transformed plants. Total RNA was extracted from young leaves using an RNAultra Extraction Kit (Qiagen). cDNA was synthesized using atzA8 and oligo(dT) as primers and SuperScript^{TM} II RNase H− (Invitrogen) as reverse transcriptase. RT-PCR was amplified using the cDNA as template and atzA7/atzA2 [20] as primers. β-actin (AB047313) was also amplified as the internal control using actin-R/actin-F as primers [21].

Germination and seedling growth in the presence of atrazine

Fifty seeds from each transgenic line and the respective control plant were directly sown on the surface of filter paper in plates containing 0, 75 or 150 μM atrazine. Seeds were placed in a growth chamber at 28°C with 16 h of 54 μmol/m^2/s light per day. Germination (based on radicles >2 mm) was recorded daily and the cumulative values at day 3 and day 7 were calculated to represent as germination potential and germination percentage. The shoot length, root length and their biomass were measured after 7 days. For seedling growth test, seven-day old seedlings germinated in absence of atrazine were placed on filter paper in pots, and incubated in Kimura B nutrition solution [22] containing 0, 75 or 150 μM atrazine in growth chamber described above. During the period, the nutrition solution containing the respective concentration of atrazine was added to keep the filter paper wet. The growth parameters as described above were measured after 10 days.

Soil-grown transgenic T$_2$ lines in the presence of atrazine

Two-week old seedlings of similar size that germinated in absence of atrazine were transplanted into pots containing 1.12 kg of soil with 0, 2, or 5 mg/kg of atrazine. Nine plants were planted in each pot, and irrigated with the same amount of water every day and with Kimura B nutrition solution [22] twice a week. Plants were incubated in the greenhouse in a 14-h light/10-h dark cycle (28/25°C) at 300 μmol m^{-2}s^{-1} light and 75% relative humidity. Plant growth and chlorophyll content were measured at

Table 1. Primers used in this study.

Primer	Sequence (5′→3′)
SL-Ubi-F	tgagctcctgcagtgcagcgtgacccggtcgt, with *Sac*I site
SL-Ubi-R	tgatatcctgcagaagtaacaccaaacaacag, with *Eco*RV site
atzA-TJ-F	tgatatcatgcaaacgctcagcatccag, with *Eco*RV site
atzA-TJ-R	tgatatcactagtctagaggctgcgccaagctg, with *Eco*RV and *Spe*I site
SL-Ubi-C-F	cccgccgtaataaatagacac
SL-Ubi-C-R	accacaccacatcatcacaac

15 -day intervals. For chlorophyll content analysis, approximately 10 mg of leaves was extracted with 5 ml acetone and then quantified the absorbance at 663.6 nm and 646.6 nm, as described by Porra et al. [23]. Chlorophyll a $= 12.25 \times A_{663.6} - 2.55 \times A_{646.6}$, Chlorophyll b $= 20.31 \times A_{646.6} - 4.91 \times A_{663.6}$.

HPLC analysis of atrazine and its metabolite hydroxyatrazine in the plant and soil

Atrazine and hydroxyatrazine in the rice leaves and soil were also determined in the soil-grown experiment. One gram of leaves, or five grams of soil samples that collected from the soil layer mixture as thick as possible, were extracted with 5 mL dichloromethane. After 30 min of incubation at room temperature, the mixture was centrifuged at 15,294 g for 10 min. The supernatant was transferred to a new tube and filtered through a 0.2-μm filter and then subjected to CoM 6000 HPLC system analysis on an analytical C18 column (5 μm, 250 mm×4.6 mm) at 30°C with linear gradients phase as follows: 0 to 6 min, 10% to 25% acetonitrile; 6 min to 21 min, 25% to 65% acetonitrile; 21 min to 23 min, 65% to 100% acetonitrile; and 23 min to 25 min, 100% acetonitrile [24] at the wavelength of 228 nm.

Spraying atrazine to transgenic T2 lines

Two-week old seedlings from the germination experiment were also transplanted into pots with soil. Each pot contained 16 plants. After 30 days' growth, the last second leaves were taken and cut into 2–3 cm section, soaked in 0, 75 and 150 μM atrazine, and incubated at 25°C under light for 2 days. The color of the leaves was observed every day. After 40 days' growth, each pot of plants was sprayed with 20 ml of 0.15% atrazine solution.

Statistical analysis

All the experiments were performed for three times. Significant difference between the specific transgenic line and the respective control was performed using independent-samples t-test at 95% or 99% confidence with IBM SPSS Statistics 11.0. Values indicated by * or ** represented significantly difference at $P<0.05$ or $P<0.01$.

Results

Selection of transgenic *atzA* plants

PCR analysis revealed that 18 (100%) Nipponbare transformants, 15 out of 18 (83%) Jindao7 transformants, 14 out of 16 (88%) Jinhui1 transformants, and 17 (100%) Jindao8 transformants were positive. RT-PCR analysis further confirmed the expression of *atzA* in the PCR-positive lines (data not shown). Subsequently, the self-pollinated seeds (T$_1$ progenies) from each plant were germinated in the presence of hygromycin. Transgenic

lines that showed a segregation pattern of 3:1 resistant/sensitive in the germination test were selected to produce the T$_2$ generations. Individual plants, whose herbicide tolerance did not segregate in the germination test, were considered as homogenous lines and selected. For simplicity, in further research, we used only two or three independent transgenic T$_2$ lines for each genotype.

Germination of transgenic *atzA* lines in the presence of atrazine

To investigate the atrazine tolerance of transgenic lines during germination, we germinated seeds of three lines of each genotype and the respective controls in the presence of atrazine (Fig. 1A, Table 2). The germination potential of the wild types decreased as the atrazine concentration increased (58.3–98%, 22.4–64.2%, and 0%, respectively, in 0 μM, 75 μM, and 150 μM atrazine); on the other hand, the germination percentage significantly decreased only in presence of 150 μM atrazine (95.2–100%, 72.1–96.7%, and 0–42.2%, respectively, in 0 μM, 75 μM, and 150 μM atrazine). In the presence of the investigated atrazine concentrations, the highest germination potential was determined for Nipponbare, followed by Jinhui1, Jindao8, and Jindao7. In the presence of 0 μM or 75 μM atrazine, the germination potentials and percentages of almost all of the transgenic lines did not differ significantly from those of the respective controls. The exceptions were the germination potential for Jindao8 in the presence of 75 μM atrazine and the germination percentage for Jinhui1 in the presence of 75 μM atrazine. On the other hand, in the presence of 150 μM atrazine, all of the transgenic lines showed significantly higher germination potentials and germination percentages than did the respective controls. Moreover, two of the wild types (Nipponbare and Jinhui1) failed to germinate or only rarely germinated at day 3 or day 7. Interestingly, in the presence of 150 μM atrazine, the highest germination percentage was determined for Jindao7, followed by Jindao8, Jinhui1, and Nipponbare; on the other hand, Jindao7 showed the lowest germination potential. When germinated in the presence of atrazine, all of the transgenic lines produced larger seedlings (with taller shoots and longer roots) than did the respective control plants (Fig. 1A). However, the presence of atrazine significantly inhibited the growth of transgenic lines and wild types (Fig. 1A).

Seedling growth of transgenic *atzA* lines in the presence of atrazine

To investigate the tolerance of transgenic lines to atrazine, we transplanted 7-day old seedlings germinated in the absence of atrazine, to pots of soil containing different concentration of atrazine (Fig. 1B and Table 3). In the absence of atrazine, we determined no differences in shoot or root growth between any of the transgenic lines and the respective controls. However, in the

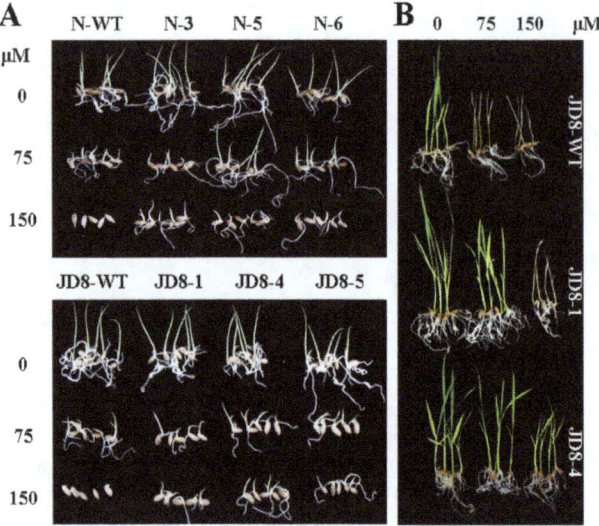

Figure 1. Germination and seedling growth of transgenic *atzA* **rice lines in the presence of atrazine.** (**A**) Representative images of seeds from the transgenic lines and the respective controls sown on plates containing 0, 75 or 150 μM atrazine for 7 days. (**B**) Representative images from seven-day old seedlings germinated in absence of atrazine were transplanted into pots containing 0, 75 or 150 μM atrazine for 10 days. For simplicity, only images of transgenic Jindao8 (JD8) and/or Nipponbare (N) were shown here.

presence of 75 μM atrazine, all of the transgenic lines (except for Jindao8) were significantly taller than were the respective controls; further, all of the transgenic lines (except for Jinhui1) produced significantly more root biomass than did the respective controls. On the other hand, in the presence of 150 μM atrazine, shoot and root growth of almost all of the transgenic lines and the respective controls were significantly inhibited. The exceptions were the shoot growth of Nipponbare and the root growth of Jinhui1–5.

Plants became yellowish, and subsequently rotted and died after 10 days (Fig. 1B).

Tolerance of soil-grown transgenic *atzA* lines to atrazine

To further investigate the tolerance of transgenic lines to atrazine, we transplanted 15-day-old seedlings germinated in the absence of atrazine, to pots of soil containing 0 mg/kg, 2 mg/kg, or 5 mg/kg of atrazine. We measured the plant height (as a non-destructive assay of plant growth) and chlorophyll content at 15-day intervals.

In the absence of atrazine, we determined no significant differences in plant height between the transgenic lines and the respective controls (Fig. 2A). In soil containing 2 mg/kg or 5 mg/kg of atrazine, all the transgenic lines were significantly taller than were the respective control plants. However, the presence of atrazine significantly inhibited the growth of transgenic lines and control plants (by 68.1–111.9% and 46.4–68.2%, respectively). In contrast to the control plants, all of the transgenic lines (except for the Jindao8) were slightly taller in soil containing 2 mg/kg of atrazine than in soil containing 5 mg/kg of atrazine. The growth of transgenic lines gradually recovered with an increase in the growth time. Further, three of the strongest transgenic lines (Jindao7–4, Jinhui1–5, and Jindao8–4) completely recovered after 45 days of growth (data not shown).

Atrazine may disrupt the electron flow of photosystem II and destroy photosynthetic pigments in plant leaves [15]. In the absence of atrazine, we observed no difference in chlorophyll content between transgenic lines and control plants (Fig. 2B). However, in soil containing 2 mg/kg or 5 mg/kg of atrazine, the transgenic lines retained more chlorophyll than did the respective control plants. With a few exceptions, the chlorophyll content of the transgenic lines decreased significantly in soil containing 5 mg/kg of atrazine, but not in soil containing 2 mg/kg of atrazine; on the other hand, the chlorophyll content of the control plants decreased significantly under both conditions. For each genotype grown in soil containing the same atrazine concentration, we observed no difference in chlorophyll content at different growth stages. After 45 days, the chlorophyll contents of the three

Table 2. Germination of transgenic *atzA* rice lines in the presence of atrazine.

Line	0 μM		75 μM		150 μM	
	3 d	**7 d**	**3 d**	**7 d**	**3 d**	**7 d**
Nᵃ-WT	98.0±2.0	100±0.0	54.0±1.6	94.5±1.4	0.0±0.0	0.0±0.0
N-3	98.0±1.6	94.2±0.0	60.0±2.2	98.2±1.1	64.3±1.6**	96.2±1.8**
N-5	96.0±2.0	100±0.0	64.0±1.3	92.1±1.5	62.3±2.6**	92.6±1.5**
JD7-WT	58.3±1.6	100±0.0	24.3±2.1	90.2±2.3	0.0±0.0	42.2±1.1
JD7-2	56.0±1.1	98.1±1.4	22.3±1.7	88.2±2.1	2.3±0.1**	88.5±1.6**
JD7-4	62.3±2.2	98.6±1.6	26.3±1.3	90.2±0.8	2.4±0.2**	86.5±1.3**
JH3-WT	86.3±2.0	95.2±2.8	64.2±2.2	72.1±1.1	0.0±0.0	4.5±0.0
JH-3	82.3±1.2	94.2±1.8	58.7±1.6	90.2±1.2*	32.3±0.9**	78.6±1.2**
JH3-5	88.3±3.2	92.2±1.3	66.3±2.7	94.1±1.0*	38.7±1.1**	80.5±1.2**
JD8-WT	82.0±4.2	98.7±0.8	22.4±1.6	96.7±1.0	0.0±0.0	26.5±1.3
JD8-1	78.3±2.5	99.2±0.8	48.3±1.1**	97.2±0.2	34.3±0.7**	94.5±2.8**
JD8-4	78.0±2.6	100±0.0	54.0±2.6**	98.3±1.0	35.2±1.6**	96.8±1.3**

Note: In each independent treatment 50 seeds were used for the experiment. Values are shown as mean ±SEM (*n* = 3). * or ** indicates significant difference (*P*<0.05) or highly significant difference (*P*<0.01) between the treatment and the control.
ᵃN, JD7, JH3 and JD8 were the abbreviations of Nipponbare, Jindao7, Jindao8 and Jinhui3, respectively. For simplicity, only two transgenic lines for each genotype were shown here.

Table 3. Seedling growth performance of transgenic *atzA* rice lines in the presence of atrazine.

Line	0 μM		75 μM		150 μM	
	Shoot length (cm)	Root biomass (mg)	Shoot length (cm)	Root biomass (mg)	Shoot length (cm)	Root biomass (mg)
N [a]-WT	9.90±1.37	32.00±3.16	6.54±0.86 (66.0)[a]	21.00±2.55 (65.6)	5.07±1.17 (51.2)	21.75±3.77 (68.0)
N-3	10.55±1.49	30.25±3.50	9.56±2.08* (90.7)	29.50±3.14** (97.5)	7.55±1.20* (71.6)	21.50±6.40 (71.1)
N-5	8.78±2.42	32.75±4.50	8.94±1.33* (101.8)	29.85±1.83** (91.2)	9.39±0.51** (106.9)	21.00±2.71 (64.1)
JD7-WT	8.47±0.37	28.50±2.56	5.28±0.57 (62.4)	20.00±3.10 (70.2)	6.54±0.61 (77.2)	17.75±5.19 (62.3)
JD7-2	9.40±2.61	31.50±3.00	7.08±0.78* (75.3)	32.50±2.65** (103.2)	5.81±0.56 (61.8)	17.00±3.46 (54.0)
JD7-4	7.99±1.65	27.50±3.35	8.16±1.27* (102.2)	27.50±2.65** (100.0)	6.78±0.53 (84.8)	17.75±4.19 (64.5)
JH3-WT	9.83±1.25	24.00±3.07	5.80±0.43 (59.0)	21.75±3.36 (90.6)	4.62±0.42 (47.0)	13.25±6.24 (55.2)
JH-3	10.44±1.47	26.75±2.75	8.27±0.59* (79.2)	24.25±2.22* (90.7)	5.12±0.61 (49.0)	15.25±2.63 (57.0)
JH3-5	8.89±1.58	25.50±3.57	8.25±0.78* (92.8)	23.25±2.87* (83.8)	6.04±0.19* (67.9)	22.00±3.37** (86.3)
JD8-WT	7.88±0.93	26.75±3.55	7.20±1.85 (91.4)	21.75±2.87 (81.3)	6.43±0.35 (81.5)	12.25±3.46 (45.8)
JD8-1	8.11±1.51	27.50±2.15	7.18±0.40 (88.6)	27.50±1.52** (100.0)	5.65±1.12 (69.6)	11.50±1.91 (41.8)
JD8-4	7.90±0.72	27.75±2.53	7.64±0.69 (96.7)	27.75±1.79** (100.0)	5.64±0.78 (71.3)	11.75±2.99 (42.3)

Note: The parameters were recorded after 10 days. Values are shown as mean ± SEM ($n = 3$). * or ** indicates significant difference ($P < 0.05$) or highly significant difference ($P < 0.01$) between the treatment and the control.
[a]N, JD7, JH3 and JD8 were the abbreviations of Nipponbare, Jindao7, Jindao8 and Jinhui3, respectively.
[b]Data in the bracket are the percentage of the treated values to the mock values.

strongest lines (Jindao7–4, Jinhui1–5, and Jindao8–4) were almost identical when grown in soil containing 5 mg/kg of atrazine and when grown in the absence of atrazine (data not shown).

Ability of soil-grown T$_2$ transgenic lines to degrade atrazine

To investigate the ability of the soil-grown transgenic plants to degrade atrazine, we determined the atrazine residue in the soil at 15-day intervals, after growth of seedlings in pots of soil containing 0 mg/kg, 2 mg/kg, or 5 mg/kg of atrazine. We observed that the atrazine residue in the soil decreased with an increase in growth time, for transgenic and control plants (Fig. 2C). However, the decrease was greater for transgenic lines than for control plants. After treatment with 2 mg/kg of atrazine, the atrazine residue remaining in the soil was 2.9–7.0% for transgenic lines, and 44.0–59.2% for control plants; after treatment with 5 mg/kg of atrazine, the atrazine residue remaining in the soil was 0.8–8.7% for transgenic lines and 28.1–30.8% for control plant. For each genotype at the same growth stage, the atrazine residue in the soil was higher after growth in 2 mg/kg atrazine than after growth in 5 mg/kg atrazine. We also observed that hydroxyatrazine, not atrazine (only trace) appeared in the leaves of the transgenic plants (Fig. S1), indicating that atrazine was metabolized rather than accumulated in the transgenic plants.

Utilization of transgenic plants in hybrid seed production

To investigate the tolerance of transgenic plants to the atrazine concentration used in weed control, we first examined the tolerance of leaf sections to atrazine solution. In the presence of 75 μM atrazine, the leaf sections of transgenic lines remained green, whereas those of control plants became bleached (Fig. 3A). In the presence of 150 μM atrazine, the leaf sections of transgenic lines became slightly bleached, but remained greener than those of control plants. We subsequently sprayed plants with 0.15% atrazine, and observed that wild-type plants became curled and withered after 6 days, whereas transgenic lines continued to grow well (Fig. 3B). Our results indicate that the transgenic rice plants

showed tolerance to the atrazine concentration used in weed control.

Discussion

Current research on the utilization of herbicide resistance genes in hybrid rice is focused on the two-line hybrid production system. Sterile lines are not stable in the two-line system, and therefore most *japonica* hybrid combinations are still produced by using three-line system. In the present study, we have developed atrazine-tolerant transgenic *japonica* rice parental lines with the potential to be used in future hybrid seed production.

Herbicide tolerance during germination is very important when utilizing transgenic *atzA* lines in hybrid seed production, because off-type plants should be eliminated as early as possible. Atrazine is a slightly water-soluble herbicide, with a saturated concentration of 153 μM. Kawahigashi [25] previously showed that an atrazine concentration of 100 μM did not affect the germination of rice plants. Therefore, we used two atrazine concentrations (75 μM and 150 μM) in the germination test, to evaluate the tolerance of transgenic rice lines. In contrast to Kawahigashi [25], we observed that the transgenic lines germinated well in the presence of atrazine whereas the respective controls did not. However, as the seedling growth of transgenic lines were also inhibited by atrazine, chemical control of contamination during germination may not be appropriate.

To evaluate the tolerance of seedlings to atrazine, we used 7-day-old seedlings germinated in the absence of atrazine as a starting material, to avoid cumulative effects derived from germination. We observed that almost all of the transgenic lines grew well in the presence of 75 μM atrazine with no difference from that in absence of atrazine. Therefore, 75 μM atrazine may be used for effective control of off-type plants during the early seedling growth stage. Our data also suggested spraying plants with 0.15% atrazine (equivalent to the standard dosage used in the field for weed control, i.e., 200–250 g 40% suspension concentrate in 30~50 kg of water) at the subsequently vegetative growth stage could serve as an alternative means of chemically controlling

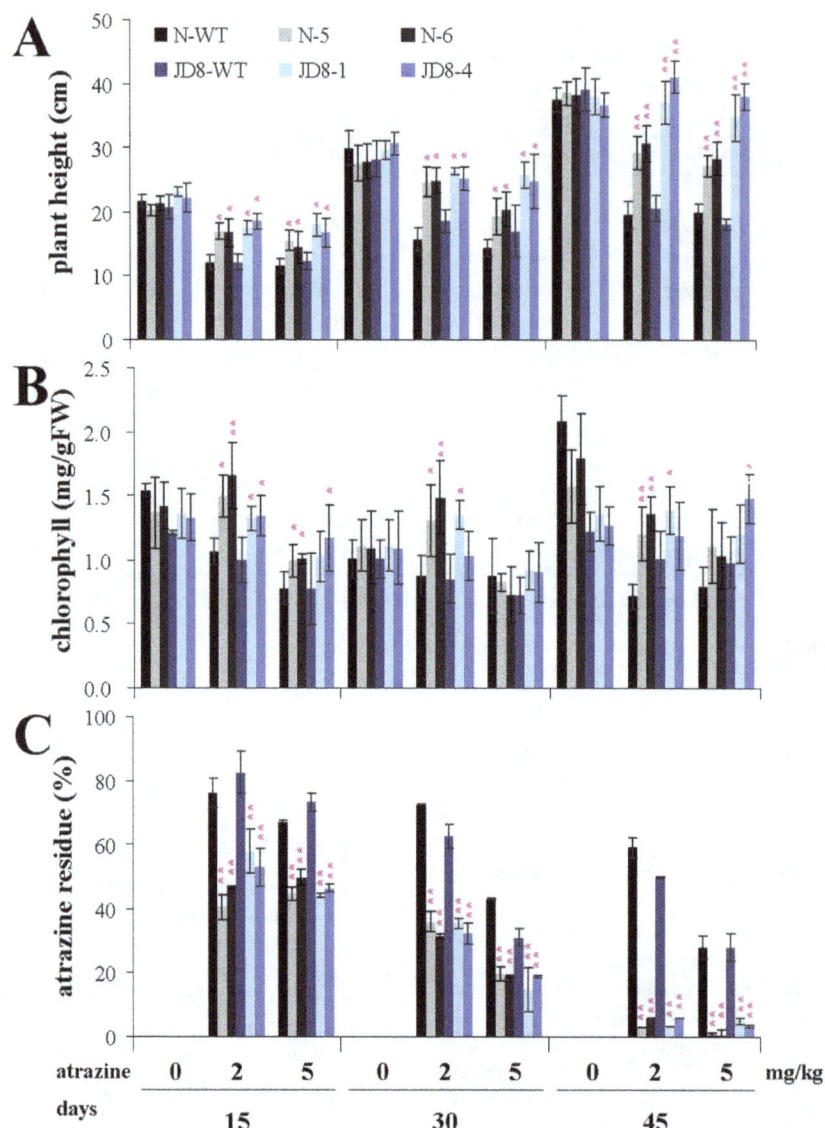

Figure 2. Growth and physiology of transgenic *atzA* rice lines in the presence of atrazine. Fifteen-day old seedlings germinated in absence of atrazine were transplanted into pots with 1.12 kg soil containing 0, 2 or 5 mg/kg of atrazine. (**A**) plant height; (**B**) chlorophyll content; and (**C**) atrazine residue were measured at 15-day intervals. Values are shown as mean ±SEM (*n* = 3). * or ** over the bar indicates significant difference (*P* < 0.05) or highly significant difference (*P* < 0.01) between the treatment and the control. For simplicity, only data of transgenic Nipponbare (N) and Jindao8 (JD8) were shown here.

off-type plants. However, we did not test the tolerance of transgenic *atzA* lines at the reproductive growth stage, which is important for the utilization in mechanical harvesting of hybrid seed production. Further studies are required to investigate whether the lethal dosage for untransformed parental lines is tolerated at the reproductive stage.

Atrazine degrades relatively slowly, with an average half-life of 4–57 weeks [26]. Therefore, application of this herbicide will inevitably lead to accumulation in the field, which may affect current crop growth or damage sensitive succession crops [27]. The transgenic rice lines were taller, and had higher chlorophyll contents than did the respective control plants when growing in the soil of 2 mg/kg or 5 mg/kg atrazine. Three of the transgenic lines (Jindao7–4, Jinhui1–5, and Jindao8–4) completely recovered after 45 days of growth in the presence of atrazine, suggesting that an atrazine concentration up to 5 mg/kg does not affect sustained

growth of transgenic lines. Our data also suggested that almost all of the atrazine was degraded after growth of the transgenic rice plants for 45 days. This was also verified by the appearance of hydroxyatrazine in the leaves of transgenic plants. Besides, atrazine metabolite hydroxyatrazine could be directly determined in transgenic plants. However, as the leave, stem and root of the plants all possibly metabolize atrazine [28], it would be more complicated if we want to quantitatively evaluate the atrazine metabolized by plants. That is why we chose to determine the atrazine residue in the soil after growth of the transgenic plants in this study.

Utilization of herbicide resistance genes in hybrid seed production may be achieved by introducing the genes into a maintainer or restorer line in the three-line system. When an herbicide-resistant maintainer is obtained, its corresponding CMS line, which possesses herbicide resistance, may be created by using

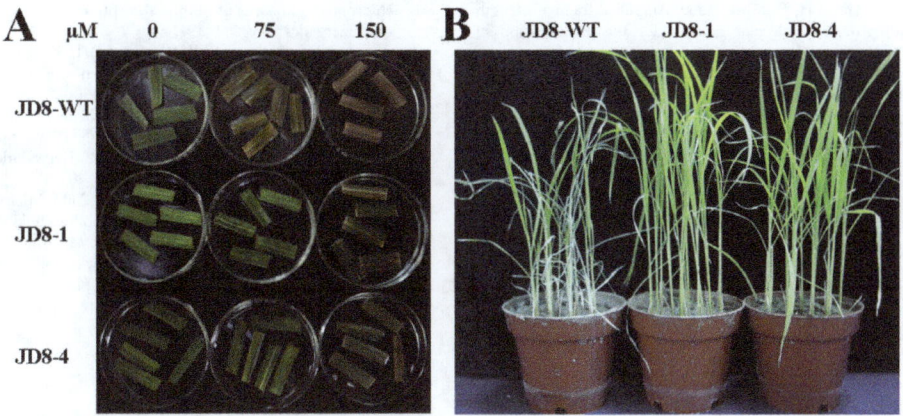

Figure 3. Tolerance of mature transgenic *atzA* plants to atrazine. (A) Representative images of leave sections in the presence of atrazine. Fifteen-day old seedlings were transplanted into pots containing soil. After 30 days' growth, the last second leaves were cut into 2–3 cm section, and soaked in 0, 75 and 150 μM atrazine, and incubated at 25°C under light for 2 days. **(B)** Representative images of mature plants after sprayed with atrazine. Fifteen-day old seedlings were transplanted into pots containing soil. After 40 days' growth, each pot of transgenic lines (9 plants) was sprayed with 20 mL of 0.15% atrazine solution. For simplicity, only data of transgenic Jindao8 (JD8) was shown here.

a backcross. The resulting herbicide-resistant CMS breeder seeds may be used to cross with the original maintainer (herbicide sensitive), to multiply the herbicide-resistant foundation seeds. During this process, the selfed maintainer and off-type plants may easily be eliminated by herbicide spraying, thus ensuring the purity of CMS seeds. The herbicide-resistant CMS line may also be used in mechanical harvesting, by spraying herbicide to kill the pollen plants after pollination, and harvesting seeds from the surviving CMS plants. A similar result may be achieved when crossing the herbicide-resistant CMS line with the herbicide-sensitive restorer line. Alternatively, the herbicide-resistant restorer line may be crossed with the herbicide-sensitive sterile line during the hybrid seed production process, by spraying herbicide to kill the off-type plants. Further, the herbicide-resistant maintainer and restorer lines may be useful for eliminating off-type plants, and for weed management when growing hybrid rice.

Field trials with genetically modified rice are strictly controlled in China, and therefore we were unable to evaluate our herbicide-resistant hybrid rice combinations on a field scale. Further studies are required to verify the validity of utilizing atrazine-tolerant transgenic lines in the field, and to determine whether herbicide-resistant hybrid rice combinations retain their original superiority in terms of yield. Nevertheless, the development of atrazine-tolerant transgenic *japonica* rice parental lines represents a valuable tool for future application in hybrid seed production.

Supporting Information

Figure S1   HPLC analysis for determination of atrazine and its metabolite hydroxyatrazine in leaves of transgenic and WT rice plants after grown in 5 mg/kg atrazine soil for 45 days. (A) Standard of hydroxyatrazine (I) and atrazine (II). **(B)** Wild type (WT). **(C–E)** Transgenic Nipponbare (N-6), Jindao7 (JD7-4) or Jindao8 (JD8-4), respectively.

Author Contributions

Conceived and designed the experiments: DFC XWC. Performed the experiments: LHZ HWC YLL YNL SJW JPS. Analyzed the data: LHZ DFC XWC. Contributed reagents/materials/analysis tools: SJW XJL DFC XWC. Wrote the paper: DFC XWC.

References

1. Peng C, Zhang H (2013) Analysis of paddy and rice markets at home and abroad in the first quarter of 2013 and its prospect. Agric Outlook (4): 4–9 (in Chinese, with English abstract).
2. Shi KB, Deng HS (2004) Chinese *japonica* hybrid rice technology innovation seminar held in Sanya. Hybrid Rice 19: 76 (in Chinese).
3. Li S, Yang D, Zhu Y (2007) Characterization and use of male sterility in hybrid rice breeding. J Integ Plant Biol 49: 791–804.
4. Tang SZ, Zhang HG, Zhu ZB, Liu C, Li P, et al. (2010) Application of HL type male sterile cytoplasm in *japonica* hybrid rice breeding. Chin J Rice Sci 24: 116–124 (in Chinese, with English abstract).
5. Yan W (2000) Crop heterosis and herbicide. US Patent 6066779.
6. Hu G, Xiao H, Yu Y, Zhu Z, Si H, et al. (2000) *Agrobacterium*-mediated transformation of the restorer lines of two-line hybrid rice with *bar* gene. Chin J Appl Environ Biol 6: 511–515 (in Chinese, with English abstract).
7. Li Y, Xu Q, Duan F, Liu G, Yan W (2000) Breeding of herbicide-resistant hybrid rice combinations. Hybrid Rice 15(6): 9–11.
8. Xue S, Zhang W, Gong X, Shen L, Huang D, et al. (2001) Breeding of isotype restorer line with *bar* gene from Miyang 46 and its combinations. J Zhejiang Agric Sci (4): 181–183 (in Chinese).
9. Ramesh S, Nagadhara D, Pasalu IC, Kumari AP, Sarma NP, et al. (2004) Development of stem borer resistant transgenic parental lines involved in the production of hybrid rice. J Biotechnol 111: 131–141.
10. Xiong X, Tang L, Deng X, Xiao G (2004) A preliminary report on the experiments of herbicide-resistant two-line hybrid rice Xiang 125S/Bar 68-1. Hybrid Rice 19(5): 41–43.
11. Wu F, Wang S, Li S, Zhang K, Li P (2006) Research progress on herbicide resistant transgenic rice and its safety issues. Mol Plant Breed 4: 846–852.
12. Xiao G, Yuan L, Sun S (2007) Strategy and utilization of a herbicide resistance gene in two-line hybrid rice. Mol Breed 20: 287–292.
13. Xiao G (2009) Recent advances in development of herbicide resistance transgenic hybrid rice in China. Rice Sci 16: 235–239.
14. Udiković-Kolić N, Scott C, Martin-Laurent F (2012) Evolution of atrazine-degrading capabilities in the environment. Appl Microbiol Biotechnol 96: 1175–1189.
15. Rutherford AW, Krieger-Liszkay A (2001) Herbicide-induced oxidative stress in photosystem II. Trends Biochem Sci 26: 648–653.
16. Cai B, Han Y, Liu B, Ren Y, Jiang S (2003) Isolation and characterization of an atrazine-degrading bacterium from industrial wastewater in China. Lett Appl Microbiol 36: 272–276.
17. Wang Y, Li X, Chen X, Chen D (2013) Directed evolution and characterization of atrazine chlorohydrolase variants with enhanced activity. Biochemistry (Moscow) 78: 1104–1111.
18. Höfgen R, Willmitzer L (1988) Storage of competent cells for *Agrobacterium* transformation. Nucleic Acids Res 16: 9877.

19. Loh JP, Kiew R, Kee A, Gan LH, Gan YY (1999) Amplified fragment length polymorphism (AFLP) provides molecular markers for the identification of *Caladium bicolor* cultivars. Ann Bot 84: 155–161.

20. Wang H, Chen X. Xing X, Hao X, Chen D (2010) Transgenic tobacco plants expressing *atzA* exhibit resistance and strong ability to degrade atrazine. Plant Cell Rep 29: 1391–1399.

21. Chen D, Chen H, Zhang L, Shi X, Chen X (2014) Tocopherol-deficient rice plants display increased sensitivity to photooxidative stress. Planta 239: 1351–1362.

22. Ma J, Takahashi E (1990) Effect of silicon on the growth and phosphorus uptake of rice. Plant Soil 126: 115–119.

23. Porra RJ, Thompson WA, Kriedemann PE (1989) Determination of accurate extinction coefficients and simultaneous equations for assaying chlorophylls *a* and *b* extracted with four different solvents: verification of the concentration of chlorophyll standards by atomic absorption spectroscopy. Biochim Biophys Acta - Bioenergetics 975: 384–394.

24. de Souza ML, Sadowsky MJ, Wackett LP (1996). Atrazine chlorohydrolase from *Pseudomonas* sp. strain ADP: gene sequence, enzyme purification, and protein characterization. J Bacteriol 178: 4894–4900.

25. Kawahigashi H, Hirose S, Ohkawa H, Ohkawa Y (2007) Herbicide resistance of transgenic rice plants expressing human CYP1A1. Biotechnol Adv 25: 75–84.

26. Erickson LE, Lee KH, Sumner DD (1989) Degradation of atrazine and related *s*-triazines. Crit Rev Environ Control 19: 1–14.

27. Rhine ED, Fuhrmann JJ, Radosevich M (2003) Microbial community responses to atrazine exposure and nutrient availability: linking degradation capacity to community structure. Microbiol Ecol 46: 145–160.

28. Wang L, Samac DA, Shapir N, Wackett LP, Vance CP, et al. (2005) Biodegradation of atrazine in transgenic plants expressing a modified bacterial atrazine chlorohydrolase (*atzA*) gene. Plant Biotech J 3: 475–486.

14

Induction of Body Weight Loss through RNAi-Knockdown of APOBEC1 Gene Expression in Transgenic Rabbits

Geneviève Jolivet[1]*, Sandrine Braud[2], Bruno DaSilva[1], Bruno Passet[4], Erwana Harscoët[1], Céline Viglietta[1], Thomas Gautier[3], Laurent Lagrost[3], Nathalie Daniel-Carlier[1], Louis-Marie Houdebine[1], Itzik Harosh[2]*

1 INRA UMR1198, Biologie du Développement et Reproduction, Jouy en Josas, France, 2 ObeTherapy Biotechnology, Evry, France, 3 INSERM UMR866, Université de Bourgogne, Dijon, France, 4 INRA UMR1313, Génétique Animale et Biologie Intégrative, Jouy-en-Josas, France

Abstract

In the search of new strategies to fight against obesity, we targeted a gene pathway involved in energy uptake. We have thus investigated the *APOB* mRNA editing protein (*APOBEC1*) gene pathway that is involved in fat absorption in the intestine. The *APOB* gene encodes two proteins, APOB100 and APOB48, via the editing of a single nucleotide in the *APOB* mRNA by the APOBEC1 enzyme. The APOB48 protein is mandatory for the synthesis of chylomicrons by intestinal cells to transport dietary lipids and cholesterol. We produced transgenic rabbits expressing permanently and ubiquitously a small hairpin RNA targeting the rabbit *APOBEC1* mRNA. These rabbits exhibited a moderately but significantly reduced level of *APOBEC1* gene expression in the intestine, a reduced level of editing of the *APOB* mRNA, a reduced level of synthesis of chylomicrons after a food challenge, a reduced total mass of body lipids and finally presented a sustained lean phenotype without any obvious physiological disorder. Interestingly, no compensatory mechanism opposed to the phenotype. These lean transgenic rabbits were crossed with transgenic rabbits expressing in the intestine the human *APOBEC1* gene. Double transgenic animals did not present any lean phenotype, thus proving that the intestinal expression of the human *APOBEC1* transgene was able to counterbalance the reduction of the rabbit *APOBEC1* gene expression. Thus, a moderate reduction of the APOBEC1 dependent editing induces a lean phenotype at least in the rabbit species. This suggests that the *APOBEC1* gene might be a novel target for obesity treatment.

Editor: Hervé Guillou, INRA, France

Funding: This study was funded by Agence Nationale de la Recherche (ANR-06-RIB-FATSTOP) to ObeTherapy as leader of the program and to INRA. Sandrine Braud and Itzik Harosh are employees of and shareholders in ObeTherapy Biotechnology. ObeTherapy Biotechnology provided support in the form of salaries for authors SB and IH, had a direct role in the study design, and analysis, decision to publish, and preparation of the manuscript.

Competing Interests: Sandrine Braud and Itzik Harosh are employees of and shareholders in ObeTherapy Biotechnology. There are no products in development or marketed products to declare.

* Email: genevieve.jolivet@jouy.inra.fr (GJ); harosh@obetherapy.com (IH)

Introduction

Obesity is becoming a major problem all over the world spreading like global epidemic with a higher prevalence in the USA [1]. Overweight and obesity are important risk factors for diabetes and cardiovascular disease. Several hundreds of genes are involved in obesity and the estimation is that one quarter of our genome is involved in weight management and energy metabolism [2,3]. In the search of new targets for obesity, we have investigated the *APOB* mRNA editing protein (*APOBEC1*) gene pathway that is involved in fat absorption in the intestine.

The *APOB* gene encodes two proteins, APOB100 and APOB48, via the editing of a single nucleotide in the mRNA by a specialized enzyme, the *APOB* mRNA editing protein (APOBEC1). This enzyme, a catalytic deaminase expressed in human and rabbit in the intestine but not in the liver, is part of a complex that deaminates a cytidine residue to an uridine one in

the intestine *APOB* mRNA (at position 6666 in the human and 6529 in the rabbit) thus generating a STOP codon; it results in the production of the shorter polypeptide designated APOB48 [4] [5] [6]. APOB48 is essential for chylomicron formation, secretion and transport of dietary cholesterol and triglyceride from the intestine [7,8]. Besides, in the liver, where the editing protein is not expressed, and editing does not occur, the unaltered mRNA gives rise to APOB100 that is an integral part of VLDL and LDL.

With the aim to show that *APOB* mRNA editing is a target mechanism for fighting against obesity, we searched to modulate APOBEC1 enzymatic activity *in vivo* in the rabbit species by modulating *APOBEC1* gene expression through transgenesis. Rabbits have the same lipid metabolism as human [9] as opposed to mice that express *APOBEC1* gene both in the liver and intestine [10], do not have CETP and have higher level of HDL and lower level of LDL, that altogether makes mice a less suitable model to study lipid metabolism than rabbits. Thus, we generated transgenic rabbits by knocking down the endogenous *APOBEC1* gene using RNA interference strategy and expressing permanently a small hairpin RNA (shRNA) targeting specifically the rabbit *APOBEC1* mRNA. We generated also transgenic rabbits expressing the human *APOBEC1* gene, and double transgenic animals by inter-crossing these two models. We observed interesting differences in the phenotypes of these rabbits, especially as regard to their body weight and total lipid content. Finally, our results suggest that APOBEC1 could be considered as a potential target for metabolic disorder treatment.

Results

Production of transgenic rabbits

We aimed to produce transgenic animals expressing a shRNA targeting the rabbit *APOBEC1* mRNA in order to knock down the expression of this gene. A construct encompassing a shRNA expressing gene (rbapobec1-shRNA, Figure 1) was therefore introduced by microinjection in the pronuclei of fertilized unicellular rabbit embryos. The sequence of the shRNA targeting the rabbit *APOBEC1* mRNA was chosen among a set of sequences designed by using the OligoWalk tool [11] after assessment of its high efficiency by using an *in vitro* test as previously described [12] (Figure S1).

Twenty-five rabbits were born after microinjection of the rbapobec1-shRNA construct in pronuclei of unicellular rabbit embryos. The screening of newborn rabbits led us to identify 5 (20%) rbapobec1-shRNA transgenic founders. Transgenic lines were successfully established from 3 (shL21, shL23, shL27) of these founders by breeding each one with a wild type animal of the facility. One copy of integrated transgene was integrated in each line. The efficiency of transgenesis and germline transmission was similar to what is currently observed in our rabbit transgenesis facility and led us to suppose that the transgenes were not deleterious for the survival of the rabbits.

rbapobec1-shRNA transgene expression

The rbapobec1-shRNA transgene was expected to produce a shRNA able to knock down the expression of the rabbit *APOBEC1* gene that is known to be specifically expressed in the

intestine [13]. The transgene expression was measured in scrapped duodenum cells. Within each line, the expression of the rbapobec1-shRNA transgene was stable over generations and not significantly different in males and females (Figure 2). Note that in shL21 line, the shRNA transgene expression was the highest compared to lines shL23 and shL27. The line shL23 was not further studied.

Expression of the rabbit *APOBEC1* gene

As presented in Figure 3, the level of the rabbit *APOBEC1* gene expression was moderately (2 to 3 times) but significantly reduced in both males and females in the rbapobec1-shRNA lines shL21 and shL27. This suggests that the shRNA produced by the rbapobec1-shRNA transgene targeted the rabbit *APOBEC1* gene probably through a RNA interference mechanism.

Unfortunately, no antibody was available to detect by Western blot the rabbit APOBEC1 protein in intestinal cell extracts. Thus we are unable to confirm that the level of rabbit APOBEC1 enzyme was lower in rbapobec1-shRNA transgenic animals than in wild type ones.

Indirect quantitative estimation of the level of APOB mRNA editing in intestinal cells

In numerous mammals, it has been already reported that the APOBEC1 induced *APOB* mRNA editing introduces a STOP codon in the *APOB* mRNA [14]. In the rabbit species, this phenomenon is responsible for the conversion of a C residue in a U one at the 2177th codon of the rabbit *APOB* mRNA [15]. We have attempted to quantify the level of editing in the various transgenic lines and in wild type animals to test whether the reduction of *APOBEC1* gene expression could modify the *APOB* mRNA editing.

This was achieved by analyzing the chromatograms of the sequence of DNA fragments encompassing the edited nucleotide and produced in each animal by PCR using reverse transcribed intestinal RNAs as template and the LapoB48F/LapoB48R set of primers (Figure 4A and Table S1). Editing was responsible for the

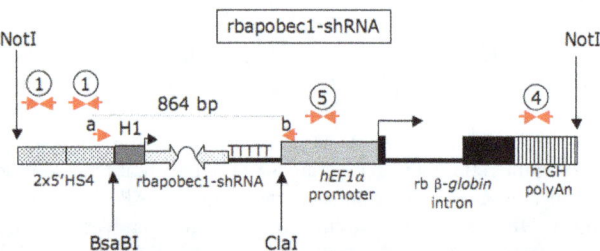

Figure 1. Structure of the rbapobec1-shRNA construct. The rbapobec1-shRNA construct encompassed the H1-rbapobec1-shRNA gene that expressed the shRNA under the activity of the H1 promoter. A gene expression insulator element (two copies of the chicken ß-*GLOBIN* gene fragment 5′HS4) and a transcription unit composed of the *hEF1alpha* – promoter, the rabbit ß-*GLOBIN* second exon and intron, and the human *GH* gene polyadenylation signal were expected to protect the shRNA expression from transcriptional extinction that occurs frequently in transgenesis. Transgenic animals were detected by PCR using sets of primers 1, 4, and 5 (Table S1). Moreover, we checked that after PCR amplification using the a/b set, a 864 bp long fragment with the expected sequence was amplified.

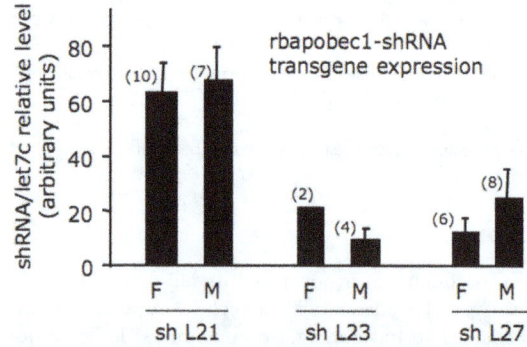

Figure 2. rbapobec1-shRNA transgene expression in rabbit intestine. The amount of shRNA targeting the rabbit *APOBEC1* mRNA was measured in RNAs prepared from duodenum cells as described in "materials and methods" section in 3 rbapobec1-shRNA lines (shL21, shL23 and shL27). Values are given in females (F) and males (M) after normalization to the level of Let7c miRNA determined simultaneously as reference gene in each sample. The number of animals in each group is indicated in brackets. Values are given with the standard error of the mean (sem). All shRNA expressing lines harbored one copy of the rbApobec1-shRNA transgene. Note that in shL21 line, the shRNA transgene expression was the hig hest compared with lines shL23 and shL27.

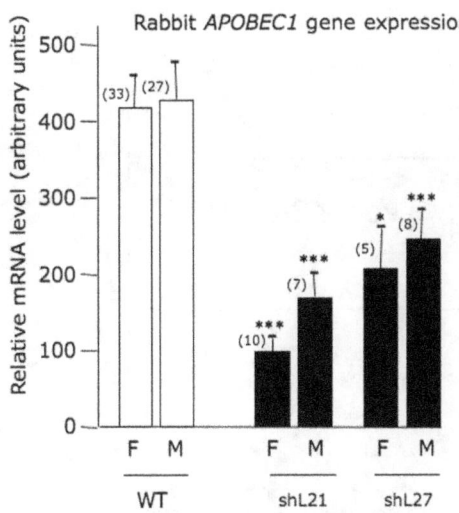

Figure 3. Expression of the rabbit *APOBEC1* gene in wild type and rbapobec1-shRNA transgenic rabbits. The amount of rabbit *APOBEC1* mRNA was measured in RNAs prepared from duodenum cells as described in "materials and methods" section in wild type animals (WT) and in two rbapobec1-shRNA lines (shL21, shL27). Values are given in females (F) and males (M) after normalization to the level of expression of three reference genes (*RPLT9, YHWAZ, HPRT*) determined simultaneously in each sample. The number of animals in each group is indicated in brackets. Values are given with the standard error of the mean (sem). Comparisons were made with control animals of the same sex (*** = p<0.001; * = p<0.05).

modification of the "C" nucleotide in a "T" one at the expected position in the amplified product. We postulated that after amplification, the yield of amplified fragment with a "T" residue was similar to the yield of edited mRNA. We deduced the later from the height of the peaks of each sequence chromatogram (Figures 4B and 4C).

Typical chromatograms are presented in Figure S2 showing that in wild type rabbits, editing occurred in the intestine and not in the liver. More than 95% of *APOB* mRNA was edited in the intestine in wild type animals (Figure 5). Interestingly, the level of editing was clearly lower in rbapobec1-shRNA expressing lines (shL21 and shL27, Figure S2 and Figure 5), with a significant reduction of the number of STOP/edited codon (UAA) encompassing *APOB* mRNA and a concomitant significant increase of the number of non-edited *APOB* mRNA. This led us to propose that the alteration of the level of intestinal editing in shRNA expressing animals was consecutive to the probable reduction of the level of APOBEC1 enzyme in this tissue.

APOB48 amount in plasma

The APOB48 protein is produced by the translation of the *APOBEC1* dependent edited *APOB* mRNA. Since the *APOBEC1* gene expression and the APOBEC1 dependent editing differed in wild type and rbapobec1-shRNA expressing rabbits, it was expected that the plasma level of APOB48 protein also differed in these rabbits.

No efficient antibody was available to detect the rabbit APOB48 protein by Western blot in intestinal extracts. However, we attempted to assay the concentration of APOB48 in the plasma of rabbits using an ELISA specific for the rabbit APOB48 [16]. Firstly, we assayed APOB48 in all plasma samples collected when animals were sacrificed. Surprisingly, all values were similar to the background level of the ELISA. In Kinoshita's paper, it was

reported that the plasma level of APOB48 was enhanced in rabbits fed for at least 8 days with a cholesterol- and triglyceride-enriched regimen. Thus, we decided to feed wild type rabbits and rbapobec1-shRNA expressing rabbits with a soybean oil (8%) and cholesterol enriched (0.2%) regimen [17]. As shown in Figure 6, the plasma level of APOB48 was significantly detected in all wild type animals after feeding for 9 days with the high fat regimen. Besides, the plasma level of APOB48 was not detected in any rbapobec1-shRNA transgenic animal on the four that have been tested. We propose that the undetectable level of plasma APOB48 in plasma samples of most rbapobec1-shRNA transgenic animals was the consequence of the reduction of *APOBEC1* gene expression in the intestine and of the modification of the *APOB* mRNA editing.

APOBEC1-mediated changes in plasma lipid levels and lipoprotein distribution

We hypothesized that chylomicron formation and secretion were impaired in rbapobec1-shRNA transgenic rabbits as a consequence of the reduction of intestinal *APOB* mRNA editing, leading to modifications of the transport of dietary cholesterol and triglyceride from the intestine. With the aim to assess the extent of this phenomenon, we analyzed the concentration of cholesterol and triglycerides in the various lipoproteic fractions of the plasma. Cholesterol (total, free and esterified) and triglycerides were assayed in rabbits fed with a normal diet. The daily food intake was not different in transgenic and wild type animals. Plasma samples were collected after 20 hours fasting, and 4 hours after re-feeding. As shown in Figure 7, after 20 hours fasting, and in all classes of lipoproteins, the concentration of lipids was not different in wild type and in transgenic rabbits (comparison of white bars in WT and shL21 rabbits in each lipid fraction). Besides, after re-feeding, the expected increase of triglycerides and cholesterol in the chylomicron + VLDL fraction was significantly reduced in rbapobec1-shRNA transgenic animals (line shL21, comparison of starved and fed rabbits in each category). Indeed, after feeding, the levels of cholesterol and triglycerides increased clearly in the chylomicron + VLDL fraction in wild type animals only, and not significantly in transgenic animals. As regard to the other classes of lipoproteins (LDL and VLDL) and after re-feeding, there were no significant differences between wild type and shRNA transgenic animals.

The ordinary diet of the rabbit is devoid of cholesterol and poor in lipids (around 2% instead of 8% in the high fat diet). To further investigate the lipoprotein distribution in the rbapobec1-shRNA transgenic rabbits and their ability to respond to a high fat diet challenge, animals were fed with a diet enriched with triglycerides and cholesterol. In this experiment, as in the alimentary challenge performed with the normal diet, the daily food intake was not different in transgenic and wild type animals. Triglycerides and cholesterol were assayed in plasma samples collected after 8 days feeding with the enriched diet, after a further 20 hours fasting, and 4 hours after re-feeding with the enriched diet. The pattern of triglycerides concentration in the plasma differed clearly in wild type and transgenic animals (Figure 8). Indeed, the plasma concentration of triglycerides was not enhanced after high fat feeding in transgenic animals as it was in wild type rabbits. More precisely, the chylomicrons + VLDL fraction was not enhanced by the diet challenge in transgenic rabbits. Taken altogether, these data support the hypothesis of an inability of the transgenic animals to produce rapidly large amounts of chylomicrons + VLDL in response to the food supply. The food intake being similar in all animals, this suggests a lower lipid absorption in transgenic animals than in wild type ones.

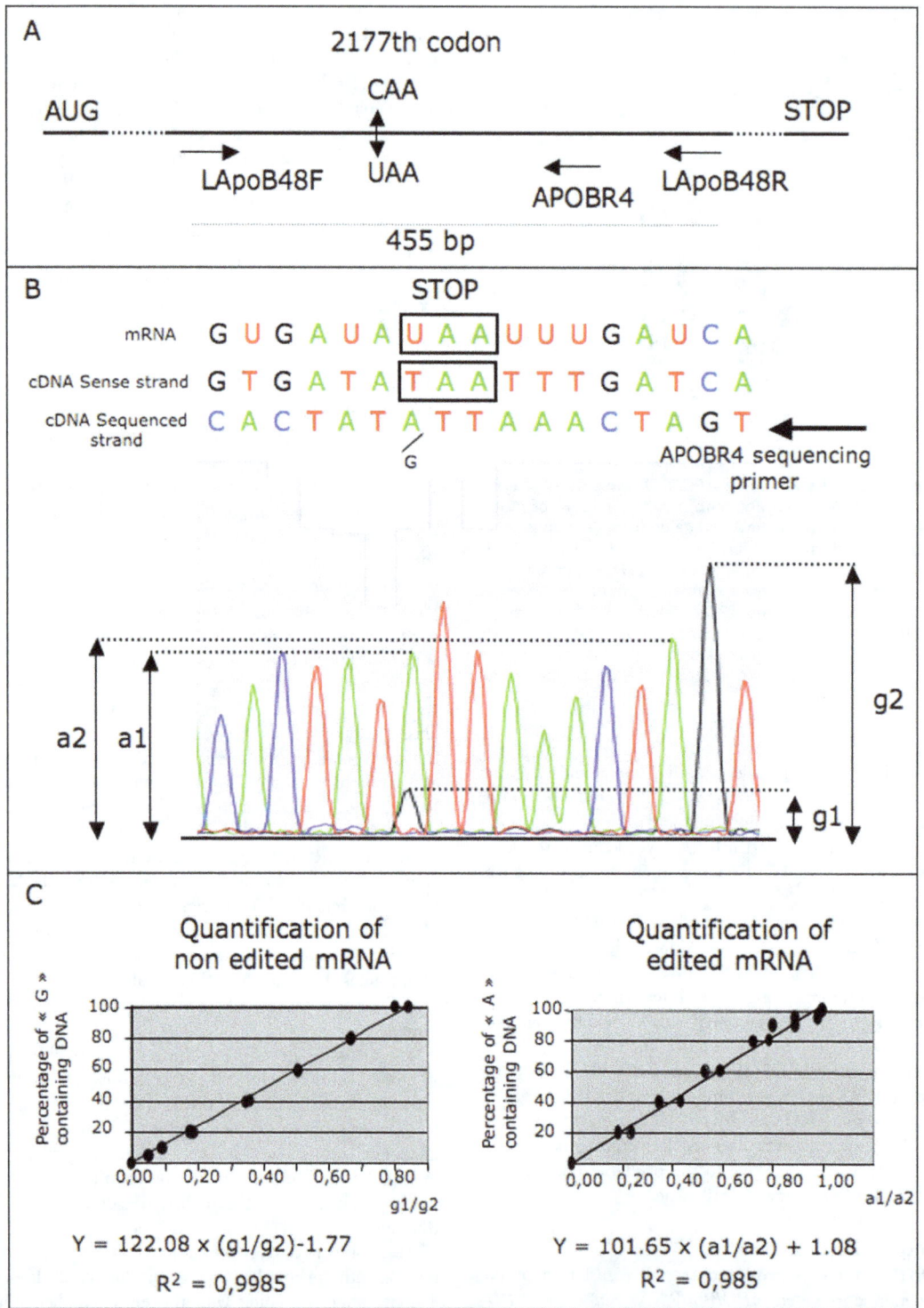

Figure 4. Indirect estimation of the level of "CAA" to "UAA" editing. A: schematic representation of the rabbit *APOB* mRNA from the AUG translation initiation codon until the STOP codon. At the 2177th codon, the "C" residue is edited in a "U" residue. Using reverse transcribed RNA as template, the LApob48F/LApoB48R set of primers amplifies a 455 bp long amplicon encompassing the 2177th codon. When using the APOBR4 primer as sequencing primer, the chromatogram shows the antisense sequence. **B:** detail of a characteristic chromatogram showing how the heights of the peaks were measured at the level of the 2177th codon. Here, the "A" residue was the major one (a1), and the "G" the minor one (g1). Consequently, a large majority of DNA strands in this mixture encompassed the edited TAA (STOP) codon at position 2177. (a2) and (g2) are measured as references. **C:** standard equations obtained by plotting the a1/a2 and g1/g2 ratios against the amount of "A" or "G" containing DNA 455 bp fragment in the sequenced sample. Amounts are given as percentage of "A" or "G" containing DNA.

APOBEC1 dependent editing in intestine

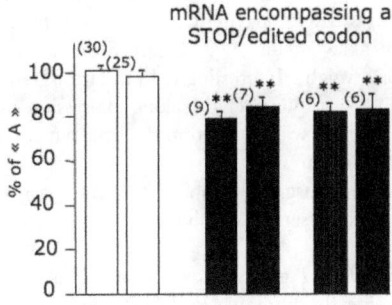

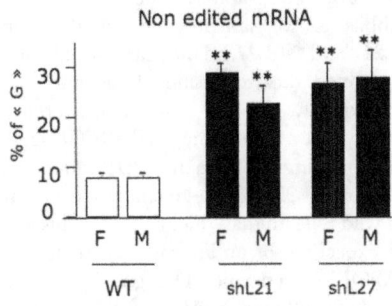

Plasma concentration of APOB48

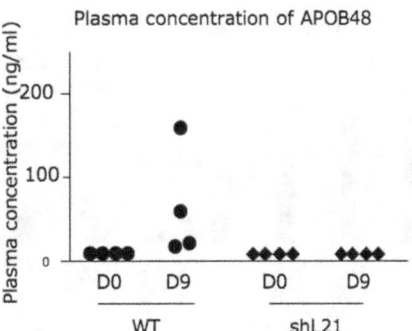

Figure 6. Plasma concentration of APOB48 in rabbits challenged by a high fat/high cholesterol regimen. Plasma concentration of APOB48 was assayed by a specific ELISA kit. Four wild type rabbits and four transgenic rabbits expressing the rbapobec1-shRNA transgene were fed ad libitum with a high fat/high cholesterol regimen for 9 days. Blood samples were collected before the high fat/high cholesterol regimen (D0) and 9 days after the starting of the regimen (D9). Each point indicates the plasma concentration of APOB48 (in ng/ml) in one animal.

Figure 5. Indirect estimation of editing in wild type and rbapobec1-shRNA transgenic rabbits. APOBEC1 dependent editing was measured in the intestine of wild type and transgenic animals. Values were deduced from sequence chromatograms of a PCR fragment encompassing the edited codon as described in "material and methods" section and in Figure 4. The amount of DNA with a "A" residue was representative of the amount of APOB mRNA with a 2177th STOP/edited codon; the amount of DNA with a "G" residue was representative of the amount of full length APOB mRNA. The number of studied animals in each group is indicated in brackets. Mean values are given as percentages with the standard error of the mean (sem). Comparisons were made with control animals (** = p<0.001).

Interestingly, in the high fat diet animals, and not in the normal diet ones, the concentration of triglycerides and cholesterol in the HDL fraction was obviously reduced in transgenic animals compared to wild type animals. This could be related to the low rate of synthesis of chylomicrons by the intestine in transgenic animals, since chylomicrons and their remnants contribute significantly to the production of HDL.

Storage of total body lipids

Since the production of chylomicrons was impaired in transgenic rabbit, one could expect that the uptake of lipids from the diet would be reduced leading to a decreased storage of lipids. To assess this hypothesis, the total mass of body fat was estimated using TOBEC analysis at around 12–16 weeks after birth. The total mass of fat was always the lowest in rbapobec1-shRNA transgenic animals (Figure 9), and the highest in wild type animals. We propose that the reduced body mass of lipids in rbapobec1-shRNA transgenic animals was the result of the reduced uptake of diet lipids consecutive from a low production of chylomicrons and low absorption of fatty acids.

Growth curves in transgenic and wild type rabbits

Our main objective was to study whether modifications of APOBEC1 gene expression in the intestine induced a lean phenotype in the rabbit species. Thus, all animals were weighed weekly from birth during 12–18 weeks. All transgenic litters were

obtained by breeding a transgenic male with a wild type female. Newborns being thus nourished by wild type mothers, this eliminated any possible incidence of the transgenic milk on growth.

At birth, the weight of newborns was not significantly different whatever animals were transgenic or not. However, after three weeks and for the whole length of the experimentation, transgenic rabbits expressing the shRNA targeting the rabbit APOBEC1 gene (shL21 and shL27) were always the lightest animals (by 10% to 20%) as shown within each litter (Figure 9). Thus, this led us to conclude that the rbapobec1-shRNA transgene expression induced actually a lean phenotype in the rabbit species. The lean phenotype could result from the low production of chylomicrons + VLDL possibly leading to a reduced uptake of diet lipids and a reduced absorption of energy deriving from fatty acids. However, additional experiments should be performed to confirm this hypothesis, and specifically to study whether the energy expenditure was affected in a different manner in transgenic and wild type animals.

Rescue of the normal phenotype in double transgenic rabbits expressing both the rbapobec1-shRNA and the human APOBEC1 gene

In order to eliminate the possibility that the lean phenotype was not consecutive to the reduction of rabbit APOBEC1 gene expression but was due to any other phenomenon induced by the rbapobec1-shRNA transgene, we decided to produce double transgenic rabbits expressing simultaneously the rbapobec1-shRNA transgene and the human APOBEC1 gene.

We first produced transgenic rabbits expressing the human APOBEC1 gene in the intestine through the tissue specific activity of the rat IFABP gene promoter [18] added in the construct (Figure 10A). Fifty-four rabbits were born after microinjection of the NotI insert, giving 4 (7.4%) rIFABP-hAPOBEC1 transgenic founders. Transgenic lines were successfully established from 2 (L01 and L02) of these founders, harboring respectively 2 and 6 copies of the transgene. A small number of double transgenic animals expressing both the human APOBEC1 gene and the shRNA targeting the rabbit APOBEC1 mRNA were produced by breeding rIFABP-hAPOBEC1 (L01 or L02) and rbapobec1-shRNA transgenic lines (shL21 or shL27). The analysis of

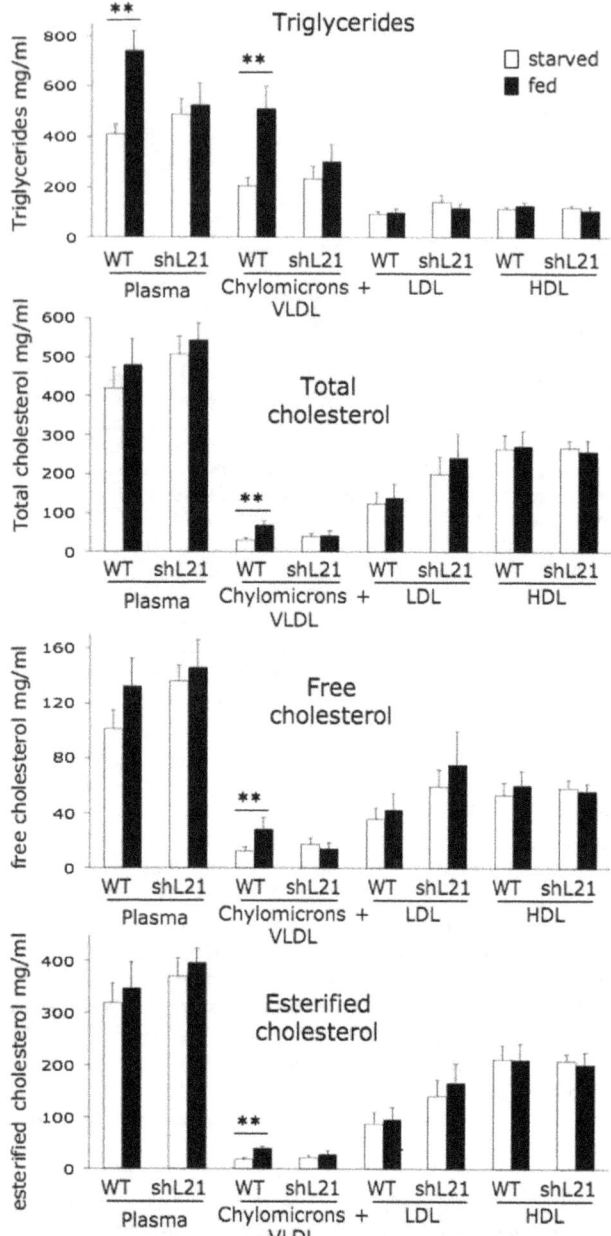

Figure 7. Plasma concentrations of triglycerides and cholesterol (total, free and esterified) in rabbits fed with a normal diet. Triglycerides and cholesterol were assayed in the plasma, and in three lipoproteic compartments separated by ultracentrifugation. Blood samples were collected in rabbits (6 wild-type and 7 transgenic rabbits from line shL21) fed with a normal diet and starved for 20 hours (white bars) and 4 hours after re-feeding with the normal diet (black bars). Values are given in mg/ml, with the standard error of the mean. Comparisons were made between starved and fed animals within each group (** = p<0.001).

transgenic lines L01, L02 and double transgenic rabbits is presented in Figures 10 and 11.

Both transgenic lines L01 and L02 expressed the human *APOBEC1* gene stably over generations, with similar levels in males and females (Figure 10B, left graph). In double transgenic lines, the level of the human *APOBEC1* gene expression was not significantly different from that in lines L01 and L02, which proves

that the shRNA produced by the rbapobec1-shRNA transgene did not alter the expression of the human *APOBEC1* transgene. The presence of the human APOBEC1 enzyme was confirmed in the intestine by western blot assay in line L02 (Figure S3) with the expected 27 kD molecular weight. Interestingly, in the transgenic rIFABP-hapobec1 line L02, an unexpected leaking expression of the rIFABP-hapobec1 transgene was detected in the liver but with a 50 times lower level than in the intestine.

As expected, in double transgenic rabbits, the rbapobec1-shRNA was significantly expressed in the intestine (Figure 10B middle graph).

The level of rabbit *APOBEC1* gene expression was similar in transgenic rabbits expressing the human *APOBEC1* gene and in wild type animals (Figure 10B, right graph). Besides, it was lower in double transgenic rabbits, as we had previously observed in rabbits from lines shL21 and shL27. Thus, as we already suggested, the shRNA targeted the expression of the intestinal rabbit *APOBEC1* gene probably through a RNA interference mechanism, without altering that of the human *APOBEC1* gene.

In transgenic animals expressing the human *APOBEC1* gene, the level of editing was at around 95% of the maximum, as it was previously determined in wild type animals (Figure 10C). This was surprising since we were expecting for an increase consecutive to the additional human APOBEC1 enzyme. Though, the human APOBEC1 enzyme was actually efficient in *APOB* mRNA editing in the rabbit as the rabbit APOBEC1 enzyme. Indeed, editing was observed in the liver of some transgenic rIFABP-hapobec1 animals (Figure 10C, line L02, middle panel) harboring a leaking expression of the human *APOBEC1* transgene in the liver, when editing is never observed in liver in wild type rabbits. Thus, the lack of any modification in *APOB* mRNA editing in transgenic animals over-expressing the APOBEC1 enzyme was not due to the inefficacy of the enzyme but probably the consequence of the saturation of the mechanism of editing.

In double transgenic animals, the level of editing was similar to that of wild type animals, despite the reduced expression of the rabbit *APOBEC1* gene in the intestine. This proves that the human APOBEC1 enzyme expressed in the intestine by the transgene was able to counterbalance the default of rabbit APOBEC1 enzyme due to the shRNA targeting the rabbit *APOBEC1* mRNA.

Interestingly, the plasma level of APOB48 was highly enhanced in the human *APOBEC1* transgenic rabbits L02 by the high fat/high cholesterol diet challenge (Figure 10C, right graph). Since editing was not modified in the intestine of these animals, it is likely that the high plasma concentration of APOB48 originated from the liver, where a significant editing of the *APOB* mRNA was measured consecutively to the leaking expression of the human *APOBEC1* transgene.

The plasma lipid levels and lipoprotein distributions were assayed in human *APOBEC1* transgenic rabbits (L02, Figure S4) submitted to the high fat/high cholesterol diet and starvation/feeding challenge. Surprisingly, the concentrations of triglycerides in the plasma and also in the chylomicrons + VLDL fraction were not enhanced by the diet, by opposition to what we were expecting for in these rabbits characterized by a high level of circulating APOB48. Clearly, in these animals, the high circulating APOB48 did not contribute to a high synthesis of chylomicrons. Other differences were further detected throughout the starvation/feeding challenge. These could be consecutive to the leaking expression of the human *APOBEC1* gene in the liver, which induced the liver editing of the *APOB* mRNA and thus the reduction of the hepatic synthesis of APOB100 protein.

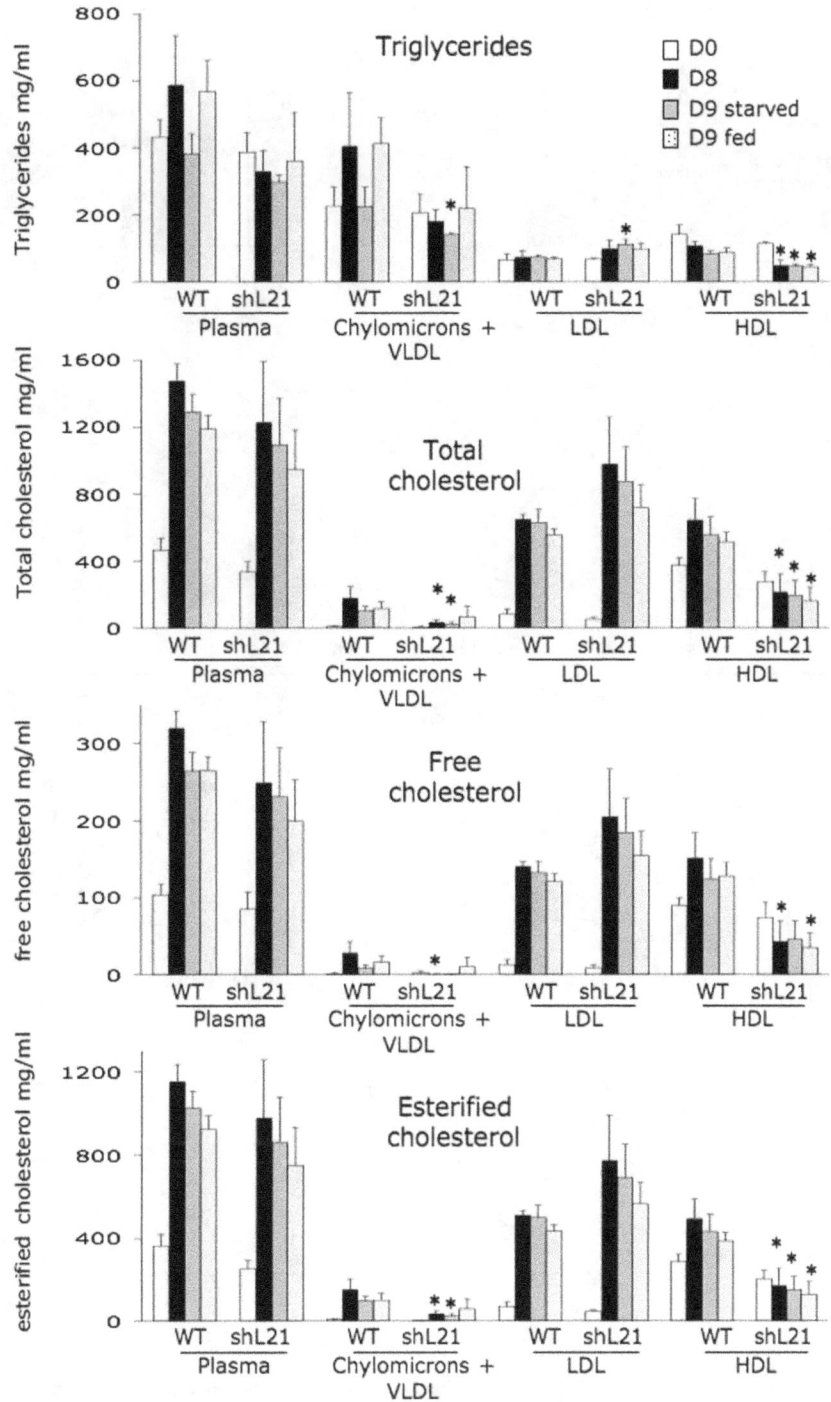

Figure 8. Plasma concentration of triglycerides and cholesterol in rabbits fed with a high fat/high cholesterol regimen. Rabbits (4 wild type, and 3 transgenic rabbits from line shL21) were fed for 8 days with a high fat/high cholesterol diet. Plasma samples were collected before the diet (D0, white bars), after feeding for 8 day with the diet (D8, black bars), after 20 hours starvation (D9 starved, grey bars) and 4 hours after re-feeding with the high fat diet (D9 fed, dotted bars). Triglycerides and cholesterol were assayed as in Figure 7. Values are given in mg/ml, with the standard error of the mean. Comparisons were made between transgenic and wild type animals for each day of the challenge (* = p<0.05).

The total mass of body lipids and growth curves were determined from a series of litters including newborns of each genotype (wild type, rbapobec1-shRNA, rIFABP-APOBEC1, and double transgenic animals, Figure 11). The transgenic animals expressing the human *APOBEC1* gene gained weight and possessed a total lipid mass as the wild type animals. This was not surprising since in these transgenic animals, the APOB mRNA editing and the production of chylomicrons were similar to those determined in wild type rabbits. A small number of animals of rIFABP-hapobec1 transgenic lines L01 and L02 were weighed for a longer time (Figure S5), in order to detect possible long-term modifications consecutive to limited but sustained modifications of

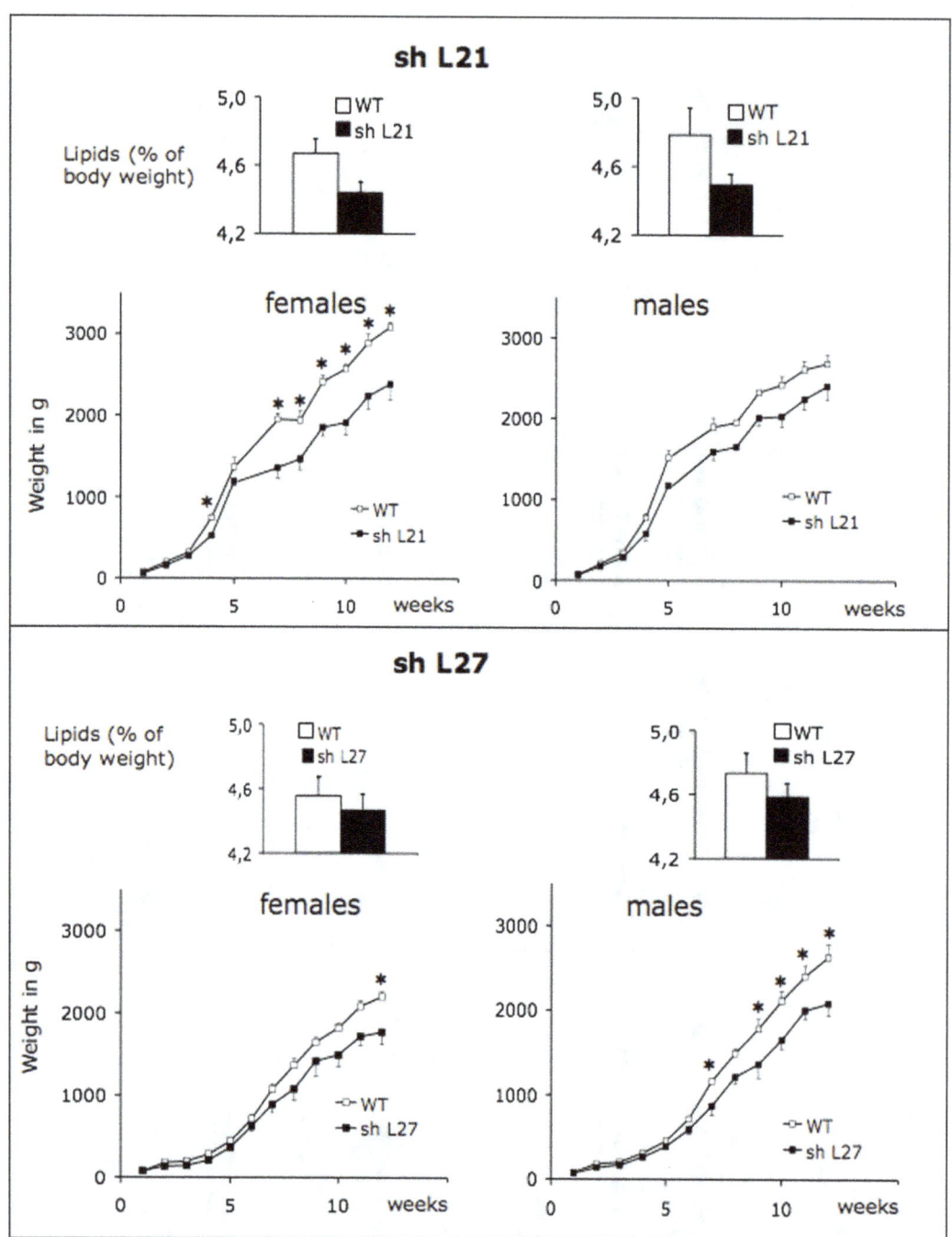

Figure 9. Total content of body lipids and growth curves of wild type and transgenic rabbits from lines shL21 and shL27. The total content of body lipids and growth curves were established on the same rabbits. All rabbits (mothers during pregnancy and lactation and their litters after weaning) were fed with the normal diet. Wild type mothers nourished all newborns (transgenic or wild type ones). The total content of body lipids, expressed as the percentage of the body weight, was measured in transgenic (shL21 and shL27, black bars) and wild type (white bars) rabbits at around 12–16 weeks after birth. Three animals at least were considered for each point. Values are means +/– sem. Note that the percentage was always the lowest in shRNA expressing animals, and the highest in wild type animals. Growth curves were established by weighing weekly each rabbit from 3–5 weeks to 12–18 weeks after birth. Males and females are shown in separate graphs. * = p<0.05 comparison of shRNA expressing animals and wild type ones.

the level of editing that we might have not been able to detect earlier. However, the weight of transgenic animals was not different from that of wild type ones, showing that even in older animals, the long term-expression of human APOBEC1 enzyme induced no significant over-weight gain.

More interestingly, in spite of the small number of animals of each genotype in the litters, the double transgenic animals were clearly heavier than the shRNA expressing animals and their total mass of body lipids was similar to that of wild type animals. This shows once more that the presence of the human APOBEC1

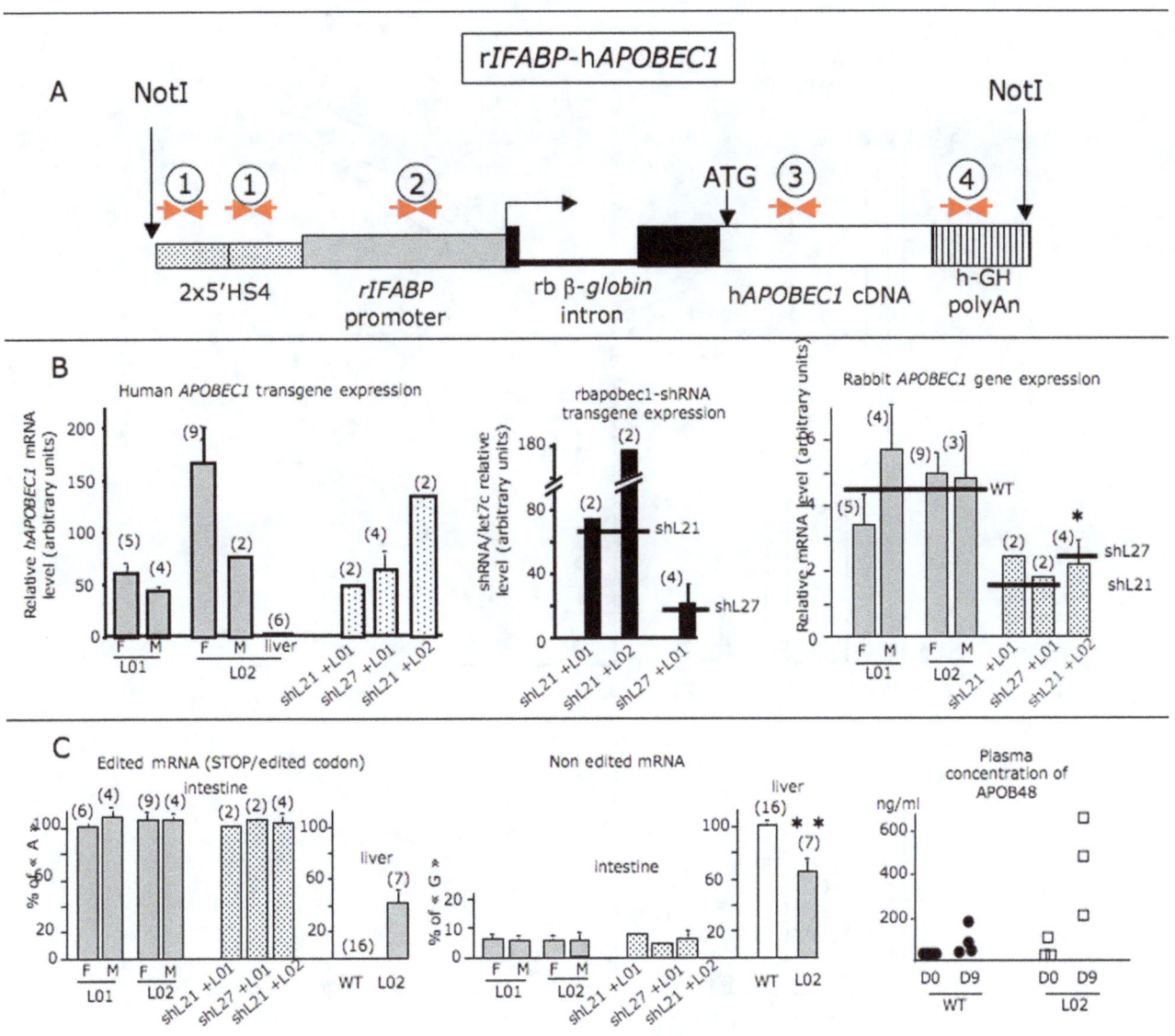

Figure 10. Analysis of rIFABP-hapobec1 transgenic and double transgenic rabbits. 10A: Structure of the recombinant gene to express the human APOBEC1 cDNA in the intestine of transgenic rabbits. The rIFABP-hAPOBEC1 construct encompassed two copies of the chicken ß-*GLOBIN* gene fragment 5'HS4 (gene expression insulator element, dotted box), the promoter of the rat intestinal fatty acid binding protein gene (r*IFABP*; grey box), the rabbit (rb) ß-*GLOBIN* second intron (black boxes and thick line), the human *APOBEC1* cDNA (white box) produced by PCR amplification from reverse transcribed RNA of HT29 cells (derived from a human colon tumor that have retained the ability to express the *APOBEC1* gene), and the human growth hormone polyadenylation sequences (box with vertical bars). The horizontal black arrow points the position of the transcription start site. ATG = translation initiation site of *hAPOBEC1* cDNA. Numbers and small horizontal arrows represent the sets of primers. All studied transgenic animals were PCR positive for the sets 1–4. **10 B: gene expression.** The levels of human *APOBEC1* mRNA (left panel) and shRNA (middle panel) were measured in RNAs prepared from duodenum cells in two rIFABP-hapobec1 lines (L01 and L02) and in double transgenic animals (shL21+L01; shL21+L02; shL27+L01). Values are given in females (F) and males (M) after normalization to the level of reference gene expression determined simultaneously in each sample: Let7c miRNA in the case of shRNA, and *RPL19, YHWAZ, HPRT* in the case of human *APOBEC1*. In double transgenic animals, males and females were not distinguished, considering the small number of animals in these groups. The number of animals in each group is indicated in brackets. Values are given with the standard error of the mean (sem). The mean level of shRNA in shL21 and shL27 as presented in Figure 2 is indicated with a horizontal line. The level of expression of the human *APOBEC1* transgene measured in the liver is given in L02. In L01, this level was not significantly detected. The level of rabbit *APOBEC1* mRNA (right panel) was measured in intestinal RNAs as described in Figure 3. The level found in wild type rabbits and in lines shL21 and shL27 is indicated with a horizontal line. **10 C: APOBEC1 dependent editing in intestine and liver and Plasma concentration of APOB48 in rabbits expressing the human APOBEC1 gene.** The estimation of editing was made as described in the legend of Figure 5. Plasma concentration of APOB48 was performed as described in legend of Figure 6 in 3 transgenic rabbits from line L02. Wild type animals are the same than those in Figure 6.

enzyme was able to counterbalance the effect of the shRNA targeting the rabbit *APOBEC1* gene. Taken altogether, our results suggest strongly that the lean phenotype observed in rbapobec1-

shRNA transgenic rabbit was the consequence of the reduced level of *APOBEC1* gene expression.

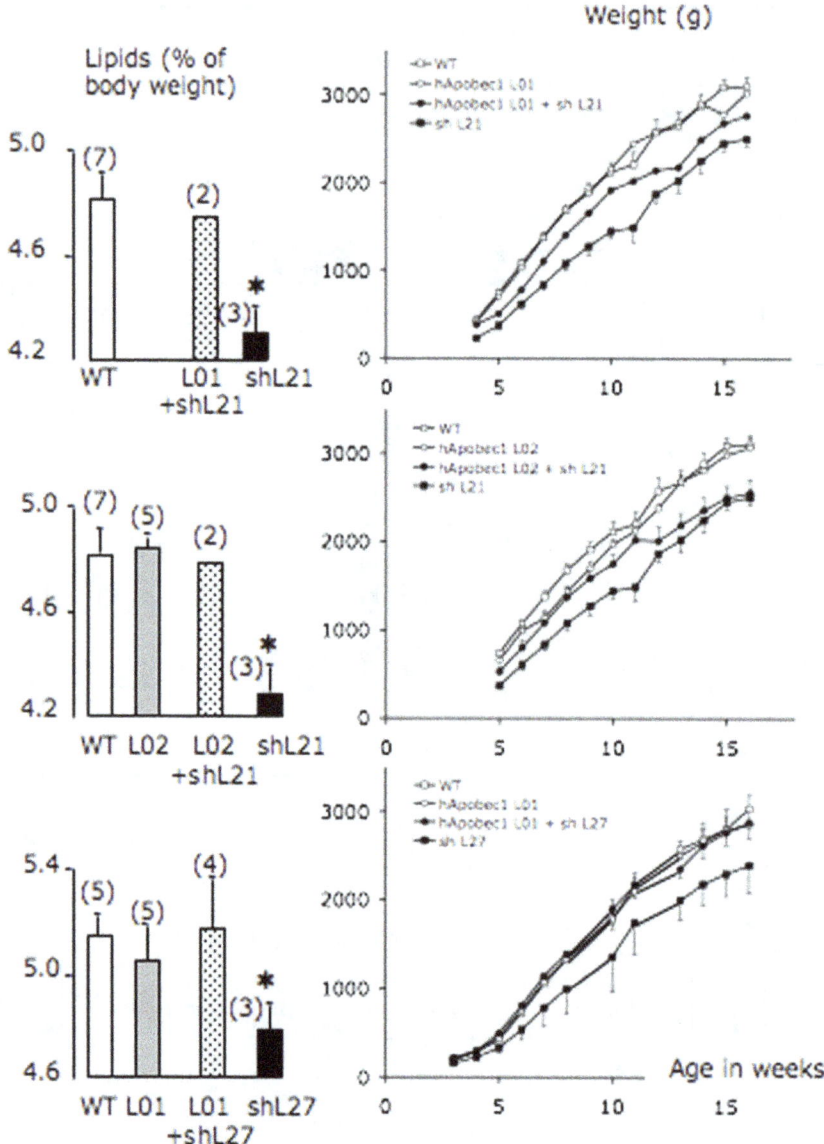

Figure 11. Total mass of body lipids and growth curves in double transgenic rabbits. Double transgenic animals (shL21+L01; shL21+L02; shL27+L01) were produced by breeding rIFABP-hAPOBEC1 (L01 or L02) and rbapobec1-shRNA transgenic lines (shL21 or shL27). In these litters, the total mass of lipids was significantly lower in shL21 or shL27 transgenic animals than in animals from all other groups (* = p<0.05). Numbers in brackets indicate the number of animals in each group. Growth curves were established by weighing weekly each rabbit from 3–5 weeks to 12–18 weeks after birth. Males and females are shown in separate graphs. * = p<0.05 comparison of shRNA expressing animals and wild type ones.

Discussion

A great number of genes are devoted to the storage of energy, and it is reasonable to propose that evolution has selected organisms able to survive in scarce conditions thanks to efficient mechanisms of energy storage. Limiting energy uptake and storage is probably a valuable strategy to fight against obesity. Thus, our approach consisted of looking for critical genes in people with lean phenotype. If a monogenic slimness disease resulting from a deficiency of fat absorption can be found, the implicated gene likely plays a critical role in the disease and is a potential target for new anti-obesity drugs. When this gene is not compensated by other mechanisms, it is therefore a powerful target for obesity treatment.

Three human genetic diseases have been described with very similar lean phenotypes: abetalipoproteinemia, hypobetalipoproteinemia, and chylomicron retention disease also known as Anderson's disease [19]. The genes involved in the first two diseases, abetalipoproteinemia and hypobetalipoproteinemia, have now been identified, but it is not yet the case in the Anderson's disease [20,21]. All three diseases are characterized by a severe reduction or total absence of APOB48 protein in intestinal cells and plasma and of chylomicrons production. This led us to investigate further the possibility of fighting against obesity through regulating APOB48 production. APOB48 resulting exclusively from the translation of the APOBEC1 dependent edited *APOB* mRNA, we decided to target the expression of the *APOBEC1* gene.

The phenotype of mice harboring a complete invalidation of the *APOBEC1* gene has been already reported by a series of laboratories [22–25]. As expected, the editing of the *APOB* mRNA was suppressed, and no APOB48 was produced in these mice. It was observed that intestinal fat absorption was less efficient in *APOBEC1⁻/⁻* mice containing only APOB100 than in wild type mice but it was not totally abolished. Probably, APOB100 could replace to some extent APOB48 in chylomicron formation and finally the plasma lipoprotein cholesterol and triglycerides profiles were not different in knock out and wild type mice [7] [8]. In contrast, in the human, APOB100 is not able to form chylomicrons and carry lipids from intestine to liver [26]. Clearly, the metabolism of lipids differs between species, and thus we decided to target the *APOBEC1* gene in another species than the mouse and closer to the human as regard to the metabolism of lipids, in order to investigate whether the *APOBEC1* dependent editing could be a valuable target for fighting against obesity through modulating the lipid uptake.

RNA interference is a natural cellular process mediated by small double strand RNA that induces knockdown of gene expression through mRNA targeting. Here, we produced transgenic rabbits expressing permanently a small interfering RNA (siRNA) targeting the rabbit intestinal *APOBEC1* gene. This was achieved through the introduction in the rabbit genome of a DNA construct expressing a small hairpin RNA by using a strategy that we had followed in a previous study [12]. This strategy had the advantage to provoke the sustained production of the siRNA, and a moderate but significant and permanent decrease of the rabbit *APOBEC1* gene expression. Our objective was to observe long term and prolonged effects of the gene knockdown that is totally different to what can be observed after a total invalidation of the gene. To validate our findings, we produced transgenic rabbits expressing the human *APOBEC1* gene in the intestine, with the aim to rescue the knockdown induced by the RNA interference mechanism.

The Figure 12 presents a model that could explain how targeting the *APOBEC1* gene induces a lean phenotype in the rabbit species. In wild type rabbits (Figure 12 A), the *APOBEC1* gene expressed in the intestine only is responsible for the editing of the *APOB* mRNA. The APOB48 protein thus produced exclusively by the intestine is processed to synthesize chylomicrons that are responsible for the lipid uptake from the diet. In the liver, the APOB mRNA is translated in the APOB100 protein that is involved in the synthesis of VLDL and LDL. In transgenic rabbits expressing the rbapobec1-shRNA (Figure 12B), the level of *APOBEC1* gene expression and the level of editing are significantly reduced in the intestine. The ability to synthesize chylomicrons in response to a diet challenge is reduced. The productions of LDL and VLDL are not significantly modified; but the production of HDL is modified since HDL is processed from chylomicrons and remnants that are reduced. The lean phenotype is thus consecutive to the reduced lipids uptake by the enterocytes through the low synthesis of chylomicrons after each food challenge.

The absence of obese phenotype in transgenic rabbits expressing the human APOBEC1 enzyme (Figure 12C) in the intestine is amazing. Indeed, we were expecting for a long-term gain of weight in these animals since the human *APOBEC1* transgene should have enhanced the level of editing of the *APOB* mRNA. It has been already published that in the rabbit species, around 90% of the *APOB* mRNA was edited in the intestine [14]. Accordingly, our results have shown that more than 95% of *APOB* mRNA were edited. However, in the present study, the expression of the human *APOBEC1* transgene was not able to enhance the level of editing, and consequently, animals did not elicit any obese phenotype. One

explanation is that the editing was already at its maximum in wild type animals, and that the over-expression of APOBEC1 was inefficient. It has been reported that the editing results from the activity of the APOB mRNA editing complex (the editosome), a multicomponent protein complex with enzymatic and regulatory activities [27]. Thus, we suggest that the editing was limited by the availability or activity of other components of the editing complex.

More surprisingly, in transgenic rabbits expressing the human *APOBEC1* gene, the plasma level of triglycerides in the chylomicrons + VLDL fraction was not enhanced after a fat-rich diet for 8 days (Figure S4) as it was in wild type rabbits. Yet, we were expecting for an enhanced production of chylomicrons, since the plasma level of APOB48 was clearly enhanced in these animals (Figure 10C). Notably, the excess of APOB48 protein was produced by the liver, in response to the leaking expression of the human *APOBEC1* transgene. Thus, it suggests that the APOB48 protein originating from the liver was not processed into chylomicrons, and that only the APOB48 produced by the intestinal cell can form chylomicrons. Finally, this study shows that in spite of a significant high expression of the human *APOBEC1* gene in the intestine, no obese phenotype can be observed in the rabbit species.

In double transgenic rabbits (Figure 12D), the expression of the human *APOBEC1* gene in the intestine counterbalances the shRNA induced knockdown of the rabbit *APOBEC1* gene expression. Consequently, the level of editing is similar to that in wild type animals, as the level of synthesized chylomicrons. The rescue of the normal phenotype in the double transgenic rabbits is a solid argument to demonstrate the relation between *APOBEC1* gene expression and the lean phenotype.

In conclusion, this study presents for the first time evidences that targeting the *APOBEC1* gene is a valuable strategy to induce a lean phenotype in the rabbit. Importantly, it has to be further confirmed that the lean phenotype is the consequence of a moderate modification of the lipid uptake through chylomicrons. Remarkably, animals did not suffer from any disease, and their breeding capacity was not apparently affected for more than two years that is a long experimental duration in this species. However, a series of additional experiments should be performed to investigate the impact of targeting the APOBEC1 gene expression on health, by specifically studying the impact of alterations of intestinal absorption of lipids and other nutrients. Finally, the success of this strategy lies probably in the fact that it concerns a gene pathway without compensatory mechanisms that affects the lipids uptake of the diet. Moreover, it suggests that looking for new genes associated to lean phenotype is probably a valuable tool to highlight novel targets for obesity treatment.

Materials and Methods

Animals

Californian rabbits (GD24 strain) were bred at the UCEA rabbit facility (Unité Commune d'Expérimentation Animale, Jouy-en-Josas, France). All experiments were performed with the approval of the local committee for animal experimentation (COMité d'ETHique appliqué à l'Expérimentation Animale (COMETHEA), Jouy-en-Josas, accreditation number 12/017). All researchers working directly with the animals possessed an animal experimentation license delivered by the French veterinary services.

All rabbits were weighed each week from birth until week 12–18, few before puberty that occurs at around 20–24 weeks in this species. Breeders were nourished with a normal diet. Since the growth rate depends on the number of newborns in each litter,

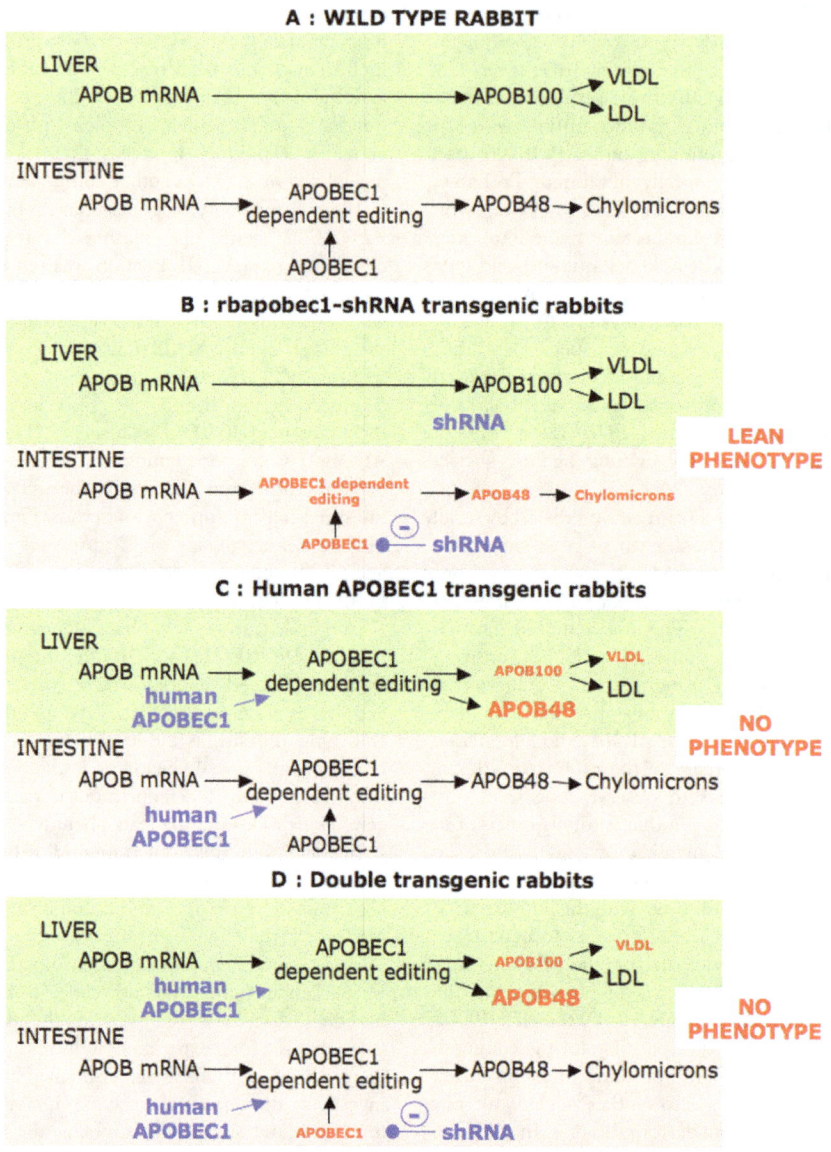

Figure 12. Intestinal and liver regulation of APOB48, APOB100, and chylomicron production. Four schematic representations are given, to simulate the regulation of *APOB* mRNA editing and the consequence upon the phenotype in wild type and transgenic rabbits. The models depict the situation after a diet challenge with normal of high fat/high cholesterol diet. Blue characters are used for the transgene expression (shRNA targeting the rabbit *APOBEC1* mRNA, and human *APOBEC1* gene); red characters indicate the measured parameters with significant modifications; the size of the letters is related to the level of production.

care was taken to compare rabbits issued from litters encompassing approximately the same number of newborns. Animals were currently weaned at around 7 weeks. After weaning, animals were fed with the normal diet except otherwise mentioned.

At around 18 weeks after birth, animals were starved for 24 hours, blood samples were collected on EDTA to prevent for coagulation, and food was immediately provided. Four hours after feeding, animals were sacrificed; blood and tissue samples were collected. Blood samples were centrifuged (10 minutes, 3000 g), then plasma and tissues were frozen at −80°C until used.

Construction of recombinant genes

The rbapobec1-shRNA-producing gene (Figure 1) encompassed two inverted repeats of the rbapobec1-shRNA and a stretch of five T residues as gene terminator. The shRNA

transcription unit was constructed from synthetic DNA fragments (Eurofins, Ebersberg, Germany). This H1-rbapobec1-shRNA gene was then inserted into the pM10 vector [12] at the enzymatic BsaBI-ClaI restriction sites as presented in Figure 1. The final construct used for microinjection encompassed the DNA fragment included between the two NotI restriction sites.

The human *hAPOBEC1* construct (rIFABP-*hAPOBEC1*, Figure 10) encompassed the human *APOBEC1* cDNA from 3 nt upstream of the site of initiation of translation (ATG) to 3 nt downstream the STOP codon linked to the rabbit (rb) ß-*globin* second intron. Transcription was driven by the promoter of the rat intestinal fatty acid binding protein gene (r*IFABP*) spanning from nucleotides −1150 to +51 as regard to the *IFABP* gene transcription start site. A tandem of the chicken ß-*globin* gene fragment 5′HS4 was added as insulator. The termination signal was brought by the terminator from the human growth hormone gene (h-GH polyAn).

The sequences of all plasmids encompassing these constructs are available upon request.

Generation of transgenic rabbits

The inserts to be used for microinjections were released from plasmids by NotI digestion, separated on 1% agarose gel in 1x TBE, purified using the Qiaquick gel extraction kit (Qiagen, Courtaboeuf, France) and then EluTipD filtration (Schleicher & Schuell, Mantes la Ville, France). The resulting DNA preparations were microinjected into Californian rabbit embryo pronuclei at a concentration of 2 ng/µl. The transgenic rabbits were identified using PCR performed on ear clip DNA extracts. Four sets of primers were designed to cover the integrity of each integrated construct: sets 1, 2, 3 and 4 for the *hAPOBEC1* expressing construct, and sets 1, 4, 5 and a/b for the shRNA expressing construct (Figures 1 and 10, table S1). Sets 1, 2, 3, 4, and 5 were used in real time PCR with the fast SYBR Green master mix (Applied Biosystems). In parallel, a set of primers (cas1, cas2, table S1) amplifying a non-coding region upstream of a control gene (the rabbit ß-*CASEIN* gene) was used as control amplification. All sets of primers were designed by the Primer Express software (Applied Biosystem) and all amplicons were 100 base-pairs long.

The number of copies of integrated gene was deduced from real time PCR amplifications by the $2^{(\Delta\,(\Delta Ct))}$ method. The sets of primers 1, 2, 3, 4 and 1, 4, 5 were used as transgene specific probes for the rIFABP-hapobec1 and rbapobec1-shRNA constructs respectively. We used the set of primers cas1/cas2 located on the rabbit ß-*CASEIN* gene as reporter probe to normalize to a two copies endogenous gene. A reference rabbit genome was produced by mixing one copy of transgene per copy of genome and used as standard. The number of copies of integrated transgene was similar for sets 1, 2, 3, 4 in the case of the rIFABP-hapobec1 transgene and for sets 1, 4, 5 in the case of the rbapobec1-shRNA transgene, thus suggesting that transgenes were intact.

shRNA assay

The concentration of shRNA produced by the shRNA constructs in transgenic rabbit tissues was estimated by RT-qPCR [28] [12]. Briefly, 5 µg of total RNA, prepared as previously described [29], were polyadenylated according to Ambion's protocol (PolyA Polymerase, Ambion, Applied Biosystems, France). The polyadenylated RNAs were reverse transcribed (High Capacity cDNA Archive kit, Applied Biosystems) using as reverse primer a polyT adapter encompassing a series of twelve "T" residues and a universal primer (Table S1). Quantification was achieved by SYBR Green quantitative PCR (Applied Biosystems) using a set of primers composed of the universal primer corresponding to the 5′ end of the polyT-adapter and a primer specific to the shRNA sequence, resulting in the amplification of a 65 bp long fragment.

The concentration of shRNA in tissue samples was estimated after normalization by the concentration of Let7c miRNA determined by the same method in each sample. It was thus given by the formula $2^{(CtLet7c-CtsiRNA)}$. A set of samples was chosen as calibrators and was assayed in all compared runs. Care was taken to consider Ct values within the linear amplification zone.

Quantification of human and rabbit *APOBEC1* gene expression

Total RNAs were extracted from tissues as previously described [29]. Reverse Transcription (RT) was performed on 1 µg of total RNA using the High Capacity cDNA Archive kit (Applied Biosystems) and the random primer mix included in the kit.

Quantification was achieved using SYBR Green quantitative PCR (fast SYBR Green master mix, Applied Biosystems) with dilutions of the RT reactions and sets of primers designed by the Primer Express software (Applied Biosystem). Whenever possible, primers were chosen on separate exons in order to avoid contaminant DNA amplification, and all amplicons were 100 base-pairs long (Table S1). The sequence of primers was chosen in order to avoid cross reactivity between human and rabbit *APOBEC1* gene measurements. Moreover, for all samples, a RT minus reaction was performed with all RT components except the reverse transcriptase enzyme, and assayed as a complete RT reaction to ensure that no amplification was due to contaminant DNA.

Three normalizing genes (*RPL19, YHWAZ, HPRT*) were tested on all samples for their stable expression in the studied tissues. The GeNorm program included in Biogazelle QBasePlus software (Biogazelle NV, Ghent, Belgium) was used to analyze the data. In order to correct for inter-run fluctuations, a set of samples was chosen as calibrators and was assayed in all compared runs. Care was taken to consider Ct values within the linear amplification zone. Gene expression was considered as significant when Ct values obtained using 2–5 ng of cDNA in each q-PCR reaction were lower than 34, and when one single DNA fragment with the expected size was amplified as template in each q-PCR reaction.

Characterization of the intestinal human APOBEC1 protein

Human APOBEC1 was characterized in intestinal protein extracts by Western blotting. Scrapped intestinal mucosa cells were homogenized with a Dounce homogenizer in RIPA buffer (50 mM Tris-HCl, pH 7.4; 1% IGEPAL; 0.5% Na-deoxycholate; 0.1% SDS; 150 mM NaCl; 2 mM EDTA; 50 mM NaF; 0.2 mM sodium orthovanadate) with protease inhibitors (Complete Protease Inhibitor Cocktail, Roche; 1 mM PMSF; 1 mM Benzamidine) extemporaneously added. After incubation on ice for 30 minutes and centrifugation at 10 000 g for 10 minutes at 4°C, the supernatant was collected and frozen in aliquots at −80°C. Protein concentration was determined by Bradford assay (BioRad, France) using BSA as standard.

Proteins were separated on a 16% acrylamide gel electrophoresis then transferred on Hybond-P membrane. The human APOBEC1 enzyme was detected after incubation with a rabbit anti-APOBEC1 antibody (Sigma, SAB2100132), a goat anti-rabbit IgG peroxidase conjugate (Sigma, A-0545) and the immunofluorescence ECL 2 detection kit (Pierce).

Indirect quantification of the proportion of edited ApoB mRNA

Reverse transcribed (RT) RNAs were obtained as for the measurement of rb*APOBEC1* gene expression. A 455 bp long DNA fragment was amplified from RT RNAs using a set of primers specific of the rabbit *APOB* cDNA (LApoB48F/ LApoB48R, Table S1). The *APOBEC1* edited 2177th codon of the *APOB* cDNA was included in this amplified DNA fragment (Figure 4A). Editing was responsible for the modification of the "CAA" codon in a "UAA" one (a TAA codon in the amplified product). The amplified fragment was purified by MSB Spin PCRapace (Stratec, Eurobio, France), then sequenced (Eurofins, MWG) using the APOBR4 oligonucleotide as sequencing primer. Using this primer, the antisense strand was sequenced. We measured the yield of 455 bp fragments with a "T" residue in the mixture by analyzing the chromatogram of the sequence of each amplified 455 bp fragment. As shown in Figure 4B, the height of the "G" (g1) and "A" (a1) peaks in the chromatogram was compared to that of the "G" (g2) and "A" (a2) peaks chosen in the vicinity to normalize for sequencing efficiency (in the antisense sequenced strand, "G" and "A" sequenced residues corresponded to the "C" and "T" residue of the edited 2177th codon). A standard curve was performed by sequencing a definite amount of the 455 bp DNA fragment containing a mixture of varying proportions of the two types of DNA strands elsewhere purified, which sequence encompassed the "C" or "T" residue at the 2177th codon (Figure 4C). Two linear equations were deduced by plotting the g1/g2 or a1/a2 ratio against the amount of "C" or "T" encompassing DNA in the mixture. These equations were then used to determine the percentage of "C" or "T" encompassing DNA in each RT RNA mixture produced from the various studied samples. Thus, we indirectly measured the yield of edited mRNA.

Plasma APOB48 levels

Plasma APOB48 levels were assayed using an ELISA [16] as indicated by the manufacturer (Shibayagi, X-Celtis GmBH, Germany). To enhance the level of APOB48 in the plasma, rabbits were fed ad libitum for 9 days with a high fat/high cholesterol regimen containing 0,2% cholesterol and 8% soybean oil [17].

Plasma cholesterol and lipid levels

The major classes of lipoproteins were isolated from plasma samples by sequential ultracentrifugation ensuring the separation of chylomicrons + VLDL, LDL and HDL [30]. Cholesterol (total, esterified, and free) and triglycerides were further determined in each fraction using commercially available enzymatic kits.

Total body content of lipids

The total content of lipids was deduced from the measure of total body electrical conductivity (TOBEC) as previously described [31] with modifications brought by L Lamothe and C Bannelier for using an EM-SCAN SA-3000-type chamber. Animals were not anesthetized during measurements. Measurements were made at around 12-16 weeks after birth, and rabbits were immediately weighed. The content of lipids was deduced from the E-value given by the TOBEC and the weight of the rabbit using a prediction equation as follows: total lipid content (% of live weight, LW) = 3.33843+0.00248 x LW − 0.00196 x E with LW = live weight and E = TOBEC measurement.

Supporting Information

Figure S1　*In vitro* assessment of the efficiency of shRNA expressing constructs. The OligoWalk web server generated a list of small hairpin RNA candidate sequences ranked by the probability of being efficient to knock down the targeted gene expression. Four sequences (named "a", "b", "c", and "d") were chosen within this list (their probability of being efficient ranged from 88% to 95.5%), and were tested *in vitro* using a cell transfection assay. These sequences targeted the 3'UTR region of the rabbit *APOBEC1* transcript. The transfections were carried out in CHO.K1 cells (ATCC number CCL-61) using ExGen500 (Euromedex, Souffelweyersheim, France), according to the manufacturer's protocol. The test aimed to measure the efficacy of rbapobec1-shRNA constructs to target the rabbit *APOBEC1* gene expression, in order to select an efficient one that will be used

to produce transgenic rabbits. Four constructs harboured the "a", or "b" or "c" or "d" shRNA sequence. A fifth construct harboured two H1-shRNA genes, one with the "a" and the other with the "d" sequence (see rbapobec1-shRNA diagrams). In the absence of rabbit intestinal cell cultures expressing the *APOBEC1* gene, we designed a chimeric target gene encompassing the *luciferase* gene fused to the targeted sequence of the rabbit *APOBEC1* gene (upper diagram). The 3'UTR of the rabbit *APOBEC1* gene (from nucleotide 756 to 905 respectively to the ATG translation initiation codon) was added at the 3' position of the *luciferase* gene. Degradation of the 3'UTR region in the target construct by shRNAs was expected to prevent translation of the *luciferase* cistron. Thus, a quantification of shRNA-induced knockdown could be achieved by measuring *luciferase* activity in the transfected cells. The reliability of this method was previously established <Hung, 2006 #30>, showing that it is possible to fuse short target sequences (such as the rabbit *APOBEC1* gene 3'UTR sequences) in the UTR of a reporter gene in order to establish a quantitative reporter-based shRNA validation system. Each shRNA-expressing constructs (0.75 µg/P35 dish) was transfected with the target *luciferase* construct (0.75 µg/P35 dish) or the control empty vector pM10, and the *β-galactosidase* vector pCH110 (Pharmacia, 1 µg/P35 dish) to correct for transfection efficiency. Luciferase and ß-galactosidase activities were measured 48 h after transfection. The results are given as percentages of luciferse activity in cells transfected by the target *Luciferase* vector and the empty pM10 vector. All luciferase values were normalized to ß-galactosidase activities. The graph is representative of two independent experiments.

Figure S2　Chromatograms of sequence of DNA amplified from RT-RNA of intestine and liver in wild type or shRNA expressing transgenic animal. The product of amplification of RT-mRNA encompassing the 2177th codon was sequenced using the APOBR4 oligonucleotide as sequencing primer. By using this oligonucleotide, the antisense strand was sequenced. Three typical chromatograms are reported showing the amplitude of A, C, G and T peaks. The sequence of the sense and the antisense strands are written below. The edited 2177th codon is boxed. The lines and small letters indicate how was measured the height of the peaks. Editing converts the "G" residue of the antisense sequence in a "A" residue". In the liver in wild type animals (**upper panel**), the APOB mRNA was not edited. A "G" residue was detected at the position of the 2177th codon and no "A" residue was possible to be detected. It was considered that all DNA strands in the amplified sample harbored a CAA codon. In the intestine in wild type animals (**middle panel**), the codon was edited. Most strands harboured a "A" residue in place of the non edited "G" residue. However, a small proportion of strands harboured the "G" residue. The **lower panel** shows that in intestine of transgenic animals expressing the shRNA (sh L21), the height of the "A" peak was reduced and that of the "G" peak was enhanced compared to the chromatogram in intestine of wild type animals. The sample was a mixture of DNA fragments harbouring the CAA or the TAA sequence. The measure of a1, a2, g1 and g2 ensured the determination of the proportion of "G" and "A" containing fragments in the mixture.

Figure S3　Western blot detection of the human APOBEC1 enzyme in intestinal cell extracts in L02 transgenic rabbits. Intestinal cell extracts (100 µg of protein in each sample) prepared from a wild type rabbit (WT) and a L02 transgenic rabbit expressing the human APOBEC1 enzyme were fractionated on

SDS-PAGE (16%). The human APOBEC1 enzyme was detected by Western blotting using the APOBEC1 antibody (1/1000 dilution). A similar amount of spleen extract was assayed on the same gel as negative control. One specific band (labelled by an arrow) was seen in the L02 transgenic animal at the expected migration rate according to the size of the human protein (27 kD). No band was possible to be detected in the wild type extract or in the spleen extract.

Figure S4 Plasma triglycerides and cholesterol concentrations in transgenic rabbits expressing the human APOBEC1 gene fed with a high fat/high cholesterol diet. The experiments and symbols are similar to those described in the legend of Figure 8. Rabbits (4 wild type, and 3 transgenic rabbits from line L02) were fed for 8 days with a high fat/high cholesterol diet. Plasma samples were collected before the diet (D0, white bars), after feeding for 8 day with the diet (D8, black bars), after 20 hours starvation (D9 starved, grey bars) and 4 hours after refeeding with the high fat diet (D9 fed, dotted bars). Triglycerides and cholesterol were assayed as in Figure 7. Values are given in mg/ml, with the standard error of the mean. Comparisons were made between transgenic and wild type animals for each day of the challenge (* = p<0.05).

Figure S5 Long-term recording of weight curves of rIFABP-hApobec1 transgenic rabbits. A series of animals (8 wild type; 3 from line L01; 5 from line L02) were weighed for up to 40 weeks after birth. Clearly, the long-term expression of the human APOBEC1 transgene did not induce an excess weigh. Values are means +/− sem.

Acknowledgments

We would like to thank Sonia Prince for the contribution to the production of transgenic rabbits; Michel Baratte, Gilbert Boyer, Jean-Pierre Albert and Gwendoline Morin (Unité Commune d'Expérimentation animale, INRA, Jouy-en-Josas, France) for their excellent care of the animals; Mathieu Leroux-Coyau (Biologie du Développement et Reproduction, INRA, Jouy-en-Josas) and Stéphanie Lemaire-Ewing for their excellent technical assistance.

Author Contributions

Conceived and designed the experiments: GJ LMH SB IH. Performed the experiments: GJ SB BP BDS EH CV TG. Analyzed the data: GJ TG LL IH. Contributed reagents/materials/analysis tools: BDS BP EH NDC. Contributed to the writing of the manuscript: GJ IH LL.

References

1. Mathus-Vliegen EMH (2012) Prevalence, Pathophysiology, Health Consequences and Treatment Options of Obesity in the Elderly: A Guideline. Obes Facts 460–483.
2. González Jiménez E (2011) Genes and obesity: a cause and effect relationship. Endocrinol Nutr 58: 492–496.
3. Reed DR, Lawler MP, Tordoff MG (2008) Reduced body weight is a common effect of gene knockout in mice. BMC Genetics 9:doi:10.1186/1471-2156-1189-1184.
4. Chan L (1995) Apolipoprotein B messenger RNA editing: An update. Biochimie 77: 75–78.
5. Teng B, Burant CF, Davidson NO (1993) Molecular cloning of an apolipoprotein B messenger RNA editing protein. Science 260: 1816–1819.
6. Davidson NO, Shelness GS (2000) APOLIPOPROTEIN B: mRNA Editing, Lipoprotein Assembly, and Presecretory Degradation. Annu Rev Nutr 20: 169–193.
7. Kendrick JS, Chan L, Higgins JA (2001) Superior role of apolipoprotein B48 over apolipoprotein B100 in chylomicron assembly and fat absorption: an investigation of apobec-1 knock-out and wild-type mice. Biochem. J. 356: 821–827.
8. Lo C-M, Nordskog BK, Nauli AM, Zheng S, vonLehmden SB, et al. (2008) Why does the gut choose apolipoprotein B48 but not B100 for chylomicron formation? Am J Physiol Gastrointest Liver Physiol 294:G344–G352.
9. Zhang X-J, Chinkes DL, Aarsland A, Herndon DN, Wolfe RR (2008) Lipid Metabolism in Diet-Induced Obese Rabbits Is Similar to That of Obese Humans. J. Nutr. 138: 515–518.
10. Nakamuta M, Oka K, Krushkal J, Kobayashi K, Yamamoto M, et al. (1995) Alternative mRNA splicing and differential promoter utilization determine tissue-specific expression of the apolipoprotein B mRNA-editing protein (Apobec1) gene in mice. J Biol Chem 270: 13042–13056.
11. Lu ZJ, Mathews DH (2008) OligoWalk: an online siRNA design tool utilizing hybridization thermodynamics. Nucl. Acids Res. 36:W104–108.
12. Daniel-Carlier N, Sawafta A, Passet B, Thépot D, Leroux-Coyau M, et al. (2013) Viral infection resistance conferred on mice by siRNA transgenesis. Transgenic Res 22: 489–500.
13. Yamanaka S, Poskay KS, Balestra ME, Zeng G-Q, Innerarity TL (1994) Cloning and mutagenesis odf the rabbit ApoB mRNA editing protein. A zinc motif is essential for catalytic activity, and noncatalytic auxiliary factor(s) of the editing complex are widely distributed. J Biol Chem 262: 21725–21734.
14. Greeve J, Altkemper I, Dieterich J-H, Greten H, Windler E (1993) Apolipoprotein B mRNA editing in 12 different mammalian species: hepatic expression is reflected in low concentrations of apoB-containing plasma lipoproteins. J Lipid Res 34: 1367–1383.
15. Greeve J, Jona VK, Chowdhury NR, Horwitz MS, Chowdhury JR (1996) Hepatic gene transfer of the catalytic subunit of the apolipoprotein B mRNA editing enzyme results in a reduction of plasma LDL levels in normal and Watanabe heritable hyperlipidemic rabbits. J Lipid Res 37: 2001–2017.
16. Kinoshita M, Matsushima T, Mashimoto Y, Kojima M, Kigure M, et al. (2010) Determination of Immuno-Reactive Rabbit Apolipoprotein B-48 in Serum by ELISA. Exp Anim 59: 459–467.
17. Picone O, Laigre P, Fortun-Lamothe L, Archilla C, Peynot N, et al. (2011) Hyperlipidic hypercholesterolemic diet in prepubertal rabbits affects gene expression in the embryo, restricts fetal growth and increases offspring susceptibility to obesity. Theriogenology 75: 287–299.
18. Yamamoto T, Yamamoto A, Watanabe M, Matsuo T, Yamazaki N, et al. (2009) Classification of FABP isoforms and tissues based on quantitative evaluation of transcript levels of these isoforms in various rat tissues. Biotechnol Lett 31: 1695–1701.
19. Dannoura A, Berriot-Varoqueaux N, Amati P, Abadie V, Verthier N, et al. (1999) Anderson's disease: exclusion of apolipoprotein and intracellular lipid transport genes. Arterioscl Throm Vasc 19: 2494–2508.
20. Okada T, Miyashita M, Fukuhara J, Sugitani M, Ueno T, et al. (2011) Anderson's disease/chylomicron retention disease in a Japanese patient with uniparental disomy 7 and a normal SAR1B gene protein coding sequence. Orphanet J Rare Dis 6: 78.
21. Georges A, Bonneau J, Bonnefont-Rousselot D, Champigneulle J, Rabes J, et al. (2011) Molecular analysis and intestinal expression of SAR1 genes and proteins in Anderson's disease (Chylomicron retention disease). Orphanet J Rare Dis 6: 1.
22. Nakamuta M, Chang BH-J, Zsigmond E, Kobayashi K, Lei H, et al. (1996) Complete Phenotypic Characterization of apobec-1 Knockout Mice with a Wild-type Genetic Background and a Human Apolipoprotein B Transgenic Background, and Restoration of Apolipoprotein B mRNA Editing by Somatic Gene Transfer of Apobec-1. J Biol Chem 271: 25981–25988.
23. Hirano K-I, Young SG, Farese RV Jr, Ng J, Sande E, et al. (1996) Targeted Disruption of the Mouse apobec-1 Gene Abolishes Apolipoprotein B mRNA Editing and Eliminates Apolipoprotein B48. J Biol Chem 271: 9887–9890.
24. Morrison JR, Paszty C, Stevens ME, Hughes SD, Forte T, et al. (1996) Apolipoprotein B RNA editing enzyme-deficient mice are viable despite alterations in lipoprotein metabolism. P Natl Acad Sci USA 93: 7154–7159.
25. Blanc V, Xie Y, Luo J, Kennedy S, Davidson N (2012) Intestine-specific expression of Apobec-1 rescues apolipoprotein B RNA editing and alters chylomicron production in Apobec1 −/− mice. J Lipid Res 53: 2643–2655.
26. Anant S, Davidson NO (2001) Molecular mechanisms of apolipoprotein B mRNA editing. Curr Opin Lipidol 12: 159–165.
27. Anant S, Davidson NO (2002) Identification and Regulation of Protein Components of the Apolipoprotein B mRNA Editing Enzyme: A Complex Event. Trends Cardiovas Med 12: 311–317.
28. Shi R, Chiang V (2005) Facile means for quantifying microRNA expression by real-time PCR. Biotechniques 39: 519–525.
29. Chomczynski P, Sacchi N (1987) Single-step method of RNA isolation by acid guanidium thiocyanate-phenol-chloroform extraction. Anal. Biochem. 162: 156–159.
30. Hatch F (1968) Practical methods for plasma lipoprotein analysis. Adv Lipid Res 6: 1–68.
31. Fortun-Lamothe L, Lamboley-Gaüzere B, Carole B (2002) Prediction of body composition in rabbit females using total body electrical conductivity (TOBEC). Livest Prod Sci 78: 133–142.

15

Cloning and Functional Characterization of a Vacuolar Na⁺/H⁺ Antiporter Gene from Mungbean (*VrNHX1*) and Its Ectopic Expression Enhanced Salt Tolerance in *Arabidopsis thaliana*

Sagarika Mishra[1], Hemasundar Alavilli[2], Byeong-ha Lee[2], Sanjib Kumar Panda[3,4], Lingaraj Sahoo[1]*

1 Department of Biotechnology, Indian Institute of Technology Guwahati, Guwahati, India, **2** Department of Life Science, Sogang University, Mapo-gu, Seoul, Korea, **3** Department of Life Sciences and Bioinformatics, Assam University, Silchar, India, **4** Department of Biochemistry & Molecular Biology, Noble Research Centre, Oklahoma State University, Stillwater, OK, United States of America

Abstract

Plant vacuolar NHX exchangers play a significant role in adaption to salt stress by compartmentalizing excess cytosolic Na⁺ into vacuoles and maintaining cellular homeostasis and ionic equilibrium. We cloned an orthologue of the vacuolar Na⁺/H⁺ antiporter gene, *VrNHX1* from mungbean (*Vigna radiata*), an important Asiatic grain legume. The *VrNHX1* (Genbank Accession number JN656211.1) contains 2095 nucleotides with an open reading frame of 1629 nucleotides encoding a predicted protein of 542 amino acids with a deduced molecular mass of 59.6 kDa. The consensus amiloride binding motif (⁸⁴LFFIYLLPPI⁹³) was observed in the third putative transmembrane domain of *VrNHX1*. Bioinformatic and phylogenetic analysis clearly suggested that *VrNHX1* had high similarity to those of orthologs belonging to Class-I clade of plant NHX exchangers in leguminous crops. *VrNHX1* could be strongly induced by salt stress in mungbean as the expression in roots significantly increased in presence of 200 mM NaCl with concomitant accumulation of total [Na⁺]. Induction of *VrNHX1* was also observed under cold and dehydration stress, indicating a possible cross talk between various abiotic stresses. Heterologous expression in salt sensitive yeast mutant AXT3 complemented for the loss of yeast vacuolar *NHX1* under NaCl, KCl and LiCl stress indicating that *VrNHX1* was the orthologue of *ScNHX1*. Further, AXT3 cells expressing *VrNHX1* survived under low pH environment and displayed vacuolar alkalinization analyzed using pH sensitive fluorescent dye BCECF-AM. The constitutive and stress inducible expression of *VrNHX1* resulted in enhanced salt tolerance in transgenic *Arabidopsis thaliana* lines. Our work suggested that *VrNHX1* was a salt tolerance determinant in mungbean.

Editor: Binying Fu, Institute of Crop Sciences, China

Funding: LS is grateful to Department of Biotechnology, Government of India for partial support from the grant (BT/PR10818/AGR/02/591/2008) to this work. BhL is grateful to Rural Development Administration, Republic of Korea for its support by Next-Generation BioGreen 21 Program (PJ009104). SM is grateful to MHRD for Research Fellowship. The funders had no role in study design, data collection and analysis, decision to publish, or preparation of the manuscript.

Competing Interests: The authors have declared that no competing interests exist.

* Email: ls@iitg.ernet.in

Introduction

Soil salinity poses increasing threat to plant growth and agricultural productivity worldwide [1]. More than 20% of the cultivated area and nearly half of the world's irrigated lands are adversely affected by salinity [2]. Enhanced crop production on salinity inflicted areas will rely on innovative agronomic practices coupled with use of genetically improved crop varieties [3]. In saline soils, Na⁺ is the predominant toxic ion. Excess accumulation of Na⁺ in cytosol is detrimental to many metabolic and physiological processes, vital for plant growth and productivity, as it causes ion imbalance, hyper osmotic stress, and oxidative damage to plants [4]. To cope with salinity stress, plants have evolved sophisticated mechanisms, including restricted uptake/ exclusion of Na⁺ from cell, and compartmentalization of Na⁺ into vacuoles. Na⁺ efflux is catalyzed by a plasma membrane Na⁺/H⁺ antiporter (NHX) encoded by *SOS1* [5,6] while, a vacuolar Na⁺/

H⁺ antiporter catalyzes the sequestration of Na⁺ into vacuoles. Compartmentalization of Na⁺ into vacuole not only provides an efficient mechanism to avert deleterious effects of Na⁺ in cytoplasm, but also allows plant to use Na⁺ as an osmoticum, for maintaining an osmotic potential for driving water into cell [4,7]. Vacuolar compartmentalization of Na⁺ is a critical process in salt adaptation, which is conserved in both halophytes and glycophytes. Na⁺ transport into vacuoles mediated through vacuolar Na⁺/H⁺ antiporter is an energy driven process involving H⁺ transporting pumps such as H⁺-ATPase and H⁺-PPase [8]. The genes encoding for Na⁺/H⁺ antiporters (NHX) have been cloned from more than 60 plant species, including gymnosperms and dicotyledonous and monocotyledonous angiosperms. The expression of most *NHXs* was induced by NaCl treatment [9]. Overexpression of vacuolar NHX genes suppressed the salt sensitive phenotype of a yeast mutant defective for endosomal and vacuolar Na⁺/H⁺ antiporters and conferred salt tolerance in transgenic plants [10,11]. Several reports on improvement of salt tolerance through overexpression of vacuolar *NHXs* in agriculturally important but glycophytic crops implicate a pivotal function of the *NHXs* in intracellular compartmentalization of Na$^+$ and salt tolerance [3,12]. In legumes, *NHX1* has been reported in *Glycine max* [13], *Medicago sativa* [14], *Trifolium repens* [15], *Lotus tenuis* [16], *Caragana korshinskii* [17] and recently by our lab, in *Vigna unguiculata* (GenBank Acc. No. JN641304.2). However, no salt-tolerant genes including *NHX* yet reported from mungbean.

Mungbean (*Vigna radiata* L. Wilczek) is an important grain legume widely cultivated in south, east and south-east Asian countries for its protein rich grains. Salinity is recognized as major constraint in the production of mungbean [18,19]. Mungbean is moderately drought tolerant [20] and therefore, this distinctive character makes it a valuable tropical crop legume for studying the molecular tolerance mechanisms for various abiotic stresses including salinity. In this paper, we report the cloning and molecular characterization of *VrNHX1* antiporter from *V. radiata*, its expression pattern under various abiotic stresses like salt, dehydration and cold stress, functional complementation of *VrNHX1* in *Saccharomyces cerevisiae* salt sensitive mutant (AXT3) and finally, increased salt tolerance by constitutive and inducible expression of *VrNHX1* in transgenic *Arabidopsis thaliana*, highlighting the potent role of *VrNHX1* in salt tolerance mechanisms.

Materials and Methods

Plant Material and Stress Treatment

Mungbean (*Vigna radiata* L. Wilczek cv. K-851) seeds were surface sterilized with 0.2% mercuric chloride and rinsed three times with distilled water. The seeds were germinated in dark chamber for 2 days, transferred to Hoagland's nutrient medium, grown hydroponically in a controlled growth chamber at 25°C, 80% relative humidity with a 16 hr/8 hr photoperiod and photosynthetic flux intensity of 300 μmol m^{-2} s^{-1} for 14 days. For salt stress treatment, these two weeks old mungbean seedlings grown under hydroponic culture were transferred to 200 mM NaCl solution for 12 hrs and roots were harvested, frozen immediately, and stored at −80°C until further use.

Molecular cloning of *VrNHX1* cDNA by RACE approach

Total RNA was isolated from salt-treated roots of mungbean using AMBION RNAqueous Kit (Ambion, Carlsbad, CA, USA). One microgram of RNA was used for cDNA synthesis using Revert Aid First Strand cDNA Synthesis Kit (Thermo Fisher Scientific, Waltham, MA, USA). The cDNA was amplified with a pair of degenerate primers (Deg FP: 5'- TAT(A/T)ATATT-CAATGC(C/A)GGGTTTCA(G/A)GT(A/G) -3' and Deg RP: 5'- GCATT(A/G)TGCCA(A/G)GT(A/G)TAATG(A/T)GA-CAT(A/G/C)AC -3') designed from the conserved region of transmembrane domains of plant NHX antiporters submitted in NCBI database. The PCR condition was: 94°C for 3 min; 94°C for 30 sec, 52°C for 30 sec, 72°C for 30 sec with 30 cycles, and a final extension at 72°C for 10 min. Based on the resulting partial fragment, gene specific primers were designed for amplification of 5'- and 3'- untranslated regions of *VrNHX1*.

The 5' RACE was performed using the 5' RACE System for Rapid Amplification of cDNA Ends Kit, Version 2.0 (Invitrogen, Carlsbad, CA, USA). Briefly, five micrograms of RNA was used for first strand cDNA synthesis using a gene specific primer (GSP1: 5'- CTGCTTCTTTTTCACCTGAAACCCAGC -3') and Superscript II reverse transcriptase (Invitrogen). cDNA was purified

using SNAP column to remove unincorporated dNTPs and primer, that might interfere in the homopolymeric tailing of cDNA. Terminal transferase enzyme was used to add dCTPs to 3' end of cDNA. The dc-tailed cDNA was amplified using abridged anchor primer (AAP: 5'- GGCCACGCGTCGACTAGTACGG-GIIGGGIIGGGIIG -3') and gene specific primer (GSP2: 5'-ACCTGAAACCCAGCATTGAATAT-3'). The PCR condition was: 94°C for 3 min; 94°C for 30 sec, 55°C for 30 sec, 72°C for 30 sec with 30 cycles, and a final extension of 72°C for 10 min. Further, nested PCR was performed using abridged universal anchor primer (AUAP: 5'- GGCCACGCGTCGACTAGTAC -3') and nested gene specific primer (GSP3: 5'- GGTATATGAA-GAAAAGATCTTC -3') using the first PCR product as template. The PCR condition was: 94°C for 3 min; 94°C for 30 sec, 52°C for 30 sec, 72°C for 30 sec with 35 cycles, and a final extension of 72°C for 10 min.

The 3' RACE was performed using 3' RACE System for Rapid Amplification of cDNA Ends Kit, Version E (Invitrogen, Carlsbad, CA, USA). Five micrograms of RNA was used to synthesize cDNA using a dT-adapter primer (AP: 5'- GGCCACGCGTCGAC-TAGTAC(T)$_{17}$ -3') and Superscript II reverse transcriptase (Invitrogen, Carlsbad, CA, USA). The first 3'-RACE-PCR was carried out using gene specific primer (GSP4: 5'- AGTGG-CATCCTCACTGTATTCTTTTGTG -3') and abridged universal anchor primer (AUAP: 5'- GGCCACGCGTCGACTAG-TAC -3'). The PCR condition was: 94°C for 3 min; 94°C for 30 sec, 60°C for 30 sec, 72°C for 1 min and 30 sec with 30 cycles, and a final extension of 72°C for 10 min. The PCR product was diluted 10 times (1:10) and used as template for nested 3' RACE-PCR. The nested 3'-RACE-PCR was carried out using gene specific primer (GSP5: 5'-GCTGTATATTGGAAGGCACTCT-3') and abridged universal anchor primer (AUAP: 5'-GGCCACGCGTCGACTAGTAC -3'). The PCR condition was: 94°C for 3 min; 94°C for 30 sec, 55°C for 30 sec, 72°C for 1 min and 30 sec with 30 cycles, and a final extension of 72°C for 10 min. The above PCR products were cloned to TA cloning vector pTZR/T (Thermo Fisher Scientific, Waltham, MA, USA) sequenced and contiguous sequences aligned to obtain full length of *VrNHX1* cDNA.

Bioinformatic analysis of *VrNHX1*

Multiple sequence alignment and phylogenetic analysis were performed using Clustal W [21]. A phylogenetic tree was constructed using neighbor joining method and reliability of the tree was analyzed with bootstrap analysis with 500 replicates using MEGA4 (Molecular Evolutionary Genetics Analysis): Tree Explorer software [22]. Hydrophobicity plot and transmembrane domain prediction was performed using TMpred software [23]. Post-translational modification of *VrNHX1* was predicted by searching for conserved motifs of N- and O- glycosylation and N-myristoylation sites using ScanProsite [24].

Southern hybridization for *VrNHX1* copy number in mungbean genome

Twenty μg of genomic DNA was used for gene copy analysis of *VrNHX1* and digested with restriction endonucleases EcoRI and HindIII. Digested genomic DNA was electrophoretically fractionated on 0.8% agarose gel and blotted onto Zeta-Probe membrane (Bio-Rad, Hercules, CA, USA). The blot was hybridized with DIG-labeled 1.6 kb PCR product, corresponding to the coding region of *VrNHX1*. Southern hybridization was carried out using solution containing 50% formamide, 5 X SSC, 5 X Denhardt's solution, 0.05 M sodium phosphate pH 6.5, 0.1% SDS, 10% dextran sulfate, 0.1 mg/ml sheared denatured salmon-sperm

DNA and 20 ng/ml probe at 42°C for 18 hrs. Washing and detection was performed according to instructions of the DIG Labeling and Detection system (Roche Diagnostics, Mannheim, Germany).

Heterologous expression of *VrNHX1* in yeast mutant

Functional complementation assay was performed in yeast strains, W303-1B (*MATα ade2-1 can1-100 his3-11,15 leu2-3,112 trp1-1 ura3-1*) and AXT3 (*Δ ena1- 4::HIS3 Δnha1::-LEU2 Δnhx1::TRP1, ura3-1*). Yeast strains were grown in YPD (1% Yeast extract, 2% Peptone and 2% Glucose), YPGal (1% Yeast extract, 2% Peptone and 2% Galactose), SC (0.67% Yeast Nitrogen Base, 2% Glucose) and APGal synthetic minimal media (10 mM arginine, 8 mM phosphoric acid, 2 mM $MgSO_4$, 1 mM KCl, 0.2 mM $CaCl_2$, 2% Galactose, trace vitamins, and minerals; pH-4.0) supplemented with appropriate amino acids as indicated.

The CDS of *VrNHX1* was cloned into yeast expression vector pYES2.0 (Invitrogen, Carlsbad, CA, USA) with restriction sites of KpnI and BamHI. The yeast strains were transformed with pYES2.0 empty vector (labeled as AXTYES2.0 strain) or pYES*VrNHX1* recombinant plasmid (labeled as AXTVrNHX1 strain) by Lithium acetate method [25] and selected on SC ura⁻ medium.

For growth assay, precultured cells were grown till OD_{600} of 1.0, diluted to an OD_{600} of 0.006, and inoculated to liquid APGal ura⁻ synthetic minimal media supplemented with different concentrations of NaCl, KCl, and LiCl and grown at 30°C for 48 hrs. For complementation assay, saturated liquid cultures (OD_{600} 0.8) of each strain were serially diluted to 10, 100 and 1000 fold and spotted on APGal solid media supplemented with or without 50, 75 and 100 mM NaCl, 0.5 M KCl, 25 mM LiCl and and YPGal media supplemented with 50 µg/ml hygromycin. Plates were maintained at 30°C. Growth was monitored after 3 days.

Intracellular measurement of Na^+ and K^+ distribution in yeast mutant

Intracellular ion was extracted from yeast strains grown in liquid APGal media, pH 4.0 supplemented without or with 75 mM NaCl [26]. Briefly, cells were harvested at an OD_{600} of 0.3–0.4, centrifuged at 3000 g/3 min, washed twice in ice-cold 10 mM $MgCl_2$, 10 mM $CaCl_2$ and 1 mM HEPES buffer and resuspended in the same buffer. The relationship between cell density (Absorbance at OD_{600}) and yeast dry weight was determined. Total intracellular ion was determined by addition of HCl to a final concentration of 0.4% and incubated at 95°C for 20 min. After removal of cell debris the supernatant was measured for presence of Na^+ and K^+. Similarly, cells were grown and washed as above and resuspended in 2% cytochrome c, 18 µg/ml antimycin, 1 mM HEPES, 10 mM $MgSO_4$, 10 mM $CaCl_2$, and 5 mM 2-Deoxy D-Glucose solution. Cytochrome c selectively permeabilizes the plasma membrane. After 20 min incubation at room temperature, cells were washed thrice with the same solution without cytochrome c. Cytoplasmic ion content was determined by pooling the supernatants. The remaining vacuolar ions were extracted with addition of HCl in a final concentration of 0.4% and incubated at 95°C for 20 min. The Na^+ and K^+ distribution in the cytoplasmic and vacuolar fractions were measured in flame photometer (Systronics, MP, India).

Vacuolar pH estimation and fluorescence imaging

Yeast cells were grown in APGal medium (pH 5.0) to an OD_{600}: 0.25–0.3, pelleted, and washed with deionized distilled water.

Further, the yeast cells were incubated with 50 µM 2′,7′-bis-(2-carboxyethyl)-5-(and-6)-carboxyfluorescein (BCECF-AM) (Molecular Probes, Eugene, Oregon) for 30 min, centrifuged, washed thrice and resuspended in APGal medium (pH 5.0) and immediately used for fluorescence measurement. Single emission fluorescence measurement at 490 nm excitation wavelength and absorbance at 600 nm were measured using LS 55 Fluorescence Spectrophotometer (Perkin Elmer, Waltham, MA, USA). The calibration curve for fluorescence intensities at different pH was obtained for each strain [27]. Briefly, the yeast strains (W303-1B, AXTYES2.0, AXTVrNHX1) were incubated in experimental medium containing 50 mM MES, 50 mM HEPES, 50 mM KCl, 50 mM NaCl, 0.2 M ammonium acetate, 10 mM NaN_3, 10 mM 2- deoxy glucose, 50 µM carbonyl cyanide m-chlorophenylhydrazone, titrated to five different pH values within the range of 4.0 to 8.0. Background subtracted I_{490} values were normalized to cell density for each strain, labeled as NI_{490} and plotted against pH values. For vacuolar pH estimation, experimental NI_{490} values corresponding to each strain was analyzed with the calibration curve specific for each strain.

For vacuolar pH imaging the yeast cells were grown, pelleted to be suspended in the same medium with 50 µM BCECF-AM pH specific dye as above. For fluorescence imaging, 100 µl of BCECF-loaded yeast suspension was plated onto glass cover slips precoated with concavalin-A (Sigma-Aldrich, St. Louis, MO, USA) and placed on glass slides. Fluorescence images were captured in Nikon eclipse Ti-U Fluorescence microscope (Nikon, Chiyoda, Tokyo, Japan).

Expression analysis of *VrNHX1* using semi-quantitative RT-PCR

Expression analysis under salt stress: Two different stages of growth in mungbean seedlings i.e. early and mid stage, were considered for expression analysis under salt stress (200 mM NaCl). Mungbean seedlings were germinated, grown in Hoagland's nutrient medium for five and ten days, in case of early and mid stage respectively, and transferred to 200 mM NaCl solution for salt stress assay. Leaves and roots of salt treated early and mid stage mungbean seedlings, were harvested at time intervals 0, 6, 12, 18, 24, and 48 hrs. Similarly, expression pattern for *VrNHX1* in response to different forms of abiotic stress such as salt (200 mM NaCl), dehydration (200 mM mannitol) and cold stress (4°C) was also studied at different time intervals (0, 6, 12, and 24 hrs) for mid-stage (10 days old) mungbean seedlings. Total RNA was extracted using RNeasy Plant Mini Kit (Qiagen, Venlo, Limburg, Netherlands) and reverse transcribed using Revert Aid First Strand cDNA Synthesis Kit. Semi-quantitative RT-PCR was performed using gene specific primers (RF: 5′- GTATTTCCACTGGCG-TAGTCATTTTGC -3′ and RR: 5′- GCATCATTCACAG-CACCCTCTCGG -3′). The PCR condition was: 94°C for 3 min; 94°C for 30 sec, 62°C for 30 sec, 72°C for 30 sec for 28 cycles, and a final extension at 72°C for 10 min. Housekeeping *VrTubulin-β* primers (FN: 5′- CTTGACTGCATCTGC-TATGTTCAG-3′ and RN: 5′-CCAGCTAATGCTCGGCA-TACTG -3′) were used as an internal control. The PCR condition was: 94°C for 3 min; 94°C for 30 sec, 58°C for 30 sec, 72°C for 30 sec for 28 cycles, and a final extension at 72°C for 10 min. Semi-quantitative RT-PCR was repeated three times. The PCR products were analyzed in 2% agarose gel stained with 10 mg/ml ethidium bromide.

Measurement of total ion content in salt stressed mungbean seedlings

Leaves and roots of untreated and salt-treated early and mid stage mungbean seedlings were harvested at different time intervals (0, 6, 12, 18, 24, 48 and 72 hrs). The samples were dried, digested with concentrated HNO_3 at 90°C for 1 hr and centrifuged at 12,000 rpm for 10 min [28]. The suspension was diluted with sterile milliQ water and analyzed for Na^+ and K^+ content in flame photometer.

Binary vector preparation and plant transformation

The 1.6 kb CDS of *VrNHX1* was cloned into standard plant binary vector pCAMBIA2301 (11.6 kb) flanked by cauliflower mosaic virus CaMV 35S promoter and terminator at *Pst*I restriction site. The resulting recombinant plant binary vector was labeled as pCAMBIA2301-35S::*VrNHX1* (13.9 kb). Further, a 0.898 kb promoter fragment of *AtRD29A* (DQ071887.1) was amplified from *A. thaliana* genomic DNA and cloned into *Eco*RI digested recombinant binary vector pCAMBIA2301-35S::*VrNHX1* (13.9 kb) by replacing the 0.4 kb 35S promoter fragment from 35SP::*VrNHX1*::35STer cassette. The resulting binary vector was named pCAMBIA2301-RD29A::*VrNHX1* (14.4 kb).

The recombinant plant binary vectors, pCAMBIA2301-35S::*VrNHX1* (13.9 kb) and pCAMBIA2301-RD29A::*VrNHX1* (14.4 kb) were transferred into *A. tumefaciens* GV3101 strain via electroporation at 1250 V with capacitance of 25 mF and resistance of 400 ohm. The constructs were used for transformation of *Arabidopsis thaliana* (ecotype Columbia) via floral dipping method [29]. The T_1 transgenic lines were screened on ½ MS medium (Duchefa, Haarlem, Netherlands) supplemented with 50 mg/l kanamycin (Duchefa, Haarlem, Netherlands). The transgenic selections were continued until T_3 generation to obtain homozygote transgenic lines with a single T-DNA locus (35S::*VrNHX1* or RD29A::*VrNHX1*).

RNA extraction and Real Time PCR of transgenic *Arabidopsis* lines

Total RNA was extracted from wild-type (WT) and T_3 independent 35S::*VrNHX1* and RD29A::*VrNHX1* transgenic lines using RNeasy Plant Mini Kit (Qiagen), quantified in Nanovue Plus Spectrophotometer (GE Healthcare, Little Chalfont, Buckinghamshire, UK) and cDNA was prepared using Revert Aid First Strand cDNA Synthesis Kit. The gene specific forward primer (VrRTF: 5′- TGATTCAATCCATCGACCAA-3′) and 35S poly-A reverse primer (TerparR: 5′-GCGAAACCC-TATAAGAACCCTAATTCC-3′) were used for amplification of a 0.283 kb fragment of *VrNHX1*::35S poly-A in transgenic *A. thaliana* plants. Housekeeping (UBQ1FP: 5′- AGAGCTGT-CAACTGCAGGAAGAA-3′ and UBQ1RP- 5′-ACAA-GAAAAACAAACCCTATCAAA GG) primers were used to amplify a 150 bp fragment of *AtUbiquitin* to be used as an internal control. Real time PCR was performed using USB VeriQuest SYBR Green qPCR Master Mix (2X) (Affymetrix, Santa Clara, CA, USA) and primers at a final concentration of 200 nM in 7500 Real-Time PCR System (Applied Biosystem, Foster City, California, USA) following the manufacturer's protocol. The experiment was repeated twice independently with three replicates. The expression values relative to the standard curve was calculated for each sample. The relative expression level of transgene *VrNHX1* in wild-type (WT) and transgenic 35S::*VrNHX1* and RD29A::*VrNHX1* *Arabidopsis* lines was estimated by normalizing *VrNHX1* expression values with respect to housekeeping *AtUBQ1* expression values in each case.

Salt tolerance assays of transgenic *Arabidopsis* lines

Wild-type (WT) and T_3 transgenic 35S::*VrNHX1* and RD29A::*VrNHX1* *Arabidopsis* seeds were germinated on ½ MS medium [30] in growth chamber maintained at 22°C and 60% relative humidity with a 16 hr/8 hr photoperiod under controlled conditions.

Studying germination efficiency under salt stress: The WT and T_3 transgenic 35S::*VrNHX1* and RD29A::*VrNHX1* lines were germinated on ½ MS medium supplemented with or without 150 mM NaCl and kept at 4°C for 3 days, prior to, transfer to growth chamber. The germination efficiency was studied after 10 days of salt stress.

Measurement of growth parameters under salt stress: The 4 days old germinated seedlings were transferred to ½ MS medium supplemented with or without 150 mM NaCl for 1 week and the difference in root length of wild-type WT and T_3 independent transgenic lines of *Arabidopsis* seedlings expressing *VrNHX1* was measured. Mean data was collected from ten replicates (n = 10) for wild-type (WT) and T_3 kanamycin selected transgenic *Arabidopsis* lines.

Measurement of physiological parameters under salt stress: The 10 days old germinated seedlings were transferred to ½ MS liquid medium supplemented with or without 200 mM NaCl for 5 days. For measurement of chlorophyll content, shoot samples were homogenized in 95% ethanol, lysate was centrifuged at 3,000 rpm for 10 min and absorbance was recorded for the extract at wavelength of 648 and 664 nm [31]. Lipid peroxidation was measured as the amount of malondialdehyde (MDA) determined by the thiobarbituric acid (TBA) reaction. Briefly, 0.2 g of fresh leaf samples were homogenized with 5 ml of 0.25% TBA containing 10% TCA (tricloroacetic acid). The homogenate was boiled for 30 min at 95°C and centrifuged at 10,000 g for 10 min Absorbance values were recorded at 532 nm and values corresponding to non-specific absorption at 600 nm were subtracted [32]. For colorimetric estimation of proline, leaf samples (0.5 g) were homogenized with 5.0 ml of sulfosalicylic acid (3%). 2 ml of homogenate was filtered through Whatman filter paper (No. 2) and incubated with 2 ml glacial acetic acid and 2 ml ninhydrin reagent at a ratio of 1:1:1 in boiling water bath at 100°C for 30 min. After cooling, 4 ml toluene was added to the reaction mixture, mixed vigorously and absorbance was measured at 520 nm [33]. Mean data was collected from three replicates (n = 3) for wild-type and T_3 kanamycin selected transgenic *Arabidopsis* lines.

Measurement of Na^+ and K^+ in transgenic *Arabidopsis* lines

The germinated seedlings were initially grown in ½ MS medium (0.5% agar) for 5 days and then subsequently transferred to soilrite and grown for 2 weeks. The WT and T_3 transgenic lines were subjected to salt stress for a period of 2 weeks by watering them with ½ MS nutrient liquid media supplemented with 250 mM NaCl. The whole plant was harvested for Na^+ and K^+ estimation using method described elsewhere [30]. Mean data was collected from three replicates (n = 3) for wild-type (WT) and T_3 kanamycin selected transgenic *Arabidopsis* lines.

Statistical analysis

Statistical comparison between the variances was determined by ANOVA (Analysis of variance) and significant differences between mean values were determined by Bonferroni analysis. Statistically significant mean values were denoted as "*" (P≤0.05).

Results

Isolation and in-silico analysis of *VrNHX1*

A *VrNHX1* cDNA of 2095 nucleotides in length (Genbank Accession number JN656211.1), with an open reading frame of 1,629 bp was obtained by RACE-PCR approach. It encodes a polypeptide of 542 amino acid residues with an estimated molecular mass 59.60 kDa and isoelectric point 6.76, predicted using ExPaSy bioinformatic tools for protein structure analysis (http://www.expasy.org/tools/). Multiple sequence alignment of deduced amino acid sequences of VrNHX1 revealed that it has 97.42% sequence identity with *Vigna unguiculata*, 92.25% with *Glycine max*, 88.48% with *Caragana korshinskii*, 87.27% with *Lotus tenuis*, 87.25% with *Trifolium repens*, 87.06% with *Medicago sativa*, and 86.72% with *Cicer arietinum* (Fig. 1 and S1). Phylogenetic relationship analysis performed using MEGA4 software indicated that VrNHX1 clustered into Class-I type IC-NHX legume *NHX* homologs, more closely to VuNHX1 and GmMHX1 (Fig. 1). The hydropathy plot of VrNHX1 protein predicted by TMpred software indicated highly hydrophobic N-terminal end with 11 putative transmembrane domains and a longer hydrophilic C-terminal end inside the vacuolar lumen (Fig. S2). The amiloride binding motif, [84]-LFFIYLLPPI-[93], a classic inhibitor of Na^+/H^+ antiporters [34] and also highly conserved among eukaryotic Na^+/H^+ exchangers, was detected in TM3 region (Fig. S1). The prediction of putative post-translational modification sites by ScanProsite software indicated presence of two potential *N*-glycosylation (ASN_glycosylation) sites, fifteen phosphorylation sites for protein kinase CK2 and protein kinase C, ten *N*-myristoylation sites, and one Leucine Zipper site (Table S1).

The Southern hybridization analysis revealed presence of single copy of *VrNHX1* in mungbean genome (Fig. 2). Two hybridization signals, one each for HindIII and EcoRI digested mungbean genome were detected, possibly due to the occurrence of a single *Hind*III site in *VrNHX1* (1.6 kb). Occurrence of a single *Eco*RI site in genome fragment of *VrNHX1* was accounted for getting two signals as probe lacked any *Eco*RI site.

Functional characterization of *VrNHX1* using salt sensitive yeast mutant

Previous work showed that heterologous expression of Na^+/H^+ antiporter genes in yeast mutant AXT3 could partly suppress its hypersensitivity to hygromycin and restore salt tolerance. The similar method was exploited to initially characterize the function of *VrNHX1*. The AXTVrNHX1 cells displayed enhanced Na^+, K^+ and Li^+ tolerance with statistically significant improvement in their survival at NaCl (75 and 100 mM) (Fig. 3 A) and 0.5 M KCl (Fig. 3 B), in contrast to AXTYES2.0 cells. Expression of *VrNHX1* in AXT3 cells under GAL1-inducible promoter restored salt tolerance upto 100 times dilution in 75 and 100 mM NaCl (Fig. 4 A), and better survival at 1000 times dilution range in 25 mM LiCl and 0.5 M KCl in AXTVrNHX1 cells on solid media (Fig. 4 B). *ScNHX1* has been suggested to ameliorate sensitivity of yeast cells by sequestering hygromycin-B, a cationic aminoglycoside antibiotic in vacuole. Therefore, yeast mutant lacking *NHX1* is more susceptible to hygromycin treatment [27]. *VrNHX1* expression showed suppression of hygromycin (50 µg/ml) sensitivity in AXTVrNHX1 cells (Fig. 4 C).

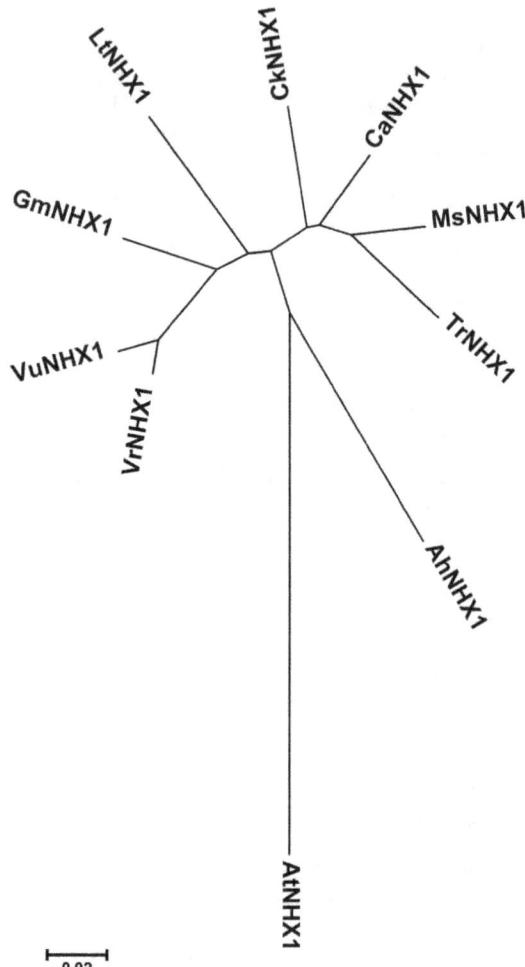

Figure 1. The phylogenetic tree for plant Na^+/H^+ antiporters was generated using MEGA4: Tree Explorer software. The evolutionary history was inferred using the neighbor-joining method and analyzed using bootstrap analysis with 500 replicates. Branches corresponding to partitions reproduced in less than 50% bootstrap replicates are collapsed. The tree is drawn to scale, with branch lengths in the same units as those of the evolutionary distances used to infer the phylogenetic tree. The evolutionary distances were computed using the Poisson correction method and are in the units of the number of amino acid substitutions per site. The GenBank Accession numbers for NHX proteins used are: VrNHX1 (AEO50758.1), VuNHX1 (AEO72079.2), GmNHX1 (AAY430061.1), CkNHX1 (ABG89337.1), MsNHX1 (AAS84487.1), CaNHX1 (ADL28385.1), TrNHX1 (ABV00895.1), LtNHX1 (ACE78322.1), AhNHX1 (ADK74832.1), AtNHX1 (NM_122597.2).

Na^+ and K^+ distribution in yeast mutants

The AXTYES2.0 cells displayed 2.3 times lower K^+ content than AXTVrNHX1 cells under normal condition owing to lack of yeast $Na^+/K^+/H^+$ antiporter activity (Fig. 5). Under salt stress, AXTVrNHX1 cells accumulated 2 times higher and 4.8 times lower vacuolar Na^+ content compared to AXTYES2.0 and W303-1B cells, respectively (Fig. 5). Similarly, vacuolar K^+ content observed for AXTVrNHX1 cells was 2.36 times higher than AXTYES2.0 cells. The cytoplsamic Na^+ content was higher in both the cell types as compared to W303-1B, due to the loss of NHA exchanger activity which cannot be solely compensated by *VrNHX1* complementation. However, cytoplasmic K^+ fractions measured were not statistically significant, though AXTVrNHX1

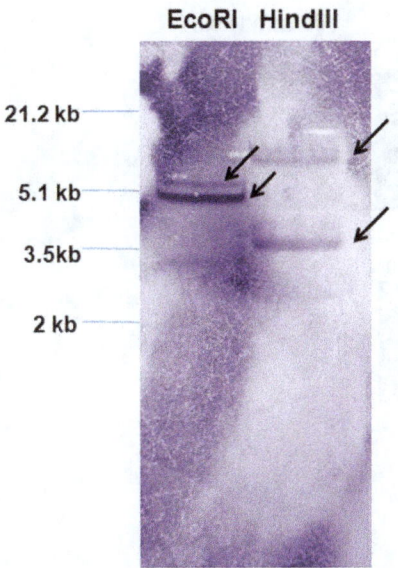

EcoRI   HindIII

21.2 kb

5.1 kb

3.5kb

2 kb

Figure 2. Copy number analysis of *VrNHX1* in mungbean genome. Mungbean genomic DNA (20 µg) was digested with EcoRI and HindIII, and hybridized with DIG-labeled probe corresponding to the CDS of *VrNHX1*. Hybridization signals are indicated as arrows.

cells exhibited higher K^+ values as compared to AXTYES2.0, indicating the improved ability of AXTVrNHX1 cells in maintaining a higher intracellular K^+/Na^+ ratio for ionic homeostasis. The total ion content in yeast cells was in accordance with distribution of Na^+ and K^+ in cytoplasm and vacuole.

Vacuolar pH estimation and imaging

2′,7′-bis-(2-carboxyethyl)-5-(and-6)-carboxyfluorescein (BCECF-AM), a widely used cell-permeant and pH-sensitive fluorescent indicator was used to measure the change in vacuolar pH of yeast mutant expressing *VrNHX1* grown under low pH environment. The study on the effect of low pH on growth efficiency of yeast cells showed that growth of AXTYES2.0 cells was highly affected with a 70.66% reduction in growth as compared to W303-1B. Moreover, AXTVrNHX1 mutant showed

improved growth under acidic condition (Fig. S3). Vacuolar pH was estimated following calibration curve plotted for each strain (Fig. S4). An acidic vacuolar pH of 5.4 was observed for AXTYES2.0 cells whereas, a pH value 5.9 and 6.2 was recorded for AXTVrNHX1 and W303-1B cells, respectively in response to low pH stress condition (Fig. 6 A). Similarly, fluorescence images provided acidic vacuolar pH values for AXTYES2.0 cells and expression of *VrNHX1* alkalinized the vacuolar compartment (Fig. 6 B).

Expression pattern of *VrNHX1* under abiotic stress by Semi-quantitative RT-PCR

The expression of *VrNHX1* was studied by semi-quantitative RT-PCR, in roots and leaves of mungbean seedlings at early (five days old) and mid (ten days old) growth stages exposed to salt stress (200 mM NaCl) for different time interval (0, 6, 12, 18, 24 and 48 hrs). The results indicated that transcript levels of *VrNHX1* were induced by NaCl in both roots and shoots of early and mid stage mungbean seedlings, indicating the potent role of *VrNHX1* in salt tolerance mechanisms in mungbean. In case of early seedling stage, higher expression level of *VrNHX1* was observed in leaves at 12, 24, and 48 hrs and in roots after 6 hrs (Fig. 7 A). The differential expression of *VrNHX1* in roots and leaves was also observed in mid stage seedlings, with a significant accumulation observed at 48 hrs in leaves whereas, some basal level of *VrNHX1* transcript was observed in roots under normal condition which further increased steadily with salt stress treatment period (Fig. 7 A).

To determine whether the expression of *VrNHX1* was also induced by dehydration (200 mM Mannitol) and cold (4°C), mid-stage (10 days old) seedlings were given the respective stress treatments for different time intervals (0, 6, 12, and 24 hrs). The *VrNHX1* expression varied with salt, cold and drought stress. The accumulation of *VrNHX1* transcript under salt, cold and dehydration stress reached its peak at 24 hours (Fig. 7 B). The results indicated that osmotic and low temperature stress is involved in the up-regulation of *VrNHX1* in addition to an ion-specific signaling component in mungbean. The *VrNHX1* expression analysis revealed involvement of cross talk between salinity, low temperature and osmotic stress in mungbean.

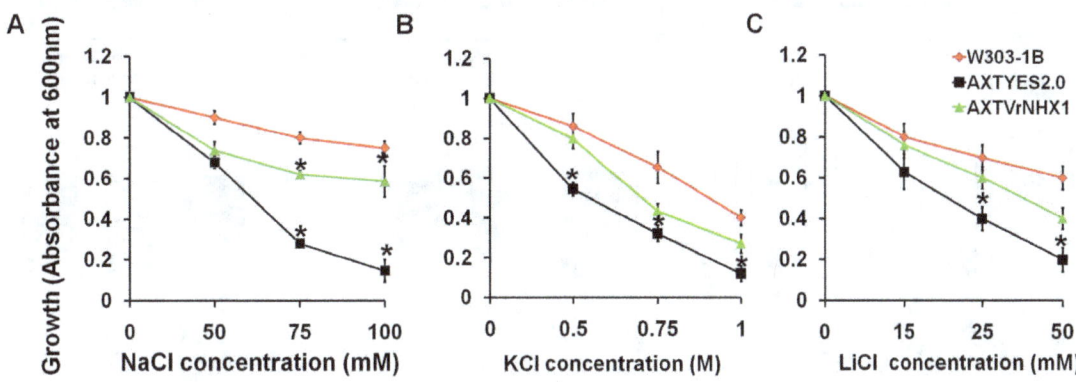

Figure 3. Cation sensitivity assay of transformed yeast strains (W303-1B, AXTYES2.0, AXTVrNHX1) under various concentrations of NaCl, KCl, and LiCl. Saturated seed cultures for each strain was diluted to an OD_{600} of 0.006 and inoculated to liquid APGal medium (pH 5.5) supplemented with or without various concentrations of (A) NaCl (0, 50, 75, 100 mM), B) KCl (0, 0.5, 0.75, 1.0 M), and (C) LiCl (0, 15, 20, 25 mM). Growth was observed at 30°C after 3 days and absorbance recorded at 600 nm. Data are means of 3 independent events (n = 3) and standard errors are plotted in the graph. Statistically significant values at P≤0.05 are indicated as "*", using Bonferroni analysis.

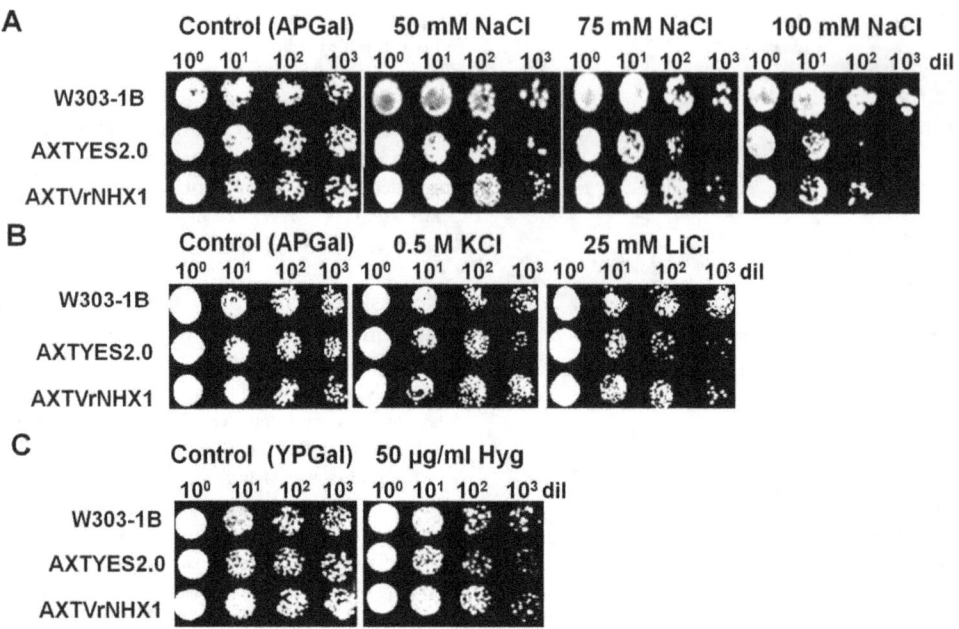

Figure 4. Heterologous expression of *VrNHX1* in yeast mutant. Wild type (W303-1B) strain was used as a control, *Δ ena1- 4 Δnha1 Δnhx1* mutant (AXT3) strain was transformed with null pYES2.0 (labeled as AXTYES2.0 strain) and pYES*VrNHX1* recombinant vector (labeled as AXTVrNHX1) were used for complementation assay. 10-fold serial dilutions of saturated seed cultures of each strain were spotted onto APGal media (pH-5.5) supplemented with or without (A) 50, 75 and 100 mM NaCl, (B) 25 mM LiCl, and (B) 0.5 M KCl. (C) Hygromycin sensitivity assay was performed by spotting 10-fold serial dilutions of saturated seed cultures of each strain onto YPGal media (pH- 5.5) supplemented with or without 50 μg/ml Hyg. The plates were incubated at 30°C for 3 days.

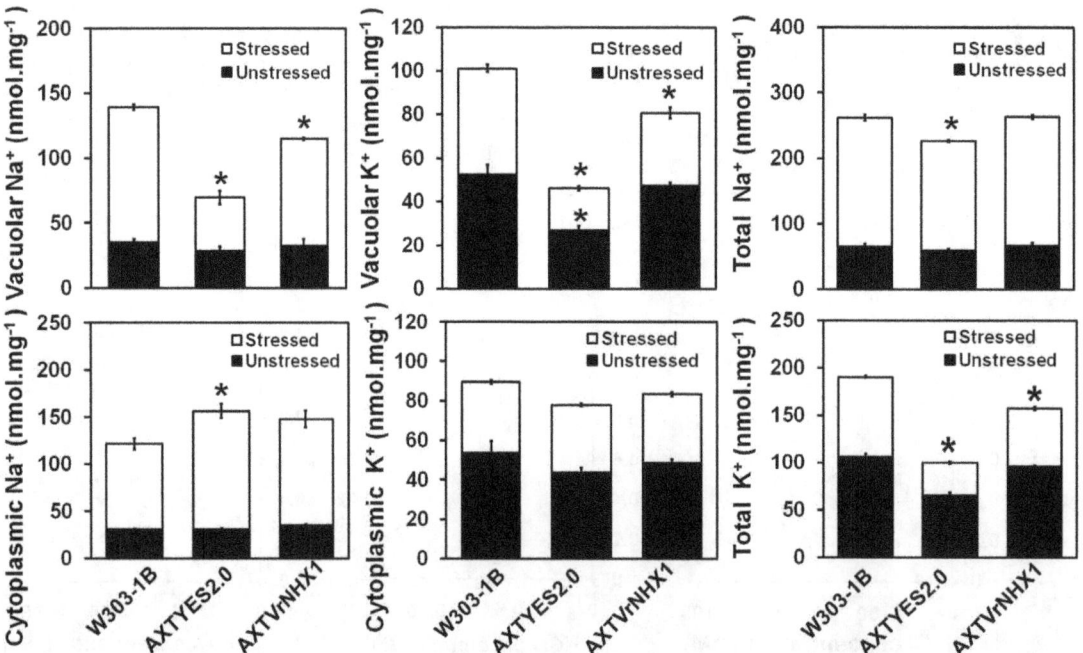

Figure 5. Total intracellular ion estimation in yeast strains W303-1B, AXTYES2.0 and AXTVrNHX1. Yeast cells were grown in APG medium (pH 4.0) with 1 mM KCl supplemented in presence (stressed) or absence (unstressed) of 75 mM NaCl (unstressed) and harvested at a cell density of 0.3. Total intracellular, vacuolar and cytoplasmic Na^+ and K^+ content was determined as described in the materials and methods section. Data are means of 3 independent events (n = 3) and standard errors are plotted in the graph. Statistically significant values at $P \leq 0.05$ are indicated as "*", using Bonferroni analysis.

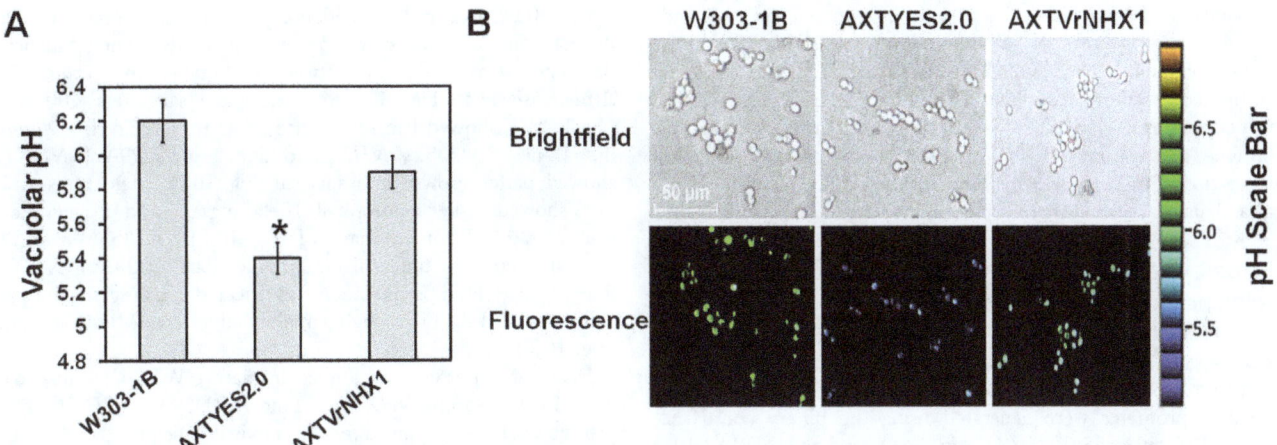

Figure 6. Measurement of vacuolar pH in yeast strains. (A) Vacuolar pH was measured for BCECF-AM loaded yeast strains W303-1B, AXTYES2.0 and AXTVrNHX1 as described in materials and methods following the calibration curve (Figure S4). Mean and SEs are plotted for three independent events (n = 3) in each case. Statistically significant values at P≤0.05 are indicated as "*", using Bonferroni analysis. (B) Accumulation of pH-sensitive fluorescent BCECF dye in yeast vacuoles was measured. The yeast strains were grown in APGal media (pH 5.0), resuspended in minimal medium with BCECF-AM dye for 30 min at 30°C. Yeast cells were visualized by Nikon eclipse Ti-U Fluorescence microscope (Nikon) at excitation wavelength of 440 nm. Bar scale: 50 µm.

Na+ and K+ measurement in salt stressed mungbean seedlings

The measurement of Na+ and K+ content in leaves and roots of untreated and salt-treated mungbean seedlings at different time intervals (0, 6, 12 18, 24, 48 and 72 hrs) showed that under salt stress, Na+ accumulation increased in leaves/roots by 1.28/2.1, 1.1/2.3, 2.1/4.36, 4.8/4.3, 4.1/4.54 times whereas, K+ accumulation decreased by 3.4/4.5, 1.6/1.78, 1.59/2.43, 2.2/3 and 2.1/3.5 times as compared to control condition at 6, 12, 18, 24 and 48 hrs, respectively in early stage mungbean seedlings (Fig. 8 A). Similarly, in mid stage seedlings, Na+ accumulation in leaves/roots

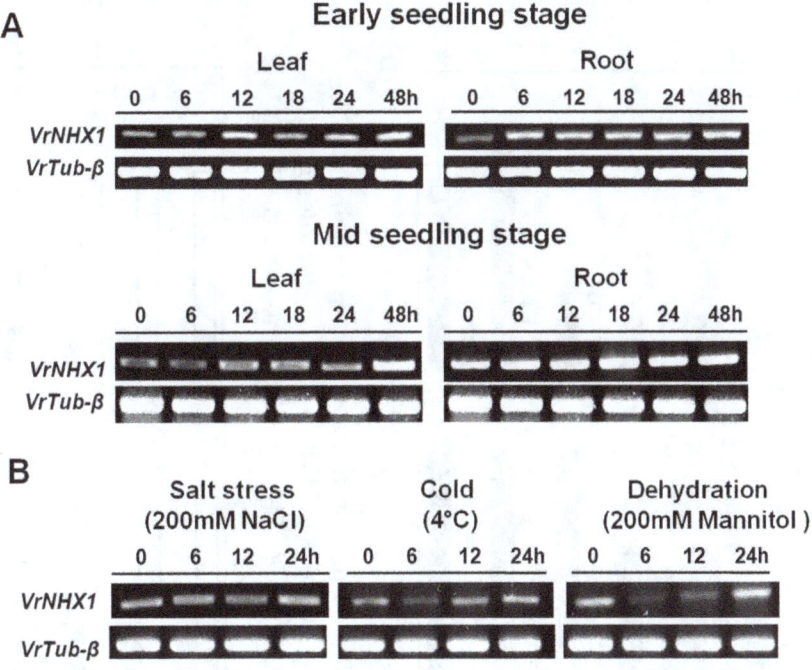

Figure 7. Expression analysis of *VrNHX1* in early and mid stage mungbean seedlings under various abiotic stresses. (A) Semi-quantitative RT-PCR for studying expression patterns of *VrNHX1* under salt stress was performed. Total RNA was isolated from leaves and roots of early (5 days) and mid stage mungbean seedlings (10 days) under 200 mM NaCl treatment at time intervals of 0, 6, 12, 18, 24, and 48 hrs. (B) Semi-quantitative RT-PCR for studying expression patterns of *VrNHX1* under different abiotic stress conditions such as salt, cold and dehydration stress was studied. Total RNA was isolated from mid stage mungbean seedlings under (A) 200 mM NaCl, (B) Cold (4°C), and (C) 200 mM Mannitol treatment at time intervals of 0, 6, 12, and 24 hrs. PCR fragments of 566 bp and 422 bp size corresponding to *VrNHX1* and *VrTubβ* were fractionated electrophoretically on 2% agarose gel stained with 10 mg/ml ethidium bromide.

also increased by 1.1/1.4, 1.4/2.4, 4/3.3, 4.5/3.5, 9.8/4.2 and 7.1/4.7 times whereas, K^+ accumulation decreased by 1.05/1.1, 1.03/1.66, 1.1/3.57, 1.34/3.2, 1.36/4.07, 1.77/4.03 times as compared to control condition at 6, 12, 18, 24, 48 and 72 hrs, respectively (Fig. 8 B). The overall higher accumulation of Na^+ (μmoles/g DW) in roots as opposed to leaves indicated the restriction of movement of toxic Na^+ to the aerial part of the plant as a plausible mechanism to confer salinity tolerance in mungbean.

Ectopic expression of *VrNHX1* resulted in enhanced salt tolerance in transgenic *Arabidopsis*

In order to characterize *VrNHX1* functionally *in planta*, T_3 homozygous *Arabidopsis* lines expressing *VrNHX1* under the control of constitutive CaMV35S promoter or a stress-responsive RD29A promoter were generated using the binary constructs pCAMBIA2301-35S::*VrNHX1* (Fig. S5 A) and pCAMBIA2301-RD29A::*VrNHX1* (Fig. S5 B), respectively, to study their performance under salt stress. The germination efficiency was studied in transgenic lines 1 (35S::*VrNHX1*) and 4 (RD29A::*VrNHX1*) after exposure to 150 mM NaCl stress for 10 days. Under normal condition, no difference was observed in WT and transgenic lines (Fig. 9 A). However, the transgenic lines exhibited better survival and germination efficiency than WT under salt stress (Fig. 9 A). Further, inhibition of root growth in WT and transgenic lines under salt stress (150 mM NaCl) was studied (Fig. 9 B). Transgenic lines 1 and 4 exhibited 2.65 and 3 times higher root length respectively, than WT (Fig. 9 C). The effect on physiological parameters was monitored in 10 days old wild-type (WT) and independent transgenic *Arabidopsis* lines expressing *VrNHX1* constitutively (Lines 1–3, 35S::*VrNHX1*) and inducibly (Lines 4–6, RD29A::*VrNHX1*) under 200 mM NaCl stress for 5 days, by analyzing the total chlorophyll, malondialde-

hyde (MDA) for lipid peroxidation and proline content. Under normal physiological condition, no qualitative and statistical difference was observed between wild-type and transgenic *Arabidopsis* lines (Fig. 10). However, under salt stress (200 mM NaCl), WT showed leaf senescence while transgenic *Arabidopsis* lines (Lines 1–3, 35S::*VrNHX1* and Lines 4–6, RD29A::*VrNHX1*) showed better growth and survival (Fig. 10 A). The transgenic lines showed higher chlorophyll (18–20 mg/ml) and proline (4.8–6 μmoles/g FW) content than WT (Fig. 10 B). The 35S::*VrNHX1* lines showed 1.35 times higher proline than RD29A::*VrNHX1* lines. A lower lipid peroxidation was detected in transgenic lines as WT showed 1.33 times higher malondialdehyde (MDA) content (Fig. 10 B).

Effect of salt stress was studied in mature WT and transgenic lines (Line 1, 35S::*VrNHX1* and Line 4, RD29A::*VrNHX1*). The transgenic lines displayed better survival efficiency while WT exhibited leaf senescence and growth inhibition upon salt stress (200 mM NaCl) (Fig. 11 A). Transgenic *Arabidopsis* 35S::*VrNHX1* plants displayed constitutively high expression of *VrNHX1* under both control (unstressed) and salt stress conditions, whereas RD29A::*VrNHX1* lines showed high induction of *VrNHX1* only after stress treatment with basal expression levels under normal conditions (Fig. 11 B). The total Na^+ and K^+ accumulated in transgenic lines was higher than WT. Further, transgenic 35S::*VrNHX1* and RD29A::*VrNHX1* lines exhibited 1.3 and 1.14 times higher Na^+/K^+ ratio, respectively, as compared to WT (Fig. 11 C, D).

Discussion

This is the first report on isolation and functional characterization of a vacuolar Na^+/H^+ antiporter (*VrNHX1*) from mungbean. Phylogenetic analysis and evolutionary relationship revealed that

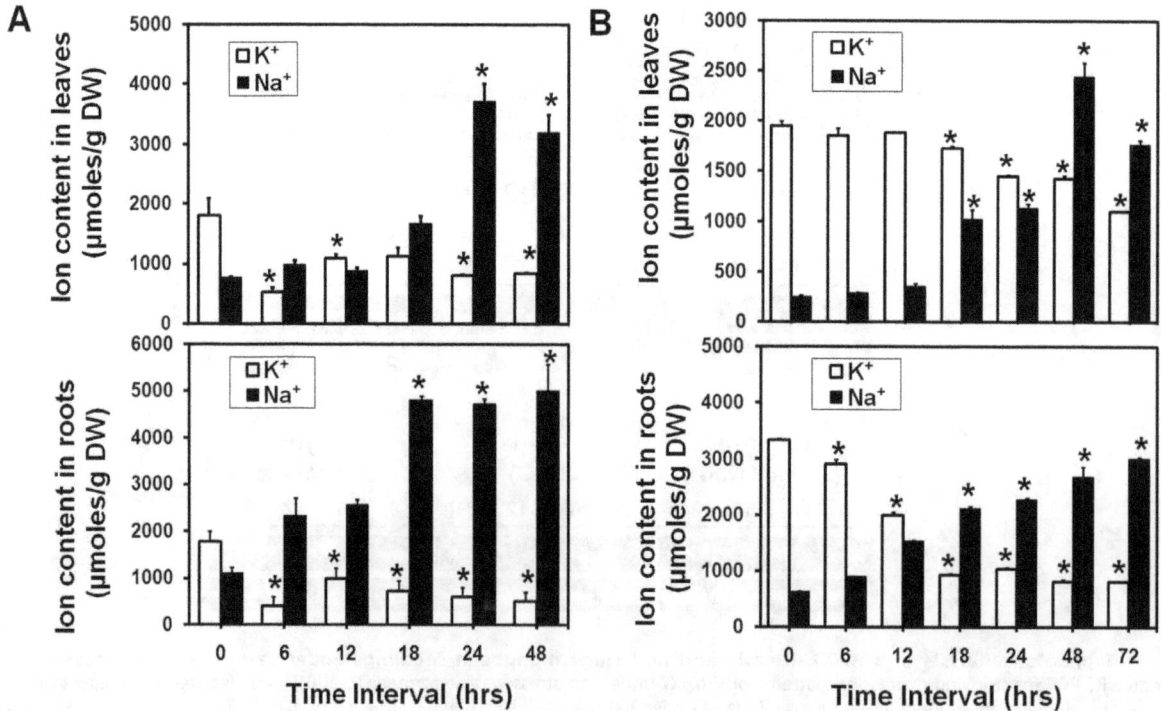

Figure 8. Total intracellular ion measurement in leaves and roots of early and mid stage mungbean seedlings. Na^+ and K^+ content in (A) leaves and (B) roots of unstressed and salt stressed mungbean seedlings harvested at time intervals of 0, 6, 12, 24, 48, and 72 hrs was measured using Flame Photometer. Values indicate means ± SE (n = 3). Statistically significant values at P≤0.05 are indicated as "*", using Bonferroni analysis.

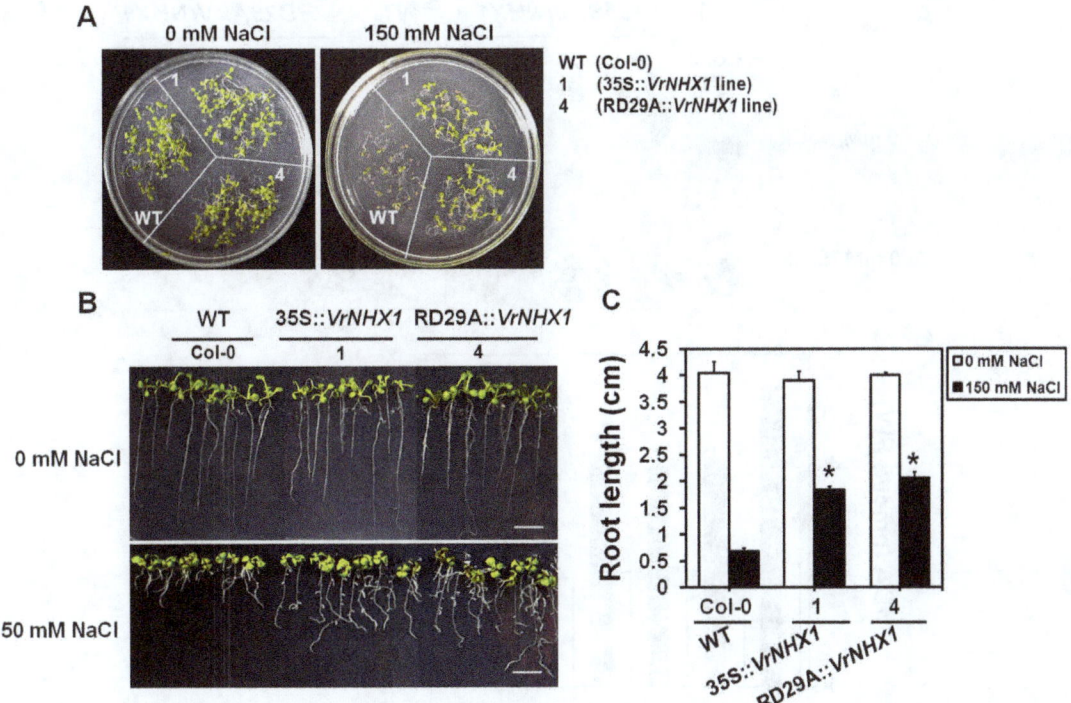

Figure 9. Effect of salt stress on germination efficiency and root growth of transgenic *Arabidopsis* **lines.** (A) The wildtype (WT, col-0) and transgenic (line 1, 35S::*VrNHX1* and line 4, RD29A::*VrNHX1*) seedlings were observed for germination score after 10 days exposure to salt stress (150 mM NaCl). (B) Root growth inhibition in wild type (WT, Col-0) and transgenic *Arabidopsis* (Line 1, 35S::*VrNHX1* and Line 4, RD29A::*VrNHX1*) plants upon salt stress (150 mM NaCl) was studied. The 4 days old germinated seedlings were transferred to 150 mM NaCl stress for a period of 7 days and (C) root length measured was plotted in graph. Values indicate means $\pm$ SE (n = 10). Statistically significant values at P≤0.05 are indicated as "*", using Bonferroni analysis.

VrNHX1 shared highest homology with reported legume Na$^+$/H$^+$ antiporters belonging to the Class-I type NHX exchanger group. The potential structural and functional similarity between yeast and plant endosomal Na$^+$/H$^+$ exchanger, serves as a valuable tool for validation of novel plant Na$^+$/H$^+$ exchangers for their role in salt tolerance [35,36]. Restored growth of AXTVrNHX1 cells in presence of high concentrations of Na$^+$, K$^+$, and Li$^+$ and suppression of hygromycin sensitivity indicated the functional complementation of *ScNHX1* by heterologous expression of *VrNHX1*. The Na$^+$ distribution pattern in vacuolar and cytoplasmic fractions of AXTVrNHX1 cells as compared to AXTYES2.0 cells, indicated the potent role of *VrNHX1* as a vacuolar Na$^+$/H$^+$ antiporter limited to vacuolar sequestration of alkali cations for establishing ion homeostasis. Similar findings were reported in functional complementation of *OsNHX1* in AXT3 mutant [37]. Moreover, *VrNHX1* expression in AXTVrNHX1 showed enhanced K$^+$ distribution within vacuolar fractions which was in accordance with the results obtained in heterologous expression of *AtNHX1* [10] and *TNHXS1* [38] in AXT3 mutant. It was also observed that cytoplasmic K$^+$ fractions were lower in AXTYES2.0 cells as compared with AXTVrNHX1 cells and W303-1B wild type cells. Alkalinization of endolytic compartments has been reported to be mediated by *ScNHX1* which serves as a leak pathway for H$^+$, thus, regulating the pH level for efficient survival against external acid stress [27,39]. In our studies, we observed that growth sensitivity of AXTVrNHX1 cells was lower than AXTYES2.0 cells under external acidic pH environment. Vacuolar acidification was reduced in AXTVrNHX1 cells under low pH indicating the role of *VrNHX1* in extrusion of excess H$^+$ by its ion specificity.

Differential regulation of Na$^+$ uptake, extrusion, compartmentalization, radial transport to stele, loading and unloading into xylem is responsible for the varied response of plants against salinity stress. Under salt stress, *VrNHX1* expression was induced in both leaves and roots of mungbean seedlings with concomitant higher expression in roots than leaves in both early and mid stage seedlings. This result was in accordance with previous reports on expression of *ZmNHX1*, *AeNHX1*, *AlNHX1*, and *ThNHX1* [40–43] and contrary to reports of expression *OsNHX1*, *AgNHX1*, *SsNHX1*, *PeNHX1*, *MsNHX1*, *TrNHX1*, *ZjNHX1*, *ZxNHX*, and *DmNHX1* [17,44–51] which had higher expression in leaves/shoots. The expression pattern of *VrNHX1* under various abiotic stress conditions in mungbean revealed gradual increase in expression under salt stress (200 mM NaCl) after 24 hrs, cold stress (4°C) at 12 hrs and dehydration stress (200 mM mannitol) after 24 hrs. The result was contrary to the previous reports on expression pattern of *PeNHX1* and *ThNHX1* [42,48] under cold stress that showed decrease in the transcript accumulation. No change in expression pattern of *AtNHX1* under cold stress has been reported [52]. Up-regulation of *VrNHX1* under cold stress can be attributed to the other unknown functional mechanisms that still remain to be deciphered. However, involvement of *NHX1* in conferring freezing tolerance has been reported in transgenic *A. thaliana* overexpressing *SsNHX1*, although the exact mechanism has not been explained [53]. Water deficit and altered water potential along with ionic imbalance are known to be primary effects of salt stress [4,8]. We found under dehydration stress the expression pattern of *VrNHX1* in mungbean seedlings similar to previous reports on expression of *GmNHX1*, *ThNHX1* and *EgNHX1*

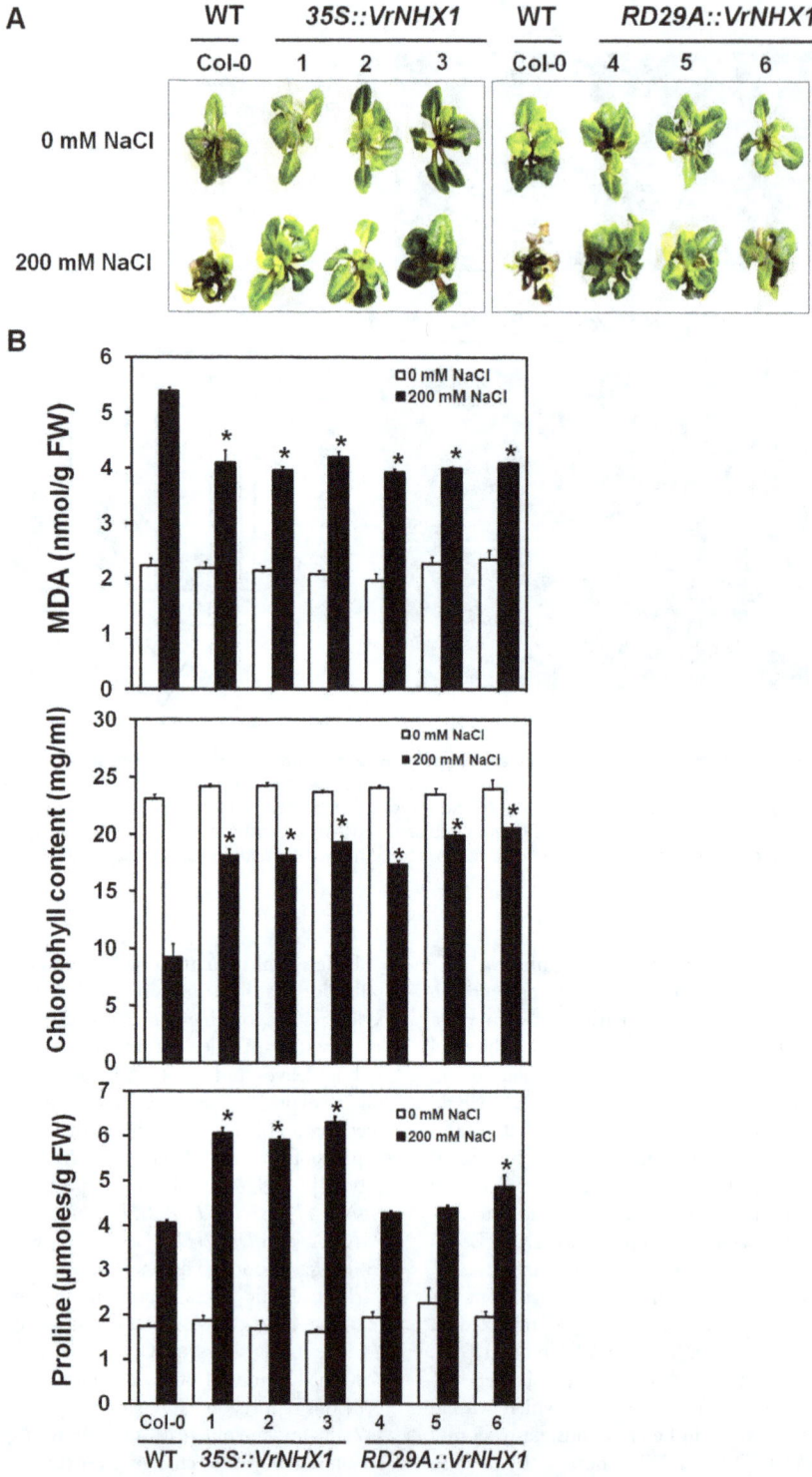

Figure 10. Studying the physiological changes in transgenic *Arabidopsis* lines under salt stress. (A) Effect of salt stress in wild type (WT, Col-0) and transgenic *Arabidopsis* lines expressing *VrNHX1* constitutively (Lines 1–3, 35S::*VrNHX1*) and inducibly (Lines 4–6, RD29A::*VrNHX1*). NaCl-induced morphological changes was visible in 10 days old WT and transgenic lines after exposure to 200 mM NaCl for 5 days. (B) Changes in chlorophyll, MDA and proline content were estimated and analyzed as explained in materials and methods section. Values indicate means ± SE (n = 3). Statistically significant values at P≤0.05 are indicated as "*", using Bonferroni analysis.

which displayed up-regulation under dehydration stress [13,42,54]. However, contrasting results have been reported for expression of *PeNHX1* and *AtNHX1* [48,52].

Physiological response under salt stress, indicated higher Na⁺ accumulation in roots than shoots in early and mid stage mungbean seedlings, in contrast to the reports in *T. repens*,

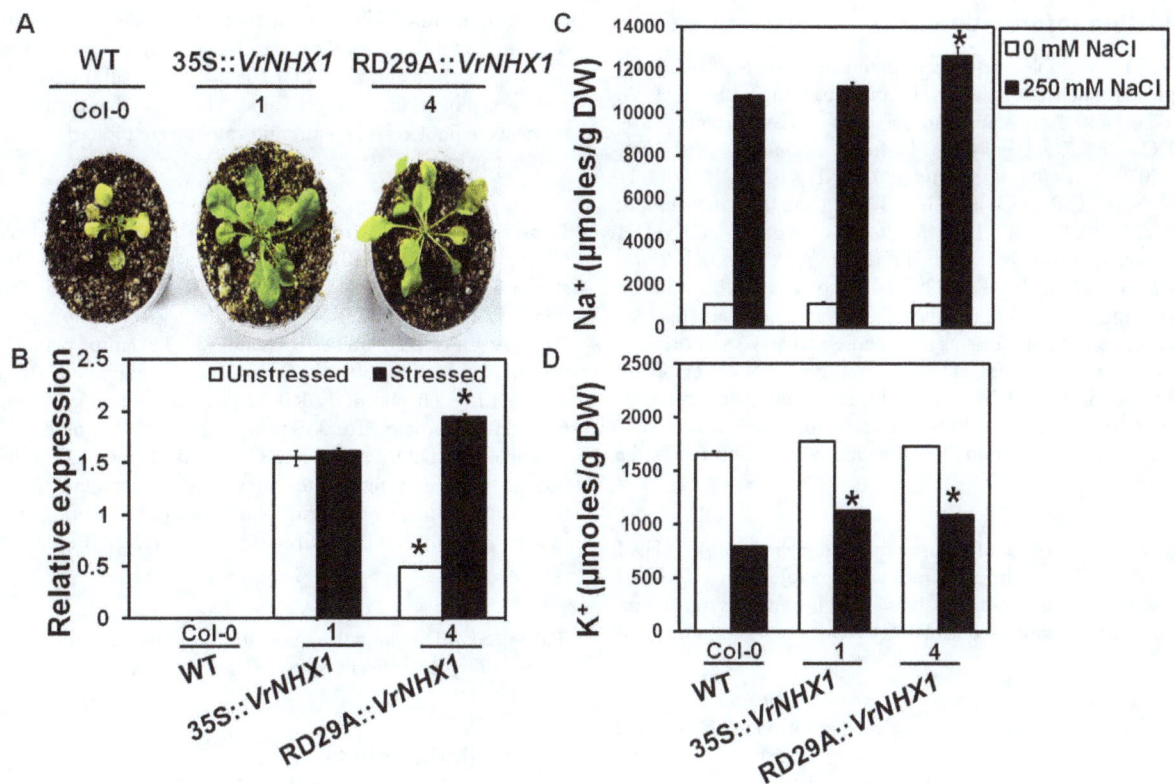

Figure 11. Salt tolerance assay in mature transgenic *Arabidopsis* lines under salt stress. (A) Effect of salt stress on wild type (WT, Col-0) and transgenic *Arabidopsis* lines expressing *VrNHX1* constitutively (Line 1, 35S::*VrNHX1*) and inducibly (Line 4, RD29A::*VrNHX1*) subjected to 250 mM NaCl treatment for 2 weeks (B) Relative transgene expression level of *VrNHX1* in transgenic *Arabidopsis* lines under unstressed and salt stressed conditions. No transgene expression was observed in WT. A 0.283 kb fragment of *VrNHX1*::35SployA and 0.150 kb fragment of *AtUBQ1* was amplified in quantitative RT-PCR analysis (C) Na+ and (D) K+ content (µmoles/g DW) was estimated in leaves of unstressed (0 mM NaCl) and salt stressed (250 mM NaCl) WT and transgenic lines, as described in materials and methods. Values indicate means ± SE (n = 3). Statistically significant values at P≤0.05 are indicated as "*", using Bonferroni analysis.

Z. japonica, H. caspica, Z. xanthoxylum, D. morifolium [17,49–51,55] that showed preferential accumulation of Na+ in leaves/shoots. This indicated that higher K+/Na+ ratio is maintained in leaves owing to sequestration of higher Na+ in root vacuoles thus, restricting their movement to the aerial part of plant. Combined together, increased *VrNHX1* transcript level coupled with higher sequestration of Na+ in roots can be attributed as the tolerance mechanism of mungbean under salt stress.

Ectopic expression of *VrNHX1* conferred salt tolerance in transgenic *Arabidopsis* lines. Both, 35S::*VrNHX1* and RD29A::*VrNHX1* homozygous T3 lines displayed better growth response in comparison to WT. Salt stress affects the photosynthetic system components including chlorophyll contents [56]. The reduction in chlorophyll content was less in transgenic lines (35S::*VrNHX1* and RD29::*VrNHX1*) as compared to WT. Lipid peroxidation is mediated by increase in accumulation of reactive oxygen species (ROS) under salinity stress [57]. Therefore, the extent of lipid peroxidation was measured using malionaldehyde (MDA), a by-product of lipid peroxidation. Transgenic lines showed lower extent of MDA generation as compared to WT indicating protection against membrane damage process. Metabolic response against salt stress, generally includes generation of proline, an osmoprotectant and compatible osmolyte, as a protective measure in plants [4]. Transgenic lines expressed higher proline content in response to salt stress. Proline is also

known as a potent ROS scavenger [58] which might also be correlated with the lower levels of generation of ROS, thus rendering reduced lipid peroxidation in transgenic plants as compared to WT. Similar result was also reported for proline content in transgenic *Arabidopsis* lines overexpressing *DmNHX1* [51]. The regulation of K+/Na+ ratio to maintain K+ homeostasis for proper cellular and enzymatic functioning is an essential mechanism against salinity stress in plants [59]. Our results demonstrated that the transgenic lines (35S::*VrNHX1* and RD29::*VrNHX1*) maintained a higher K+/Na+ ratio than WT plants under salt stress indicating effective tolerance in transgenic lines under salt stress. The phenotypical, physiological and expressional analysis using quantitative real-time PCR concluded that the transgenic RD29::*VrNHX1* line displayed comparable higher survival and growth than 35S::*VrNHX1* lines under salt stress and can be further exploited in crop plants.

The expression of *VrNHX1* under constitutive and inducible promoter enhanced salt tolerance in transgenic *Arabidopsis*. *AtNHX1* is one of the most effective genes in improving plant salt tolerance, however, it played a dominant role mainly in leaf. Our result suggested that *VrNHX1* might play an important role in the root resistance to Na+ toxicity. Therefore, we could assume that overexpression of *VrNHX1* in crop plants might generate enhanced salt tolerance.

Supporting Information

Figure S1 Multiple sequence alignment was performed for amino acid sequences of plant NHX proteins using CLUSTAL W. The GenBank Accession numbers for NHX proteins are: VrNHX1 (AEO50758.1), *Vigna radiata*; VuNHX1 (AEO72079.2), *Vigna unguiculata*; GmNHX1 (AAY430061.1), *Glycine max*; CkNHX1 (ABG89337.1), *Caragana korshinskii*; MsNHX1 (AAS84487.1), *Medicago sativa*; CaNHX1 (ADL28385.1), *Cicer arietinum*; TrNHX1 (ABV00895.1), *Trifolium repens*; LtNHX1 (ACE78322.1), *Lotus tenuis*. "*" indicates identical amino acid (AA) residues. ":" indicates conservative AA substitutions and "." represents semi-conservative AA substitutions in the sequence alignment. The transmembrane region of VrNHX1 as indicated by TM 1–11 and conserved amiloride binding motif, [84]-LFFIYLLPPI-[93], a classic inhibitor of the Na^+/H^+ antiporters detected in TM3 region is also shown in the alignment.

Figure S2 Prediction of transmembrane helices of VrNHX1 (AEO50758.1).The hydropathy plot was generated using TMPred online software. The positive values indicate putative transmembrane domains as indicated as TM 1–11.

Figure S3 Growth measurement of yeast strains under low pH. Yeast strains were grown in synthetic medium APGal (pH 4.0) and absorbance was measured at 600 nm. The data shown above are normalized to growth under normal condition (APGal, pH 7.0). W303-1B:- Wild type strain, AXTYES2.0:- AXT3 mutant harboring null pYES2.0 plasmid, AXTVrNHX1:- AXT3 mutant harboring pYES*VrNHX1* recombinant plasmid. Data represent mean from three independent events (n = 3) and standard error plotted in the graph. Statistically significant values at P≤0.05 are indicated as "*", using Bonferroni analysis.

Figure S4 Calibration curve for pH sensitive BCECF fluorescent dye was plotted using standards ranging from pH 4.0–8.0. Yeast strains (W303-1B, AXTYES2.0, AXTVrNHX1) grown in APGal medium (pH 4.0) were loaded with BCECF dye as

described in materials and methods, fluorescence intensity was measured at 440 and 490 nm, background values (measured with only cell extract and only BCECF dye) were subtracted and the ratio was plotted for each pH value. The data from the three yeast stains were pooled and mean ratio values were plotted with a fitted non-linear graph.

Figure S5 T-DNA region of pCAMBIA2301-35S::*VrNHX1* (13.9 kb) and pCAMBIA2301-RD29A::*VrNHX1* (14.4 kb). Restrction enzyme PstI and EcoRI used for cloning 35SP::*VrNHX1*::35STer cassette (2.3 kb) and RD29A::*VrNHX1*::35STer cassette (2.8 kb) into plant binary vector pCAMBIA 2301 (11.6 kb) is also highlighted. Abbreviations: LB, left border; RB, right border; 35S Promoter, Cauliflower mosaic virus 35S promoter; RD29A promoter, Stress indicible AtRD29A promoter; CaMV 35S poly-A, Cauliflower mosaic virus 35S terminator; nos poly-A, nopaline transferase terminator; *nptII*, neomycin phosphotransferase; *intron-gus-A*, intron interrupted β-glucuronidase; *VrNHX1*, *Vigna radiata* NHX1.

Table S1 The putative post-translational modification sites predicted by ScanProsite software for VrNHX1.

Acknowledgments

We express our sincere thanks to Prof. Edward Blumwald and Dr. Olivier Cagnac for the yeast strains, W303 and AXT3 respectively. We also thank Dr. Luciana Loureiro Penha for providing vector for yeast expression analysis, Department of Civil Engineering, IIT Guwahati for ion estimation analysis using Flame Photometer.

Author Contributions

Conceived and designed the experiments: LS BhL SKP. Performed the experiments: SM HA. Analyzed the data: LS BhL SM. Contributed reagents/materials/analysis tools: LS SM HA BhL. Wrote the paper: SM LS. Supervised the study: LS BhL. Provided critical revision of the manuscript for important intellectual content: BhL.

References

1. Kronzucker HJ, Britto DT (2011) Sodium transport in plants: a critical review. New Phytol 189: 54–8.
2. Mahajan S, Tuteja N (2005) Cold, salinity and drought stresses: an overview. Arch Biochem Biophys 444: 139–158.
3. Schroeder JI, Delhaize E, Frommer WB, Guerinot ML, Harrison MJ, et al. (2013) Using membrane transporters to improve crops for sustainable food production. Nature 497: 60–66.
4. Hasegawa PM, Bressan RA, Zhu JK, Bohnert HJ (2000) Plant Cellular and Molecular Responses to High Salinity. Annu Rev Plant Physiol Plant Mol Bio 51: 463–99.
5. Shi H, Ishitani M, Kim C, Zhu JK (2000) The *Arabidopsis thaliana* salt tolerance gene *SOS1* encodes a putative Na⁺/H⁺ antiporter. Proc Natl Acad Sci USA 97: 6896–6901.
6. Shi H, Quintero FJ, Pardo JM, Zhu JK (2002) The Putative Plasma Membrane Na⁺/H⁺ Antiporter SOS1 Controls Long-Distance Na⁺ Transport in Plants. Plant Cell 14: 465–477.
7. Blumwald E, Aharon GS, Apse MP (2000) Sodium transport in plant cells. Biochim Biophys Acta 1465: 140–151.
8. Blumwald E (2000) Sodium transport and salt tolerance in plants. Curr Opin Cell Biol 12: 431–434.
9. Pardo JM, Cubero B, Leidi EO, Quintero FJ (2006) Alkali cation exchangers: roles in cellular homeostasis and stress tolerance. J Exp Bot 57: 1181–1199.
10. Yokoi S, Quintero FJ, Cubero B, Ruiz MT, Bressan RA, et al. (2002) Differential expression and function of *Arabidopsis thaliana* NHX Na⁺/H⁺ antiporters in the salt stress response. Plant J 30: 529–539.
11. Zhang HX, Blumwald E (2001) Transgenic salt-tolerant tomato plants accumulate salt in foliage but not in fruit. Nat Biotechnol 19: 765–768.
12. Rodríguez-Rosales MP, Gálvez FJ, Huertas R, Aranda MN, Baghour M, et al. (2009) Plant NHX cation/proton antiporters. Plant Signal Behav 4:265–276.
13. Li WYF, Wong FL, Tsai SN, Phang TH, Shao G, et al. (2006) Tonoplast-located *GmCLC1* and *GmNHX1* from soybean enhance NaCl tolerance in transgenic bright yellow (BY)-2 cells. Plant Cell Environ 29: 1122–1137.
14. An BY, LuoY Li JR, Qiao WH, Zhang XS, et al. (2008) Expression of a vacuolar Na⁺/H⁺ antiporter gene enhances salinity tolerance in transgenic *Arabidopsis*. Acta Agron Sin 34: 557–564.
15. Tang R, Li C, Xu K, Du Y, Xia T, et al. (2010) Isolation, functional characterization, and expression pattern of a vacuolar Na⁺/H⁺ Antiporter gene *TrNHX1* from *Trifolium repens* L. Plant Mol Biol Rep 28: 102–111.
16. Teakle NL, Amtmann A, Real D, Colmer TD (2010) *Lotus tenuis* tolerates combined salinity and waterlogging: maintaining O₂ transport to roots and expression of an *NHX1*-like gene contribute to regulation of Na⁺ transport. Physiol Plant 139: 358–374.
17. Yang DH, Song LY, Hu J, Yin WB, Li ZG, et al. (2012) Enhanced tolerance to NaCl and LiCl stresses by over-expressing *Caragana korshinskii* sodium/proton exchanger 1 (*CkNHX1*) and the hydrophilic C terminus is required for the activity of *CkNHX1* in *Atsos3-1* mutant and yeast. Biochem Biophys Res Commun 417: 732–737.
18. Hasanuzzaman M, Ali MR, Hossain M, Kuri S, Islam MS (2013) Evaluation of total phenolic content, free radical scavenging activity and phytochemical screening of different extracts of *Averrhoa bilimbi* (fruits). Int Curr Pharmaceut J 2: 92–96.
19. Jacoby B (1999) Mechanisms involved in salt tolerance of plants. In: Pessarakli M, editor. Plant and Crop Stress. pp. 97–123.

20. Nair RM, Yang RY, Easdown WJ, Thavarajah D, Thavarajah P, et al. (2013) Biofortification of mungbean (*Vigna radiata*) as a whole food to enhance human health. J Sci Food Agric 93: 1805–1813.

21. Thompson JD, Gibson TJ, Plewniak F, Jeanmougin F, Higgin DG (1997) The CLUSTAL_X windows interface: flexible strategies for multiple sequence alignment aided by quality analysis tools. Nucleic Acids Res 25: 4876–4882.

22. Tamura K, Dudley J, Nei M, Kumar S (2007) MEGA4: Molecular Evolutionary Genetics Analysis (MEGA) Software Version 4.0. Mol Biol Evol 24: 1596–1599.

23. Hofmann K, Stoffel W (1993) A database of membrane spanning proteins segments. Biol Chem 374: 166.

24. Gattiker A, Gasteiger E, Bairoch A (2002) ScanPROSITE: a reference implementation of a PROSITE scanning tool. Appl Bioinformatics 1: 107–108.

25. Gietz D, St. Jean A, Woods RA, Schiestl RH (1992) Improved method for high efficiency transformation of intact yeast cells. Nucleic Acids Res 20: 1425.

26. Venema K, Belver A, Marin-Manzano MC, Rodgriguez-Rosales MP, Donaire JP (2003) A novel intracellular K^+/H^+ antiporter related to Na^+/H^+ antiporters is important for K^+ ion homeostasis in plants. J Biol Chem 278: 22453–22459.

27. Brett CL, Tukaye DN, Mukherjee S, Rao R (2005) The yeast endosomal $Na^+(K^+)/H^+$ exchanger Nhx1 regulates cellular ph to control vesicle trafficking. Mol Biol Cell 16: 1396–1405.

28. Munns R, Wallace PA, Teakle NL, Colmer TD (2010) Measuring soluble ion concentrations (Na^+, K^+, Cl^-) in salt-treated plant. In: Sunkar R, editor. Plant Stress Tolerance: Methods in molecular biology. pp. 371–382.

29. Clough SJ, Bent AF (1998) Floral dip: a simplified method for Agrobacterium-mediated transformation of *Arabidopsis thaliana*. Plant J 16: 735–743.

30. Murashige T, Skoog F (1962) A revised medium for rapid growth and bio assays with tobacco tissue cultures. Physiol Plantarum 15: 473–497.

31. Lichtenthaler HK (1987) Chlorophyll fluorescence signatures of leaves during the autumnal chlorophyll breakdown. J Plant Physiol 131: 101–110.

32. Heath RL, Packer L (1968) Photoperoxidation in isolated chloroplasts. I. Kinetic and stoichiometry of fatty acid peroxidation. Arch Biochem Biophys 125: 189–198.

33. Bates LS, Waldren RP, Teare ID (1973) Rapid determination of free proline for water-stress studies. Plant Soil 39: 205–207.

34. Harris C, Fliegel L (1999) Amiloride and the Na^+/H^+ exchanger protein: mechanism and significance of inhibition of the Na^+/H^+ exchanger. Int J Mol Med 3: 315–321.

35. Darley CP, van Wuytswinkel OCM, van der Woude K, Mager WH, de Boer AH (2000) *Arabidopsis thaliana* and *Saccharomyces cerevisiae* NHX1 genes encode amiloride sensitive electroneutral Na^+/H^+ exchangers. Biochem J 351: 241–249.

36. Quintero FJ, Blatt MR, Pardo JM (2000) Functional conservation between yeast and plant endosomal Na^+/H^+ antiporters. FEBS Lett 471: 224–228.

37. Kinclova-Zimmermannova O, Flegelova H, Sychrova H (2004) Rice Na^+/H^+-antiporter Nhx1 partially complements the alkali-metal-cation sensitivity of yeast strains lacking three sodium transporters. Folia Microbiol 49: 519–525.

38. Gouiaa S, Khoudi H, Leidi EO, Pardo JM, Masmoudi K (2012) Expression of wheat Na^+/H^+ antiporter *TNHXS1* and H^+-pyrophosphatase *TVP1* genes in tobacco from a bicistronic transcriptional unit improves salt tolerance. Plant Mol Biol 79:137–155.

39. Ali R, Brett CL, Mukherjee S, Rao R (2004) Inhibition of sodium/proton exchange by a Rab-GTPase- activating protein regulates endosomal traffic in yeast. J Biol Chem 279: 4498–4506.

40. Zhang GH, Su Q, An LJ, Wu S (2008) Characterization and expression of a vacuolar Na^+/H^+ antiporter gene from the monocot halophyte *Aeluropus littoralis*. Plant Physiol Biochem 46: 117–126.

41. Zorb C, Noll A, Karl S, Leib K, Yan F, et al. (2005) Molecular characterization of Na^+/H^+ antiporters (*ZmNHX*) of maize (*Zea mays* L.) and their expression under salt stress. J Plant Physiol 162: 55–65.

42. Wu C, Gao X, Kong X, Zhao Y, Zhang H (2009) Molecular Cloning and Functional Analysis of a Na^+/H^+ Antiporter Gene *ThNHX1* from a Halophytic Plant *Thellungiella halophila*. Plant Mol Biol Rep 27: 1–12.

43. Qiao WH, Zhao XY, Li W, Luo Y, Zhang XS (2007) Overexpression of *AeNHX1*, a root-specific vacuolar Na^+/H^+ antiporter from *Agropyron elongatum*, confers salt tolerance to *Arabidopsis* and *Festuca* plants. Plant Cell Rep 26: 1663–1672.

44. Fukuda A, Nakamura A, Tagiri A, Tanaka H, Miyao A, et al. (2004) Function, intracellular localization and the importance in salt tolerance of a vacuolar Na^+/H^+ antiporter from rice. Plant Cell Physiol 45: 146–159.

45. Hamada A, Shono M, Xia T, Ohta M, Hayashi Y, et al. (2001) Isolation and characterization of a Na^+/H^+ antiporter gene from the halophyte *Atriplex gmelini*. Plant Mol Biol 46: 35–42.

46. Ma XL, Zhang Q, Shi HZ, Zhu JK, Zhao YX, et al. (2004) Molecular cloning and different expression of a vacuolar Na^+/H^+ antiporter gene in *Suada salsa* under salt stress. Biol Plantarum 48: 219–225.

47. An BY, LuoY LiJR, Qiao WH, Zhang XS, et al. (2008) Expression of a vacuolar Na^+/H^+ antiporter gene of alfalfa enhances salinity tolerance in transgenic *Arabidopsis*. Acta Agron Sin 34:557–564.

48. Rajagopal D, Agarwal P, Tyagi W, Singla-Pareek SL, Reddy MK, et al. (2007) Pennisetum *glaucum* Na^+/H^+ antiporter confers high level of salinity tolerance in transgenic *Brassica juncea*. Mol Breed 19: 137–151.

49. Du Y, Hei Q, Liu Y, Zhang H, Xu K, et al. (2010) Isolation and characterization of a putative vacuolar Na^+/H^+ Antiporter gene from *Zoysia japonica* L. J Plant Biol 53: 251–258.

50. Wu GQ, Xi JJ, Wang Q, Bao AK, Ma Q, et al. (2011) The *ZxNHX* gene encoding tonoplast Na^+/H^+ antiporter from the xerophytes *Zygophyllum xanthoxylum* plays important roles in response to salt and drought. J Plant Physiol 168: 758–767.

51. Zhang H, Liu Y, Xu Y, Chapman S, Love AJ, et al. (2012) A newly isolated Na^+/H^+ antiporter gene, *DmNHX1*, confers salt tolerance when expressed transiently in *Nicotiana benthamiana* or stably in *Arabidopsis thaliana*. Plant Cell Tiss Organ Cult 110: 189–200.

52. Shi H, Zhu JK (2002) Regulation of expression of the vacuolar Na^+/H^+ antiporter gene *AtNHX1* by salt stress and abscisic acid. Plant Mol Biol 50: 543–550.

53. Li J, Jiang G, Huang P, Ma J, Zhang F (2007) Overexpression of the Na^+/H^+ antiporter gene from *Suaeda salsa* confers cold and salt tolerance to transgenic *Arabidopsis thaliana*. Plant Cell Tiss Organ Cult 90: 41–48.

54. Baltierra Q, Castillo M, Gamboa MC, Rothhammer M, Krauskopf E (2012) Molecular characterization of a novel Na^+/H^+ antiporter cDNA from *Eucalyptus globules*. Biochem and Biophys Res Commun 430: 535–540.

55. Guan B, Hu YZ, Zeng YL, Wang Y, Zhang FC (2010) Molecular characterization and functional analysis of a vacuolar Na^+/H^+ antiporter gene (*HcNHX1*) from *Halostachys caspica*. Mol Biol Rep 38: 1889–1899.

56. Demetriou G, Neonaki C, Navakoudis E, Kotzabasis K (2007) Salt stress impact on the molecular structure and function of the photosynthetic apparatus—the protective role of polyamines. BBA Bioenergetics 1767: 272–280.

57. Bor M, Özdemir F, Türkan I (2003) The effect of salt stress on lipid peroxidation and antioxidants in leaves of sugar beet *Beta vulgaris* L. and wild beet *Beta maritima* L. Plant Sci 164: 77–84.

58. Szabados L, Savouré A (2010) Proline: a multifunctional amino acid. Trends Plant Sci 15: 89–97.

59. Maathuis FJ, Amtmann ANNA (1999) K^+ nutrition and Na^+ toxicity: the basis of cellular K^+/Na^+ ratios. Ann Bot 84: 123–133.

16

Multiple Different Defense Mechanisms Are Activated in the Young Transgenic Tobacco Plants Which Express the Full Length Genome of the Tobacco Mosaic Virus, and Are Resistant against this Virus

Balaji Jada[1], Arto J. Soitamo[1], Shahid Aslam Siddiqui[2], Gayatri Murukesan[3], Eva-Mari Aro[1], Tapio Salakoski[3], Kirsi Lehto[1]*

1 Department of Biochemistry, Laboratory of Molecular Plant Biology, University of Turku, Turku, Finland, 2 Department of Agricultural sciences, University of Helsinki, Helsinki, Finland, 3 Department of Information Technology, University of Turku, Turku, Finland

Abstract

Previously described transgenic tobacco lines express the full length infectious Tobacco mosaic virus (TMV) genome under the 35S promoter (Siddiqui et al., 2007. Mol Plant Microbe Interact, 20: 1489–1494). Through their young stages these plants exhibit strong resistance against both the endogenously expressed and exogenously inoculated TMV, but at the age of about 7–8 weeks they break into TMV infection, with typical severe virus symptoms. Infections with some other viruses (Potato viruses Y, A, and X) induce the breaking of the TMV resistance and lead to synergistic proliferation of both viruses. To deduce the gene functions related to this early resistance, we have performed microarray analysis of the transgenic plants during the early resistant stage, and after the resistance break, and also of TMV-infected wild type tobacco plants. Comparison of these transcriptomes to those of corresponding wild type healthy plants indicated that 1362, 1150 and 550 transcripts were up-regulated in the transgenic plants before and after the resistance break, and in the TMV-infected wild type tobacco plants, respectively, and 1422, 1200 and 480 transcripts were down-regulated in these plants, respectively. These transcriptome alterations were distinctly different between the three types of plants, and it appears that several different mechanisms, such as the enhanced expression of the defense, hormone signaling and protein degradation pathways contributed to the TMV-resistance in the young transgenic plants. In addition to these alterations, we also observed a distinct and unique gene expression alteration in these plants, which was the strong suppression of the translational machinery. This may also contribute to the resistance by slowing down the synthesis of viral proteins. Viral replication potential may also be suppressed, to some extent, by the reduction of the translation initiation and elongation factors eIF-3 and eEF1A and B, which are required for the TMV replication complex.

Editor: Xiao-Wei Wang, Zhejiang University, China

Funding: Research was supported by Turku University Foundation, and by the Academy of Finland, grant numbers 127203, 128943 and 118637. The funders had no role in study design, data collection and analysis, decision to publish, or preparation of the manuscript.

Competing Interests: The authors have declared that no competing interests exist.

* Email: klehto@utu.fi

Introduction

Viruses are obligate intracellular molecular parasites which depend on host's cellular machinery and on multiple host factors to complete their infectious life cycle. They utilize a large variety of host-encoded proteins and molecular structures as components of their replication complex or cell-to-cell movement machinery [1–6] and various cellular compartments (typically various membranous structures) as their replication sites [1,7–10]. For instance, many RNA-viruses use the host's translation elongation factor 1A (eEF1A) as a component of their replication complex, tobamoviruses use also the factors eEF1B and eIF3 in this complex, and potyviral VPg molecules interact with the host's initiation factor 4E (eIF4E) for promoting their translation [11–19].

Viruses can also alter the functions and composition of their host cells to benefit their own proliferation. For instance, they can enhance the expression of their needed host factors, bind or suppress various resistance factors, induce changes in the lipid composition of infected cells, and interfere with host's hormonal pathways [1,8,13,20–22].

Viruses can initiate infection process only in susceptible host species that provide compatible host factors, needed for the viral replication and spread. However, many of these potential host species also recognize the invading viruses and mount different defense mechanisms to stop their proliferation or spread. For instance, activation of the R-gene mediated resistance leads to hypersensitive reaction (HR) and virus localization, enhanced expression of various defense-related genes and induction of systemic acquired resistance (SAR) [23–25]. Accumulation of virus-specific double-stranded RNAs also induces the RNA-silencing mediated immune-system in plants [26], which leads to sequence-specific degradation of the viral RNAs. To counteract these silencing-mediated defense reactions, viruses produce specific silencing suppressor proteins (Viral RNA-silencing suppressors, VRSS) which interfere with different steps of the silencing pathways [22,27–31]. Some of these VRSS-factors have been identified as viral host determinants or pathogenicity factors within specific host species [32–35], demonstrating the importance of this defense/counter-defense interaction between viruses and their hosts. The VRSS factors may also interfere with the silencing-mediated endogenous regulatory pathways in the cells [36–39]. This may happen as a mere side-effect of the viral counter-defense, or as an active means to weaken hosts' cellular status.

The virus-host interactions thus consist of a very complex molecular interplay. It involves depletion of various host factors and energy compounds through viral parasitism, altered expression of the viral-induced host factors, active defense mechanisms mounted by the host, active viral counter-defense mechanisms, disturbance of the silencing-mediated regulatory network, and the general infection related stress-reaction in the host [17,23,40]. Some of these interactions depend on functions of individual virus-encoded genes, while others are related to the replication of the viral RNA or to the consorted action of various viral products. They may lead to either plant resistance, or to virus proliferation and symptom development in the host plant.

Now, different system biology approaches are available to investigate the plant responses induced in different stages of virus infections, or in transgenic plants that express individual viral genes. Expression levels of many hundreds or even several thousands of genes have been found to be altered in these plants, and although these alterations have some common features (enhanced expression of defense- and stress-related genes), they are mostly unique and specific to each virus/host combination [20,21,41–47]. This illustrates that the molecular interactions are unique and specific in each compatible or incompatible virus-host combination.

Through the last few decades, Tobacco mosaic virus (TMV) or its constituent genes have often been used as a model system to investigate the plant-virus interactions and to dissect the details of viral replication, movement, host resistance and physiological alterations [6,19,42,48–52]. Here we are using the functional genomics to study the molecular response of transgenic tobacco plants expressing the infectious TMV genome under the constitutive 35S promoter [53]. Interestingly, during the early growth stages (up to about 7–8 weeks after germination) these plants accumulate only a very low level of the TMV RNA, and also exhibit strong resistance against external TMV infections. After this period the resistance breaks, plants become infected from the transformed TMV sequence, accumulate high levels of TMV RNA and show typical TMV symptoms. To identify the gene functions that are associated to this early resistance we have conducted a microarray analysis of these transgenic plants just before resistance break (BRB), and compared their gene expression patterns to those of corresponding healthy wild type (wt) plants. The observed trasncriptome alterations were compared to those observed in the same plants after resistance break (ARB) and in TMV-infected (TMVi) wt tobacco plants. Gene expression alterations were also compared to those observed in other transgenic tobacco plants expressing various virus-derived VRSS genes [54] that are known not to be resistant against TMV, to reveal the genes or processes that would be specific to the resistance condition.

Results

Transgenic tobacco plants that express the wt TMV genome have been previously produced and characterized in our laboratory [53]. All transgenic lines derived from separate transformation events, and sibling lines from the same transformations all had the same, consistent, stunted phenotype (Figure 1A), were initially resistant against TMV, and broke into a strong TMV infection typically at about 7–9 weeks after germination. This emerging infection verified the positive transcription of the transgene. Positive transgene expression status, even before the resistance break, was also shown by positive, although very low detection of the viral RNA by northern blotting and by RT-PCR. At the resistance breaking stage, the TMV-coat protein positive cells, as detected by *in situ* immunolabeling, first appeared as isolated infection foci in the vascular tissues of the upper leaves (Figure 1B). Typical viral symptoms also first appeared on upper leaves of the plants and then slowly progressed towards the lower older leaves, i.e. showing similar symptom pattern as a normal TMV-infection in wt tobacco plants (data not shown).

To investigate the molecular mechanisms underlying this resistant condition, transcriptomic profiles of three young resistant plants (BRB), and of the same plants after the resistance break (ARB), and of three wt TMV-infected (TMVi) plants were analyzed by the microarray approach, and compared each to the transcriptomes of healthy control plants of corresponding age. The observed transcriptome alterations were compared between these three types of plants to reveal gene functions that would be specific to each condition.

The microarray analysis was performed using Tobacco 4×44K microarray (Agilent), according to Agilent's standard procedures (see Methods). Our previous microarray analyses of different transgenic tobacco lines have revealed that the transcript profiles of control plants transformed with the empty pBin61 transformation vector are approximately equal to those of the wild type healthy plants [41], and therefore only the wild type plants were used here as controls. The raw microarray data was normalized and subjected to statistical analysis, with BH false discovery rate of less than 0.05 (Student t-test with adjusted P-value <0.05). Subsequently, the transcripts that were 2-fold up- or down-regulated, as compared to the corresponding healthy control plants, were considered as differentially expressed in the test plants. Some of the expression levels of randomly selected genes were verified by using RT-qPCR, with essentially same results as was attained with the microarray method (Table 1).

The microarray analyses indicate that total of 1362 transcripts were up- and 1422 transcripts were down-regulated in the BRB transgenic plants (Figure 2, Table S1), total of 1150 transcripts were up- and 1200 were down-regulated in the ARB transgenic plants (Figure 2, Table S2), and 550 transcripts were up- and 480 transcripts down-regulated in the TMVi plants, respectively (Figure 2, Table S3). The transcriptional alterations were very distinct and different between these three types of plants. The details of these alterations are compared and discussed in further sections.

Transcripts of the protein synthesis machinery were strongly reduced in BRB plants

Many viruses infecting eukaryotic hosts use different mechanisms to reduce the host-specific protein synthesis, and to increase the synthesis of the virus-specific proteins [55]. Interestingly, the three different types of plants examined in this study (BRB and ARB transgenic plants and TMVi tobacco plants) showed strong –

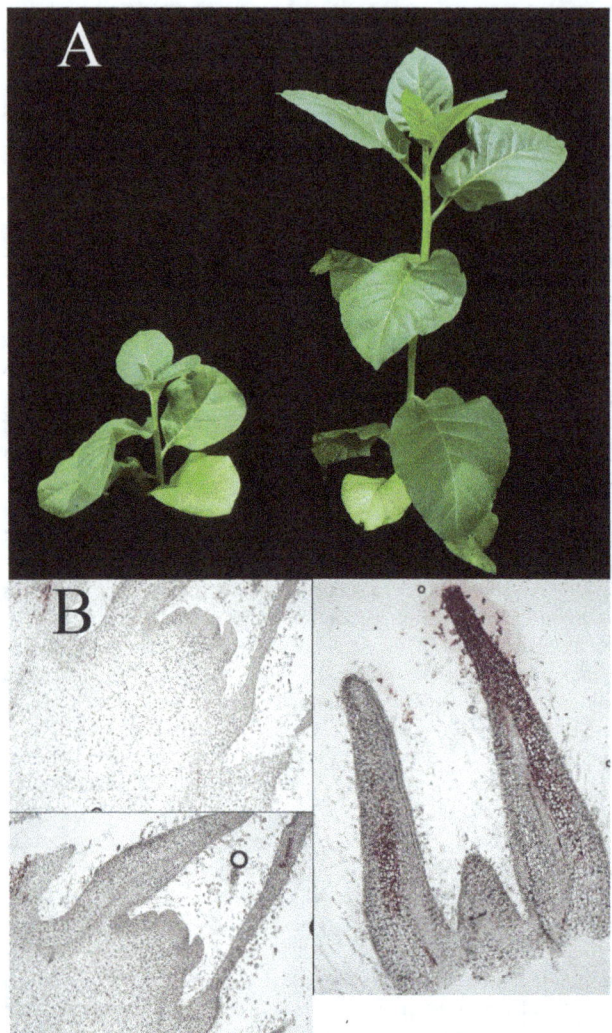

genes), and for elongation-related genes like nascent polypeptide complex, LOS1 and elongation factors eEF1A, B, P, Δ and TuA. Also, several transcripts related to the post-translational modifications, protein targeting (such as TIC40- and TOC75-complexes), amino acid biosynthesis and various protein folding chaperons were down-regulated (Table S4) in the BRB-plants. In addition, about 150 protein processing and degradation-related transcripts were clearly up-regulated in the BRB-plants, including ubiquitin ligases and conjugating proteins, autophagy 8c proteins and a variety of peptidases and proteases. Also, some transcripts coding for translation initiation factors, amino acid synthesis, protease inhibitors, ribosomal proteins and protein kinases were up-regulated (Table S5).

Interestingly, nearly opposite expression pattern of translation-related genes was observed in the ARB transgenic plants. In these plants, only 120 translation-related transcripts were down-regulated, and a total of 135 translation-related transcripts were up-regulated (Table S5). Among these, several transcripts related to the ribosomes and translational initiation (total of 13), amino acid biosynthesis (total of 25), and protein secreting pathways (6) were induced. In contrast to the BRB-plants, 60 transcripts coding for proteases, peptidases and ubiquitin-mediated protein degradation pathways were down-regulated (Table S4), including the transcripts for the S41 and M48 peptidases, believed to be involved in the processing of the D1 protein of the photosystem II (PSII) in plants [56–58]. Also, several genes related to transcription (e.g. sigma factors and RNA polymerases), or to the posttranslational modifications were down-regulated. Only 16 genes coding for different proteases and 29 genes coding for ubiquitin-mediated protein degradation pathway were up-regulated. Some of the up-regulated transcripts in the ARB transgenic plants were related to metallocarboxypeptidase, kinases and amino acid degradation.

In the TMVi plants, the translation and protein processing related transcripts were predominantly up-regulated, with a total of 197 translation-related transcripts being up-regulated and 44 being down-regulated (Table S4 & S5). Most interestingly, many of the up-regulated transcripts in the TMVi plants were related to 60S (114) and 40S (55) ribosomes and their subunits, and to translation elongation, which were down-regulated in the BRB transgenic plants; these transcripts were enhanced by 2–23.2 fold. Also, some transcripts related to, nucleus targeting, protein folding and protein phosphatases, and protein and amino acid degradation pathways were up-regulated in the TMVi plants, including ubiquitin mediated pathway and several proteases and peptidases. Only 7 transcripts related to ribosomal proteins and 26S ribosomal RNA, and 11 transcripts related to post-translational modification pathways were down-regulated in TMVi plants.

Many stress-related genes were induced in TMV-transgenic plants but not in wild type TMV infected tobacco plants

In compatible plant-virus interactions the virus must be able to infect the host plant without mounting excessive or fast defense reactions, or to suppress the basal host defense mechanisms [50,59,60]. The interaction of TMV with tobacco is known to be compatible in nature [50]. In this work such full compatibility was found only in TMVi wild type tobacco plants where only few biotic and abiotic stress response-related transcripts were induced. Instead, the TMV- expressing BRB and ARB transgenic plants showed strongly activated defense-responses.

In the BRB transgenic plants, total of 174 stress response-related transcripts were up-regulated and 76 were down-regulated (Figure S2, Table S7 & S6). Many of the up-regulated transcripts were coding for heat shock proteins (20), chitinases (17), elicitor

Figure 1. Phenotype of the transgenic plants that harbor the full length infectious TMV cDNA, under 35S promoter. (A) BRB-TMV transgenic plant (on the left), at the age of six weeks after germination. The plants show no viral symptoms, and do not contain any detectable viral RNA or CP, but they are severely stunted as compared to the wild type plants of the same age. (B), Anti-TMV CP labeled thin sections of the apical domain of the TMV transgenic plant during the early stage of the resistance break (at 8 weeks after germination). Isolated TMV-positive foci are detected in the vascular tissue, and in the tips of leaf initials. On the left, two adjacent sections are shown to illustrate the small size of the foci.

but strikingly different - alterations in the expression levels of the translation-related transcripts.

In BRB transgenic plants, the largest group of down-regulated genes (total of 750) was composed of translation-related transcripts (Figures 2, S1, Table S4). Most of them were pertained to the cytosolic ribosomal genes, including transcripts coding for various 60S ribosomal proteins (total of 391 transcripts, e.g. the L5–L15, L17–L39, L44, P0, P1, P3 and P4), and for various 40S ribosomal proteins (total of 222 transcripts, e.g. S3–S29), with reduction levels ranging up to 5,5-fold level. Also, some transcripts related to the 18S ribosomal RNA and ribosomal biogenesis regulators were down-regulated, as were several (total of 33) transcripts coding for translation initiation factors (including eIF2γ, eIF-4A, eIF-3

Table 1. Microarray results verification by using quantitative real-time PCR (RT-qPCR).

EST/mRNA	Description	Log value Microarray	Log value RT-qPCR	s.e. of C_t of RT-qPCR
BRB transgenic plants	**up-regulated transcripts**			
EB683763	P-rich protein NtEIG-C29	2.9	3.0	0.46
BP128776	DNAJ heat shock protein	4.6	6.8	0.21
BRB transgenic plants	**down-regulated transcripts**			
CV018266	60 s acidic ribosomal protein-like protein	−2.0	−2.4	0.15
DV158570	40S ribosomal protein S8	−2.1	−2.6	0.11
EB683199	60S ribosomal protein L35	−2.1	−2.8	0.03
ARB transgenic plants	**up-regulated transcripts**			
EB438730	Dicer-like 2 protein (DCL901)	3.1	3.3	0.53
EH620111	Pathogenesis-related protein 1B precursor	2.6	3.8	0.24
EH617029	WRKY transcription factor-30	1.0	1.1	0.03
ARB transgenic plants	**down-regulated transcripts**			
CV017513	Chlorophyll a-b binding protein 3A	−5.1	−5.3	0.47
EH620344	F box related protein	−1.2	−1.1	0.09
TMV infected plants	**up-regulated transcripts**			
EH620111	Pathogenesis-related protein 1B precursor	3.5	3.6	0.15
EB643469	60 s Acidic ribosomal protein	4.5	6.0	0.57
TMV infected plants	**down-regulated transcripts**			
EH620344	F box related protein	−1.5	−1.7	0.22
CV017513	Chlorophyll a–b binding protein 3A	−1.8	−1.9	0.22
TA12913_4097	Pollen coat like protein	−3.5	−4.7	0.29

Accumulation of some up- or down-regulated transcripts of the the wild type, BRB- and ARB- transgenic, and of TMV infected wt plants were tested by RT-qPCR, and compared to their accumulation levels observed by the microarray assay. The depicted log values are normalized mean intensive value (n = 3) differences of the wt control plants and the different plant types of TMV-transgenic and TMV-infected wt plants. Statistical significance of the results was tested using Student's t-test (p< 0.05). The Standard error of mean (s.e) is calculated for the C_t values of the RT-qPCR results.

inducible proteins (16), HR-proteins (12), glycine rich proteins (13), osmotin precursors (8), wound responsive proteins (16) and disease resistance proteins (6). Interestingly, also the non-functional allele of the TMV-resistance gene N, 13 SAR- and 16 HR-related transcripts were also enhanced in the BRB plants [59,61], even that no visible signs of HR was seen on the plants. Several transcripts related to various abiotic stresses such as cold acclimation, dehydration, reactive oxygen species such as peroxidases, and to various redox-reactions, as well as transcripts coding for cytochrome p450 and glutathione transferases were also up-regulated in these plants (Table S7). On the other hand, several other transcripts pertaining to heat shock proteins, defense proteins, ATP binding proteins, arabinogalactan proteins, peroxidases and cytochrome p450 were down-regulated in BRB plants (Table S6).

Even more (total of 192) of stress-related genes were up-regulated in the ARB transgenic plants (Table S7), including transcripts coding for heat shock proteins (25), PR-proteins (17), chitinases (12), dehydration response factors (8), different protease inhibitors (10), peroxidases (18), redoxins (11) and various defense proteins (23) such as defensins, thaumatin, thionins and germins. Also, the transcripts of the TMV viral coat protein transcript, which was completely absent in BRB transgenic plants, accumulated to very high levels in the ARB transgenic plants. Various stress-related transcripts (total of 131), e.g. coding for heat shock proteins, wound induced proteins, elicitor inducible proteins, methanol inducible proteins, NDR1-like, cytochrome p450,

thioredoxins and glutathione transferases were down-regulated in the ARB transgenic plants (Table S6). Interestingly, SAR-proteins (6) and HR- (3) related transcripts were down-regulated in the ARB transgenic plants in contrast to their up-regulation in the BRB transgenic plants.

In TMVi plants, only total of 61 of stress-related transcripts were up-regulated (Table S7). These included several transcripts coding for heat shock proteins (12), chitinases (9), pathogenesis related proteins (11) and senescence associated proteins (2). The transcript for the viral coat protein was highly expressed in the TMV infected plants. A total of 40 stress-related transcripts, e.g. coding for heat shock proteins, wound induced proteins, glutathione-S-transferase and peroxidases were down-regulated in the TMVi plants as well (Table S6).

Hormone and development related transcripts

Plant hormones play a major role in various defense signaling pathways, and some viruses interact with these signaling pathways to enhance their infection process [14,17,18]. Our microarray data also indicates differential hormonal regulation between the BRB and ARB transgenic plants, and the TMVi plants. In the BRB transgenic plants, total of 60 transcripts related to hormones and development were up-regulated. Particularly the auxin repressed/ dormancy associated, and auxin- and ethylene responsive transcripts were up-regulated by 10–15 fold (Tables S8 & S9). Also several other hormones and development-related transcripts were either down- or up-regulated (Tables S8 & S9) in these plants,

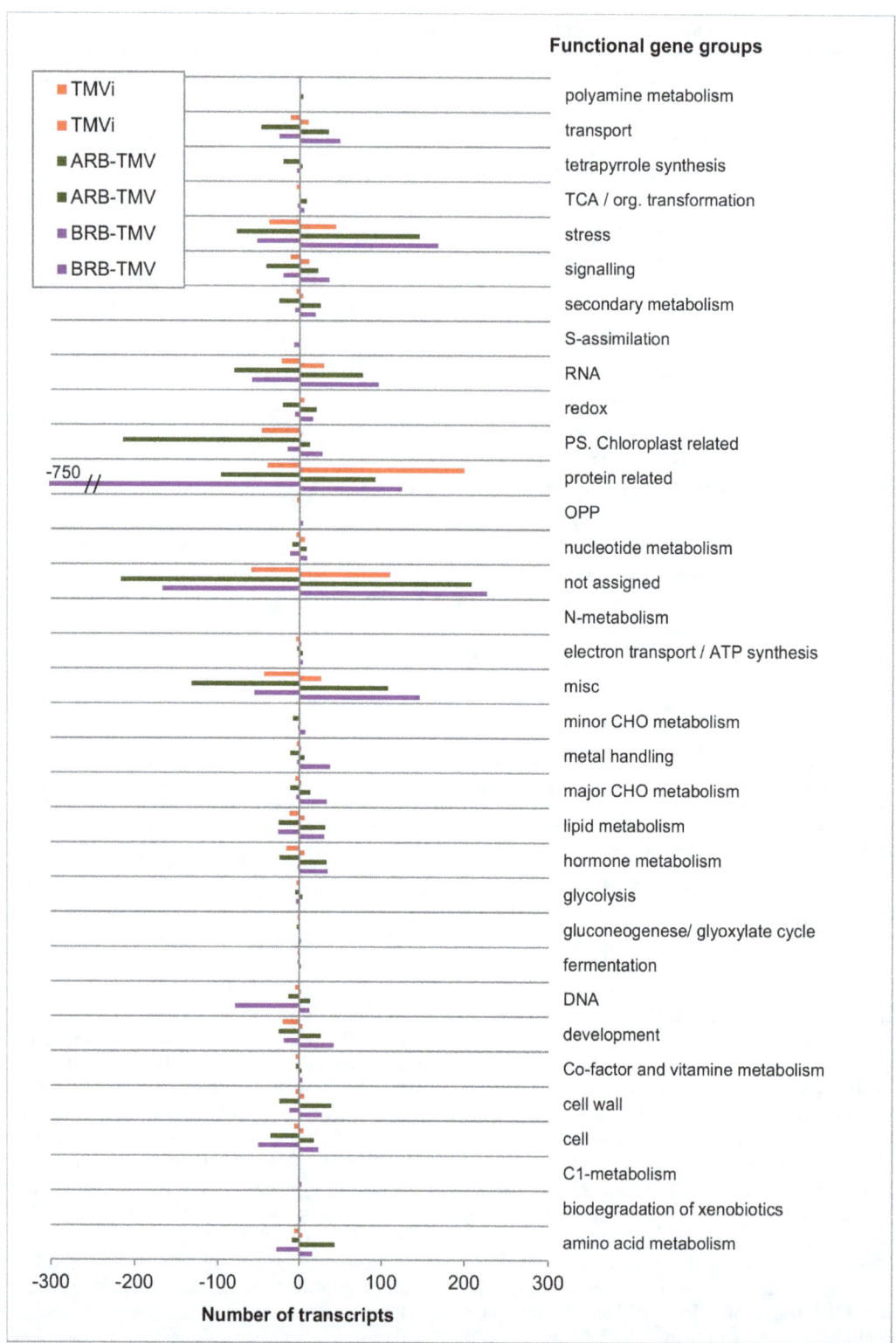

Figure 2. A overview chart showing the distribution in different functional groups of the up- and down-regulated transcripts and their numbers in the BRB- and ARB-TMV transgenic, and in TMVi tobacco plants.

including transcripts related to ethylene, senescence, and abscisic acid synthesis or to various development associated genes.

In ARB plants, a total of 59 transcripts related to defence hormones (ethylene and jasmonic acid) or coding for development-related embryo-specific proteins were up-regulated (Table S9). The jasmonic acid-related genes were induced by up to 26-fold in these plants. Interestingly, 8 transcripts related to auxin repressed/dormancy, auxin associated and SAUR-families were down-regulated, in contrast to their up-regulation in the BRB transgenic plants. Some transcripts related to ethylene synthesis, abscisic acid, gibberellin 20-oxidase, LEA proteins and pentatricopeptide repeat-containing proteins were down- regulated in these plants (Table S8).

In TMVi plants, a total of 33 transcripts related to hormones and development were down-regulated (Table S8), including some transcripts coding for auxin repressed genes, ethylene signal transduction, gibberellin oxidase, jasmonic acid and senescence related genes. A few transcripts related to hormones and development (total of 5) including transcripts coding for ethylene biosynthesis, protodermal factor and pale cress related were up-regulated (Table S9).

Photosynthesis and carbohydrate metabolism related transcripts

Several studies indicate that virus infections or expression of virus-derived genes in transgenic plants reduce the photosynthesis process and cause alterations of the carbohydrate metabolism and translocation in the plants [17,41,43,44,62,63]. Furthermore, some chloroplast proteins (i.e. Rubisco activase, ATP-synthase γ-subunit, and the 33K subunit of the oxygen evolving complex) interact directly with the TMV-encoded replicase protein. These proteins mediate some level of suppression of virus replication, while the virus infection causes some suppression of their expression [51,64]. Interestingly, in the three types of our studied plants, fewer photosynthesis and carbohydrate metabolism-related transcripts were down-regulated in the BRB plants as compared to the ARB and TMVi plants. Only total of 13 transcripts coding for chloroplast proteins, Rubisco activase, NADP-dependent g-3-p dehydrogenase, plastocyanin, ferredoxin and tetrapyrrole synthesis were down-regulated in the BRB plants (Table S10). Similarly, only few photosynthesis-related transcripts (total 28), coding for PSI and II subunits L, O and R, OEC proteins, PGR5-1A and alternative oxidase, or related to chlororespiration and NADPH dehydrogenase complex were up-regulated in these plants (Table S11).

Contrastingly, total of 239 photosynthesis-related genes were down-regulated in the ARB transgenic plants (Table S10). Many of these (total of 227) were related to the photosynthetic machinery, as they coded for the chlorophyll binding and synthesis related proteins (125), or for the subunits of PSI and II, and of the OEC (including the 33 kDa subunit), and for plastocyanin and PGR5-1A. A total of 48 transcripts related to carbon metabolism (i.e. related to starch degradation, sucrose synthesis, Calvin cycle or photorespiration, or coding for electron carriers, carbonic anhydrase or enolase) were up-regulated in ARB plants (Table S11).

Similar to the ARB transgenic plants, photosynthesis related transcripts were predominantly down-regulated in TMVi plants (Table S10). Total of 55 down-regulated transcripts were related to photosynthesis and carbohydrate metabolism (i.e. coding for the chlorophyll a, b binding proteins, components of the PSI and PSII complexes, PGR5-like, carbonic anhydrase, glycolysis, Ribulose biphosphate carboxylase, and for the OEC components, including the 33 kDa protein). A few transcripts (total of 5) related to carbohydrate metabolism (including β-amylase, trehalose-6-phospahte phosphatase and L-lactate dehydrogenase encoding transcripts) were up-regulated in the TMV-infected plants (Table S11).

Cell division, cell organization and DNA binding and repair related transcripts

The gene expression related to cell cycle and cell organization was differentially altered in BRB and ARB transgenic plants and in TMVi plants. Specific to the BRB transgenic plants, 30 and 18 transcripts related to cell cycle and organization were up- and down- regulated, respectively (Table S13 & S12). Specifically, several transcripts of B-type cyclins, peptidyl-prolyl cis-trans isomerases, mitotic spindle check proteins, XKLP2 targeting protein, Knolle protein, and various nucleolus and transport-related proteins were down-regulated, while some transcripts related to cell division and cell organization, including annexins, HIPL2, myosin-13 and tubulins were up-regulated in these plants.

In the ARB transgenic plants total of 36 transcripts related to cell cycle and cell organization were down-regulated (Table S12), including some transcripts coding for cell division control protein 48, tubulins, ankyrin repeat proteins, kinesin and vesicle transport realated proteins. Meanwhile, a total of 17 transcripts coding e.g. for cyclins, cell cycle check point control proteins, motor proteins and cell division inhibited proteins were up-regulated in these plants (Table S13). The cell division check point control protein RAD9A, which is abnormally expressed in several cancer types in animal cells [65,66] was induced by 138× fold, as compared to the wild type controls.

Very few of cell division and organization-related transcripts were altered in the TMVi plants. Only 5 transcripts, coding for the CDC kinase and annexin were down-regulated (Table S12), and 5 transcripts related to cell cycle check point control protein and other miscellaneous cell cycle proteins were up-regulated in these plants (Table S13). Interestingly, the transcript of the cell division check point control protein RAD9A accumulated to 138× fold level also in the TMVi plants, as it did in the ARB transgenic plants.

Similar to the cell cycle and organization gene expression pattern, also the transcripts for histones and DNA repair proteins were affected differently in BRB and ARB transgenic plants and in TMVi plants. In the BRB transgenic plants, a total of 54 histone-encoding transcripts and 28 DNA binding and repair protein-encoding transcripts were down-regulated (Table S12), while 11 transcripts related to DNA repair and binding were up-regulated in these plants (Table S13). In the ARB transgenic plants, only 13 DNA binding protein transcripts were down-regulated, and 16 transcripts coding for various DNA binding proteins were up-regulated (Tables S12 & S13).

In TMVi plants, only four transcripts coding for DNA modifying proteins, i.e. one coding for a transposase, one coding for nuclease and two coding for DNA binding proteins were down-regulated, and only one NAP1-related transcript was up-regulated (Tables S12 & S13).

Gene expression alterations unique for the BRB-TMV plants

To reveal the gene expression alterations that were unique to the resistant stage of the BRB transgenic plants, their RNA expression profile was compared with that of the same transgenic plants at the ARB stage, with TMVi plants and with our previously published transcriptomes of transgenic plants express-ing various viral silencing suppressors, i.e. HC-Pro from *Potato virus Y* [41], AC2 from *African cassava mosaic virus* [44], and P25 from *Potato virus × [43]*, which all are known not to be resistant against TMV (data not shown). In total, 1305 up-regulated transcripts of the BRB transgenic plants were compared against 3453 up-regulated transcripts from the other transgenic and

TMVi plants. This comparison revealed 695 unique and 610 commonly up-regulated transcripts in the TMV-resistant BRB transgenic plants (Figure 3). Many of the uniquely up-regulated transcripts in the BRB transgenic plants were related to stress (98), translation (88), photosynthesis (53), molecular transporters (27), lipids (27), hormones and development (40) and transcription factors (46) (Table S14).

Similarly, comparison of the total of 1462 transcripts that were down-regulated in the BRB transgenic plants against the 2884 down-regulated transcripts of other plants revealed that 1232 of these were unique to the BRB plants (Figure 3). Most of these unique down-regulated transcripts were related to translation (676), but some were related to chromatin (63), cell division (44), stress (48), and RNA processing (31) or were unknown for their function (159) in the BRB transgenic plants (Table S15). The strong suppression of protein synthesis machinery (see Figure S1) is a very special response observed only in this plant material, suggesting that it may be related to the unique virus resistance occurring in these plants. Still, the large number of transcripts that were distinctly altered (either up or down) in the resistance stage of the BRB plants suggests that many other functions may also contribut to the resistance.

Not only transcripts but also total protein levels and profiles were different between BRB, ARB and TMVi plants

To find out how the strong reduction of the proteins synthesis-related transcripts, and of the increase of protein degradation-related transcripts in the BRB-plants affects the total protein content of these plants, their total soluble protein was extracted, quantitated by the Lowry method, and compared to the soluble protein content of healthy control plants of the same age. This revealed that the total soluble protein content of the BRB-TMV transgenic plants was 28% lower that that of the wt plants (Figure 4). From previous literature it is known that the wt TMV infection does not significantly change the total protein content of tobacco leaves, although it significantly changes their protein

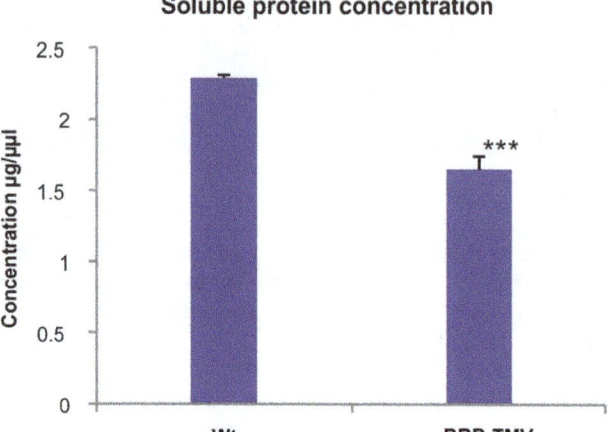

Figure 4. Lowry-quantitation of the soluble protein content of the BRB-TMV plants and of the corresponding wt control plants. Standard error of mean is presented as bars above the columns (consisting of three biological replicates). The confidence level determined by the Student's T-test, with confidence level higher than 95% is indicated by *, higher than 99% indicated by **, and higher than 0,001 indicated by ***.

composition [42,51,62,64,67]. We are not aware of any other virus infection condition where the total protein content would be reduced.

The total protein composition of the three types (BRB, ARB and TMVi) of test plants was analysed by using the 1D and 2D-PAGE, and compared with their corresponding wild type controls. The analyses were repeated with three biological replicates for each plant type. Upon loading of equal amounts of protein samples, the 2D-PAGE analysis showed that the protein profile of the BRB samples contained multiple altered (either enhanced or reduced) bands or spots as compared to the wild type control

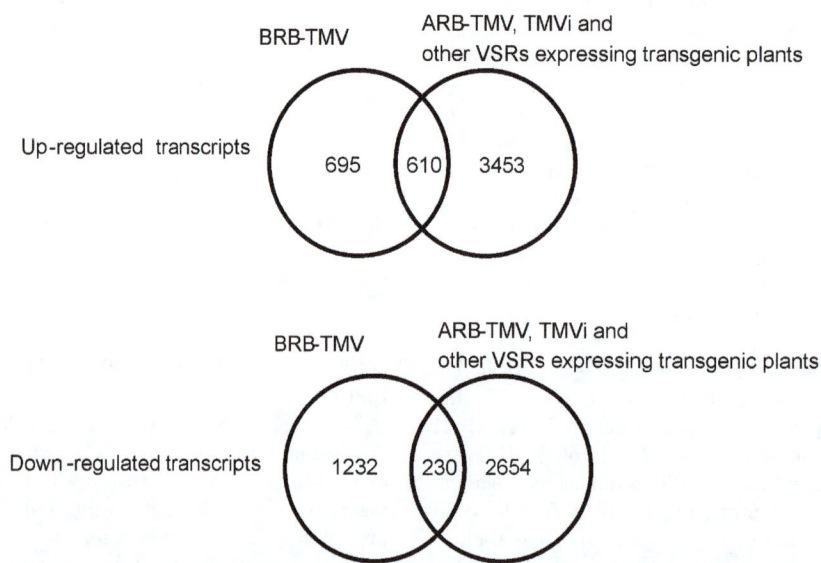

Figure 3. A Venn-diagram showing the numbers of up- and down-regulated transcripts that are either uniquely altered in the BRB-TMV transgenic plants, or in the ARB-TMV transgenic plants or in the TMVi plants, or in the transgenic tobacco lines that express various viral silencing suppressors (as described in the text), and numbers of genes that are detected both in the BRB-TMV plants and in some of the other plants.

plants, but the altered spots were not identified (Figure S3). Also the samples of the ARB transgenic plants and TMV infected plants revealed reduction of multiple spots in the protein profiles compared to the healthy wild type plants, with strong accumulation of the TMV coat protein either as monomer (17,5 kDa) or as a dimer protein (35 kDa) Figure S3.

TMV infection reduces photosynthetic oxygen evolution

Differential photosynthetic gene expression and appearance of the chlorotic TMV symptoms at different stages of plant growth indicated that photosynthetic activity was differentially altered in BRB and ARB transgenic plants and in TMVi plants. To analyze this, we measured their oxygen evolution per ug chlorophyll, and compared this against the oxygen evolution rate of control plants of the same age. The results indicated that oxygen evolution was not changed much in the BRB transgenic plants (Figure 5 A), whereas in ARB and TMVi plants it was strongly reduced (Figure 5 B). Thus, reduction of the photosynthetic activity was correlated to TMV virus accumulation.

Response of the TMV-resistant plants to other viruses

To check whether the observed TMV resistance in the young TMV-transgenic plants was TMV-specific, or active against a broader range of viruses, the plants were inoculated with PVX,

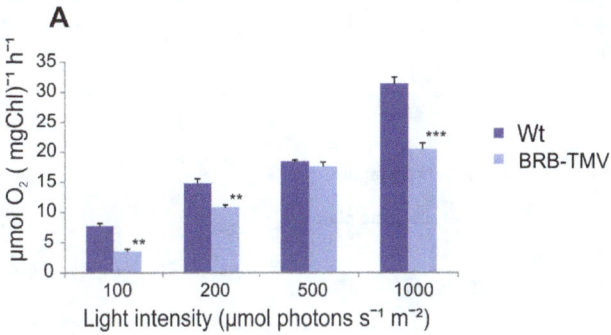

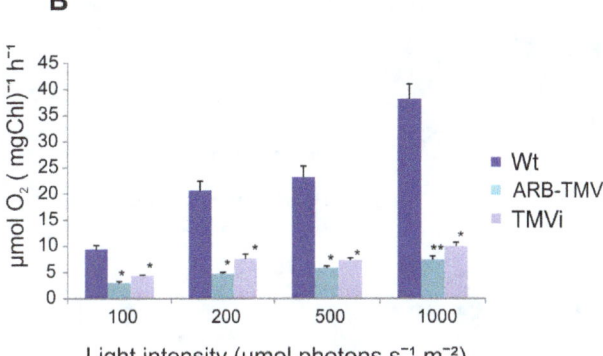

Figure 5. Light-responsive O_2-evolution of photosystem II of the different test plants, as compared to the wild type control plants. BRB- (A), and ARB-TMV transgenic and TMVi plants (B). O_2-evolution was measured of freshly isolated thylakoid membranes using DCBQ as an electron acceptor. Four light intensities is shown in x-axis (μmol photons s^{-1} m^{-2}) and the O_2-evolution in y-axis (μmol O_2 $(mgChl)^{-1}$ h^{-1}). Standard error of mean is presented as bars above the columns (n=4, consisting of five biological replicates). The confidence level determined by the Student's T-test, with confidence level higher than 95% is indicated by *, higher than 99% indicated by **, and higher than 99,9% indicated by ***.

PVY and PVA. ELISA results indicated that these viruses reacted somewhat differently to theTMV- transgenic host. The PVY level was somewhat increased, and PVX level somewhat decreased in the inoculated leaves of the transgenic plants, as compared to the inoculated leaves of the wt plants at 7 days post inoculation (dpi). Any of these viral levels did not differ significantly in the systemically infected leaves of TMV-transgenic and wt plants at 7 dpi, but PVA infection was strongly enhanced in the transgenic plants at 10 dpi (Figure 6). Meanwhile, these second infections also caused the breaking of the TMV-resistance and induced strong accumulation of the TMV in the transgenic plants, prior to the time of typical resistance break (Figure 6 B).

Discussion

Typically, viruses can infect and complete their life cycle in a susceptible host and also easily spread systemically in such a host plant. This kind of compatible interaction involves effective use of necessary host factors and suppression of the host defense mechanisms [20,50,59]. In natural infections TMV has a very compatible interaction with its tobacco host. In inoculated mature tobacco leaves it typically spreads and accumulates evenly in all cells, while in systemically infected young leaves it spreads unevenly, causing mosaic where some tissues become fully infected, while others remain virus-free. These dark-green islands (DGI) are protected against the spreading infection by RNA-silencing, which becomes activated at the marginal regions of the initial infection foci [52].

Our transgenic tobacco plants represent an artificial inoculation system, where the infectious TMV genome is expressed under the constitutive 35S promoter, supposedly already in all cells of the germinating embryo. This leads to a very strong resistance condition that prevails in the plants until they reach a certain stage of maturity (about seven weeks from germination). This resistance was mostly specific to its endogenous inducer virus, although it provided low level transient protection also against two exogenously inoculated viruses (PVY and PVA), maybe due to the multiple defence genes that are activated in these tissues (Figure 6). However, as these viruses infected the plants, also the TMV-resistance was broken, indicating that the resistance mechanism was not durable under the stress factors, or was compromised by the silencing suppressors produced by the second viruses.

This suggests that the resistance may be related to the TMV-specific RNA silencing, and similar in nature to the resistance surrounding the DGIs in the young systemically infected leaves. Still, this was not supported by our earlier results, which showed that the TMV-derived transgene was not silenced by methylation, and that virus-specific siRNAs were not detected in the resistant tissues. The methylation level increased and the siRNAs became detectable only after the resistance break, indicating that the RNA silencing became activated at this stage [53].

The resistance was very strong in the young transgenic plants, and the viral RNA level remained very low: it was not detectable by Northern blotting [53] or by the microarray hybridization, but was barely detectable by qRT-PCR analysis, being reduced by a factor of 2^{10} as compared to normal productive infection (data not shown). No viral CP was detected during this period in the plants. Still, the transgene expression, and the resistance reaction against the expressed infectious viral RNA caused a severe stress condition of the plants, manifested by the strong reduction of the plant growth (Figure 1).

When the resistance was broken, at about 1.5–2 months after germination, the plants became fully infected from the endogenous inoculum. The first sign of the resistance break was the

A

(i)

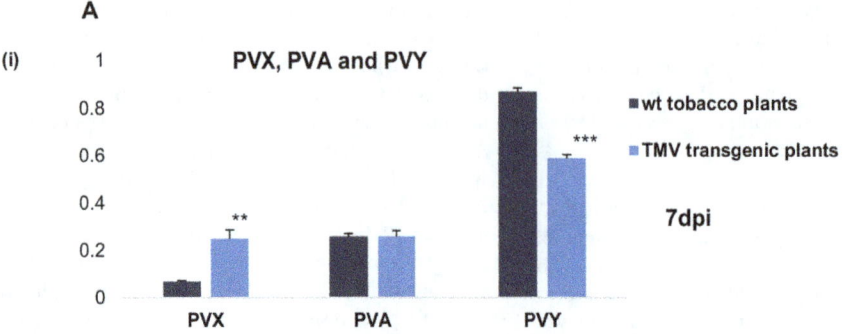

(ii)

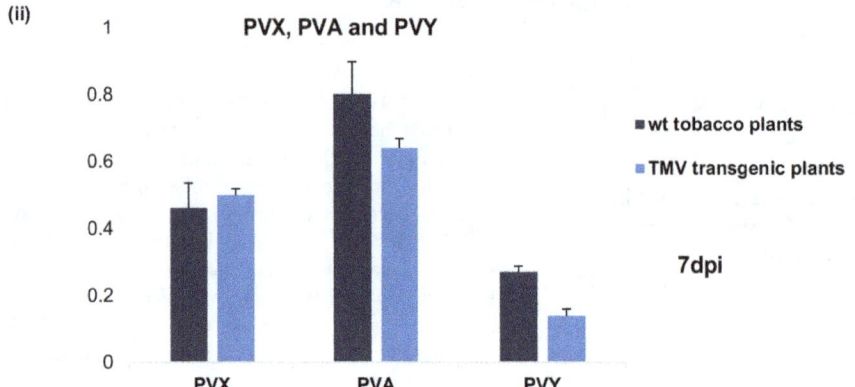

(iii)

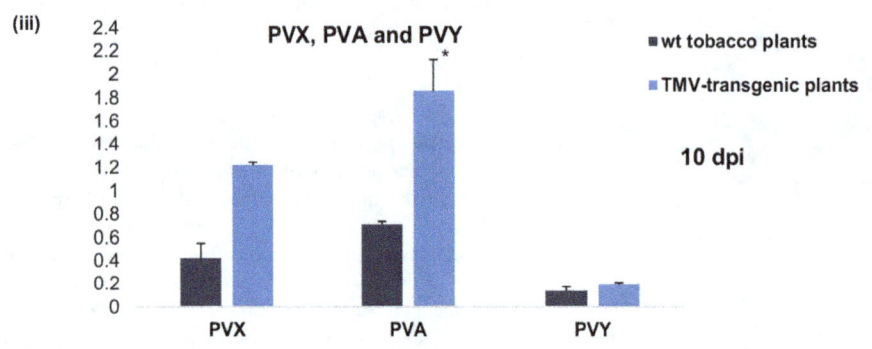

B

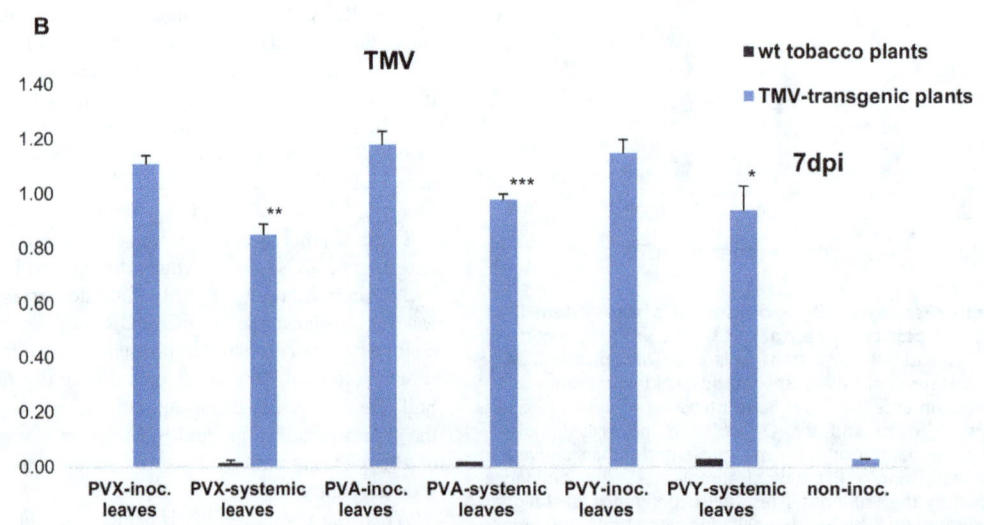

Figure 6. ELISA-mediated detection of the exogenously inoculated PVX, PVY and PVA viruses, and of the endogenously infecting TMV, from the BRB-TMV transgenic plants. A. Detection of PVX, PVY and PVA the from the inoculated leaves (i), and form the systemic leaves (ii) at 7 dpi (ii), and from the systemic leaves at 10 dpi (iii), and (B) detection of the TMV from the wt tobacco plants, and from the TMV-transgenic plants at 7 days after inoculation either with PVX, PVA and PVY. Standard error of mean is presented as bars above the columns (consisting of three biological replicates). The confidence level determined by the Student's T-test, with confidence level higher than 95% is indicated by *, higher than 99% indicated by **, and higher than 99,9% indicated by ***.

appearance of typical patchy mosaic symptoms in the small apical leaves. Also, the viral coat protein was first detected by immunolabeling as small patches in the apical leaf initials, and in the vascular tissue, but not in the apical meristems (Figure 1B). This patchy pattern of the resistance break particularly in the uppermost leaves suggests that it was related to the changing developmental and maturity level of the plants.

Similar, but not identical, virus resistance has been reported in transgenic *N. benthamiana* plants that express the full length genome of *Plum pox potyvirus* (PPV) [68]. In that case the virus resistance occurred only in some of the PPV-expressing transgenic lines, and apparently several different mechanisms contributed to its induction, including RNA silencing.

To identify physiological and molecular processes that are associated with the resistance status of the young plants, we investigated the transcriptome of the transgenic plants before and after resistance break. The microarray analysis revealed multiple gene expression alterations in these plants, and several of them - for example the strong suppression of the translational machinery, enhanced biotic stress responses including the activation of the SAR and HR pathways, hormonal changes or cell division alterations - may all contribute to the resistance.

Protein translation machinery was strongly suppressed in the BRB transgenic plants

An interesting feature in the transcript profile was the strong down-regulation of the multiple (more than 700) transcripts coding for different components of the translation machinery. The strong reduction of the 40S and 60S ribosomal RNAs, and of other ribosomal genes in the BRB plants, compared to their strong up-regulation in the TMV infected wt plants, and also to their normal expression in the ARB transgenic plants, suggest that the availability of host translational machinery is actively restricted in the BRB transgenic plants. This may directly suppress the accumulation of viral proteins. Furthermore, the reduction of the eIF3 and eEF1A and eEF1B translation initiation and elongation factors, which are known to be needed for the TMV-specific replicase complex [1,15,16,19,30,49,69–71], may, to some extent, directly suppress the TMV replication.

Many viruses modify the host translational machinery to increase the viral protein synthesis but not host protein synthesis [55,72–74]. In the case of TMV, the viral genomic and coat protein RNAs are stronger translational templates than host mRNAs [55]. TMV genomic RNA has a long 5′-leader sequence, so called omega sequence, which promotes its translation by efficiently recruiting the 40S and 60S ribosomal subunits to form the 80S-preinitiation complex [74,75]. The 5′-leader sequence also interacts with the heat shock protein 101 to recruit eIF4F [76]. Thus, TMV should strongly compete for the host cell's transla-tional capacity, even under the situation where the total translation machinery is reduced. Thus the significance of this response, in terms of the induced viral resistance, is not clear.

Many protein degradation pathways and several proteases were induced at the resistant stage but reduced in the TMVi plants, indicating that the BRB transgenic plants may promote the resistance also through enhanced turnover of the viral proteins.

The reduced protein synthesis, and enhanced protein degradation were reflected in the reduced soluble protein content, and also in the altered protein profile of the BRB plants.

TMV-resistant BRB transgenic plants exhibited less photosynthetic damage and higher defense responses than did the ARB transgenic and TMVi plants

Chloroplasts are the main center for many important metabolic functions, and many biotic/abiotic stresses, including virus infections, influence their environment [63]. For instance, TMV infections affects the composition of chloroplast proteins, and efficient TMV accumulation depends on the silencing of the 33k subunit of the OEC and ferridoxin I proteins, which are involved in defense against TMV [51,62,64,77]. Up-regulation of OEC complex proteins in the BRB-TMV plants, and their down-regulation in the the ARB plants may thus relate to the induced defense condition in the BRB-TMV plants.

In the TMVi plants the transcriptome was altered much less than in the transgenic BRB and ARB plants. Interestingly, while induction of defense-related genes is not typical to TMVi plants due to the compatible interaction, this reaction was quite opposite in the BRB and ARB transgenic plants. Induced expression of SAR- and HR-associated proteins in the BRB transgenic plants, and their down-regulation after resistance break or virus infection (ARB and TMVi plants) indicates the incompatible host-virus interactions and induction of the active resistance pathways [61] in the BRB plants. This was also indicated by the expression of the non-functional allele of the N-resistance gene and of other R-genes, observed in the BRB transgenic plants. All these activated defence-related genes and pathways are likely to contribute to the strong TMV resistance condition in the BRB plants.

Cell division

Several animal studies have revealed that viruses can hijack the host cell division machinery to control the anti-cancer mechanisms [78] and thus provide a suitable environment for virus replication process. One of the cell division check point control protein RAD9A is expressed on very high level (~140 x) in our ARB transgenic and TMVi plants but not in the BRB transgenic plants. RAD9A protein is known for its high expression levels during high DNA damage conditions. Several cancer studies indicate that cells accumulate RAD9A protein during carcinogenesis and also the RAD9A down-regulation by siRNA reduces the tumorogenesis in the cells [65,66]. It is not clear how the RAD9A protein increases tumorogenesis, but one hypothesis is that it may induce the expression of adjacent carcinogenic genes. Abnormally high expression of RAD94 in the ARB transgenic and TMVi plants may indicate that it plays some role in the TMV infection process or cell's stress reaction under the virus infection.

Many cyclins (A and B type) were strongly down-regulated in the BRB transgenic plants. Cyclins are involved in the cell cycle regulation by the cyclin dependent protein kinases phosphoryla-tion process (CDKs) [79,80], and their down-regulation may indicate that the cell cycle is stalled during G2 and M phase in the BRB transgenic plants, leading to their stunted phenotype.

Conclusions

The microarray analysis reveals that the expression of the infectious TMV genome in the germinating and developing tobacco tissue induces strong alterations in an unique pool of transcripts, many of which may contribute to the TMV resistance status of the plants. One important factor in this condition appears to be the strong enhancement of the SAR- and HR-type defence pathways. Another interesting response is the suppression of the translational machinery, which is a totally unique reaction observed in these plants. This may slow down the synthesis of viral proteins, and also deprive the cells of different host factors (e.g. translation factors eIF-3 and eEF1A and B) which are needed for TMV replication, but the role of these reactions in terms of the virus resistance remains unclear. How all these responses are induced and mutually integrated in the young transgenic plants remains to be solved.

Methods

Plant material

The wild type tobacco (*Nicotiana tabacum* cv. xanthi nn) and transgenic tobacco plants, which express the whole TMV genome (strain U1) were grown in greenhouse at 60% relative humidity and in 22°C temperature, with a light/night regime of 16 h light (150 μmol photons $m^{-2}s^{-1}$) and 8 h dark. Three replicate samples were collected from the selected transgenic tobacco line, from three BRB plants at 6 weeks, and from the same ARB plants at 8 weeks after germination. At the same time, corresponding sets of the control samples were collected from wild type (wt) plants, which were grown in the same conditions and of the same age as the test plants at the time of sampling. During this collection period all plants remained in the vegetative growth stage. Always the third leaf from the apex (about half of the mature leaf size) was collected, and the collection took place always at the same time of the day (11 am). Wt tobacco plants were mechanically inoculated with TMV at 8 weeks after germination, and samples of systemically infected leaves were collected from these plants at 8 days after inoculation, in parallel with corresponding control samples from healthy wt plants. All the leaf samples were directly frozen in liquid nitrogen and used for RNA extraction. Positive expression of transgene (TMV) RNA was detected in the leaf samples by qRT-PCR prior to the array analysis; TMV RNA was not detectable in these samples by Northern blotting [53] but was detectable on very low level by qRT-PCR. The first set of protein extractions were done from the same samples what were used for the RNA extraction, and additional protein samples were later collected from separate sets of plants of the same line, at the same growth stage and growth conditions.

RNA extraction, cDNA labeling and microarray hybridization

Total RNA was extracted from healthy wild type (controls), BRB and ARB transgenic, and TMVi wild type tobacco plants by using TRIsure reagent (Bio line, UK) according to manufacturer's instructions. The extracted total RNA was purified with the RNA purification kit (Nucleospin RNA clean-up, Macherey-Nagel) and then subjected to DNaseI treatment (Promega RQ1 RNase free-DNaseI) according to the manufacturer's recommendations. Subsequently, the total RNA was concentrated with Amicon Ultra-0.5 centrifugal filter devices. The cDNA labeling, and Agilent 4×44k microarray hybridization were done according to the manufactures instructions (Center for Biotechnology, Turku, Finland) and the raw numerical data handling and its statistical analysis were done by using Chipster (CSC, Finland) program [81] as previously described [41].

Annotation of differentially regulated genes in microarray data

The probe information provided by the manufacturer for the Agilent 4×44k microarray was limited and mainly based on EST and cDNA sequences. Therefore, most of this descriptive annotation information was obtained from the http://mapman. gabipd.org/web/guest/mapman-annotationexperts website, with additional information obtained from the websites like JCVI http://plantta.jcvi.org/ and BLAST http://blast.ncbi.nlm.nih. gov/Blast.cgi. The functional grouping of probes was also attained from the mapman.gabipd.org website, with some additional manual adjustments.

Verification of differentially expressed genes

The microarray results were verified by using the quantitative real-time PCR (RT-qPCR) method by following the MIQE guidelines [82]. A total of 1 μg purified leaf total RNA was used to make cDNA synthesis by using the Revert Aid reverse transcriptase enzyme (product # EPO441, Fermentas) according to the manufacture instructions. The RT-qPCR reactions were performed using 10 ng (3 μl) of diluted cDNA samples (1:15), gene specific primers (Table S16) and Maxima SYBR Green/Fluorescein RT-qPCR Master Mix (2X) (Product #K0242, Fermentas) with a total volume of 25 μl. For each biological replicate, 3–4 technical replicates were run to reduce the pipetting errors. The standard error of mean was measured from the three biological replicates. The Bio-Rad iQ5 machine was used to perform the RT-qPCR in 96–well plate and the results were calculated by employing the quantification cycle (Cq) method (delta delta Cq). Primer specificity was tested by checking the single peak in the DNA melting curves.

In situ labeling of the TMV coat protein

Sample sections (7 μm) were prepared from the shoot apical domains of BRB/ARB TMV transgenic plants at 7 weeks after germination, i.e. just at the time of resistance break, by excising and immediate fixing as described previously [83]. The sample sections were initially incubated in phosphate buffered saline (PBS) containing 4% bovine serum albumin at room temperature for 30 min. Later, the samples were subjected to incubation by TMV-specific alkaline phosphatase–conjugated polyclonal antibodies (dilution 1:50) at 4°C for overnight. Next morning, the samples were washed and stained with freshly prepared fuchsin substrate solution and examined with a Leitz, Laborlux S light microscope (Leica Microsystems AG) at 40× and 100× magnifications.

Photosynthetic Measurements

Equal amount of leaf samples (1.0 g) from the healthy control, BRB and ARB transgenic plants, and from TMVi tobacco plants were taken and ground in 4 ml of thylakoid isolation buffer (0.3 M sorbitol, 50 mM HEPES/KOH pH 7.4, 5 mM $MgCl_2$, 1 mM EDTA, and 1% BSA) with ice cold mortar and pestle. The ground mixture was filtered through the Miracloth and 2 ml of the filtrate was taken and centrifuged at 12000×g for 2 minutes. The chlorophyll concentration measurements were done according to the procedure stated in [84]. The supernatant was removed and the thylakoid pellet was resuspended in oxygen electrode buffer (0.3 M sorbitol, 50 mM HEPES/KOH pH 7.4, 5 mM $MgCl_2$ and 1 mM KH_2PO_4). The oxygen evolution measurements were carried out by a Clark type electrode by using 0.5 mM DCBQ as electron donor.

Protein Isolation and 2D gel electrophoresis

Protein extraction for the measurement of the total soluble protein content from the BRB plants and healthy control plants was done from by 0.5 g of leaf tissue, ground inthe ice cold oxygen electrode buffer (0.3 M sorbitol, 50 mM HEPES/KOH pH 7.4, 5 mM $MgCl_2$ and 1 mM KH_2PO_4). Subsequently, supernantent was isolated and used for protein quantity measurement by using lowry assay.

Protein samples for the electrophoresis analysis from leaves of the healthy control, BRB and ARB transgenic plants and of the TMVi plants were isolated by using TRIsure reagent (Bio line, U.K) according to the manufacture recommendations with some adaptive steps from TRIzol protocol (Invitrogen Inc. USA). The concentrations of isolated protein samples were measured by using Lowry assay. A total of 20 μg protein samples were loaded and run on 1D SDS-page electrophoresis to verify the equal loading of the protein samples. 250 μg of protein samples were taken and mixed with rehydration buffer (8 M urea, 4% CHAPS, 2 M thiourea, 20 mM Tris-HCl, 0.05% bromophenol blue, 100 mM DTT and 5 μl/ml of Bio-lyte ampholyte solution). The protein samples were incubated for 2 hours at room temperature and separated by isoelectric focusing using Bio-Rad 7 cm IPG pH 3–10 strips. Subsequently, the IPG strips were subjected to the second dimension separation with the 15% PAGE gels by using protein II apparatus (Bio-Rad). The protein gels were fixed with isopropanol and acetic acid treatments for 15 minutes each and incubated overnight with coomassie blue stain (Page Blue staining kit, Fermentas). In the next morning, gels were destained and photographed. Optionally, the gels were stained again with silver staining kit (Page silver staining kit, Fermentas) according to manufacturer's instructions to analyze even the low abundance proteins.

ELISA analysis of the virus titers

Different viruses used in this study (PVY, PVX, PVA and TMV) were detected by using double antibody sandwich enzyme-linked immunosorbent assay (DAS-ELISA) according to the manufacturers guidelines (Bioreba, Reinach, Switzerland) with slightly modified protocol. The commercial polyclonal, alkaline phosphatases conjugated antibodies (Bioreba, Reinach, Switzerland) against all viruses were diluted to 1:10000 for use. The ELISA reactions were developed by using the p-Nitrophenyl phosphate as a substrate, and measured at 405 nm absorbance by using the ELISA plate reader (Benchmark, Bio-Rad, Hercules, CA, U.S.A.). 100 ng of purified virions of the corresponding viruses were used as internal standards.

Supporting Information

Figure S1 Graphic presentation of the altered protein synthesis-related transcripts (log2 value>1), as detected and portrayed by the MapMan software from the microarray data of the BRB-, ARB-TMV transgenic and TMVi plants. The blue- and red squares and bars represent the up- and down-regulated transcripts, respectively, and the picture frame depicts their functional location in the cell. The boxes portrayed inside the nuclear circles indicate the alterations in the transcription- and mRNA processing-related transcripts, respectively. Alterations of the translation-related transcripts are portrayed in the cytoplasm, separately for the plastidic, mitochondrial and cytoplasmic ribosomes.

Figure S2 Graphic presentation of the altered stress-related transcripts (log2 value>1) as detected and portrayed by using the MapMan software from the microarray data of the BRB- and ARB-TMV transgenic and TMVi plants. The blue and red squares represent the up- and down-regulated transcripts, respectively, that are involved in the different signaling pathways and in different stress responces.

Figure S3 1D-SDS-PAGE gels (left panel), and 2D-polyacrylamide gel electrophoresis (2D-PAGE) (right panel) showing, respectively, the equal loading of samples and the levels of various individual proteins. Upper two panels show BRB-TMV transgenic plants protein samples analysis at before resistance break stage, the middle two panels show the ARB-TMV transgenic plants protein samples analysis at after resistance break stage, and the bottom two panels show samples from the TMV-infected wild type tobacco plants (TMVi). All gels were stained with coomassie blue. Molecular weight ladder (Thermo Scientific) contains markers for the 250, 130, 100, 70, 55, 35, 25, 15, and 10 kDa proteins – the most important ones are marked in the panels. The TMV CP is indicated with the * in the ARB and TMVi gels.

Table S1 Normalized microarray data showing both down- (sheet1) and up-regulated (sheet 2) transcripts in the leaves of BRB-TMV transgenic tobacco plants.

Table S2 Normalized microarray data showing both down- (sheet1) and up-regulated (sheet 2) transcripts in the leaves of ARB-TMV transgenic tobacco plants.

Table S3 Normalized microarray data showing both down- (sheet1) and up-regulated (sheet 2) transcripts in the leaves of TMVi wild type tobacco plants.

Table S4 Down-regulated transcripts related to the protein synthesis, degradation and amino acid metabolism in the leaves of BRB-, ARB- transgenic and TMVi plants.

Table S5 Protein synthesis, degradation and amino acid metabolism related up-regulated transcripts detected in the leaves of BRB-, ARB- transgenic and TMVi plants.

Table S6 Biotic stress related down-regulated transcripts detected in the leaves of BRB-, ARB- transgenic and TMVi plants.

Table S7 Biotic stress related up-regulated transcripts detected in the leaves of BRB-, ARB- transgenic and TMVi plants.

Table S8 Hormones and development related down-regulated detected in the leaves of BRB-, ARB- transgenic and TMVi plants.

Table S9 Hormones and development related up-regulated detected in the leaves of BRB-, ARB- transgenic and TMVi plants.

Table S10 Photosynthesis and carbohydrate metabolism related down-regulated transcripts detected in the leaves of BRB-, ARB- transgenic and TMVi plants.

Table S11 Photosynthesis and carbohydrate metabolism related up-regulated transcripts detected in the leaves of BRB-, ARB- transgenic and TMVi plants.

Table S12 Cell division and DNA-binding related down-regulated transcripts detected in the leaves of BRB-, ARB- transgenic and TMVi plants.

Table S13 Cell division and DNA-binding related up-regulated transcripts detected in the leaves of BRB-, ARB- transgenic and TMVi plants.

Table S14 Different up-regulated transcripts in the BRB-TMV transgenic plants after subtracting the up-regulated transcripts from other ARB-TMV, TMVi and different VRS expressing transgenic tobacco plants (HcPro, AC2 and P25).

Table S15 Different down-regulated transcripts in the BRB-TMV transgenic plants after subtracting the up-regulated transcripts from other ARB-TMV, TMVi and different VRS expressing transgenic tobacco plants (HcPro, AC2 and P25).

Table S16 Primers used in RT-qPCR experiment for validation of microarray data.

Acknowledgments

The Finnish Microarray and sequencing center (FMSC) at Turku Center for Biotechnology is acknowledged for labeling the cDNAs, hybridizations, scanning the chips and producing raw microarray data. The Center of Scientific Calculating (CSC, Espoo, Finland) is acknowledged for Chipster program.

Author Contributions

Conceived and designed the experiments: BJ AS SS KL. Performed the experiments: BJ AS GM SS. Analyzed the data: BJ AS GM SS E-MA TS KL. Contributed reagents/materials/analysis tools: BJ AS GM SS E-MA TS KL. Contributed to the writing of the manuscript: BJ AS GM SS E-MA TS KL.

References

1. Ahlquist P, Noueiry AO, Lee WM, Kushner DB, Dye BT (2003) Host factors in positive-strand RNA virus genome replication. J Virol 77: 8181–8186.
2. Barends S, Bink HH, van den Worm SH, Pleij CW, Kraal B (2003) Entrapping ribosomes for viral translation: tRNA mimicry as a molecular Trojan horse. Cell 112: 123–129.
3. Harries PA, Park JW, Sasaki N, Ballard KD, Maule AJ, et al. (2009) Differing requirements for actin and myosin by plant viruses for sustained intercellular movement. Proc Natl Acad Sci U S A 106: 17594–17599.
4. Harries PA, Schoelz JE, Nelson RS (2010) Intracellular transport of viruses and their components: utilizing the cytoskeleton and membrane highways. Mol Plant Microbe Interact 23: 1381–1393.
5. Whitham SA, Wang Y (2004) Roles for host factors in plant viral pathogenicity. Curr Opin Plant Biol 7: 365–371.
6. Wright KM, Wood NT, Roberts AG, Chapman S, Boevink P, et al. (2007) Targeting of TMV movement protein to plasmodesmata requires the actin/ER network: evidence from FRAP. Traffic 8: 21–31.
7. Ahlquist P (2006) Parallels among positive-strand RNA viruses, reverse-transcribing viruses and double-stranded RNA viruses. Nat Rev Microbiol 4: 371–382.
8. Schwartz M, Chen J, Lee WM, Janda M, Ahlquist P (2004) Alternate, virus-induced membrane rearrangements support positive-strand RNA virus genome replication. Proc Natl Acad Sci U S A 101: 11263–11268.
9. Hatta T, Bullivant S, Matthews RE (1973) Fine structure of vesicles induced in chloroplasts of Chinese cabbage leaves by infection with turnip yellow mosaic virus. J Gen Virol 20: 37–50.
10. McCartney AW, Greenwood JS, Fabian MR, White KA, Mullen RT (2005) Localization of the tomato bushy stunt virus replication protein p33 reveals a peroxisome-to-endoplasmic reticulum sorting pathway. Plant Cell 17: 3513–3531.
11. Leonard S, Plante D, Wittmann S, Daigneault N, Fortin MG, et al. (2000) Complex formation between potyvirus VPg and translation eukaryotic initiation factor 4E correlates with virus infectivity. J Virol 74: 7730–7737.
12. Duprat A, Caranta C, Revers F, Menand B, Browning KS, et al. (2002) The Arabidopsis eukaryotic initiation factor (iso)4E is dispensable for plant growth but required for susceptibility to potyviruses. Plant J 32: 927–934.
13. Zhu S, Gao F, Cao X, Chen M, Ye G, et al. (2005) The rice dwarf virus P2 protein interacts with ent-kaurene oxidases in vivo, leading to reduced biosynthesis of gibberellins and rice dwarf symptoms. Plant Physiol 139: 1935–1945.
14. Padmanabhan MS, Kramer SR, Wang X, Culver JN (2008) Tobacco mosaic virus replicase-auxin/indole acetic acid protein interactions: reprogramming the auxin response pathway to enhance virus infection. J Virol 82: 2477–2485.
15. Li Z, Pogany J, Tupman S, Esposito AM, Kinzy TG, et al. (2010) Translation elongation factor 1A facilitates the assembly of the tombusvirus replicase and stimulates minus-strand synthesis. PLoS Pathog 6: e1001175.
16. Osman TA, Buck KW (1997) The tobacco mosaic virus RNA polymerase complex contains a plant protein related to the RNA-binding subunit of yeast eIF-3. J Virol 71: 6075–6082.
17. Culver JN, Padmanabhan MS (2007) Virus-induced disease: altering host physiology one interaction at a time. Annu Rev Phytopathol, 45: 221–243.
18. Bari R, Jones JD (2009) Role of plant hormones in plant defence responses. Plant Mol Biol 69: 473–488.
19. Hwang J, Oh CS, Kang BC (2013) Translation elongation factor 1B (eEF1B) is an essential host factor for Tobacco mosaic virus infection in plants. Virology 439: 105–114.
20. Lodha TD, Basak J (2012) Plant-pathogen interactions: what microarray tells about it? Mol Biotechnol 50: 87–97.
21. Whitham SA, Yang C, Goodin MM (2006) Global impact: elucidating plant responses to viral infection. Mol Plant Microbe Interact 19: 1207–1215.
22. Song L, Gao S, Jiang W, Chen S, Liu Y, et al. (2011) Silencing suppressors: viral weapons for countering host cell defenses. Protein Cell 2: 273–281.
23. Soosaar JL, Burch-Smith TM, Dinesh-Kumar S (2005) Mechanisms of plant resistance to viruses. Nat Rev Microbiol 3: 789–798.
24. Lukasik E, Takken FL (2009) STANDing strong, resistance proteins instigators of plant defence. Curr Opin Plant Biol 12: 427–436.
25. Coll NS, Epple P, Dangl JL (2011) Programmed cell death in the plant immune system. Cell Death Differ 18: 1247–1256.
26. Ding SW (2001) RNA-based antiviral immunity. Nat Rev Immunol 2010, 10(9): 632–644.
27. Vance V, Vaucheret H: RNA silencing in plants—defense and counterdefense. Science 292: 2277–2280.
28. Voinnet O (2005) Induction and suppression of RNA silencing: insights from viral infections. Nat Rev Genet 6: 206–220.
29. Wu Q, Wang X, Ding SW (2010) Viral suppressors of RNA-based viral immunity: host targets. Cell Host Microbe 8: 12–15.
30. Rodrigo G, Carrera J, Jaramillo A, Elena SF (2011) Optimal viral strategies for bypassing RNA silencing. J R Soc Interface 8: 257–68.
31. Burgyan J, Havelda Z (2011) Viral suppressors of RNA silencing. Trends Plant Sci 16: 265–72.
32. Brigneti G, Voinnet O, Li WX, Ji LH, Ding SW, et al. (1998) Viral pathogenicity determinants are suppressors of transgene silencing in Nicotiana benthamiana. EMBO J 17: 6739–6746.
33. Voinnet O, Pinto YM, Baulcombe DC (1999) Suppression of gene silencing: a general strategy used by diverse DNA and RNA viruses of plants. Proc Natl Acad Sci U S A 96: 14147–14152.
34. Voinnet O (2001) RNA silencing as a plant immune system against viruses. Trends Genet 17: 449–459.
35. Dunoyer P, Lecellier CH, Parizotto EA, Himber C, Voinnet O (2004) Probing the microRNA and small interfering RNA pathways with virus-encoded suppressors of RNA silencing. Plant Cell 16: 1235–1250.
36. Kasschau KD, Xie Z, Allen E, Llave C, Chapman EJ, et al. (2003) P1/HC-Pro, a viral suppressor of RNA silencing, interferes with Arabidopsis development and miRNA unction. Dev Cell 4: 205–217.
37. Pfeffer S, Dunoyer P, Heim F, Richards KE, Jonard G, et al. (2002) P0 of beet Western yellows virus is a suppressor of posttranscriptional gene silencing. J Virol 76: 6815–6824.

38. Thomas CL, Leh V, Lederer C, Maule AJ (2003) Turnip crinkle virus coat protein mediates suppression of RNA silencing in Nicotiana benthamiana. Virology 306: 33–41.

39. Voinnet O, Rivas S, Mestre P, Baulcombe D (2003) An enhanced transient expression system in plants based on suppression of gene silencing by the p19 protein of tomato bushy stunt virus. Plant J 33: 949–956.

40. Laliberte JF, Sanfacon H (2010) Cellular remodeling during plant virus infection. Annu Rev Phytopathol 48: 69–91.

41. Soitamo AJ, Jada B, Lehto K (2011) HC-Pro silencing suppressor significantly alters the gene expression profile in tobacco leaves and flowers. BMC Plant Biol 11: 68.

42. Golem S, Culver JN (2003) Tobacco mosaic virus induced alterations in the gene expression profile of Arabidopsis thaliana. Mol Plant Microbe Interact 16: 681–688.

43. Jada B, Soitamo AJ, Lehto K (2013) Organ-specific alterations in tobacco transcriptome caused by the PVX-derived P25 silencing suppressor transgene. BMC Plant Biol 13: 8.

44. Soitamo AJ, Jada B, Lehto (2012) Expression of geminiviral AC2 RNA silencing suppressor changes sugar and jasmonate responsive gene expression in transgenic tobacco plants. BMC Plant Biol 12: 204.

45. Postnikova OA, Nemchinov LG (2012) Comparative analysis of microarray data in Arabidopsis transcriptome during compatible interactions with plant viruses. Virol J 9: 101.

46. Dardick C (2007) Comparative expression profiling of Nicotiana benthamiana leaves systemically infected with three fruit tree viruses. Mol Plant Microbe Interact 20: 1004–1017.

47. Whitham SA, Quan S, Chang HS, Cooper B, Estes B, et al. (2003) Diverse RNA viruses elicit the expression of common sets of genes in susceptible Arabidopsis thaliana plants. Plant J 33: 271–283.

48. Ishibashi K, Nishikiori M, Ishikawa M (2010) Interactions between tobamovirus replication proteins and cellular factors: their impacts on virus multiplication. Mol Plant Microbe Interact 23: 1413–1419.

49. Liu C, Nelson RS (2013) The cell biology of Tobacco mosaic virus replication and movement. Front Plant Sci 4: 12.

50. Wang X, Goregaoker SP, Culver JN (2009) Interaction of the Tobacco mosaic virus replicase protein with a NAC domain transcription factor is associated with the suppression of systemic host defenses. J Virol 83: 9720–9730.

51. Bhat S, Folimonova SY, Cole AB, Ballard KD, Lei Z, et al. (2012) Influence of host chloroplast proteins on Tobacco mosaic virus accumulation and intercellular movement. Plant Physiol 161: 134–147.

52. Hirai K, Kubota K, Mochizuki T, Tsuda S, Meshi T (2008) Antiviral RNA silencing is restricted to the marginal region of the dark green tissue in the mosaic leaves of tomato mosaic virus-infected tobacco plants. J Virol 82: 3250–3260.

53. Siddiqui SA, Sarmiento C, Valkonen S, Truve E, Lehto K (2007) Suppression of infectious TMV genomes in young transgenic tobacco plants. Mol Plant Microbe Interact 20: 1489–1494.

54. Siddiqui SA, Sarmiento C, Truve E, Lehto H, Lehto K (2008) Phenotypes and functional effects caused by various viral RNA silencing suppressors in transgenic Nicotiana benthamiana and N. tabacum. Mol Plant Microbe Interact 21: 178–187.

55. Walsh D, Mohr I (2011) Viral subversion of the host protein synthesis machinery. Nat Rev Microbiol 9: 860–875.

56. Sokolenko A, Pojidaeva E, Zinchenko V, Panichkin V, Glaser VM, et al. (2002) The gene complement for proteolysis in the cyanobacterium Synechocystis sp. PCC 6803 and Arabidopsis thaliana chloroplasts. Curr Genet 41: 291–310.

57. Diner BA, Ries DF, Cohen BN, Metz JG (1988) COOH-terminal processing of polypeptide D1 of the photosystem II reaction center of Scenedesmus obliquus is necessary for the assembly of the oxygen-evolving complex. J Biol Chem 263: 8972–8980.

58. Liao DI, Qian J, Chisholm DA, Jordan DB, Diner BA (2000) Crystal structures of the photosystem II D1 C-terminal processing protease. Nat Struct Biol 7: 749–753.

59. Elvira MI, Galdeano MM, Gilardi P, Garcia-Luque I, Serra MT (2008) Proteomic analysis of pathogenesis-related proteins (PRs) induced by compatible and incompatible interactions of pepper mild mottle virus (PMMoV) in Capsicum chinense L3 plants. J Exp Bot 59: 1253–1265.

60. Baebler S, Stare K, Kovac M, Blejec A, Prezelj N, et al. (2011) Dynamics of responses in compatible potato-Potato virus Y interaction are modulated by salicylic acid. PLoS One 6: e29009.

61. Durrant WE, Dong X (2004) Systemic acquired resistance. Annu Rev Phytopathol 42: 185–209.

62. Lehto K, Tikkanen M, Hiriart JB, Paakkarinen V, Aro EM (2003) Depletion of the photosystem II core complex in mature tobacco leaves infected by the flavum strain of tobacco mosaic virus. Mol Plant Microbe Interact 16: 1135–1144.

63. Bilgin DD, Zavala JA, Zhu J, Clough SJ, Ort DR, et al. (2010) Biotic stress globally downregulates photosynthesis genes. Plant Cell Environ 33: 1597–1613.

64. Abbink TE, Peart JR, Mos TN, Baulcombe DC, Bol JF, et al. (2002) Silencing of a gene encoding a protein component of the oxygen-evolving complex of photosystem II enhances virus replication in plants. Virology 295: 307–319.

65. Zhu A, Zhang CX, Lieberman HB (2008) Rad9 has a functional role in human prostate carcinogenesis. Cancer Res 68: 1267–1274.

66. Lieberman HB, Bernstock JD, Broustas CG, Hopkins KM, Leloup C, et al. (2011) The role of RAD9 in tumorigenesis. J Mol Cell Biol 3: 39–43.

67. Šindelářová M, Šindelář L (2001) Changes in Composition of Soluble Intercellular Proteins Isolated from Healthy and TMV-Infected Nicotiana tabacum L. cv. Xanthi-nc. Biologia Plantarum 44: 567–572.

68. Calvo M, Dujovny G, Lucini C, Ortuno J, Alamillo JM, et al. (2010) Constraints to virus infection in Nicotiana benthamiana plants transformed with a potyvirus amplicon. BMC Plant Biol 10: 139.

69. Quadt R, Kao CC, Browning KS, Hershberger RP, Ahlquist P (1993) Characterization of a host protein associated with brome mosaic virus RNA-dependent RNA polymerase. Proc Natl Acad Sci U S A 90: 1498–1502.

70. Thivierge K, Nicaise V, Dufresne PJ, Cotton S, Laliberte JF, et al. (2005) Plant virus RNAs. Coordinated recruitment of conserved host functions by (+) ssRNA viruses during early infection events. Plant Physiol 138: 1822–1827.

71. Noueiry AO, Chen J, Ahlquist P (2000) A mutant allele of essential, general translation initiation factor DED1 selectively inhibits translation of a viral mRNA. Proc Natl Acad Sci U S A 97: 12985–12990.

72. Toribio R, Ventoso I (2010) Inhibition of host translation by virus infection in vivo. Proc Natl Acad Sci U S A 107: 9837–9842.

73. Kerekatte V, Keiper BD, Badorff C, Cai A, Knowlton KU, et al. (1999) Cleavage of Poly(A)-binding protein by coxsackievirus 2A protease in vitro and in vivo: another mechanism for host protein synthesis shutoff? J Virol 73: 709–717.

74. Tyc K, Konarska M, Gross HJ, Filipowicz W (1984) Multiple ribosome binding to the 5'-terminal leader sequence of tobacco mosaic virus RNA. Assembly of an 80S ribosome X mRNA complex at the AUU codon. Eur J Biochem 140: 503–511.

75. Sogorin EA, Shirokikh NE, Ibragimova AM, Vasiliev VD, Agalarov SC, et al. (2012) Leader sequences of eukaryotic mRNA can be simultaneously bound to initiating 80S ribosome and 40S ribosomal subunit. Biochemistry (Mosc) 77: 342–345.

76. Wells DR, Tanguay RL, Le H, Gallie DR (1998) HSP101 functions as a specific translational regulatory protein whose activity is regulated by nutrient status. Genes Dev 12: 3236–3251.

77. Jimenez I, Lopez L, Alamillo JM, Valli A, Garcia JA (2006) Identification of a plum pox virus CI-interacting protein from chloroplast that has a negative effect in virus infection. Mol Plant Microbe Interact 19: 350–358.

78. Nevins JR (1994) Cell cycle targets of the DNA tumor viruses. Curr Opin Genet Dev 1994, 4: 130–134.

79. Bloom J, Cross FR (2007) Multiple levels of cyclin specificity in cell-cycle control. Nat Rev Mol Cell Biol 8: 149–160.

80. Morgan DO (1995) Principles of CDK regulation. Nature 374: 131–134.

81. Kallio MA, Tuimala JT, Hupponen T, Klemela P, Gentile M, et al. (2011) Chipster: user-friendly analysis software for microarray and other high-throughput data. BMC Genomics 12: 507.

82. Bustin SA, Benes V, Garson JA, Hellemans J, Huggett J, et al. (2009) The MIQE guidelines: minimum information for publication of quantitative real-time PCR experiments. Clin Chem 55: 611–622.

83. Siddiqui SA, Valkonen JP, Rajamaki ML, Lehto K (2011) The 2b silencing suppressor of a mild strain of Cucumber mosaic virus alone is sufficient for synergistic interaction with Tobacco mosaic virus and induction of severe leaf malformation in 2b-transgenic tobacco plants. Mol Plant Microbe Interact 24: 685–693.

84. Porra RJ, Thompson WA, Kriedemann PE (1989) Determination of accurate extinction coefficients and simultaneous equations for assaying chlorophyll a and b with four different solvents: verification of the concentration of chlorophyll by atomic absorption spectroscopy. Biochim Biophys Acta 975: 384–394.

17

Tomato *ABSCISIC ACID STRESS RIPENING (ASR)* Gene Family Revisited

Ido Golan[1], Pia Guadalupe Dominguez[2], Zvia Konrad[1], Doron Shkolnik-Inbar[1], Fernando Carrari[2], Dudy Bar-Zvi[1]*

1 Department of Life Sciences and Doris and Bertie Black Center for Bioenergetics in Life Sciences, Ben-Gurion University of the Negev, Beer-Sheva, Israel, **2** Instituto de Biotecnología, Instituto Nacional de Tecnología Agropecuaria, Buenos Aires, Argentina

Abstract

Tomato *ABSCISIC ACID RIPENING 1 (ASR1)* was the first cloned plant *ASR* gene. *ASR* orthologs were then cloned from a large number of monocot, dicot and gymnosperm plants, where they are mostly involved in response to abiotic (drought and salinity) stress and fruit ripening. The tomato genome encodes five *ASR* genes: *ASR1, 2, 3* and *5* encode low-molecular-weight proteins (ca. 110 amino acid residues each), whereas *ASR4* encodes a 297-residue polypeptide. Information on the expression of the tomato *ASR* gene family is scarce. We used quantitative RT-PCR to assay the expression of this gene family in plant development and in response to salt and osmotic stresses. *ASR1* and *ASR4* were the main expressed genes in all tested organs and conditions, whereas *ASR2* and *ASR3/5* expression was two to three orders of magnitude lower (with the exception of cotyledons). *ASR1* is expressed in all plant tissues tested whereas ASR4 expression is limited to photosynthetic organs and stamens. Essentially, *ASR1* accounted for most of *ASR* gene expression in roots, stems and fruits at all developmental stages, whereas *ASR4* was the major gene expressed in cotyledons and young and fully developed leaves. Both *ASR1* and *ASR4* were expressed in flower organs, with *ASR1* expression dominating in stamens and pistils, *ASR4* in sepals and petals. Steady-state levels of *ASR1* and *ASR4* were upregulated in plant vegetative organs following exposure to salt stress, osmotic stress or the plant abiotic stress hormone abscisic acid (ABA). Tomato plants overexpressing *ASR1* displayed enhanced survival rates under conditions of water stress, whereas *ASR1*-antisense plants displayed marginal hypersensitivity to water withholding.

Editor: Sara Amancio, ISA, Portugal

Funding: This study was supported in part by the I-CORE Program of the Planning and Budgeting Committee and the Israel Science Foundation (Center No. 757). DBZ is the incumbent of The Israel and Bernard Nichunsky Chair in Desert Agriculture, Ben-Gurion University. The funders had no role in study design, data collection and analysis, decision to publish, or preparation of the manuscript.

Competing Interests: The authors have declared that no competing interests exist.

* Email: barzvi@bgu.ac.il

Introduction

The first member of the tomato *ABSCISIC ACID STRESS RIPENING (ASR)* gene family, *ASR1*, was identified by screening a tomato fruit cDNA library with cDNA from stressed leaves [1], hence its name. Since then, a large number of *ASR* orthologs have been cloned from many plant species, including gymnosperms and angiosperms (reviewed by [2]). *ASR* gene families are found in the genomes of both monocots and dicots, but they are missing in the model plant *Arabidopsis* [2]. Interestingly, no *ASR* orthologs have been found in organisms outside the plant kingdom. *ASR* genes have been shown to be induced by abscisic acid (ABA) and abiotic stress, mainly salinity and drought [1,3–18]. They are also highly expressed in ripening fruit [1,4,14,19–23], and during potato-tuber development [24,25].

The tomato ASR gene family consists of five genes localized in one cluster on chromosome 4. Four members of the *ASR* gene family have been cloned from tomato [4,26–30]. These genes encode highly homologous proteins and possess a single intron of different size, but conserved location [4,27,28]. Upon completion of the tomato genome sequence [31], a fifth *ASR* gene was annotated, whose exon nucleotide sequence is highly similar to that of *ASR3* (88% identity in coding sequences, see also [32]). The loci of *ASR1*–*ASR5* genes in the tomato genome are Solyc04g071610, Solyc04g071580, Solyc04g071590, Solyc04g071620 and Solyc04g071600, respectively. Four of the genes (*ASR1*–*ASR3*, *ASR5*) encode low-molecular-weight proteins, whereas the polypeptide encoded by *ASR4* is approximately double the size of the other proteins [30]. Wild tomato species also encode this five member ASR gene family. suggesting that this family was not lost during tomato domestication and breeding [30,33]. Furtheremore, *ASR1, 2* and *4* genes from wild tomato species were also induced by drought and cold. The majority of plant ASR genes encode low molecular weight proteins. In addition, genomes of a number of plant species contain in addition a gene encoding higher molecular ASR polypeptides [25,34,35].

ASR proteins have been proposed to belong to the hydrophylin group of proteins [36]. Tomato ASR1 was shown to be a natively unordered protein [37] that possesses chaperone-like activity [38], and was localized to both the cytosol and nucleus [39]. Dual subcellular localization was also shown for the lily pollen ASR protein [40]. Rice ASR1 was also shown to possess chaperone-like activity [41,42]. ASR1 has Zn^{2+}-dependent DNA-binding activity [39,43]. Upon binding of Zn^{2+}, ASR1 becomes structured and dimerizes [26,37], and is translocated to the nucleus [37,44,45]. Zn^{2+} and Fe^{3+} ions affected the structure of a soybean ASR [46]. Nuclear ASR proteins modulate gene expression via binding to specific promoter sequences [14,39,47,48].

ASR genes have been shown to play a central role in drought and salinity stress. Overexpression of ASR genes resulted in increased tolerance of the transgenic plants to water/osmotic [42,48–51], salinity [48,50,52,53] and cold [45,54] stresses. However, until now these responses have been seen only in heterologous systems. Transgenic Arabidopsis plants expressing ASR proteins from other plant species [42,54] demonstrated increased tolerance to abiotic stresses. Arabidopsis plants do not encode ASR proteins. Ectopic expression of tomato ASR1 in Arabidopsis was shown to affect the plant's response to ABA, glucose and tolerance to abiotic stress via competition for DNA binding with the transcription factor ABI4 [55]. Thus, expression studies in heterologous organisms that do not naturally have the studied gene(s) should be analyzed with caution, especially for regulatory proteins, as results may not directly reflect the biological role of the analyzed gene. In addition, ASR gene involvement in carbohydrate signaling, sugar trafficking and metabolism has been shown [14,24,56,57], as has its influence on the biogenesis of branched-chain amino acids [58].

In tomato, the best-studied member of the ASR family is ASR1, followed by ASR2. Information on ASR3, ASR4 and ASR5 is scarce. Although a few studies have compared the expression of some members of the tomato ASR gene family [59,60], results are from northern blot, semi-quantitative PCR, or histological staining studies analyzing two or three of the family's genes, under rather restricted conditions. In this work, we revisited the tomato ASR gene family using the highly accurate quantitative RT-PCR technology to determine the expression patterns of its members in vegetative and reproductive organs. Because of their high sequence homology, we could not design gene-specific primers for ASR3 and ASR5, we determined the summed expression of these genes (ASR3/5). We found that ASR1 and ASR4 are highly expressed in vegetative tissues of nonstressed plants, whereas the steady-state levels of ASR2 and ASR3/5 transcripts are two to three orders of magnitude lower. ASR1 was the major member expressed in roots, stems, stamens, pistils and fruits. ASR4 accounted for more than two-thirds of total ASR transcripts in leaves, shoot vegetative meristem, sepals and petals. Both of these genes were induced by ABA, osmotic stress and salt stress. Tomato plants overexpressing ASR1 (ASR1-OE) survived better under

water stress than wild-type (WT) plants, where ASR1-antisense (ASR1-AS) plants had slightly lower survival rates than the WT.

Materials and Methods

Plant material and growth conditions

Generation and selection of transgenic lines. The 348-bp coding region of the tomato ASR1 gene (GenBank U86130.1) was cloned in sense or antisense orientation into the multiple cloning site of the vector [61] between the *Cauliflower Mosaic Virus* (*CaMV*) 35S promoter and the *octopine synthase* (*ocs*) terminator. Tomato plants (*Solanum esculentum* cv. Moneymaker) were transformed by *Agrobacterium* as previously described [62]. Emerging shoots were excised and selected on Murashige and Skoog (MS) medium containing kanamycin (100 mg/l), and then transferred to the greenhouse for selection by qPCR in T1 plants. Transgenic plants were numbered XX-YY, where XX represents an independent transformation event, and YY represents a subline of the founder of T2 generation seeds. Plants were self-pollinated and seeds were collected. T3 generation plants were used in this study. Western blot analyses showed that the ASR1-OE lines have higher levels of ASR1 than WT plants (Figure S1 in File S1).

Plant growth. Plants were grown in pots in the greenhouse or hydroponically in the growth room in aerated half-strength Hoagland mineral solution at 28°C and 70% relative humidity, under a diurnal cycle of 18 h light, or in the greenhouse at an average temperature of 28°C, and >50% relative humidity. Seeds were germinated in water-soaked vermiculite. Ten-day-old seedlings were transferred to aerated containers with half-strength Hoagland mineral solution [63], or to pots containing equal volumes of planting mix and vermiculite. Hydroponic growth medium was replaced 1 week after transfer and every 3–4 days thereafter.

Water-stress tolerance and survival assays. Seedlings were grown in pots under optimal conditions for 3 weeks. Water was then withheld for 22 days, followed by rewatering. Plant survival was scored 17 days later. At least 20 plants from each line were used. Plants were grown in random order and pot location was changed every few days.

NaCl, polyethylene glycol (PEG) and ABA treatment

WT seedlings were germinated on vermiculite and transferred to aerated 0.5X Hoagland's solution as described above. After 1 week acclimation, plants were transferred to fresh mineral solution containing, where indicated, 40 µM ABA, 0.2 M NaCl, or 8% (w/v) PEG 8000. Plants were harvested 24 h after treatments.

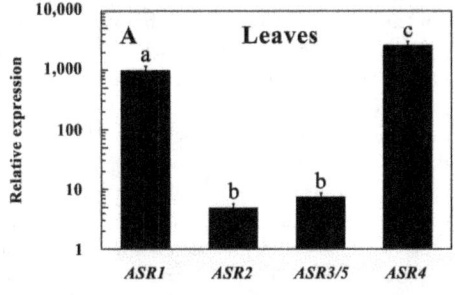
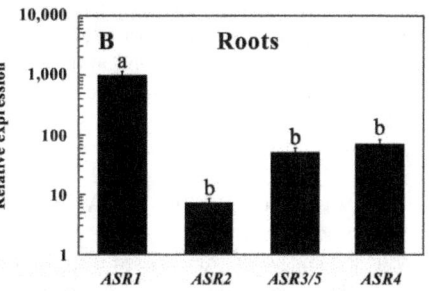

Figure 1. Relative expression of the tomato *ASR* genes. Steady-state levels of the indicated genes were determined in leaves and roots of 10-day-old hydroponically grown tomato seedlings. Expression of *ASR1* in each tissue was defined as 1000. Data shown are average ± SE. Bars with different letters represent statistically different values by Tukey's HSD post-hoc test (P≤0.05).

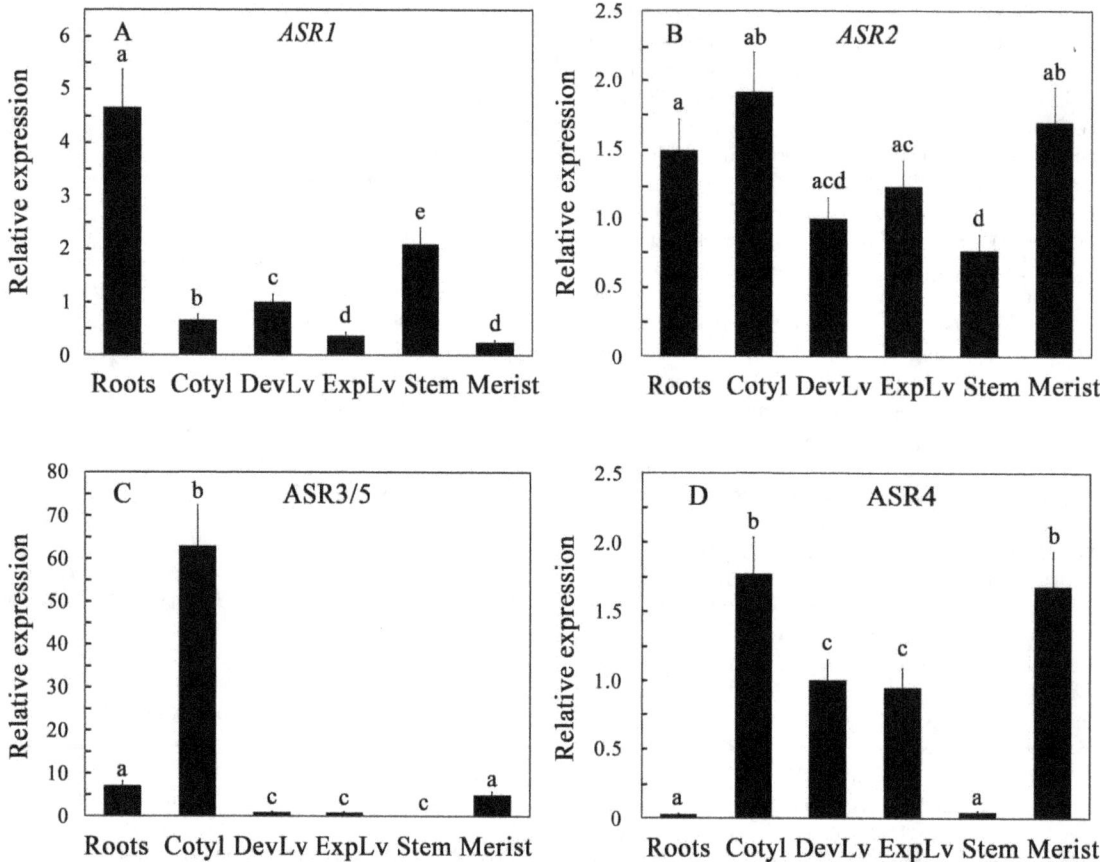

Figure 2. Expression of the tomato *ASR* genes in vegetative tissues. Steady-state levels of the indicated genes were determined in hydroponically grown tomato plants at the 8-true-leaf stage. The expression levels were normalized for each of the genes to their expression in young developing leaves, defined as 1. Cotyl, cotyledons; DevLv, developing leaves; ExpLv, fully expanded leaves; Merist, shoot meristem. Data shown are average ± SE. Bars with different letters represent statistically different values by Tukey's HSD post-hoc test (P≤0.05).

mRNA level assay

Total RNA was extracted from the indicated plant tissues using EZ-RNA (Biological Industries, Israel) according to the manufacturer's protocol. This protocol uses improved RNA-extraction methods as described by Chomczynski and Sacci [64]. RNA quality integrity was checked spectrally (at 230 nm, 260 nm and 280 nm) and by running samples on denaturing agarose gels electrophoresis. Relative steady-state transcript levels were assayed by RT-qPCR as described previously [50,55,65,66]. cDNA was synthesized from DNase-treated RNA using high-capacity cDNA reverse transcription kit (Applied Biosystems). Primers were designed by Primer-Express software Vers. 2.0 (Applied Biosystems). When possible, one of the primers in each set was anchored at an exon–exon border to reduce possible amplification from contaminating genomic DNA. All amplicon lengths were between 75 and 90 bp. Primer sequences are presented in Table S1 in File S1. Transcript levels were assayed using the 7300 Real-Time PCR System (Applied Biosystems), with 18S rRNA as the internal standard. PCR efficiency was close to 100%. RNA relative quantification analyses were performed using 7300 System SDS software (Applied Biosystems). The list of primers used is shown in Table S1 in File S1. The data represent the mean ± SE of $n = 3$ independent experiments. Each data point was determined in triplicates in each of the three biological replicates and presented as mean ± SE. Data presented in a single graph were carried in a single run. Differences between groups were analyzed by Tukey's HSD post-hoc test (P≥0.05).

Results and Discussion

Relative expression levels of the tomato *ASR* genes

Relative steady-state levels of the members of the *ASR* gene family were determined in leaves and roots of hydroponically grown seedlings. In general, *ASR1* and *ASR4* were the most highly expressed genes in this family (Figure 1). In young leaves, *ASR4* levels were 2.6 times higher than those of *ASR1*, whereas transcript levels of *ASR2* and *ASR3/5* were more than two orders of magnitude lower. In tomato roots, *ASR1* was the most abundantly expressed gene, whereas transcript levels of *ASR3/5* and *ASR4* were approximately one order of magnitude lower than that of *ASR1*, and that of *ASR2* two orders of magnitude lower than the steady-state levels of *ASR1*. These results are in agreement with Frankel et al. [30] who reported that *ASR2* and *ASR3/5* transcripts could not be detected in tomato leaves by northern blot analysis. On the other hand, previous studies from the same laboratory reported that *ASR2* transcripts are highly abundant in roots of stressed tomato plants [60] and that the *ASR2* promoter can drive transcription in both tomato and other Solanaceae plant cells [60,67]. RNA Seq also suggest that *ASR1* and *ASR4* transcripts in tomato leaves and fruits are relatively

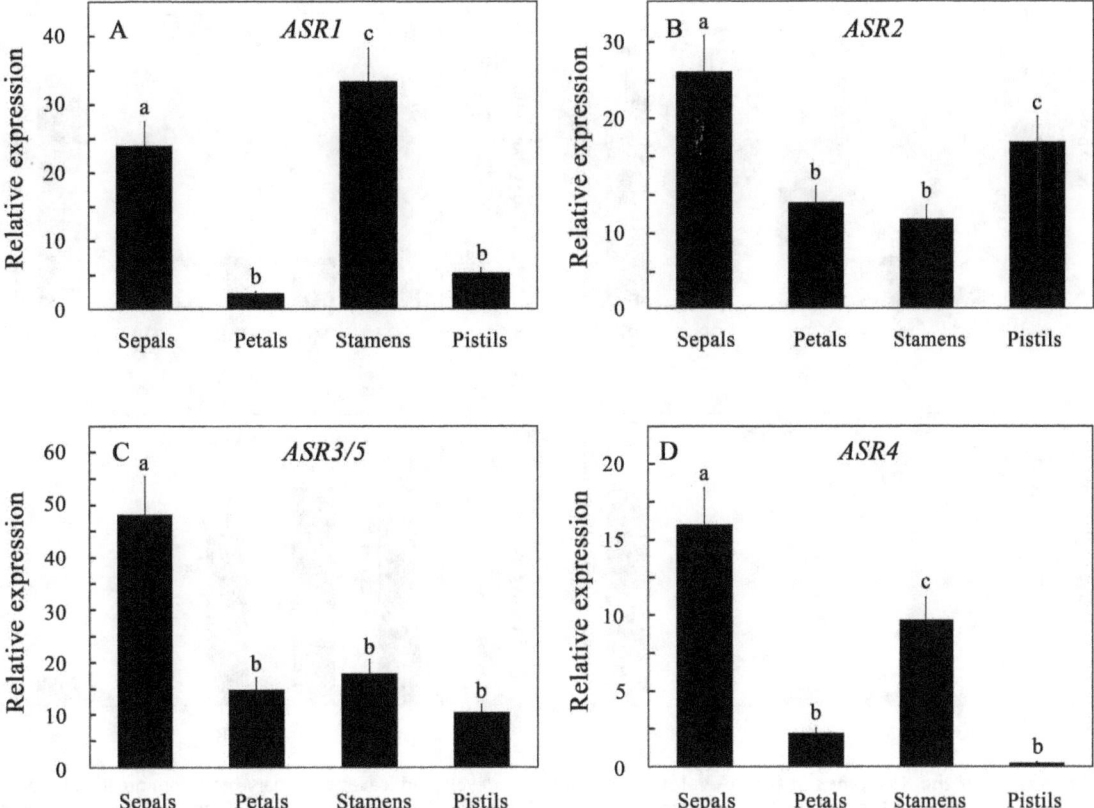

Figure 3. Expression of the tomato *ASR* genes in flower organs. Flowers and developing leaves were harvested from greenhouse-grown tomato plants. Flowers were dissected, and steady-state levels of the indicated genes were determined. The expression levels were normalized for each of the genes to their expression in young developing leaves, defined as 1. Data shown are average ± SE. Bars with different letters represent statistically different values by Tukey's HSD post-hoc test (P≤0.05).

abundant, whereas transcripts from *ASR2*, *ASR3* and *ASR5* are hardly found [31].

(data presented at the tomato Sol Genomic Network database (http://solgenomics.net)).

Expression of tomato *ASR* genes in vegetative tissues

Expression of tomato *ASR* genes was determined in vegetative tissues of hydroponically grown tomato (Figure 2). Transcript levels of each of the genes were normalized to the expression of the same gene in developing leaves. This reference tissue was selected since it is present in all plant ages used in the study. Highest levels of *ASR*1 transcript were found in roots and stems (ca. 4.5 and 2 times that in the leaves, respectively). *ASR1* transcript levels decreased with leaf development (Figure 2A), in agreement with Amitai-Zeigerson et al. [68]. *ASR1* expression in stems and roots suggests its role in the plant's vascular system [59]. *ASR2* transcript levels were only slightly different in vegetative organs (Figure 2B), and its steady-state levels in all vegetative organs were marginal (see Figure 1). *ASR3/5* levels were highest in the cotyledons (Figure 2C), reaching transcript levels in the same order of magnitude of that of *ASR1* in cotyledons, suggesting that *ASR3* and/or *ASR5* my play a role in advanced stages of seed development. ASR proteins were detected immunologically in tomato seeds using anti-ASR1 antiserum [37]. Since there is high amino acid conservation between the different members of the tomato ASR proteins, is it likely that this antibody crossreacts with other ASR proteins such as ASR3 and ASR5. *ASR4* was expressed mainly in leaves, cotyledons and meristem, with

relatively low expression rates in roots and stems (Figure 2D). After normalization of the relative expression of each gene in young leaf tissues (Figure 2A), *ASR1* seemed to be the main expressed gene in the roots and stems. On the other hand, *ASR4* was the highest expressed *ASR* gene in cotyledons, and young and fully expanded leaves. *ASR1* accounted for up to one quarter of of *ASR* gene family expression in these tissues-. Expression of *ASR2* and *ASR3/5* in leaves was negligible. Interestingly, and Our results indicate that *ASR1* is the primarily expressed gene in tomato roots, stems and fruits, whereas both *ASR1 and ASR4* are both expressed in cotyledons, leaves and meristems, where *ASR4* levels exceed that of *ASR1*. Transcript levels of *ASR2* and *ASR3/5* were marginal in vegetative tissues, with the exception of *ASR3/5* in cotelydons. Our analyzes determine averaged transcript levels in the entire organ, rather than cell specific expression. Thus, it will be interesting to find out if in these organs, *ASR1* and *ASR4* coexpress in the same cells, or in different cell types.

Expression of the tomato ASR genes in reproductive tissues

The *ASR* gene family showed differential expression in flower organs. Although *ASR2* and *ASR3/5* expression varied in the different flower organs (Figure 3B, C), their expression levels can be estimated to be two order of magnitude order lower than that of *ASR1* and *ASR4* of *ASR1* and *ASR4* were highly expressed in the sepals and stamens, and to a lower extent in petals and pistils (Figure 3A, D), where *ASR4* estimated transcript levels were higher than *ASR1* in the sepals and petals. One the other hand,

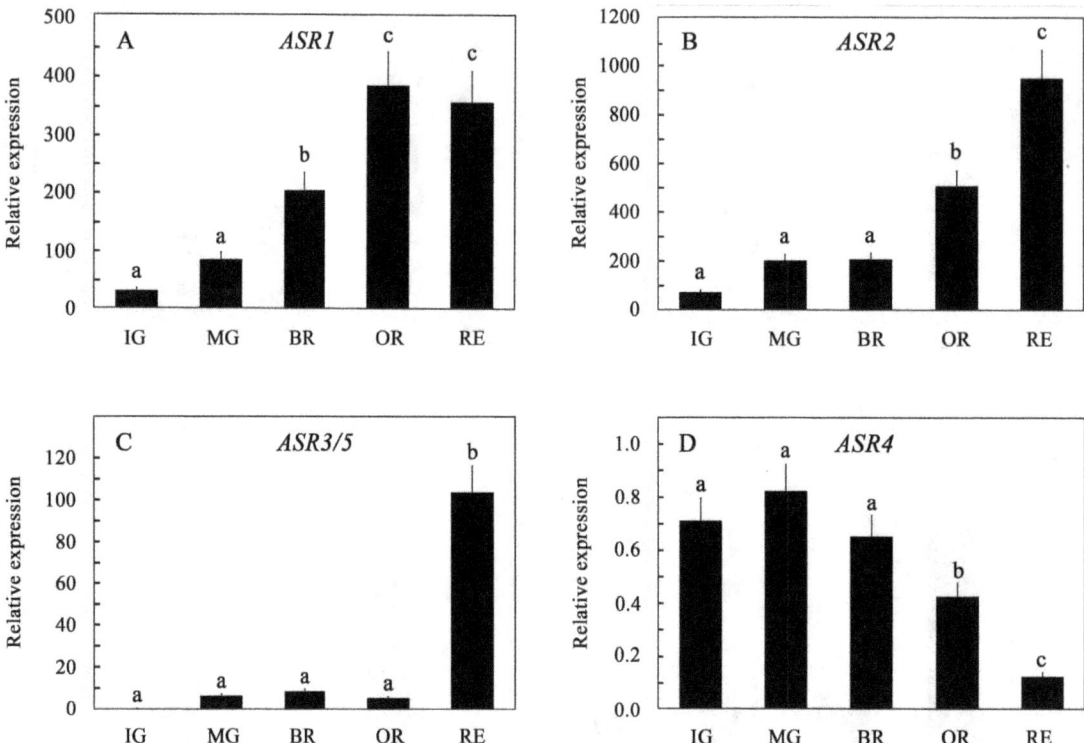

Figure 4. Expression of the *ASR* genes in fruit development. Fruits and developing leaves were harvested from greenhouse-grown tomato plants, and steady-state levels of the indicated genes were determined. The expression levels were normalized for each of the genes to their expression in young developing leaves, defined as 1. Fruit stages were defined according to [75]: IG, immature green; MG, mature green; BR, breaker; OR, orange; RE, red. Data shown are average ± SE. Bars with different letters represent statistically different values by Tukey's HSD post-hoc test (P≤ 0.05).

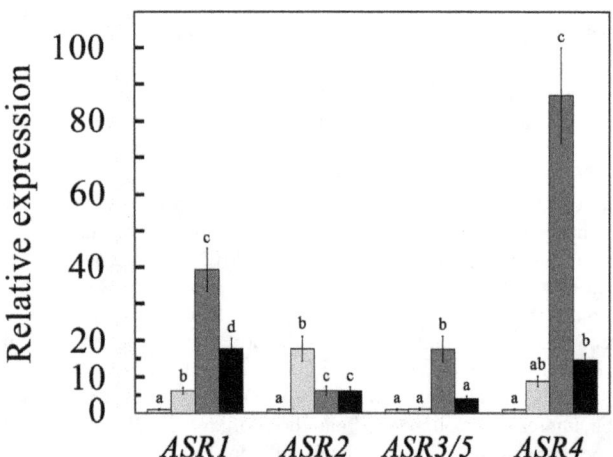

Figure 5. Effects of ABA, NaCl and PEG on the steady-state transcript levels of *ASR* genes. One-week-old hydroponically grown seedlings were transferred to fresh Hoagland solution containing: no addition (white bars), 40 μM ABA (light gray bars), 0.2 M NaCl (dark gray bars), or 8% PEG 8000 (black bars). Leaves were harvested 24 h later and expression levels were determined as described in Materials and Methods. Data were normalized for each of the genes to their expression in young developing leaves, defined as 1. Data shown are average ± SE. Expression of each gene in response to plant treatment was analyzed separately using Tukey's HSD post-hoc test. Bars with different letters represent statistically different values by (P≤0.05).

ASR1 is the most highly expressed *ASR* gene in the stamen and pistils.

Although *ASR1* expression increased during fruit development (Figure 4A), it was already higher than that in all other tissues in young immature green fruits. *ASR1* was essentially the main fruit-expressed *ASR* gene at all fruit developmental (Figure 4). Steady-state levels of *ASR2 and ASR3/5* transcripts in fruit tissues were also the highest measured in any plant tissue. Nevertheless, their levels in fruits are approximately two order of magnitude lower of that of *ASR1*. An increase in *ASR1* transcript levels in tomato fruit development and ripening is in agreement with Gilad et al. [4] and Iusem et al. [1], but not with Maskin et al. [60]. Increase in *ASR1* during tomato fruit ripeing also correlates with increase in its protein levels [20]. Increased steady state transcript levels during fruit ripening of *SlASR1* orthologous genes were reported in a number of plant species [21–23,69,70].

Tomato ASR genes are differentially responsive to ABA and abiotic stress

Steady-state levels of *ASR1* and *ASR4* transcript increased following plant exposure to salt stress, osmotic stress (PEG) or to the hormone ABA (Figure 5). The relative induction levels by these treatments were rather similar for these two highly expressed *ASR* genes. On the other hand, the less expressed genes *ASR2* and *ASR3/5* showed different responses: *ASR2* responded most strongly to ABA treatment and to a lesser extent to salt or osmotic stress, whereas *ASR3/5* was not affected by ABA and was relatively highly expressed after NaCl treatment, suggesting that they are induced by an ABA-independent pathway. A large

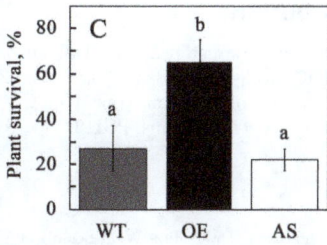

Figure 6. ASR1-overexpressing tomato plants have enhanced tolerance to water stress. Plants were grown in pots in the greenhouse using optimal irrigation for 17 days. Water was withheld for 22 days, following by rewatering for 17 days. Panels A and B show representative plants (left to right): wild type, *ASR1*-OE-31, *ASR1*-OE-12, *ASR1*-OE-16 after 22 days of dehydration (A) and 17 days of rewatering (B). Panel C, quantitative data of survival of three lines of *ASR1*-overexpressing (*ASR1*-OE) plants and four lines of *ASR1*-antisense (*ASR1*-AS) plants, measured at the end of the rewatering stage. Data shown are average ± SE. Bars with different letters represent statistically different values by Tukey's HSD post-hoc test (P≤ 0.01).

number of *ASR* genes have been identified in different plant species based on their response to abiotic stresses such as water, salinity and osmotic stress [9,10,14,71,72]. Expression results for the tomato *ASR1* gene are in agreement with previous studies [4,59,60,68]. In contrast, activity of the tomato *ASR2* promoter was enhanced by ABA when expressed in papaya and tobacco, but not in tomato or potato [67].

Genetic manipulation of *ASR1* in tomato affects water-stress tolerance

Tomato plants overexpressing *CaMV 35S:ASR1* (*ASR1*-OE) or *CaMV 35S:Reverse ASR1* (*ASR1*-AS) were tested for tolerance to water and salt stresses. The modified plants did not perform significantly differently from WT plants when treated with NaCl (not shown). However, *ASR1*-OE plants survived water stress better than WT plants (Figure 6A and 6B). The response of *ASR1*-AS plants was highly variable: many lines showed hypersensitivity to water stress, whereas others showed no difference, or even better tolerance (Figure S2 in File S1). Averaging the recovery rates of all lines showed that *ASR1*-OE plants were significantly more tolerant to water stress than WT controls and *ASR1*-AS plants, whereas the latter were slightly more sensitive to lack of water than WT plants (Figure 6C). Transgenic plants overexpressing *ASR* genes have been reported to be more tolerant to abiotic stresses such as water/osmotic stress [42,48–51], salinity stress [48,50,52,53] and cold stress [44,54]. However, most of those studies expressed the studied gene in a heterologous system [42,48–50,52–54]. In some of those studies, the biological species of the gene of origin and the transgenic plant belonged to the same botanical genus or family [24,53,56], but only a few studies have been performed within a single species [45,51]. The biological relevance of studying the role of a regulatory protein on a genetic background that naturally lacks it has been questioned [55]: the phenotype of *Arabidopsis* plants expressing tomato *ASR1* was shown to result from competition for DNA binding between the ectopically expressed tomato gene and the Arabidopsis transcription factor ABI4, essentially resulting in an *abi4*-mutant-like phenotype [55]. Thus, expressing the studied gene on the same genetic background ([45,51] and this study), or in closely related species [24,53,56], is more likely to shed light on the actual role of the studied gene and its protein product. The increased survival of transgenic tomato plants overexpressing tomato *ASR1* following water stress (Figure 6) is in agreement with reports on the expression of plant *ASR* genes on similar or different genetic backgrounds in response to stress [42,48–51]. Results obtained using the *ASR1*-AS lines were more variable: lines ASR1-AS5-4 and ASR1-AS18-4 were significantly hypersensitive to water stress

as compared to WT plants (Figure S2 in File S1), whereas other lines were not significantly different. On the other hand, no significant differences were found in the sensitivity of *ASR1*-OE and *ASR1*-AS plants to NaCl stress (not shown), most probably due to the relative tolerance of the tomato WT line. ASR proteins are localized in the cytosol and nucleus. In the latter organelle, they are associated with DNA [14,39,40,48,73]. Tomato and grape ASR proteins have been shown to bind specific DNA sequences [14,53], suggesting that they regulate the expression of genes involved in abiotic stress responses or sugar metabolism [4,14,24,54,57,74]. In addition, ASR proteins have been shown to possess protein-chaperone-like activity [38,41], increasing protein stability. Thus, the increased water-stress tolerance of *ASR1*-OE plants most likely results from an increase in the transcription of ASR1-regulated genes and from ASR1 chaperon activity.

Conclusions

Two of the five genes in the *ASR* gene family (*ASR1* and *ASR4*) are significantly expressed in tomato plants, whereas the expression levels of the other three member of the gene family are less pronounced. *ASR1* and *ASR4* encode 115- and 297-amino acid polypeptides, respectively, thus most likely encoding proteins whose activity may not be fully redundant. ASR1 is expressed in all plant tissues tested: it is most highly expressed in the stem, roots and reproductive organs–stamen, pistils and fruit at all developmental stages. *ASR4* is mainly expressed in photosynthetic organs, in in sepals and petals. The steady-state levels of *ASR1* and *ASR4* increased following salt stress, osmotic stress and treatment with ABA. *ASR2* expression is negligible in all tested tissues. *ASR3* and *ASR5*, being highly similar genes, were assayed together, with significant expression detected only in the cotyledons. These results suggest that *ASR2/3/5* may be very low expressing genes, or that their expression is limited to specific low abundant cells, thus resulting in low transcript activity when assayed in the whole tissue. Tomato plants overexpressing *ASR1* show increased tolerance to water stress, whereas *ASR1*-AS plants show a certain degree of hypersensitivity to water withholding. Our data suggest that *ASR1* and *ASR4* may be expressed in different cell types. The differential expression patterns of the tomato ASR gene family under non-stress and stress conditions may be used for genetic manipulation of tomato (as well as other crop plants) to affect vegetative and fruit parameters under non-stress and stress conditions.

Author Contributions

Conceived and designed the experiments: IG DSI DBZ. Performed the experiments: IG ZK DSI. Analyzed the data: IG ZK DSI DBZ. Contributed to the writing of the manuscript: IG PGD DSI FC DBZ. Constructed the transgenic tomato plants: PGD FC.

References

1. Iusem ND, Bartholomew DM, Hitz WD, Scolnik PA (1993) Tomato (*Lycopersicon esculentum*) transcript induced by water deficit and ripening. Plant Physiol 102: 1353–1354.
2. Carrari F, Fernie AR, Iusem ND (2004) Heard it through the grapevine? ABA and sugar cross-talk: The ASR story. Trends Plant Sci 9: 57–59.
3. Chang S, Puryear JD, Dias MADL, Funkhouser EA, Newton RJ, et al. (1996) Gene expression under water deficit in loblolly pine (*Pinus taeda*): Isolation and characterization of cDNA clones. Physiol Plantar 97: 139–148.
4. Gilad A, Amitai-Zeigerson H, Bar-Zvi D, Scolnik PA (1997) *ASR1*, a tomato water-stress regulated gene: Genomic organization, developmental regulation and DNA-binding activity. Acta Hortic 447: 447–453.
5. Padmanabhan V, Dias DMAL, Newton RJ (1997) Expression analysis of a gene family in loblolly pine (*Pinus taeda* L.) induced by water deficit stress. Plant Mol Biol 35: 801–807.
6. Schneider A, Salamini F, Gebhardt C (1997) Expression patterns and promoter activity of the cold-regulated gene *ci21A* of potato. Plant Physiol 113: 335–345.
7. Wang CS, Liau YE, Huang JC, Wu TD, Su CC, et al. (1998) Characterization of a desiccation-related protein in lily pollen during development and stress. Plant Cell Physiol 39: 1307–1314.
8. De Vienne D, Leonardi A, Damerval C, Zivy M (1999) Genetics of proteome variation for QTL characterization: Application to drought-stress responses in maize. J Exp Bot 50: 303–309.
9. Vaidyanathan R, Kuruvilla S, Thomas G (1999) Characterization and expression pattern of an abscisic acid and osmotic responsive gene from rice. Plant Sci 140: 21–30.
10. Dóczi R, Csanaki C, Bánfalvi Z (2002) Expression and promoter activity of the desiccation-specific *Solanum tuberosum* gene, StDS2. Plant Cell Environ 25: 1197–1203.
11. Salekdeh GH, Siopongco J, Wade LJ, Ghareyazie B, Bennett J (2002) A proteomic approach to analyzing drought- and salt-responsiveness in rice. Field Crops Res 76: 199–219.
12. Sugiharto B, Ermawati N, Mori H, Aoki K, Yonekura-Sakakibara K, et al. (2002) Identification and characterization of a gene encoding drought-inducible protein localizing in the bundle sheath cell of sugarcane. Plant Cell Physiol 43: 350–354.
13. Wang JT, Gould JH, Padmanabhan V, Newton RJ (2002) Analysis and localization of the water-deficit stress-induced gene (*lp3*). J Plant Growth Regul 21: 469–478.
14. Çakir B, Agasse A, Gaillard C, Saumonneau A, Delrot S, et al. (2003) A grape ASR protein involved in sugar and abscisic acid signaling. Plant Cell 15: 2165–2180.
15. Frankel N, Hasson E, Iusem ND, Rossi MS (2003) Adaptive evolution of the water stress-induced gene *Asr2* in *Lycopersicon* species dwelling in arid habitats. Mol Biol Evol 20: 1955–1962.
16. Liu HY, Dai JR, Feng DR, Liu B, Wang HB, et al. (2010) Characterization of a novel plantain *Asr* gene, *MpAsr*, that is regulated in response to infection of *Fusarium oxysporum* f. sp *cubense* and abiotic stresses. J Integ Plant Biol 52: 315–323.
17. Philippe R, Courtois B, McNally KL, Mournet P, El-Malki R, et al. (2010) Structure, allelic diversity and selection of *Asr* genes, candidate for drought tolerance, in *Oryza sativa* L. and wild relatives. Theor Applied Genet 121: 769–787.
18. Joo J, Lee YH, Kim YK, Nahm BH, Song SI (2013) Abiotic stress responsive rice *ASR1* and *ASR3* exhibit different tissue-dependent sugar and hormone-sensitivities. Mol Cells 35: 421–435.
19. Srivastava A, Handa AK (2005) Hormonal regulation of tomato fruit development: A molecular perspective. J Plant Growth Regul 24: 67–82.
20. Rocco M, D'Ambrosio C, Arena S, Faurobert M, Scaloni A, et al. (2006) Proteomic analysis of tomato fruits from two ecotypes during ripening. Proteomics 6: 3781–3791.
21. Xu BY, Su W, Liu JH, Wang JB, Jin ZQ (2007) Differentially expressed cDNAs at the early stage of banana ripening identified by suppression subtractive hybridization and cDNA microarray. Planta 226: 529–539.
22. Chen JY, Liu DJ, Jiang YM, Zhao ML, Shan W, et al. (2011) Molecular characterization of a strawberry *FaASR* gene in relation to fruit ripening. PLoS One 6: e24649.
23. Luo C, Dong L, He XH, Yu HX, Ou SJ, et al. (2014) Molecular cloning and characterisation of a cDNA encoding an abscisic acid-, stress-, and ripening-induced (ASR) protein in mango (*Mangifera indica* L.). J Hort Sci Biotechnol 89: 352–358.
24. Frankel N, Nunes-Nesi A, Balbo I, Mazuch J, Centeno D, et al. (2007) *ci21A/Asr1* expression influences glucose accumulation in potato tubers. Plant Mol Biol 63: 719–730.
25. Silhavy D, Hutvagner G, Barta E, Banfalvi Z (1995) Isolation and characterization of a water-stress-inducible cDNA clone from *Solanum chacoense*. Plant Mol Biol 27: 587–595.
26. Iusem ND, Maskin L, Frankel N, Gudesblat G, Demergasso MJ, et al. (2007) Dimerization and DNA-binding of *ASR1*, a small hydrophilic protein abundant in plant tissues suffering from water loss. Biochem Biophys Res Commun 352: 831–835.
27. Rossi M, Lijavetzky D, Bernacchi D, Hopp HE, Iusem N (1996) *Asr* genes belong to a gene family comprising at least three closely linked loci on chromosome 4 in tomato. Mol Gen Genet 252: 489–492.
28. Amitai-Zeigerson H, Scolnik PA, Bar-Zvi D (1994) Genomic nucleotide sequence of tomato *Asr2*, a second member of the stress/ripening-induced *Asr1* Gene Family. Plant Physiol 106: 1699–1700.
29. Rossi M, Iusem ND (1995) Sequence of *Asr2*, a member of a gene family from *Lycopersicon esculentum* encoding chromosomal proteins: homology to an intron of the polygalacturonase gene. DNA Seq 5: 225–227.
30. Frankel N, Carrari F, Hasson E, Iusem ND (2006) Evolutionary history of the *Asr* gene family. Gene 378: 74–83.
31. The Tomato Genome Consortium (2012) The tomato genome sequence provides insights into fleshy fruit evolution. Nature 485: 635–641.
32. Fischer I, Camus-Kulandaivelu L, Allal F, Stephan W (2011) Adaptation to drought in two wild tomato species: The evolution of the *Asr* gene family. New Phytol 190: 1032–1044.
33. Fischer I, Steige KA, Stephan W, Mboup M (2013) Sequence evolution and expression regulation of stress-responsive genes in natural populations of wild tomato. PLoS One 8: e78182.
34. Hara M, Kumagai K, Kuboi T (2002) Characterization and expression of a water stress responsive gene from a seashore plant *Calystegia soldanella*. Plant Biotechnol 19: 277–281.
35. Yamada T, Ichimura K, Kanekatsu M, Doorn W (2007) Gene expression in opening and senescing petals of morning glory (*Ipomoea nil*) flowers. Plant Cell Reports 26: 823–835.
36. Battaglia M, Olvera-Carrillo Y, Garciarrubio A, Campos F, Covarrubias AA (2008) The enigmatic LEA proteins and other hydrophilins. Plant Physiol 148: 6–24.
37. Goldgur Y, Rom S, Ghirlando R, Shkolnik D, Shadrin N, et al. (2007) Desiccation and zinc binding induce transition of tomato Abscisic Acid Stress Ripening 1, a water stress- and salt stress-regulated plant-specific protein, from unfolded to folded state. Plant Physiol 143: 617–628.
38. Konrad Z, Bar-Zvi D (2008) Synergism between the chaperone-like activity of the stress regulated ASR1 protein and the osmolyte glycine-betaine. Planta 227: 1213–1219.
39. Kalifa Y, Gilad A, Konrad Z, Zaccai M, Scolnik PA, et al. (2004) The water- and salt-stress-regulated *Asr1* (abscisic acid stress ripening) gene encodes a zinc-dependent DNA-binding protein. Biochemical Journal 381: 373–378.
40. Wang HJ, Hsu CM, Guang YJ, Wang CS (2005) A lily pollen ASR protein localizes to both cytoplasm and nuclei requiring a nuclear localization signal. Physiol Plantar 123: 314–320.
41. Kim IS, Kim YS, Yoon HS (2012) Rice ASR1 protein with reactive oxygen species scavenging and chaperone-like activities enhances acquired tolerance to abiotic stresses in *Saccharomyces cerevisiae*. Mol Cells 33: 285–293.
42. Dai JR, Liu B, Feng DR, Liu H, He Y, et al. (2011) *MpAsr* encodes an intrinsically unstructured protein and enhances osmotic tolerance in transgenic Arabidopsis. Plant Cell Report 30: 1219–1230.
43. Rom S, Gilad A, Kalifa Y, Konrad Z, Karpasas MM, et al. (2006) Mapping the DNA- and zinc-binding domains of ASR1 (abscisic acid stress ripening), an abiotic-stress regulated plant specific protein. Biochimie 88: 621–628.
44. Ricardi MM, Guaimas FF, Gonzalez RM, Burrieza HP, Lopez-Fernandez MP, et al. (2012) Nuclear Import and dimerization of tomato ASR1, a water stress-inducible protein exclusive to plants. PLoS One 7: e41008.
45. Kim SJ, Lee SC, Hong SK, An K, An G, et al. (2009) Ectopic expression of a cold-responsive *OsAsr1* cDNA gives enhanced cold tolerance in transgenic rice plants. Mol Cells 27: 449–458.
46. Li RH, Liu GB, Wang H, Zheng YZ (2013) Effects of Fe^{3+} and Zn^{2+} on the structural and thermodynamic properties of a soybean ASR protein. Biosci Biotechnol Biochem 77: 475–481.
47. Saumonneau A, Agasse A, Bidoyen MT, Lallemand M, Cantereau A, et al. (2008) Interaction of grape ASR proteins with a DREB transcription factor in the nucleus. FEBS Lett 582: 3281–3287.
48. Yang CY, Chen YC, Jauh GY, Wang CS (2005) A lily ASR protein involves abscisic acid signaling and confers drought and salt resistance in Arabidopsis. Plant Physiol 139: 836–846.

49. Hu W, Huang C, Deng XM, Zhou SY, Chen LH, et al. (2013) *TaASR1*, a transcription factor gene in wheat, confers drought stress tolerance in transgenic tobacco. Plant Cell Environ 36: 1449–1464.

50. Shkolnik-Inbar D, Bar-Zvi D (2010) ABI4 mediates abscisic acid and cytokinin inhibition of lateral root formation by reducing polar auxin transport in Arabidopsis. Plant Cell 22: 3560–3573.

51. Jeanneau M, Gerentes D, Foueillassar X, Zivy M, Vidal J, et al. (2002) Improvement of drought tolerance in maize: towards the functional validation of the *Zm-Asr1* gene and increase of water use efficiency by over-expressing C4-PEPC. Biochimie 84: 1127–1135.

52. Jha B, Lal S, Tiwari V, Yadav SK, Agarwal PK (2012) The *SbASR-1* gene cloned from an extreme halophyte *Salicornia brachiata* enhances salt tolerance in transgenic tobacco. Marine Biotechnol 14: 782–792.

53. Kalifa Y, Perlson E, Gilad A, Konrad Z, Scolnik PA, et al. (2004) Over-expression of the water and salt stress-regulated *Asr1* gene confers an increased salt tolerance. Plant Cell Environ 27: 1459–1468.

54. Hsu YF, Yu SC, Yang CY, Wang CS (2011) Lily ASR protein-conferred cold and freezing resistance in Arabidopsis. Plant Physiol Biochem 49: 937–945.

55. Shkolnik D, Bar-Zvi D (2008) Tomato ASR1 abrogates the response to abscisic acid and glucose in Arabidopsis by competing with ABI4 for DNA binding. Plant Biotechnol J 6: 368–378.

56. Dominguez PG, Frankel N, Mazuch J, Balbo I, Iusem N, et al. (2013) ASR1 mediates glucose-hormone cross talk by affecting sugar trafficking in tobacco plants. Plant Physiol 161: 1486–1500.

57. Saumonneau A, Laloi M, Lallemand M, Rabot A, Atanassova R (2012) Dissection of the transcriptional regulation of grape *ASR* and response to glucose and abscisic acid. J Exper Bot 63: 1495–1510.

58. Virlouvet L, Jacquemot MP, Gerentes D, Corti H, Bouton S, et al. (2011) The *ZmASR1* protein nfluences branched-chain amino acid biosynthesis and maintains kernel yield in maize under water-limited conditions. Plant Physiol 157: 917–936.

59. Maskin L, Maldonado S, Iusem N (2008) Tomato leaf spatial expression of stress-induced *Asr* genes. Mol Biol Report 35: 501–505.

60. Maskin L, Gudesblat GE, Moreno JE, Carrari FO, Frankel N, et al. (2001) Differential expression of the members of the *Asr* gene family in tomato (*Lycopersicon esculentum*). Plant Sci 161: 739–746.

61. Liu X, Prat S, Willmitzer L, Frommer W (1990) cis regulatory elements directing tuber-specific and sucrose-inducible expression. Mol Gen Genet 223: 401–406.

62. Obiadalla-Ali H, Fernie AR, Lytovchenko A, Kossmann J, Lloyd JR (2004) Inhibition of chloroplastic fructose 1,6-bisphosphatase in tomato fruits leads to decreased fruit size, but only small changes in carbohydrate metabolism. Planta 219: 533–540.

63. Hoagland DR, Arnon DI (1950) The water-culture method for growing plants without soil. California Agricultural Experiment Station Circular 347: 1–32.

64. Chomczynski P, Sacchi N (1987) Single-step method of RNA isolation by acid guanidinium thiocyanate phenol chloroform extraction. Anal Biochem 162: 156–159.

65. Adler G, Blumwald E, Bar-Zvi D (2010) The sugar beet gene encoding the sodium/proton exchanger 1 (BvNHX1) is regulated by a MYB transcription factor. Planta 232: 187–195.

66. Shkolnik-Inbar D, Adler G, Bar-Zvi D (2013) ABI4 downregulates expression of the sodium transporter *HKT1;1* in Arabidopsis roots and affects salt tolerance. Plant J 73: 993–1005.

67. Rossi M, Carrari F, Cabrera-Ponce JL, Vazquez-Rovere C, Herrera-Estrella L, et al. (1998) Analysis of an abscisic acid (ABA)-responsive gene promoter belonging to the Asr gene family from tomato in homologous and heterologous systems. Mol Gen Genet 258: 1–8.

68. Amitai-Zeigerson H, Scolnik PA, Bar-Zvi D (1995) Tomato *Asr1* mRNA and protein are transiently expressed following salt stress, osmotic stress and treatment with abscisic acid. Plant Sci 110: 205–213.

69. Hong SH, Kim IJ, Yang DC, Chung WI (2002) Characterization of an abscisic acid responsive gene homologue from *Cucumis melo*. J Exper Bot 53: 2271–2272.

70. Liu J, Jia C, Dong F, Wang J, Zhang J, et al. (2013) Isolation of an abscisic acid senescence and ripening inducible gene from litchi and functional characterization under water stress. Planta 237: 1025–1036.

71. Doczi R, Kondrak M, Kovacs G, Beczner F, Banfalvi Z (2005) Conservation of the drought-inducible, DS2 genes and divergences from their ASR paralogues in solanaceous species. Plant Physiol and Biochem 43: 269–276.

72. Riccardi F, Gazeau P, de Vienne D, Zivy M (1998) Protein changes in response to progressive water deficit in maize - Quantitative variation and polypeptide identification. Plant Physiol 117: 1253–1263.

73. Maskin L, Frankel N, Gudesblat G, Demergasso MJ, Pietrasanta LI, et al. (2007) Dimerization and DNA-binding of ASR1, a small hydrophilic protein abundant in plant tissues suffering from water loss. Biochem Biophys Res Commun 352: 831–835.

74. Ricardi MM, Gonzalez RM, Zhong S, Dominguez PG, Duffy T, et al. (2014) Genome-wide data (ChIP-seq) enabled identification of cell wall-related and aquaporin genes as targets of tomato ASR1, a drought stress-responsive transcription factor. BMC Plant Biol 14: 29.

75. Gillaspy G, Bendavid H, Gruissem W (1993) Fruits - A Developmental perspective. Plant Cell 5: 1439–1451.

18

Arabidopsis CAPRICE (MYB) and GLABRA3 (bHLH) Control Tomato (*Solanum lycopersicum*) Anthocyanin Biosynthesis

Takuji Wada[1], Asuka Kunihiro[2], Rumi Tominaga-Wada[1]*

1 Graduate School of Biosphere Sciences, Hiroshima University, Higashi-Hiroshima, Hiroshima, Japan, **2** Faculty of Applied Biological Science, Hiroshima University, Higashi-Hiroshima, Hiroshima, Japan

Abstract

In *Arabidopsis thaliana* the MYB transcription factor CAPRICE (CPC) and the bHLH transcription factor GLABRA3 (GL3) are central regulators of root-hair differentiation and trichome initiation. By transforming the orthologous tomato genes *SlTRY* (*CPC*) and *SlGL3* (*GL3*) into *Arabidopsis*, we demonstrated that these genes influence epidermal cell differentiation in Arabidopsis, suggesting that tomato and *Arabidopsis* partially use similar transcription factors for epidermal cell differentiation. CPC and GL3 are also known to be involved in anthocyanin biosynthesis. After transformation into tomato, *35S::CPC* inhibited anthocyanin accumulation, whereas *GL3::GL3* enhanced anthocyanin accumulation. Real-time reverse transcription PCR analyses showed that the expression of anthocyanin biosynthetic genes including *Phe-ammonia lyase* (*PAL*), the flavonoid pathway genes *chalcone synthase* (*CHS*), *dihydroflavonol reductase* (*DFR*), and *anthocyanidin synthase* (*ANS*) were repressed in *35S::CPC* tomato. In contrast, the expression levels of *PAL, CHS, DFR*, and *ANS* were significantly higher in *GL3::GL3* tomato compared with control plants. These results suggest that *CPC* and *GL3* also influence anthocyanin pigment synthesis in tomato.

Editor: Takaya Moriguchi, NARO Institute of Fruit Tree Science, Japan

Funding: This work was supported by a Cooperative Research Grant of the Gene Research Center, the University of Tsukuba, and JSPS KAKENHI (Grant numbers 24658032, 23570057 and 25114513). The funders had no role in study design, data collection and analysis, decision to publish, or preparation of the manuscript.

Competing Interests: The authors have declared that no competing interests exist.

* Email: rtomi@hiroshima-u.ac.jp

Introduction

Anthocyanins are important chemical compound of polyphenolic pigments derived from the phenylpropanoid biosynthetic pathway. Anthocyanins belong to the group of flavonoids, of which they are noticeable in the wide range of chemical structures [1]. Anthocyanins provide appealing color to leaves, flowers, fruits and seeds in plants. In addition to this obvious feature, they have other essential functions. Anthocyanin synthesis was induced by the stressful occasions, such as low temperature or strong irradiation of the sunlight, against which they protect the plant as scavengers for radical species or a light-screen [2]. Anthocyanins are produced through several enzymatic step [3]. The enzymes which are involved in anthocyanin synthesis are fully analyzed by both biochemical and genetic approaches.

Thus, it is important to identify the regulatory factors governing this enzymatic steps. In *Arabidopsis*, anthocyanin biosynthesis is regulated by the TTG1-bHLH-MYB protein complex [4–10]. In *Arabidopsis*, overexpressions of *PAP1/MYB75*, *PAP2/MYB90*, *MYB113* and *MYB114*, which are R2R3-type MYB transcription factors, accelerate the anthocyanin accumulations in *Arabidopsis* [10,11]. Two homologous bHLH proteins, GLABRA3 (GL3) and ENHANCER OF GLABRA3 (EGL3) enhance anthocyanin biosynthesis together with PAP1 and PAP2 [7]. In contrast, *CAPRICE* (*CPC*), one of R3-type MYB genes, compete with the

binding of PAP1/2 to GL3/EGL3 and disrupt the TTG1-GL3/EGL3-PAP1/2 protein complex, thus inhibiting the activity of anthocyanin biosynthesis [12].

CPC has been initially identified as a key regulator of root-hair differentiation in *Arabidopsis thaliana* [13]. *Arabidopsis* has six additional *CPC*-like MYB genes in its genome, including *TRYPTICHON* (*TRY*), *ENHANCER OF TRY AND CPC1* (*ETC1*), *ENHANCER OF TRY AND CPC2* (*ETC2*), *ENHANCER OF TRY AND CPC3/CPC-LIKE MYB3* (*ETC3/CPL3*), *TRICHOMELESS1* (*TCL1*), and *TRICHOMELESS2/CPC-LIKE MYB4* (*TCL2/CPL4*) [14–22]. These *CPC*-like MYB family genes cooperatively regulate Arabidopsis epidermal cell differentiation including root-hair and trichome formation [14–23].

GL3 is also important for root-hair and trichome differentiation in *Arabidopsis* [24]. The gene products of *GL3*, *EGL3* [25], *WEREWOLF* (*WER*), which encodes an R2R3 type MYB protein [26] and *TRANSPARENT TESTA GLABRA1* (*TTG1*), which encodes a WD-40 protein [27] form a transcriptional complex [7,24,28]. This protein complex, including the WER, GL3/EGL3 and TTG1 proteins, controls transcription of the *GLABRA2* (*GL2*) gene [29]. The *GL2* gene encodes a homeodomain leucine zipper protein and is thought to act farthest downstream in the Arabidopsis root-hair and trichome differentiation regulatory pathway [13,26,27,30,31]. CPC moves form non-hair cells to hair cells where it disrupts TTG1-GL3/EGL3-WER transcriptional complex by competing the binding of WER [32].

In the previous study, we identified Arabidopsis *CPC* and *GL3* homologous genes from tomato and named them *Solanum lycopersicum TRYPTICHON* (*SlTRY*) and *Solanum lycopersicum GLABRA3* (*SlGL3*), respectively [33]. The *SlTRY*-encoded protein was most closely related to TRY among the CPC-like MYBs [33]. Transformants expressing the tomato *TRY* homologous gene (*SlTRY*) in Arabidopsis had a greater number of root-hairs and no trichomes, a phenotype similar to that seen in over-expressors of *CPC*-like MYB genes. On the other hand, transformants expressing the tomato *GL3* homologous gene (*SlGL3*) in Arabidopsis had no obvious *GL3*-like phenotypes related to non-hair and trichome cell differentiation [33]. We concluded that tomato and Arabidopsis use similar transcription factors for root-hair and trichome cell differentiation and that the *SlTRY*-like R3 MYB may be a key common regulator of plant root-hair and trichome development [33]. In prior work, we also analyzed the anthocyanin content of *SlTRY* and *SlGL3* transgenic Arabidopsis [34]. We showed that anthocyanin accumulation was repressed in the *CPC::SlTRY* and *GL3::SlGL3* transgenic Arabidopsis plants, suggesting that the tomato genes of *SlTRY* and *SlGL3* are involved in anthocyanin biosynthesis [34].

In this study, we have expressed the Arabidopsis *CPC* and *GL3* genes in tomato to show the effect of these genes on tomato anthocyanin biosynthesis, indicating that *GL3* is a positive regulator for anthocyanin biosynthesis, but *CPC* is a negative regulator.

Materials and Methods

Plant materials and growth conditions

Tomato, *Solanum lycopersicum* L. cv. Micro-Tom, was used. Seeds were surface-sterilized with 10% commercial bleach including a detergent (Kitchen Haiter, Kao, Tokyo, Japan), for 20 min and then rinsed with sterilized water three times for 5 min each and sown on 1.5% agar plates containing 0.5× MS medium [35]. Seeded plates were held at 4°C for 2 d and then incubated at 25°C under constant white light (50–100 μmol m^{-2} s^{-1}) for 7 days to produce seedlings for RNA extraction. Some 7-day-old seedlings were transplanted into soil and grown in a photoperiod of 16 h light (50–100 μmol m^{-2} s^{-1}) at 25°C for an additional week to produce mature plant tissues for anthocyanin extraction.

Transgenic plants

Gene constructs of *35S::CPC* [13] and *GL3::GL3* [36] were introduced into tomato (Micro-Tom) according to a highly efficient transformation protocol for Micro-Tom [37]. *Agrobacterium tumefaciens* C58C1 was grown for 24 h at 28°C. Cotyledon explants were sectioned, dipped in the bacterial suspension to allow adsorption, and transferred to callus induction medium containing 100 mg L^{-1} kanamycin, 1.5 mg L^{-1} zeatin and 375 mg L^{-1} Augmentin (GlaxoSmithKline, Uxbridge, UK) [37]. Transgenic shoots were selected and rooted on a medium containing 50 mg/L kanamycin.

Homozygous transgenic lines were selected based on kanamycin resistance. We obtained ten and four T2 transgenic tomato lines and selected eight and three homozygous lines of *35S::CPC* and *GL3::GL3*, respectively. The presence of *35S::CPC* and *GL3::GL3* in the transgenic plants was confirmed by PCR using *CPC* or *GL3* forward and reverse primers (Table 1) (Figure S1). Only those plants with the expected PCR products (*CPC* and *GL3*) were used in the analyses.

Real-time reverse transcription PCR analysis

The sequences of all primers used in this study are listed in Table 1. Total RNA from tomato tissues was extracted with MagDEA RNA 100 (GC) (PSS, Chiba, Japan) using a Magtration System 12 GC (PSS, Chiba, Japan). To remove contaminating genomic DNA, RNA samples were treated with RNase-free DNase I (Ambion, Austin, TX, USA) according to the Magtraction System protocol. Plant tissue (100 mg) was homogenized using a TissueLyser II (Qiagen, Valencia, CA, USA) with 100 μL of RLT buffer (Qiagen, Valencia, CA, USA). Sample supernatants were applied to the instrument, and RNA was eluted with 50 mL of sterile distilled water.

First-strand cDNA was synthesized from 1 μg total RNA in a 20 μL reaction mixture using the Prime Script RT Master Mix (Perfect Real Time) (Takara, Tokyo, Japan). Real-time PCR was performed using a Chromo4 Real-Time IQ5 PCR Detection System (Bio-Rad, Hercules, CA, USA) with SYBR Premix Ex Taq II (Takara, Tokyo, Japan). PCR amplification employed a 30 s denaturing step at 95°C, followed by 5 s at 95°C and 30 s at 60°C with 40 cycles for *CPC*, *GL3*, *PAL*, *CHS*, *DFR*, *ANS* and *LeActin*. Real-time PCR was used to analyze the mRNA expression level of each transcript encoding *CPC*, *GL3*, *PAL*, *CHS*, *DFR* and *ANS* in transgenic tomato. The relative expression of each transcript was calculated by the $\Delta\Delta$CT method [38]. The expression levels of *CPC*, *GL3*, *PAL*, *CHS*, *DFR* and *ANS* were estimated after being normalized to the endogenous control gene *LeActin* (TC116322) [39]. The primers were: *CPC-F* and *CPC-R* for *CPC*; *GL3-F* and *GL3-R* for *GL3*; *PAL-F* and *PAL-R* for *PAL*; *CHS-F* and *CHS-R* for *CHS*; *DFR-F* and *DFR-R* for *DFR*; *ANS-F* and *ANS-R* for *ANS*; *LeActin-F* and *LeActin-R* for *LeActin* [39–41].

Extraction and analysis of anthocyanins

Anthocyanin levels were measured according to previously reported protocols [42,43]. Control and transgenic plants were grown together in a growth chamber as described above. Anthocyanins were extracted from cotyledons of 7-day-old seedlings, leaves and stems of three-week-old plants, and fresh weights were determined. Total plant pigments were extracted overnight in 0.3 mL acidic methanol (1% (v/v) HCl). After the addition of 0.2 mL water and an equal volume of chloroform, anthocyanins were separated from the chlorophylls by partitioning into the aqueous methanol phase, and the absorption was measured at 530–657 nm in a spectrometer (GENios, TECAN). Anthocyanin levels were then normalized to the total fresh weight of tissue used in each sample.

Light microscopy

To observe anthocyanin pigment localization in hypocotyls of the control, *35S::CPC* and *GL3::GL3* transgenic plants, we prepared hand-cut sections from 3-week-old plants and observed them by light microscopy using a Zeiss (Jena, Germany) Axio Imager. Z1 microscope.

Results

Anthocyanin pigmentation of the *35S::CPC* and *GL3::GL3* transgenic plants

To establish whether Arabidopsis CPC and GL3 transcription factors function in tomato, we introduced these genes into one of tomato cultivars (*Solanum lycopersicum* L. cv. Micro-Tom). Previously, we showed that *35S::CPC* transgenic Arabidopsis plants have an unusually large number of root-hairs and no leaf trichomes [13]. Thus, we chose to introduce the *35S::CPC*

Table 1. Primer sequences used in this study.

Primer Name	Sequence (5′ to 3′)
CPC-F	5′-GGATGTATAAACTCGTTGGCGACAG-3′
CPC-R	5′-GCCGTGTTTCATAAGCCAATATCTC-3′
GL3-F	5′-GATAACCATCGCAGGACTAAGC-3′
GL3-R	5′-CCCACTCAAGACTACTCACTTCTG-3′
PAL-F	5′-ATTGGGAAATGGCTGCTGATT-3′
PAL-R	5′-TCAACATTTGCAATGGATGCA-3′
CHS-F	5′-TGGTCACCGTGGAGGAGTATC-3′
CHS-R	5′-GATCGTAGCTGGACCCTCTGC-3′
DFR-F	5′-CAAGGCAGAGGGAAGATTCATTTG-3′
DFR-R	5′-GCACCATCTTAGCCACATCGTA-3′
ANS-F	5′-GAACTAGCACTTGGCGTCGAA-3′
ANS-R	5′-TTGCAAGCCAGGCACCATA-3′
LeActin-F	5′-TGTCCCTATTTACGAGGGTTATGC-3′
LeActin-R	5′-CAGTTAAATCACGACCAGCAAGAT-3′

construct into tomato in this experiment. In contrast, the root-hair number of *35S::GL3* transgenic Arabidopsis plants is not significantly different from the wild-type [25], suggesting that the 35S promoter is not suitable for *GL3* gene overexpression. The expression of *GL3* should be precisely controlled by the *GL3* promoter [31]. Therefore, we decided to use the *GL3::GL3* construct (a genome fragment of *GL3* driven by the *GL3* promoter) for transformation of tomato in this study [36].

The *35S::CPC* and *GL3::GL3* transgenic tomato plants were phenotypically similar to the control plants (Figure 1; Figure S1). We did not detect any remarkable differences between *35S::CPC* or *GL3::GL3* transgenic tomato plants and the control tomato plant in root-hair and trichome phenotypes (Figure 1; Figure S1). On the other hand, we observed qualitatively less and more reddish-purple coloration in the stems and leaves of *35S::CPC* and *GL3::GL3* plants, respectively (Figure 2A, 2D and 2G). The first true leaves of two-week-old *35S::CPC* transgenic plants had clearly lower amounts of anthocyanin pigmentation on the adaxial and abaxial sides of the leaves compared with that of the control plants (Figure 2B, 2C, 2E and 2F). Control plant leaves accumulated reddish-purple anthocyanin mainly in the leaf veins on the adaxial side and nearly the entire surface of the abaxial side of the leaves (Figure 2B, 2C). Leaf veins of the *35S::CPC* plants were pale green and no anthocyanin accumulation was observed on either side of the leaves (Figure 2E and 2F). On the other hand, leaves of the *GL3::GL3* plants accumulated greater amounts of

Figure 1. Leaf and root epidermal phenotypes of *35S::CPC* and *GL3::GL3* transgenic tomato plants. (A) The first true leaf from the two-week old control plant. (B) Close-up view of the adaxial side of the leaf shown in A. (C) Five-day-old seedling roots of control plants. (D) The first true leaf from the two-week old *35S::CPC* plant. (E) Close-up view of the adaxial side of the leaf shown in E. (F) Five-day-old seedling roots of *35S::CPC* plants. (G) The first true leaf from the two-week old *GL3::GL3* plant. (H) Close-up view of the adaxial side of the leaf shown in G. (I) Five-day-old seedling roots of *GL3::GL3* plants. Scale bars: 1 mm in A, C, D, F, G and I; 20 μm in B, E and H.

Figure 2. Phenotypes of *35S::CPC* and *GL3::GL3* transgenic tomato plants. (A) Two-week old control plant. (B) Adaxial side of the first true leaf from the plant shown in A. (C) Abaxial side of the first true leaf from the plant shown in A. (D) Two-week-old *35S::CPC* transgenic plant. (E) Adaxial side of the first true leaf from the plant shown in D. (F) Abaxial side of the first true leaf from the plant shown in D. (G) Two-week-old *GL3::GL3* transgenic plant. (H) Adaxial side of the first true leaf from the plant shown in G. (I) Abaxial side of the first true leaf from the plant shown in G. Scale bars: 1 cm in A for A, D and G; 5 mm in B for B, C, E, F, H and I.

reddish-purple anthocyanin in the leaf veins compared with the control plants (Figure 2H, 2I).

To determine the tissue distribution of anthocyanin pigments in the *35S::CPC* and *GL3::GL3* transgenic tomato plants, we examined hand-cut sections prepared from stem samples with a light microscope as shown in Figure 2A, 2D and 2G. In hypocotyls of two-week-old control tomato seedlings, anthocyanin pigments were observed in a few cells, as was previously reported in tomato hypocotyls (Figure 3A) [44]. Anthocyanins did not accumulate in the hypocotyls of young *35S::CPC* tomato seedlings (Figure 3B). In the hypocotyls of *GL3::GL3* seedlings, anthocyanin pigments were present in two to three layers of an epidermal cell and subepidermal cells (Figure 3C). These results suggest that *CPC* expression did not induce any remarkable changes in root-hair and trichome formation but reduced anthocyanin accumulation in transgenic tomato. *GL3* also did not affect the epidermal phenotype but induced anthocyanin accumulation in transgenic tomato.

Analysis of anthocyanin levels in the cotyledons, leaves and stems of transgenic plants

We examined the effects of *CPC* and *GL3* on anthocyanin accumulation in the different tissues. Expression levels of the introduced *CPC* gene were checked by PCR, and we selected three lines (*35S::CPC#10, 35S::CPC #18* and *35S::CPC #21*) among eight transgenic lines for analysis (Figure S2A). Expression levels of the introduced *GL3* gene were also checked by PCR.

Among three *GL3::GL3* transgenic lines, only one line, *GL3::GL3#12*, showed stable expression of *GL3*. Therefore, we used the *GL3::GL3#12* line for further analyses (Figure S2B). To compare the levels of anthocyanin accumulation in *35S::CPC* and *GL3::GL3* with those in control tomato, the anthocyanin content in extracts of two-week-old seedlings was determined (Figure 4). Compared with the control tomato cotyledons, all three lines of *35S::CPC* transgenic tomato cotyledons had significantly reduced levels of anthocyanin (Figure 4A). On the other hand, cotyledons of *GL3::GL3* accumulated higher levels of anthocyanin compared with that of the control plants (Figure 4A). Consistent with the observations shown in Figure 2, very low levels of anthocyanin accumulation were observed in leaves of all three *35S::CPC* lines (Figure 4B). Compared with control tomato leaves, significantly larger amounts of anthocyanin were measured in *GL3::GL3* leaves (Figure 4B). Consistent with the observations shown in Figure 2 and 3, anthocyanin accumulation was also significantly reduced in the stems of all three *35S::CPC* lines and increased in *GL3::GL3* stems compared with those in the control plants (Figure 4C). We confirmed that introduction of the *CPC* gene under the control of the 35S promoter significantly inhibited anthocyanin accumulation in cotyledons, leaves and stems of tomato as observed in Arabidopsis [12]. Introduction of the *GL3* gene under the control of the *GL3* promoter significantly increased anthocyanin accumulation also in mature leaves and stems of tomato as observed in Arabidopsis [45].

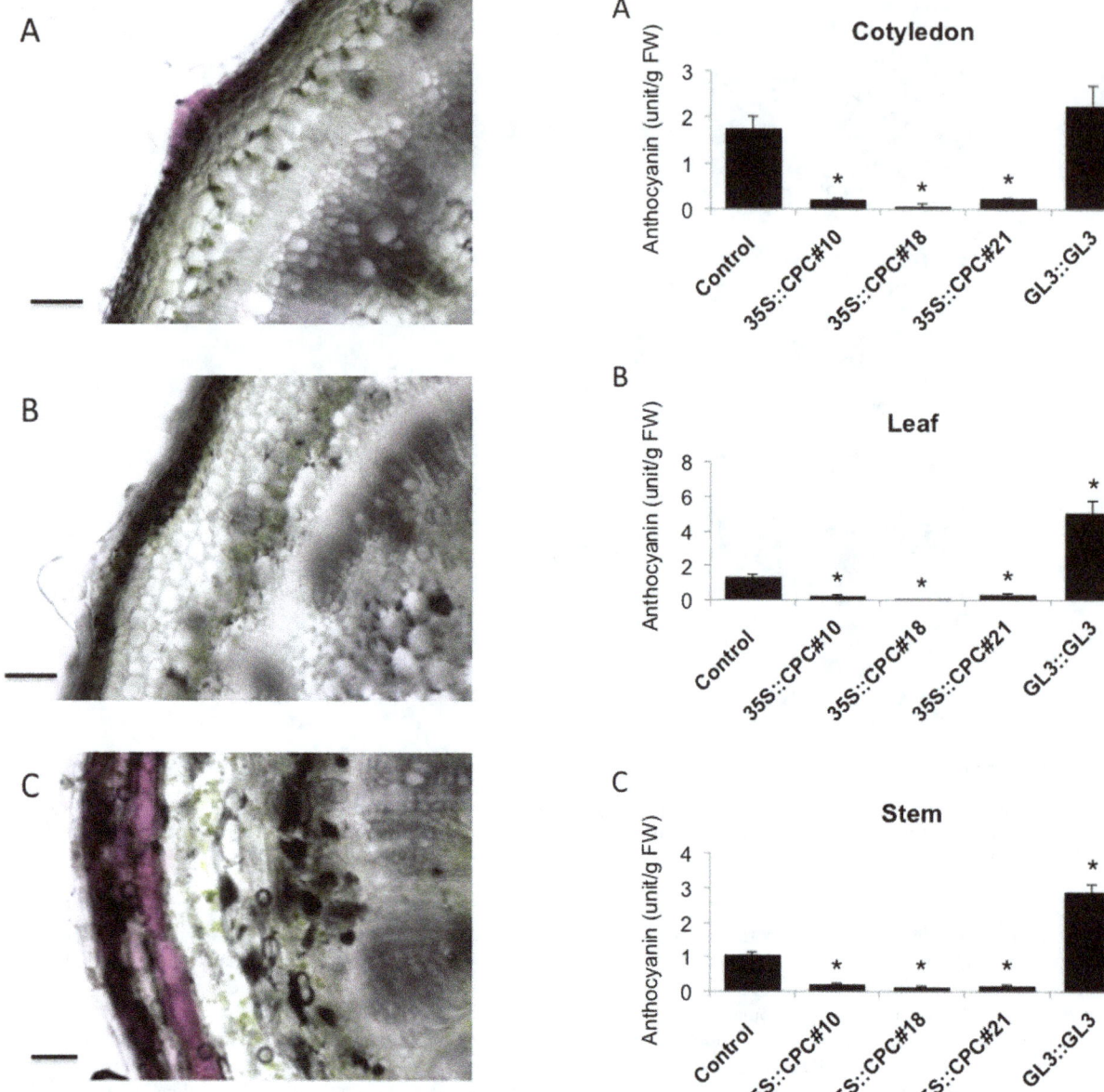

Figure 3. Stem phenotypes of *35S::CPC* and *GL3::GL3* transgenic tomato plants. (A) Transverse section of a hypocotyl of a two-week-old control plant. (B) Transverse section of a hypocotyl of a two-week-old *35S::CPC* transgenic plant. (C) Transverse section of a hypocotyl of a two-week-old *GL3::GL3* transgenic plant. Scale bars: 100 μm.

Figure 4. Anthocyanin content in control, *35S::CPC* and *GL3::GL3* transgenic tomato plants. (A) The anthocyanin content of cotyle-dons from control, *35S::CPC#10*, *35S::CPC#18*, *35S::CPC#21* and *GL3::GL3* plants are shown. (B) The anthocyanin content in leaves of control, *35S::CPC#10*, *35S::CPC#18*, *35S::CPC#21* and *GL3::GL3* plants are shown. (C) The anthocyanin content in stems of control, *35S::CPC#10*, *35S::CPC#18*, *35S::CPC#21* and *GL3::GL3* plants are shown. Error bars indicate the standard deviations. Bars marked with asterisks indicate a significant difference between the control and the *35S::CPC* or the *GL3::GL3* transgenic lines by Student's *t*-test (P<0.050).

Effect of *CPC* and *GL3* on the expression of anthocyanin pathway genes

To characterize more fully the involvement of the introduced CPC and GL3 transcription factors on the regulation of anthocyanin biosynthesis in tomato, we examined the expression levels of genes that encode anthocyanin biosynthetic enzymes. The effects of *CPC* and *GL3* on the expression of anthocyanin biosynthesis genes were examined by real-time RT-PCR, as described in the Materials and Methods section. First and second true-leaf samples of representative *35S::CPC*, *GL3::GL3* and control plants, harvested from two-week-old seedlings, were homogenized, and total RNA was isolated from each tissue sample. Anthocyanins are synthesized through the flavonoid

biosynthetic pathway [46]. Therefore, expression levels of tomato genes for *Phe-ammonia lyase* (*PAL*), the flavonoid pathway genes *chalcone synthase* (*CHS*), *dihydroflavonol reductase* (*DFR*), and *anthocyanidin synthase* (*ANS*) were determined and expressed relative to the *LeActin* gene, a tomato gene that encodes an actin protein [39]. Consistent with the reduced anthocyanin accumulation in *35S::CPC* transgenic tomato (Figure 4B), *PAL*, *CHS*, *DFR* and *ANS* expression levels were significantly lower in

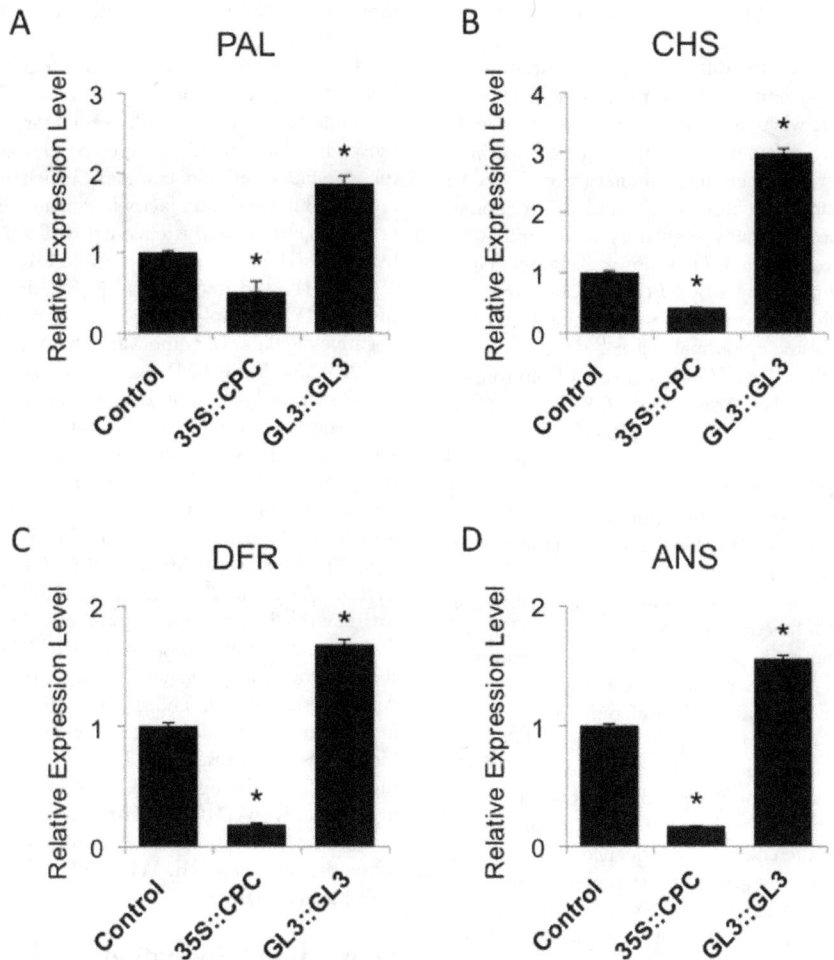

Figure 5. Expression analysis of genes associated with the anthocyanin biosynthetic pathway in tomato leaves. Enzyme names of the anthocyanin biosynthetic pathway are abbreviated as follows: phenyl alanine ammonia-lyase (PAL), chalcone synthase (CHS), dihydroflavonol 4-reductase (DFR), leucoanthocyanidin dioxygenase (ANS). (A) Real-time reverse transcription PCR analysis of *PAL* gene expression in *35S::CPC* and *GL3::GL3* transgenic tomato leaves. (B) Real-time reverse transcription PCR analysis of *CHS* gene expression in *35S::CPC* and *GL3::GL3* transgenic tomato leaves. (C) Real-time reverse transcription PCR analysis of *DFR* gene expression in *35S::CPC* and *GL3::GL3* transgenic tomato leaves. (D) Real-time reverse transcription analysis of *ANS* gene expression in *35S::CPC* and *GL3::GL3* transgenic tomato leaves. Total RNA was isolated from the indicated leaves from two-week-old tomato plants. Expression levels of *PAL*, *CHS*, *DFR* and *ANS3* in each sample relative to those in the control plants are shown. The experiments were repeated three times. Error bars indicate the standard error. Bars marked with asterisks indicate a significant difference between control and indicated transgenic plants by Student's *t*-test ($P < 0.050$).

35S::CPC transgenic tomato compared with the control plants (Figure 5). In contrast, consistent with the large amount of anthocyanin accumulation in *GL3::GL3* transgene tomato (Figure 4B), *PAL*, *CHS*, *DFR* and *ANS* expression levels were significantly higher in *GL3::GL3* transgenic tomato compared with control plants (Figure 5). These results suggest that Arabidopsis *CPC* and *GL3* can regulate gene expression of the anthocyanin biosynthetic pathway in tomato and affect the anthocyanin accumulation.

Discussion

In this study, we introduced the Arabidopsis *CPC* and *GL3* genes into tomato under the control of the *35S* promoter and the *GL3* promoter, respectively. Overexpression of *CPC* is known to induce root-hair cell differentiation and inhibits trichome forma-

tion in Arabidopsis [13]. Overexpression of *GL3* is known to reduce root-hair cell differentiation and induce trichome formation in Arabidopsis [25,31]. However, overexpression of *CPC* and *GL3* in tomato did not result in visible differences in the root-hair and trichome phenotypes (Figure 1; Figure S1). The reasons for the differences in CPC and GL3 function between tomato and Arabidopsis may arise from fundamental differences in the way epidermal organs develop in the two plants. Root epidermal development in vascular plants is classified into three types [47]. Tomato root epidermal development belongs to type 1, in which root-hairs can be produced from any root epidermal cell [48]. Conversely, Arabidopsis root epidermal development belongs to type 3 in which root-hair cell files and non-hair cell files are organized in the root epidermis [47]. Regulation of root-hair cell and non-hair cell fate determination by the TTG1-GL3/EGL3-

WER complex and CPC might be specific for Arabidopsis but not for tomato.

Trichome phenotypes are also different between Arabidopsis and tomato. Arabidopsis trichomes are normally large single cells with three branches [49], whereas tomato trichomes are chemically and morphologically divergent [50–52]. Tomato trichomes are classified into seven types, including glandular (types I, IV, VI and VII), and non-glandular (types II, III and V) trichomes [51,53]. The participation of many regulatory genes might be necessary to form tomato trichomes. Hence, it is likely difficult to change tomato trichome phenotypes by CPC or GL3 overexpression only. Tomato might need other transcriptional factors to change the morphology of the epidermal cell.

In a previous study, we isolated SlTRY and SlGL3 from tomato as orthologous genes of the Arabidopsis CPC and GL3, respectively [33]. The full length SlTRY protein shares 50% amino acid identity with CPC [33]. Phylogenic analysis suggested that SlTRY and CPC originated from a single common ancestor [33]. SlTRY was shown to function quite similarly to the Arabidopsis CPC, including in the formation of ectopic root-hairs, in the induction of a no-trichome phenotype and in its action as a repressor of anthocyanin accumulation in Arabidopsis [34]. In summary, SlTRY functions in a similar way as CPC for the epidermal cell differentiation and the anthocyanin accumulation in Arabidopsis. On the other hand, there was no obvious effect on trichome or non-hair cell differentiation in the Arabidopsis GL3::SlGL3 transformants [33]. Rather, anthocyanin accumulation was reduced in the GL3::SlGL3 transgenic Arabidopsis compared with the wild-type [34]. In contrast, GL3 functions as a positive regulator for the anthocyanin accumulation in Arabidopsis [7]. The difference of the sequence between GL3 and SlGL3 might contribute to the opposite functions although they share 45% amino acid identity at the entire region [33]. Taken together, the functions of SlGL3 are completely different from those of GL3.

In this study, we demonstrated that Arabidopsis CPC and GL3 genes regulate anthocyanin biosynthesis in tomato. We made 35S::CPC transgenic tomatoes that accumulated significantly less anthocyanin in comparison with the control plants (Figure 4). In contrast, anthocyanin accumulation in GL3::GL3 transgenic tomato was greater than the control plants (Figure 4). CPC and GL3 are known to regulate anthocyanin biosynthesis in Arabidopsis [12,54]. Our study suggests that the regulatory system for anthocyanin biosynthesis by CPC and GL3 is maintained in both Arabidopsis and tomato.

Genes encoding enzymes of the anthocyanin biosynthetic pathway are divided into two groups: early biosynthetic genes including PAL and CHS, and late biosynthetic genes including DFR and ANS. The two groups have independent activation mechanisms in dicotyledonous species [55,56]. Whereas PAL and CHS are involved in the synthesis of precursors and flavonoids, DFR and ANS are more specific for the synthesis of anthocyanins. Analysis of the biosynthetic pathway genes in tomato showed that genes of both groups were regulated by CPC and GL3. Expression levels of PAL, CHS DFR and ANS were significantly lower in 35S::CPC transgenic tomato compared with the control plants (Figure 5). In contrast, expression levels of PAL, CHS DFR and ANS were significantly higher in GL3::GL3 transgenic tomato compared with the control plants (Figure 5). GL3 and CPC were strong up- and down-regulators of the entire anthocyanin biosynthesis pathway in tomato, respectively (Figure 5), which reflect the results form Arabidopsis [7,12]. These results suggest the presence of a TTG1-TT8/GL3-PAP1/2 like protein complex

that specifically regulates anthocyanin biosynthesis in tomato [45,57–59].

Many studies contributed to the elucidation of the anthocyanin biosynthetic pathway using Arabidopsis [10,60–64]. As a result, the molecular genetics of the regulatory system for anthocyanin biosynthesis has greatly progressed [1,46,65–67]. In Arabidopsis, the regulatory protein complex, which includes WD40, bHLH and MYB transcription factors, regulates anthocyanin biosynthesis [10,58,68,69]. WD40 is encoded by TTG1, bHLHs are encoded by TT8, GL3 and EGL3, and MYBs are encoded by PAP1, PAP2, MYB113 and MYB114 [65]. In addition to the WD40-bHLH-MYB complex, CPC, a single repeat R3-MYB, is a negative regulator of anthocyanin biosynthesis in Arabidopsis [12]. MYBL2, another R3-MYB gene, functions as a negative regulator of anthocyanin biosynthesis in Arabidopsis seedlings [70,71]. Our study suggests the existence of a WD40-bHLH-MYB complex that regulates anthocyanin biosynthesis in tomato. CPC may disrupt this putative WD40-bHLH-MYB protein complex, thus inhibiting the activity of downstream anthocyanin biosynthetic genes in tomato. In Arabidopsis, there are a total of seven CPC family R3-type MYB genes, including CPC, TRY, ETC1 ETC2, ETC3/CPL3, TCL1 and TCL2/CPL4 [14–22]. In contrast, only SlTRY was identified as a putative tomato ortholog of CPC so far [33]. Although the total number of tomato CPC orthologous gene(s) is still unknown, fewer genes are expected than are present in the Arabidopsis genome. The small number of R3-type MYB gene(s) in tomato might reflect their specific functions in anthocyanin biosynthesis. Because SlGL3 did not induce anthocyanin accumulation in Arabidopsis [34], SlGL3 probably does not participate in the putative WD40-bHLH-MYB protein complex in tomato as is the case in Arabidopsis. A model for regulating anthocyanin biosynthesis in tomato by WD40-bHLH-MYB will be forthcoming with further analyses.

Supporting Information

Figure S1 Root and leaf epidermal phenotypes of 35S::CPC and GL3::GL3 transgenic tomato plants. Five-day-old seedlings (left panels) and two-week-old plants (right panels) from control (top), 35S::CPC (middle) and GL3::GL3 (bottom) transgenic plants.

Figure S2 *CPC* or *GL3* expression in the transgenic tomato plants. (A) Real-time reverse transcription PCR analyses of the CPC gene in eight 35S::CPC (#6, #10, #15, #18, #20, #21, #24 and #26) transgenic plants. Expression levels of CPC in each line are reported relative to that of transgenic line #10. (B) Real-time reverse transcription PCR analyses of the GL3 gene in three GL3::GL3 (#4, #12 and #22) transgenic plants. Expression levels of GL3 in each line are reported relative to that of transgenic line #4. Expression levels were normalized to Act2 expression. The experiment was repeated three times. Error bars indicate the standard deviations.

Acknowledgments

We thank Shusei Sato for useful suggestions and Yuka Nukumizu for technical support.

Author Contributions

Conceived and designed the experiments: TW RT. Performed the experiments: TW AK RT. Analyzed the data: TW RT. Wrote the paper: TW RT.

References

1. Holton TA, Cornish EC (1995) Genetics and Biochemistry of Anthocyanin Biosynthesis. Plant Cell 7: 1071–1083.

2. Gould KS (2004) Nature's Swiss army knife: The diverse protective roles of anthocyanins in leaves. Journal of Biomedicine and Biotechnology: 314–320.

3. Li S (2014) Transcriptional control of flavonoid biosynthesis: Fine-tuning of the MYB-bHLH-WD40 (MBW) complex. Plant Signal Behav 9.

4. Quattrocchio F, Wing JF, van der Woude K, Mol JN, Koes R (1998) Analysis of bHLH and MYB domain proteins: species-specific regulatory differences are caused by divergent evolution of target anthocyanin genes. Plant J 13: 475–488.

5. Larkin JC, Walker JD, Bolognesi-Winfield AC, Gray JC, Walker AR (1999) Allele-specific interactions between ttg and gl1 during trichome development in Arabidopsis thaliana. Genetics 151: 1591–1604.

6. Walker AR, Davison PA, Bolognesi-Winfield AC, James CM, Srinivasan N, et al. (1999) The TRANSPARENT TESTA GLABRA1 locus, which regulates trichome differentiation and anthocyanin biosynthesis in Arabidopsis, encodes a WD40 repeat protein. Plant Cell 11: 1337–1350.

7. Zhang F, Gonzalez A, Zhao M, Payne CT, Lloyd A (2003) A network of redundant bHLH proteins functions in all TTG1-dependent pathways of Arabidopsis. Development 130: 4859–4869.

8. Carey CC, Strahle JT, Selinger DA, Chandler VL (2004) Mutations in the pale aleurone color1 regulatory gene of the Zea mays anthocyanin pathway have distinct phenotypes relative to the functionally similar TRANSPARENT TESTA GLABRA1 gene in Arabidopsis thaliana. Plant Cell 16: 450–464.

9. Schwinn K, Venail J, Shang Y, Mackay S, Alm V, et al. (2006) A small family of MYB-regulatory genes controls floral pigmentation intensity and patterning in the genus Antirrhinum. Plant Cell 18: 831–851.

10. Gonzalez A, Zhao M, Leavitt JM, Lloyd AM (2008) Regulation of the anthocyanin biosynthetic pathway by the TTG1/bHLH/Myb transcriptional complex in Arabidopsis seedlings. Plant J 53: 814–827.

11. Borevitz JO, Xia Y, Blount J, Dixon RA, Lamb C (2000) Activation tagging identifies a conserved MYB regulator of phenylpropanoid biosynthesis. Plant Cell 12: 2383–2394.

12. Zhu HF, Fitzsimmons K, Khandelwal A, Kranz RG (2009) CPC, a single-repeat R3 MYB, is a negative regulator of anthocyanin biosynthesis in Arabidopsis. Mol Plant 2: 790–802.

13. Wada T, Tachibana T, Shimura Y, Okada K (1997) Epidermal cell differentiation in Arabidopsis determined by a Myb homolog, CPC. Science 277: 1113–1116.

14. Schellmann S, Schnittger A, Kirik V, Wada T, Okada K, et al. (2002) TRIPTYCHON and CAPRICE mediate lateral inhibition during trichome and root hair patterning in Arabidopsis. EMBO J 21: 5036–5046.

15. Kirik V, Simon M, Huelskamp M, Schiefelbein J (2004) The ENHANCER OF TRY AND CPC1 gene acts redundantly with TRIPTYCHON and CAPRICE in trichome and root hair cell patterning in Arabidopsis. Dev Biol 268: 506–513.

16. Kirik V, Simon M, Wester K, Schiefelbein J, Hulskamp M (2004) ENHANCER of TRY and CPC 2 (ETC2) reveals redundancy in the region-specific control of trichome development of Arabidopsis. Plant Mol Biol 55: 389–398.

17. Esch JJ, Chen MA, Hillestad M, Marks MD (2004) Comparison of TRY and the closely related At1g01380 gene in controlling Arabidopsis trichome patterning. Plant J 40: 860–869.

18. Simon M, Lee MM, Lin Y, Gish L, Schiefelbein J (2007) Distinct and overlapping roles of single-repeat MYB genes in root epidermal patterning. Dev Biol 311: 566–578.

19. Tominaga R, Iwata M, Sano R, Inoue K, Okada K, et al. (2008) Arabidopsis CAPRICE-LIKE MYB 3 (CPL3) controls endoreduplication and flowering development in addition to trichome and root hair formation. Development 135: 1335–1345.

20. Wang S, Kwak SH, Zeng Q, Ellis BE, Chen XY, et al. (2007) TRICHOME-LESS1 regulates trichome patterning by suppressing GLABRA1 in Arabidopsis. Development 134: 3873–3882.

21. Gan L, Xia K, Chen JG, Wang S (2011) Functional Characterization of TRICHOMELESS2, a New Single-Repeat R3 MYB Transcription Factor in the Regulation of Trichome Patterning in Arabidopsis. BMC Plant Biol 11: 176.

22. Tominaga-Wada R, Nukumizu Y (2012) Expression Analysis of an R3-Type MYB Transcription Factor CPC-LIKE MYB4 (TRICHOMELESS2) and CPL4-Related Transcripts in Arabidopsis. Int J Mol Sci 13: 3478–3491.

23. Hulskamp M, Misra S, Jurgens G (1994) Genetic dissection of trichome cell development in Arabidopsis. Cell 76: 555–566.

24. Payne CT, Zhang F, Lloyd AM (2000) GL3 encodes a bHLH protein that regulates trichome development in arabidopsis through interaction with GL1 and TTG1. Genetics 156: 1349–1362.

25. Bernhardt C, Lee MM, Gonzalez A, Zhang F, Lloyd A, et al. (2003) The bHLH genes GLABRA3 (GL3) and ENHANCER OF GLABRA (EGL3) specify epidermal cell fate in the Arabidopsis root. Development 130: 6431–6439.

26. Lee MM, Schiefelbein J (1999) WEREWOLF, a MYB-related protein in Arabidopsis, is a position-dependent regulator of epidermal cell patterning. Cell 99: 473–483.

27. Galway ME, Masucci JD, Lloyd AM, Walbot V, Davis RW, et al. (1994) The TTG gene is required to specify epidermal cell fate and cell patterning in the Arabidopsis root. Dev Biol 166: 740–754.

28. Esch JJ, Chen M, Sanders M, Hillestad M, Ndkium S, et al. (2003) A contradictory GLABRA3 allele helps define gene interactions controlling trichome development in Arabidopsis. Development 130: 5885–5894.

29. Koshino-Kimura Y, Wada T, Tachibana T, Tsugeki R, Ishiguro S, et al. (2005) Regulation of CAPRICE Transcription by MYB Proteins for Root Epidermis Differentiation in Arabidopsis. Plant Cell Physiol 46: 817–826.

30. Rerie WG, Feldmann KA, Marks MD (1994) The GLABRA2 gene encodes a homeo domain protein required for normal trichome development in Arabidopsis. Genes Dev 8: 1388–1399.

31. Bernhardt C, Zhao M, Gonzalez A, Lloyd A, Schiefelbein J (2005) The bHLH genes GL3 and EGL3 participate in an intercellular regulatory circuit that controls cell patterning in the Arabidopsis root epidermis. Development 132: 291–298.

32. Tominaga-Wada R, Wada T (2014) Regulation of root hair cell differentiation by R3 MYB transcription factors in tomato and Arabidopsis. Front Plant Sci 5: 91.

33. Tominaga-Wada R, Nukumizu Y, Sato S, Wada T (2013) Control of Plant Trichome and Root-Hair Development by a Tomato (Solanum lycopersicum) R3 MYB Transcription Factor. PLoS One 8: e54019.

34. Tominaga-Wada R, Nukumizu Y, Wada T (2013) Tomato (Solanum lycopersicum) Homologs of TRIPTYCHON (SlTRY) and GLABRA3 (SlGL3) are involved in anthocyanin accumulation. Plant Signal Behav 8.

35. Murashige T, Skoog F (1962) A Revised Medium for Rapid Growth and Bio Assays with Tobacco Tissue Cultures. Physiologia Plantarum 15: 473–497.

36. Yoshida Y, Sano R, Wada T, Takabayashi J, Okada K (2009) Jasmonic acid control of GLABRA3 links inducible defense and trichome patterning in Arabidopsis. Development 136: 1039–1048.

37. Sun HJ, Uchii S, Watanabe S, Ezura H (2006) A highly efficient transformation protocol for Micro-Tom, a model cultivar for tomato functional genomics. Plant Cell Physiol 47: 426–431.

38. Livak KJ, Schmittgen TD (2001) Analysis of relative gene expression data using real-time quantitative PCR and the 2(−Delta Delta C(T)) Method. Methods 25: 402–408.

39. Girardi CL, Bermudez K, Bernadac A, Chavez A, Zouine M, et al. (2006) The mitochondrial elongation factor LeEF-Tsmt is regulated during tomato fruit ripening and upon wounding and ethylene treatment. Postharvest Biology and Technology 42: 1–7.

40. Povero G, Gonzali S, Bassolino L, Mazzucato A, Perata P (2011) Transcriptional analysis in high-anthocyanin tomatoes reveals synergistic effect of Aft and atv genes. J Plant Physiol 168: 270–279.

41. Bovy A, de Vos R, Kemper M, Schijlen E, Almenar Pertejo M, et al. (2002) High-flavonol tomatoes resulting from the heterologous expression of the maize transcription factor genes LC and C1. Plant Cell 14: 2509–2526.

42. Beggs CJ, Kuhn K, Böcker R, Wellmann E (1987) Phytochrome-induced flavonoid biosynthesis in mustard (Sinapis alba L.) cotyledons. Enzymic control and differential regulation of anthocyanin and quercetin formation. Planta 172: 121–126.

43. Rabino I, Mancinelli AL (1986) Light, Temperature, and Anthocyanin Production. Plant Physiology 81: 922–924.

44. Mustilli AC, Fenzi F, Ciliento R, Alfano F, Bowler C (1999) Phenotype of the tomato high pigment-2 mutant is caused by a mutation in the tomato homolog of DEETIOLATED1. Plant Cell 11: 145–157.

45. Feyissa DN, Lovdal T, Olsen KM, Slimestad R, Lillo C (2009) The endogenous GL3, but not EGL3, gene is necessary for anthocyanin accumulation as induced by nitrogen depletion in Arabidopsis rosette stage leaves. Planta 230: 747–754.

46. Winkel-Shirley B (2001) Flavonoid biosynthesis. A colorful model for genetics, biochemistry, cell biology, and biotechnology. Plant Physiol 126: 485–493.

47. Dolan L (1996) Pattern in the root epidermis: an interplay of diffusible signals and cellular geometry. Annals of Botany 77: 547–553.

48. Pemberton LMS, Tsai SL, Lovell PH, Harris PJ (2001) Epidermal patterning in seedling roots of eudicotyledons. Annals of Botany 87: 649–654.

49. Glover BJ, Martin C (2000) Specification of epidermal cell morphology. Advances in Botanical Research 31: 193–217.

50. Kang JH, Liu G, Shi F, Jones AD, Beaudry RM, et al. (2010) The tomato odorless-2 mutant is defective in trichome-based production of diverse specialized metabolites and broad-spectrum resistance to insect herbivores. Plant Physiol 154: 262–272.

51. Kang JH, Shi F, Jones AD, Marks MD, Howe GA (2010) Distortion of trichome morphology by the hairless mutation of tomato affects leaf surface chemistry. J Exp Bot 61: 1053–1064.

52. Schilmiller A, Shi F, Kim J, Charbonneau AL, Holmes D, et al. (2010) Mass spectrometry screening reveals widespread diversity in trichome specialized metabolites of tomato chromosomal substitution lines. Plant J 62: 391–403.

53. Luckwill LC (1943) The genus Lycopersicon: a historical, biological and taxonomic survey of the wild and cultivated tomato. Aberd Univ Stud 120: 1–44.

54. Feyissa DN, Lovdal T, Olsen KM, Slimestad R, Lillo C (2009) The endogenous GL3, but not EGL3, gene is necessary for anthocyanin accumulation as induced by nitrogen depletion in Arabidopsis rosette stage leaves. Planta 230: 747–754.

55. Martin C, Gerats T (1993) Control of Pigment Biosynthesis Genes during Petal Development. Plant Cell 5: 1253–1264.

56. Povero G, Gonzali S, Bassolino L, Mazzucato A, Perata P (2011) Transcriptional analysis in high-anthocyanin tomatoes reveals synergistic effect of Aft and atv genes. Journal of Plant Physiology 168: 270–279.

57. Gonzalez A, Zhao M, Leavitt JM, Lloyd AM (2008) Regulation of the anthocyanin biosynthetic pathway by the TTG1/bHLH/Myb transcriptional complex in Arabidopsis seedlings. Plant Journal 53: 814–827.

58. Ramsay NA, Glover BJ (2005) MYB-bHLH-WD40 protein complex and the evolution of cellular diversity. Trends in Plant Science 10: 63–70.

59. Zhang F, Gonzalez A, Zhao MZ, Payne CT, Lloyd A (2003) A network of redundant bHLH proteins functions in all TTG1-dependent pathways of Arabidopsis. Development 130: 4859–4869.

60. Bloor SJ, Abrahams S (2002) The structure of the major anthocyanin in Arabidopsis thaliana. Phytochemistry 59: 343–346.

61. Cominelli E, Gusmaroli G, Allegra D, Galbiati M, Wade HK, et al. (2008) Expression analysis of anthocyanin regulatory genes in response to different light qualities in Arabidopsis thaliana. Journal of Plant Physiology 165: 886–894.

62. Peng MS, Hudson D, Schofield A, Tsao R, Yang R, et al. (2008) Adaptation of Arabidopsis to nitrogen limitation involves induction of anthocyanin synthesis which is controlled by the NLA gene. Journal of Experimental Botany 59: 2933–2944.

63. Rowan DD, Cao MS, Lin-Wang K, Cooney JM, Jensen DJ, et al. (2009) Environmental regulation of leaf colour in red 35S:PAP1 Arabidopsis thaliana. New Phytologist 182: 102–115.

64. Shi MZ, Xie DY (2010) Features of anthocyanin biosynthesis in pap1-D and wild-type Arabidopsis thaliana plants grown in different light intensity and culture media conditions. Planta 231: 1385–1400.

65. Zhou LL, Shi MZ, Xie DY (2012) Regulation of anthocyanin biosynthesis by nitrogen in TTG1-GL3/TT8-PAP1-programmed red cells of Arabidopsis thaliana. Planta 236: 825–837.

66. Grotewold E (2006) The genetics and biochemistry of floral pigments. Annual Review of Plant Biology 57: 761–780.

67. Lloyd AM, Walbot V, Davis RW (1992) Arabidopsis and Nicotiana anthocyanin production activated by maize regulators R and C1. Science 258: 1773–1775.

68. Gonzalez A, Mendenhall J, Huo Y, Lloyd A (2009) TTG1 complex MYBs, MYB5 and TT2, control outer seed coat differentiation. Dev Biol 325: 412–421.

69. Shi MZ, Xie DY (2011) Engineering of red cells of Arabidopsis thaliana and comparative genome-wide gene expression analysis of red cells versus wild-type cells. Planta 233: 787–805.

70. Dubos C, Le Gourrierec J, Baudry A, Huep G, Lanet E, et al. (2008) MYBL2 is a new regulator of flavonoid biosynthesis in Arabidopsis thaliana. Plant Journal 55: 940–953.

71. Matsui K, Umemura Y, Ohme-Takagi M (2008) AtMYBL2, a protein with a single MYB domain, acts as a negative regulator of anthocyanin biosynthesis in Arabidopsis. Plant Journal 55: 954–967.

19

Engineering and Two-Stage Evolution of a Lignocellulosic Hydrolysate-Tolerant *Saccharomyces cerevisiae* Strain for Anaerobic Fermentation of Xylose from AFEX Pretreated Corn Stover

Lucas S. Parreiras[1], Rebecca J. Breuer[1], Ragothaman Avanasi Narasimhan[1], Alan J. Higbee[1,2], Alex La Reau[1], Mary Tremaine[1], Li Qin[1], Laura B. Willis[3], Benjamin D. Bice[1], Brandi L. Bonfert[2], Rebeca C. Pinhancos[2], Allison J. Balloon[2], Nirmal Uppugundla[4,5], Tongjun Liu[4,6], Chenlin Li[7], Deepti Tanjore[7], Irene M. Ong[1], Haibo Li[1], Edward L. Pohlmann[1], Jose Serate[1], Sydnor T. Withers[1], Blake A. Simmons[8], David B. Hodge[4,9,10,11], Michael S. Westphall[2], Joshua J. Coon[2], Bruce E. Dale[4,5], Venkatesh Balan[4,5], David H. Keating[1], Yaoping Zhang[1], Robert Landick[1,3,12], Audrey P. Gasch[1,13], Trey K. Sato[1]*

1 DOE Great Lakes Bioenergy Research Center, University of Wisconsin-Madison, Madison, Wisconsin, United States of America, 2 Department of Chemistry, University of Wisconsin-Madison, Madison, Wisconsin, United States of America, 3 Department of Bacteriology, University of Wisconsin-Madison, Madison, Wisconsin, United States of America, 4 DOE Great Lakes Bioenergy Research Center, Michigan State University, East Lansing, Michigan, United States of America, 5 Biomass Conversion Research Laboratory, Department of Chemical Engineering and Materials Science, Michigan State University, Lansing, Michigan, United States of America, 6 School of Food and Bioengineering, Qilu University of Technology, Jinan, China, 7 Advanced Biofuels Process Demonstration Unit, Lawrence Berkeley National Laboratory, Emeryville, California, United States of America, 8 Deconstruction Division, Joint BioEnergy Institute, Emeryville, California, United States of America, 9 Department of Chemical Engineering & Materials Science, Michigan State University, East Lansing, Michigan, United States of America, 10 Department of Biosystems & Agricultural Engineering, Michigan State University, East Lansing Michigan, United States of America, 11 Division of Sustainable Process Engineering, Luleå University of Technology, Luleå, Sweden, 12 Department of Biochemistry, University of Wisconsin-Madison, Madison, Wisconsin, United States of America, 13 Laboratory of Genetics, University of Wisconsin-Madison, Madison, Wisconsin, United States of America

Abstract

The inability of the yeast *Saccharomyces cerevisiae* to ferment xylose effectively under anaerobic conditions is a major barrier to economical production of lignocellulosic biofuels. Although genetic approaches have enabled engineering of *S. cerevisiae* to convert xylose efficiently into ethanol in defined lab medium, few strains are able to ferment xylose from lignocellulosic hydrolysates in the absence of oxygen. This limited xylose conversion is believed to result from small molecules generated during biomass pretreatment and hydrolysis, which induce cellular stress and impair metabolism. Here, we describe the development of a xylose-fermenting *S. cerevisiae* strain with tolerance to a range of pretreated and hydrolyzed lignocellulose, including Ammonia Fiber Expansion (AFEX)-pretreated corn stover hydrolysate (ACSH). We genetically engineered a hydrolysate-resistant yeast strain with bacterial xylose isomerase and then applied two separate stages of aerobic and anaerobic directed evolution. The emergent *S. cerevisiae* strain rapidly converted xylose from lab medium and ACSH to ethanol under strict anaerobic conditions. Metabolomic, genetic and biochemical analyses suggested that a missense mutation in *GRE3*, which was acquired during the anaerobic evolution, contributed toward improved xylose conversion by reducing intracellular production of xylitol, an inhibitor of xylose isomerase. These results validate our combinatorial approach, which utilized phenotypic strain selection, rational engineering and directed evolution for the generation of a robust *S. cerevisiae* strain with the ability to ferment xylose anaerobically from ACSH.

Editor: Y-H Percival Zhang, Virginia Tech, United States of America

Funding: This work was funded in part by the Department of Energy (DOE) Great Lakes Bioenergy Research Center (DOE BER Office of Science BER DE-FC02-07ER64494). The portion of work conducted by the Joint BioEnergy Institute was supported by the Office of Science, Office of Biological and Environmental Research, of the United States Department of Energy under Contract No. DE-AC02-05CH11231. The portion of the work conducted by the Advanced Biofuels Process Demonstration Unit was funded by support from Office of Biomass Program within the United States DOE's Office of Energy Efficiency and Renewable Energy, and also the funding support from the American Recovery and Reinvestment Act. The funders had no role in study design, data collection and analysis, decision to publish, or preparation of the manuscript.

Competing Interests: The authors have declared that no competing interests exist.

* Email: tksato@glbrc.wisc.edu

Introduction

As the world's human population increases, so does the demand for energy. Renewable biofuels offer a route to replace a portion of the finite amounts of liquid petroleum and natural gas-based fuels. Although bioethanol produced from grain has been employed as a partial replacement for gasoline, this process is generally viewed as unsustainable [1]. An alternative to grain-based ethanol, which has been traditionally produced by microbial fermentation of starch sugars, is bioethanol generated from lignocellulosic (LC) sugars derived from renewable and sustainable plant feedstocks [2]. Despite well over two decades of research, microbial-based production of cellulosic ethanol at an industrial scale remains largely unpracticed throughout the world. Part of the reason for this is a number of molecular barriers that have profound impact on the metabolic catabolism of LC sugars, thereby limiting their efficient and cost-effective conversion into ethanol. In particular, the yeast *Saccharomyces cerevisiae*, in its native form, does not convert most LC sugars into ethanol due to insufficient biochemical activities and metabolic inhibition.

Although *S. cerevisiae* excels at fermenting glucose from cornstarch and sugar cane juice, the fermentation of pentose sugars from the hemicellulose component of lignocellulosic biomass is challenging. In particular, xylose, a pentose sugar and a major component of hemicellulose, composes 30–40% of total cell-wall carbohydrate in grasses and some woody biomass [3]. Conversion of xylose to ethanol is crucial to maximize the economic return from fuel production in excess of feedstock and production costs. However, native *S. cerevisiae* cannot efficiently ferment xylose, as most strains have either lost or downregulated the activities of xylose catabolism enzymes [4] and lack specific xylose transporters [5]. To overcome this deficiency, yeast have been engineered to express a minimal enzyme set from native xylose-metabolizing organisms that allow conversion of xylose into xylulose-5-phosphate (X5P), which can then be catabolized by the pentose phosphate pathway into ethanol. Specifically, engineering of *S. cerevisiae* to express xylose reductase (XR) and xylitol dehydrogenase (XDH), or xylose isomerase (XI) alone, has permitted the limited conversion of xylose into xylulose, which can then be phosphorylated to X5P by overexpression of native or exogenous xylulokinase (XK) (for reviews, see [6–8]). With additional rational engineering approaches, yeast strains with improved xylose fermentation in lab medium have been created (reviewed in [9–11]). Some of these approaches have been employed with varying degrees of success, including metabolic reengineering of *S. cerevisiae* strains through overexpression of native pentose phosphate pathway enzymes [12,13], deletion of genes such as *PHO13* to improve xylose metabolism [14], and heterologous expression of putative xylose transporters [5,15]. Experimental directed evolution is another well-utilized means to improve desired phenotypic traits (reviewed in [16]). A combination of rational engineering followed by directed evolution on xylose-containing medium under aerobic [17,18] or oxygen-limited [10,12] conditions has generated yeast strains with increased anaerobic xylose consumption rates relative to their parental strains. Most recently, two sequential anaerobic selections of an XR-XDH engineered *S. cerevisiae* strain on xylose resulted in an evolved isolate with a significantly faster anaerobic consumption rate of xylose than its ancestor, although most of the xylose appears to be converted to xylitol and glycerol [19]. These approaches have allowed for effective consumption of xylose in innocuous and nutrient-rich lab medium; however, conversion of complex, LC-derived xylose from lignocellulosic hydrolysates into biofuels is much more challenging.

Before being deconstructed into fermentable sugars, plant biomass requires chemical, thermal and/or mechanical pretreatments that alter cellulose, hemicellulose and lignin organization, thereby allowing hydrolytic enzymes greater access to sugar polymers for faster rates of enzymatic hydrolysis. Numerous pretreatment methods have been developed and they include the use of dilute acid, bases and ionic liquids (IL) (reviewed in [20]). Although it significantly increases the rate and effectiveness of biomass deconstruction, lignocellulose pretreatment generates a number of common degradation products released from plant cell walls. These chemical compounds include hemicellulose-derived acetate and lignin-derived aromatic aldehydes that induce microbial stress by draining reducing cofactors, limiting ATP generation and causing cellular damage (reviewed in [21] and [22]). Additionally, each pretreatment process can generate its own set of dominant stress-inducing compounds, such as furans generated from dilute acid pretreatment [23,24] and hydroxycinnamic acids in alkaline hydrogen peroxide (AHP) pretreatment [25]. In some cases, the pretreatment compound itself can be a major biological inhibitor, as is the case for the IL, 1-ethyl-3-methylimidazolium acetate ([C2mim][OAc]) [26]. Ammonia Fiber Expansion (AFEX) is a highly effective pretreatment for herbaceous biomass. In contrast to dilute acid, which degrades the hemicellulose, AFEX pretreatment retains hemicellulose as intact polymers [27] that can then be hydrolyzed into fermentable sugars for additional fuel production. However, AFEX pretreatment of corn stover generates diverse inhibitory small molecules, including phenolic amides, which impair xylose fermentation by *S. cerevisiae* [28,29]. These effects are often compounded during xylose fermentation in the absence of oxygen, likely due to reduced ATP yield from pentose sugars compared to hexoses, combined with decreased energetic yield under anaerobic conditions. Given that ATP drives numerous detoxification processes [21], the cellular stress induced by these compounds has profound impacts on xylose fermentation. To cope with these diverse inhibitory compounds present in LC hydrolysates and their impacts on cellular physiology and metabolism, many researchers have opted to employ industrial or environmental *S. cerevisiae* strains with innate stress tolerant properties [12,25,30–32]. A combination of directed engineering and evolution approaches with *S. cerevisiae* for xylose metabolism in the presence of defined spruce [30] or AHP [25] hydrolysate inhibitors or undefined raw spruce hydrolysates [12] have resulted in a small number of *S. cerevisiae* strains with improved xylose fermentation properties in lignocellulosic hydrolysates compared to their parental strains.

At present, little is known about how the inhibitors found in AFEX-pretreated corn stover hydrolysate (ACSH) impact xylose fermentation by *S. cerevisiae*, particularly under strict, industrially relevant, anaerobic conditions where ethanol production is maximized. As the foundation for investigating this knowledge gap, and the goal of determining what genetic factors are important for improving anaerobic xylose fermentation, we sought to develop and compare closely related *S. cerevisiae* strains with varying anaerobic xylose fermentation phenotypes. Previously, we identified a natural *S. cerevisiae* strain, GLBRCY0 (originally named as NRRL YB-210 [33]) with growth tolerance to individual LC-derived inhibitors [25] as well as ACSH at elevated temperature [34]. Despite engineering NRRL YB-210 (YB-210) with known XR-XDH genes, the strain fermented xylose in lab medium and ACSH at slow rates, even in the presence of limited oxygen. Here, we report the development of an engineered and evolved derivative of YB-210, which displayed robust cell growth in a variety of pretreated lignocellulosic hydrolysates (LCH) relative to other strains, with the ability to rapidly ferment xylose from lab medium and ACSH under completely anaerobic conditions. Combined genetic and metabolomic analyses indicate that the evolved strain, named GLBRCY128 (Y128), incurred a missense mutation in *GRE3*, which, combined with additional unknown mutations, allowed for faster anaerobic xylose consumption rates relative to its parent. Together, these results identify Y128 as a novel *S. cerevisiae* strain with evolved genetic traits for robust anaerobic fermentation of xylose from ACSH. They also illustrate how careful selection of genetic background can accelerate development of biocatalysts with the ability to ferment xylose anaerobically in an inhibitor-laden LCH.

Materials and Methods

AFEX pretreated corn stover hydrolysate (ACSH) preparation

Zea mays (Pioneer hybrid 36H56) corn stover from Field 570-C Arlington Research Station, University of Wisconsin was harvested in 2008 for use in 96-well plate phenotyping or in 2009 for use in anaerobic fermentation experiments. AFEX pretreatment of corn stover was performed as described previously [27]. AFEX pretreated corn stover was hydrolyzed at 6% or 9% glucan loading in a 1.5 L reaction volume in a 3 L Applikon fermenter (Applikon Biotechnology Inc. USA) with Spezyme CP (15 FPU/g glucan loading, DuPont Danisco), Multifect Xylanase (10% of Spezyme CP, DuPont Danisco), and Novozyme 188 (64 pNPGU/g glucan, Sigma-Aldrich) at 50°C for 5 days. Tetracycline (40 mg/L) was used to prevent microbial contamination and pH 4.8 was maintained during the hydrolysis process. Biomass was added to the reaction mixture in 4 batches within 4 h from the start of hydrolysis to facilitate better mixing at 1000 rpm. After 120 h, the hydrolysis mixture was centrifuged (2500×g for 30 min.) and sterile filtered (0.22 μm pore size; Millipore Stericup). For 6% glucan loading ACSH, the final sugar concentrations were 53 g/L glucose and 21.7 g/L xylose. For 9% glucan loading ACSH, the final sugar concentrations were 80 g/L glucose and 36 g/L xylose. For ACSH used in bioreactor fermentations, hydrolysates were prepared as described previously [35] with one additional modification; prior to fermentation, the hydrolysate was adjusted to pH 5.0 and again filtered through a 0.2 μm filter to remove precipitates and to ensure sterility.

Alkaline hydrogen peroxide pretreated hydrolysates

Pioneer hybrid 36H56 corn stover described above and switchgrass (*Panicum virgatum* cv. Cave-In-Rock) described elsewhere [36] were milled (Circ-U-Flow model 18-7-300, Schutte-Buffalo Hammermill, LLC) to pass through a 5 mm screen. AHP pretreatment was performed as reported previously [37] at a hydrogen peroxide loading of 0.125 g H_2O_2/g biomass in an incubator with shaking at 150 rpm at 30°C for 24 h with periodic pH readjustment to 11.5 during pretreatment using 5 N NaOH. For switchgrass, pretreatment was conducted at biomass concentration of 20% w/w. For corn stover, pretreatment was conducted at biomass concentration of 15% w/w. Following pretreatment, the whole slurry was adjusted to pH 4.8 using 72% H_2SO_4. Accelerase 1000 (Novozymes A/S), Multifect xylanase, and Multifect pectinase (DuPont Danisco) were used at the protein ratio of 0.62:0.24:0.14 with a total protein loading of 30 mg/g glucan as assayed by the Bradford Assay. Hydrolysis was performed at 50°C with shaking speed of 180 rpm for 24 h. After enzymatic hydrolysis, the whole slurry was centrifuged at 18,000×g for 30 min. The supernatant was used as undetoxified raw hydrolysate or for detoxified hydrolysate, activated carbon (Fisher Scientific #05-690A) was mixed with undetoxified hydrolysate at 5% concentration (5 g activated carbon with 100 mL hydrolysate) and incubated at 50°C for 1 h in an unbaffled shake flask at 150 rpm. After centrifugation at 18000×g for 30 min, the supernatant was used as the detoxified hydrolysate. All hydrolysates were filter-sterilized (0.22 μm pore size; Millipore Stericup). Final sugar concentrations for AHP hydrolysates were 30 g/L glucose and 20 g/L xylose for raw AHP corn stover hydrolysate, 35 g/L glucose and 23 g/L xylose for detoxified AHP CSH, and 27 g/L glucose and 19 g/L xylose for both raw and detoxified AHP SGH.

Dilute acid pretreated lignocellulosic hydrolysates

An industrial collaborator provided two different versions of dilute acid pretreated wheat straw that was hydrolyzed using a proprietary blend of cellulase enzymes at pH 5.0 and 50°C. Both

Table 1. Engineered and evolved *S. cerevisiae* strains used in this study.

Strain name	Genotype	Reference
GLBRCY73	NRRL YB-210 MATa/α *HOΔ::SsXYL1-SsXYL2-SsXYL3-loxP-KanMX-loxP*, aerobically evolved on YPX	[40]
GLBRCY22-3	NRRL YB-210 MATa spore *HOΔ::ScTAL1-CpxylA-SsXYL3-loxP-KanMX-loxP*	This study
GLBRCY127	GLBRCY22-3 MATa, aerobically evolved isolate on YPDX	This study
GLBRCY128	GLBRCY127 MATa, anaerobically evolved isolate on YPDX	This study
GLBRCY36	GLBRCY22-3 with *loxP-KanMX-loxP* marker excised by Cre	This study
GLBRCY132	GLBRCY127 with *loxP-KanMX-loxP* marker excised by Cre	This study
GLBRCY133	GLBRCY128 with *loxP-KanMX-loxP* marker excised by Cre	This study
GLBRCY132 *xylAΔ*	GLBRCY132 *xylAΔ::loxP-KanMX-loxP*	This study
GLBRCY132 *tal1Δ*	GLBRCY132 synthetic *tal1Δ::loxP-KanMX-loxP*	This study
GLBRCY133 *xylAΔ-A*	GLBRCY133 *xylAΔ::loxP-KanMX-loxP* transformant A	This study
GLBRCY133 *xylAΔ-B*	GLBRCY133 *xylAΔ::loxP-KanMX-loxP* transformant B	This study
GLBRCY36 *gre3Δ*	GLBRCY36 *gre3Δ::loxP-KanMX-loxP*	This study
GLBRCY132 *gre3Δ*	GLBRCY132 *gre3Δ::loxP-KanMX-loxP*	This study
GLBRCY133 *gre3Δ*	GLBRCY132 *gre3Δ::loxP-KanMX-loxP*	This study

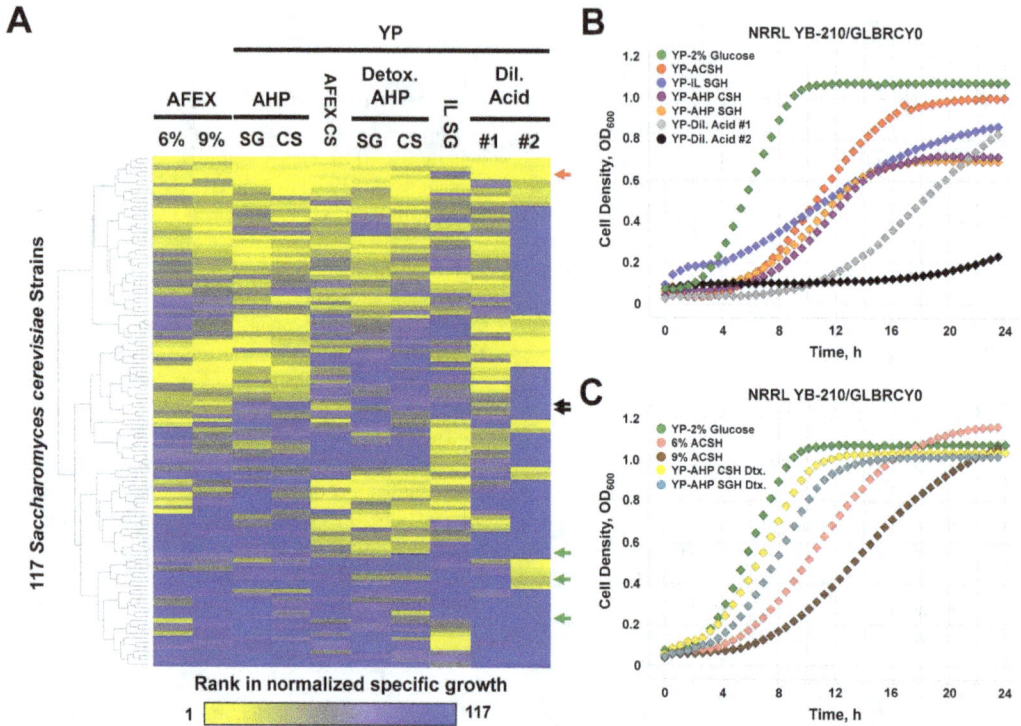

Figure 1. Phenotypic screening of wild and domesticated *S. cerevisiae* strains identifies NRRL YB-210 with tolerance to hydrolysates made from a variety of pretreated lignocellulose. In (**A**), 117 *S. cerevisiae* strains (including some in duplicate) were cultured in 96-well plates and monitored for changes cell density and growth rates calculated as described in Materials and Methods. All strains in each condition were then ranked from 1 (highest growth rate in yellow) to 117 (lowest growth rate, or no growth, in blue) and hierarchically clustered. Arrows indicate clustered rows for BY4741 (green), CEN.PK2 (black) in duplicate microtiter wells, and NRRL YB-210/GLBRCY0 (red). Representative growth data for the YB-210/GLBRCY0 strain in the indicated media from Fig. 2A are plotted (**B–C**). CS, corn stover; SG, switchgrass; YP; Yeast Extract and Peptone supplementation, 6%; 6% glucan loading ACSH, 9%; 9% glucan loading ACSH, Dtx.; Detoxified.

hydrolysates were diluted 4:5 in sterile water supplemented with 10 g/L yeast extract and 20 g/L peptone.

Ionic liquid pretreated switchgrass hydrolysate (IL-SGH)

Switchgrass was pretreated with [C2mim][OAc] (1-ethyl-3-methylimidazolium acetate) at 15% solids loading as described elsewhere [38]. IL-pretreated switchgrass was hydrolyzed with CTec2 (54 mg/g glucan) and HTec2 (6 mg/g glucan) enzymes (Novozyme) for 72 h in a 2 L IKA bioreactor. [C2mim][OAc]-pretreated SGH was generated at Advanced Biofuels Process Demonstration Unit (batch ABPDU 110201S02). Final sugar concentrations in the IL-SGH were 41 g/L glucose and 10 g/L xylose.

Lab media

Standard undefined yeast lab medium was prepared as previous described [39]. Briefly, liquid and plate-based medium contained 10 g/L yeast extract and 20 g/L peptone (YP), and various sugar concentrations (X = 20 g/L xylose, D = 20 g/L dextrose or glucose, DX = 60 g/L glucose and 30 g/L xylose). Where indicated, hydrolysates were supplemented with 10 g/L yeast extract and 20 g/L peptone. For bioreactor experiments, this YPDX medium also contained 50 mM potassium phosphate, pH 5.0.

Saccharomyces cerevisiae strains, 96-well growth assays and hierarchical clustering

Native *S. cerevisiae* strains used in this study (see **Table S1**) were obtained from Dr. Cletus Kurtzmann (USDA ARS, Peoria, IL), National Collection of Yeast Cultures (Norwich, UK), and Dr. Justin Fay (Washington University, Saint Louis, MO). Aerobic growth assays in microtiter plates were performed as previously described [40] [25] [34], with one exception; 10 μL of saturated culture was inoculated into 190 μL of YPD or a single type of pretreated lignocellulosic hydrolysate. Cell growth was measured by optical density at 595 nm every 10 min for 24 h in Tecan F500 or M1000 multimode plate readers with an interior temperature of 30°C. Background-subtracted cell density readings for each strain were analyzed by the program GCAT [25]. Normalized specific growth rates for each strain from three independent biological replicates in pretreated hydrolysates were normalized to their average growth rate in YPD alone, and then ranked ordered from 1 to 117 (including control strains – Y389, BY4741, CEN.PK113-5D and CEN.PK2-1D in duplicate) for highest average specific growth to lowest, respectively. For all strains with no detectable specific growth rates, strains were assigned a rank of 117. Strain ranks in each medium condition were hierarchically clustered and displayed with Spotfire (TIBCO).

DNA constructs and strain engineering

Genotypes of engineered strains used in this study are described in **Table 1**. Construction of the GLBRCY73 strain has been described elsewhere [40]. The expression cassette containing

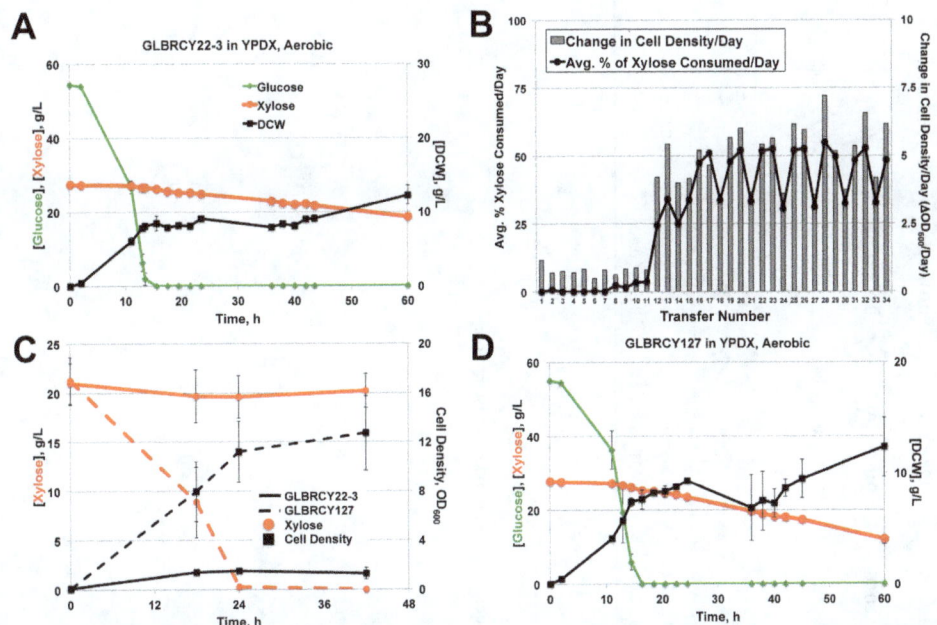

Figure 2. The GLBRCY127 strain developed by directed engineering with xylose isomerase coupled with batch evolution can rapidly consume xylose aerobically. Average sugar consumption and cell growth of unevolved GLBRCY22-3 strain engineered with *ScTAL1*, *CpxylA* and *SsXYL3* cultured in bioreactors containing YPDX media and sparged with air from biological duplicates is shown (**A**). Indicated components were quantified from media samples at times from initial inoculation. In (**B**), the average percentage of xylose consumed and change in cell density per day are plotted for each transfer during the adaption of the Y22-3 strain in YP media containing 0.1% glucose and 2% xylose. The pattern of lower % of xylose consumed and change in cell density per day during every third transfer is due to reaching saturated growth prior to transfer. Average extracellular xylose concentrations and cell density measurements from parental Y22-3 and evolved Y127 strains grown aerobically in culture tubes with YPX media from three independent biological replicates are plotted in (**C**). In (**D**), evolved isolate Y127 was cultured in the same conditions as in (**A**), and samples measurements taken in an identical manner.

TAL1 from *S. cerevisiae* S288c (*ScTAL1*), *xylA* from *Clostridium phytofermentans* ISDg (*CpxylA*) and *XYL3* from *Scheffersomyces stipitis* CBS 6054 (*SsXYL3*) was generated in a similar manner with some modifications. Codon-optimized versions of each gene were synthesized (GeneArt, Life Technologies) and inserted via homologous recombination in the following promoter-open reading frame-terminator combinations in order from 5' to 3': *ScPGK1* promoter-*ScTAL1*-*ScTDH3* terminator, *ScTDH3* promoter-*CpxylA*-*ScTEF2* terminator, *ScTEF2* promoter-*SsXYL3*-*ScCYC1* terminator. This cassette, which also contains a loxP-KanMX-loxP selection marker [41], was flanked by *ScHO* sequences [42] for targeted recombination at the genomic *HO* locus. The complete *CpxylA* cassette was amplified by standard polymerase chain reaction (PCR) using primers that anneal to the 5' ends of the *HO* flanking sequences, gel purified and transformed into the NRRL YB-210 strain. Following selection on YPD plates containing 200 µg/mL Geneticin (Life Technologies), verification of cassette insertion was determined by PCR using combinations of primers that anneal outside of the HO flanking sequence and specific to synthesized DNA cassette. The engineered YB-210 diploid strain was then subjected to sporulation and tetrad dissection. One spore, which was derived from a tetrad with three other inviable spores, was confirmed for a single MATa mating type and subsequently named GLBRCY22-3 (Y22-3). LoxP-KanMX-loxP marker rescues from Y22-3, Y127 and Y128 were carried out by expression of Cre recombinase as published elsewhere [41] to generate the respective Y36, Y132 and Y133 strains. Deletion of engineered *CpxylA* and *ScTAL1*, and endogenous *ScGRE3* were performed by integration of PCR product using a loxP-KanMX-loxP cassette [41] into the marker-rescued versions of GLBRCY36, GLBRCY132 and GLBRCY133

strains. Sanger sequencing of PCR products and DNA plasmids was performed by the University of Wisconsin-Madison Biotechnology Center.

Quantitative RT-PCR

GLBRCY128 and GLBRCY128 *xylAΔ* strains were aerobically cultured at 30°C in baffled shake flasks containing YPD media. When the culture reached log phase growth (OD$_{600}$ = 0.8), 7 mL of cell culture was then harvested by centrifugation at 3,000 ×*g* for 3 min, the supernatant was decanted and then cell pellets were flash frozen on dry ice/ethanol. Frozen cell pellets were resuspended and then vortexed in 0.8 mL phenol, pH 4.3 and 0.8 mL lysis buffer (10 mM Tris-Cl, pH 7.4, 10 mM EDTA, 0.5% SDS). Cell lysates were incubated at 65°C for 30 min and then centrifuged at 20,000 ×*g* at 4°C. The aqueous phase was then transferred to a new 2 mL centrifuge tube and further extracted with two additional rounds of phenol and then chloroform. The final extracted aqueous phase was then transferred to RNase-free minicentrifuge tubes and RNA precipitated by addition of 0.1 volumes of 0.3 M NaOAc, pH 5.2 and 2.5 volumes of 100% ethanol at −20°C for 1 h. Precipitated RNA was pelleted by centrifugation at 20,000 ×*g* for 30 min at 4°C. The RNA pellet was washed with 2 mL 70% ethanol, dried in a Speed-Vac, and dissolved in 0.2 mL RNase-free TE buffer (10 mM Tris-HCl, 1 mM EDTA, pH 8.0). RNA was further purified by RNeasy Mini Kit (Qiagen) according to the manufacturer protocol. cDNA synthesis from 10 µg total RNA was performed with Superscript III reverse transcriptase (Life Technologies) according to manufacturer protocol. Generated cDNA was purified and concentrated with a PCR Minelute Purification kit (Qiagen) into 12 ml elution buffer according to manufacturer protocol. For quantitative

Table 2. Fermentation kinetic profiles for engineered and evolved *S. cerevisiae* strains.

Media	Aerobic YPDX			Anaerobic YPDX			Anaerobic ACSH		
Strain	Y73	Y22-3	Y127	Y73	Y127	Y128	Y73	Y127	Y128
Absolute xylose consumption rate[1]	0.47±0.02	0.17±0.003	0.31±0.01	0.30±0.02	0.094±0.031	1.68±0.06	0.28±0.01	0.04±0.03	0.52±0.01
Specific xylose consumption rate[2]	0.039±0.001	0.019±0.000	0.036±0.01	0.066±0.010	0.016±0.007	0.27±0.06	0.059±0.01	0.013±0.01	0.18±0.02
% of theoretical ethanol yield for consumed sugars[3]	ND	ND	ND	80.1±1.4	86.0±2.5	87.5±1.1	72.1±10.4	78.9±14.3	77.2±6.4
% of theoretical ethanol yield for consumed xylose[4]	ND	ND	ND	9.8±5.2	ND*	86.2±15.6	20.9±18.9	ND*	71.5±7.0
$Y_{x/glc}$[5]	ND	ND	ND	0.08±0.00	0.09±0.00	0.11±0.02	0.066±0.012	0.045±0.004	0.05±0.01
$Y_{glycerol/glc}$[6]	ND	ND	ND	0.10±0.00	0.07±0.00	0.08±0.01	0.051±0.002	0.038±0.006	0.04±0.00

ND, Not Determined for aerobic conditions; ND*, Not Determined – no ethanol produced.

[1] In g xylose consumed/L/h.

[2] In g xylose consumed/g DCW/h.

[3] Calculated from the maximum ethanol concentration produced divided by the consumed xylose concentration at that time.

[4] Calculated from the ethanol concentration produced between two time points after glucose depletion.

[5] Yield of g DCW/g glucose consumed calculated at or near the time of glucose depletion and prior to xylose consumption. No cell growth was observed during xylose consumption.

[6] Yield of g glycerol/g glucose consumed calculated at or near the time of glucose depletion and prior to xylose consumption.

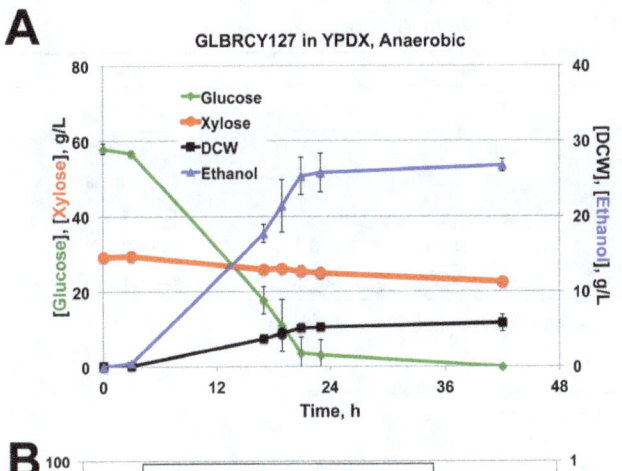

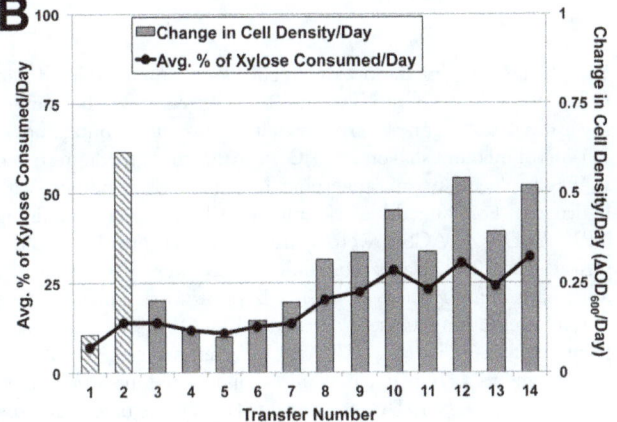

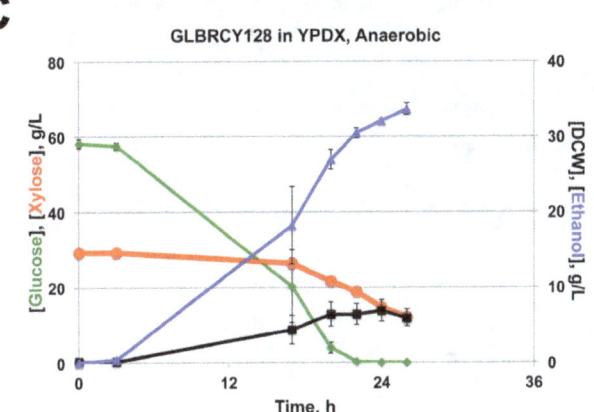

Figure 3. Second stage anaerobic adaptation on xylose enabled rapid xylose fermentation by evolved GLBRCY128 isolate. Average fermentation kinetic profiles of the GLBRCY127 strain cultured in bioreactors containing YPDX media and sparged with nitrogen from biological duplicates are shown (**A**). Average concentrations with standard deviations of indicated compounds were quantified from media samples at times from initial inoculation. In (**B**), the percentage of xylose consumed and change in cell density per day is plotted for each transfer during the anaerobic adaptation of Y127 in YP media containing 0.1% glucose and 2% xylose. In the first two transfers (hatched bars), Tween-80 and ergosterol were added to the media. In (**C**), evolved isolate Y128 was cultured in biological duplicate under the same conditions as in (**A**), and samples measurements taken in an identical manner.

reverse-transcriptase polymerase chain reaction (qPCR) of *xylA*, 10 ng of cDNA was mixed with 500 nM SynCpXylA FWD (5′-

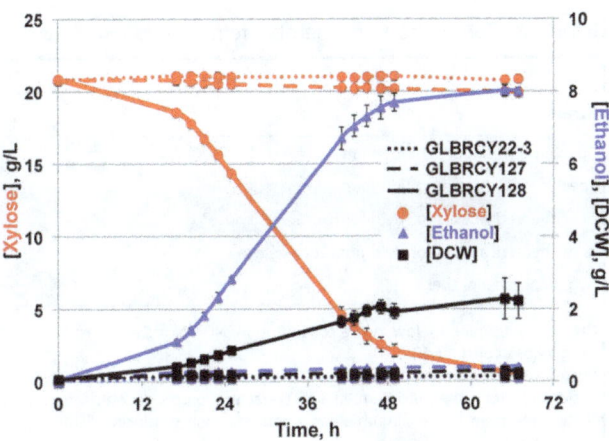

Figure 4. GLBRCY128 can anaerobically ferment xylose from YPX medium into ethanol faster than its parental strains. Fermentation kinetic profiles comparing Y22-3, Y127 and Y128 strains cultured anaerobically in bioreactors containing YPX medium and sparged with nitrogen are shown. Average concentrations with standard deviations of xylose, ethanol and DCW concentrations in the media were calculated from independent biological triplicates.

GGTGGATGCTAGGTTGTCTTT-3′) and 250 nM SynCp-XylA REV (5′-CACGCCTTCTTGCTCAAATAAC-3′), or ScERV25qPCRfor and ScERV25 FWD (5′-GTCGCGGA-TATTCACTCAGATG-3′) and ScERV25 REV (5′-CCTGCAAAGTCCCTCTTTCTAC-3′) primers and SYBR JumpStart Taq Ready Mix with Rox internal standard (Sigma-Aldrich) according to manufacturer protocol. Relative quantities of *CpxylA* and *ScERV25* RNA were determined using an ABI7500 Real-Time PCR instrument (Applied Biosystems). Relative concentrations of *CpxylA* and *ScERV25* transcripts were determined by ΔC_t method.

Directed Evolution

Cell density measurements were determined from OD_{600} measurements from the cultures diluted 1:10 in 1 cm path length cuvettes by a Beckman Coulter DU720 spectrophotometer. For aerobic adaptation, GLBRCY22-3 was inoculated at $OD_{600} = 0.05–0.1$ in 250 mL YP medium containing 2% xylose and 0.1% glucose. The first 15 transfers took place over 53 days with serial 1:10 dilutions in fresh medium occurring every 3–4 days. After transfer 15, the adaptation culture was diluted every 2–3 days over the course of 44 additional days, ending after transfer 34. For anaerobic adaptation, GLBRCY127 was inoculated at $OD_{600} = 0.05–0.1$ in a flask containing 50 mL YP medium with 2% xylose, 0.1% glucose and 50 μg/mL Geneticin, and then placed in an anaerobic chamber. For the first two anaerobic transfers, the medium also contained 40 μg/L ergosterol (Sigma-Aldrich) and 4 g/L Tween-80 (Sigma-Aldrich). Anaerobic cultures were maintained in suspension using a stir bar and magnetic stir plate, and passaged every 7 days during the first 5 transfers. After the 5th transfer, the culture was passaged every 3–4 days with the final 14th transfer finishing 66 days after the start of the anaerobic adaptation. Xylose concentrations in the medium at the end of each transfer cycle were measure by YSI 2700 Select instrument. At the end of each adaptation, the culture at 1:10,000 dilution was spread onto multiple YPD-Geneticin plates and incubated at 30°C for 48 h. Single clonal isolates were picked and evaluated for growth in YPX medium either aerobically or anaerobically in an anaerobic chamber as described previously [25,40].

Table 3. Comparison of anaerobic fermentation kinetics for Y22-3, Y127 and Y128 in YPX medium.

Strain	Y22-3	Y127	Y128
Absolute xylose consumption rate[1]	ND	0.101±0.004	0.596±0.028
Specific xylose consumption rate[2]	ND	0.515±0.009	0.587±0.038
% of theoretical ethanol yield for consumed xylose[3]	ND*	ND*	77.7±0.6
% of theoretical ethanol yield from total xylose[4]	1.2±0.8	3.9±0.2	74.7±0.5
Avg. final ethanol titer[5]	0.11±0.03	0.42±0.02	8.0±0.1

ND, Not Determined – low xylose consumption; ND*, Not Determined – no ethanol produced and low xylose consumption.
[1]In g xylose consumed/L/h.
[2]In g xylose consumed/g DCW/h.
[3]Calculated from the maximum ethanol concentration produced divided by the consumed xylose concentration at that time.
[4]Calculated from the maximum ethanol concentration produced divided by the starting xylose concentration.
[5]In g ethanol/L.

Aerobic and anaerobic fermentations

Aerobic and anaerobic batch fermentations in 3 L bioreactors (Applikon Biotechnology) were conducted using 2.1 L of ACSH or 1.7 L of YPX or YPDX with 50 mM potassium phosphate medium. Vessels were sparged in the headspace with N_2 (150 sccm) for anaerobic fermentation or in the medium with air (200 sccm) for aerobic fermentation. Inocula were prepared from single colonies grown in YPD-Geneticin medium aerobically for ~9 h, and then were diluted in ACSH or YPD medium with Geneticin at an initial $OD_{600} = 0.1$ (approximately 0.08–0.1 g DCW/L), and then grown aerobically for approximately 20 h. Starter cultures were then inoculated at a starting $OD_{600} = 0.1$ in bioreactor vessels maintained at 30°C and pH 5.0 with NaOH or HCl, and stirred at 500 rpm. For aerobic and anaerobic YPX growth assays, inoculum cultures were started from single colonies grown in

YPD-Geneticin medium overnight and then diluted to $OD_{600} = 0.1$ in YPX (no Tween-80 or ergosterol) at the start of the assay. Yeast cultures were grown in culture tubes containing 5–7.5 mL of medium shaken at 30°C, or in 30 mL of medium stirred in flasks placed in an anaerobic chamber (Coy) purged with hydrogen. For anaerobic experiments, bioreactors containing YPX, YPDX or ACSH were sparged with N_2 into the medium for at least 2 h. Flasks containing YPX were placed in the anaerobic chamber for at least 2 h prior to inoculation. Cell density measurements were determined by OD_{600} measurements from cultures diluted 1:10 or 1:25 in water. All OD_{600} measurements were blanked against uninoculated medium diluted in the same manner. Dry cell weight (DCW) was determined by vacuum filtering 50 mL of culture at 4 time points onto pre-weighed filters, washing with water and microwaving on 10%

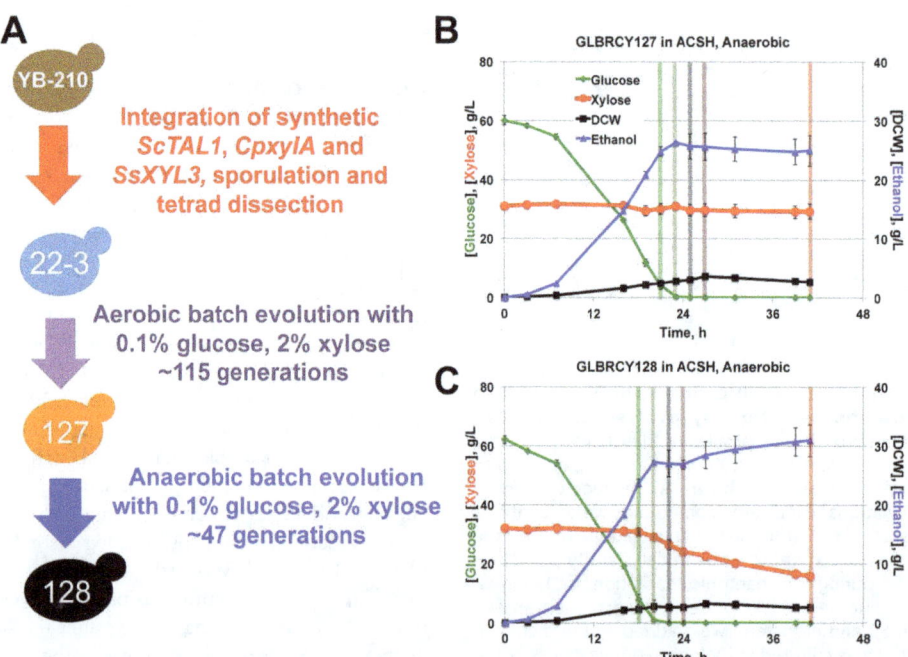

Figure 5. GLBRCY128 can anaerobically ferment xylose from ACSH. A diagram summarizing the engineering and evolution of the YB-210 strain into the evolved Y128 strain is provided in (**A**). Fermentation kinetic profiles of the Y127 (**B**) and Y128 (**C**) strains cultured in bioreactors containing ACSH and sparged with nitrogen from biological duplicates are shown. Average concentrations and standard deviations of indicated components were quantified from media samples at times from initial inoculation. Vertical colored bars indicate time points at which samples were taken for metabolomic analysis described in Fig. 7A–D.

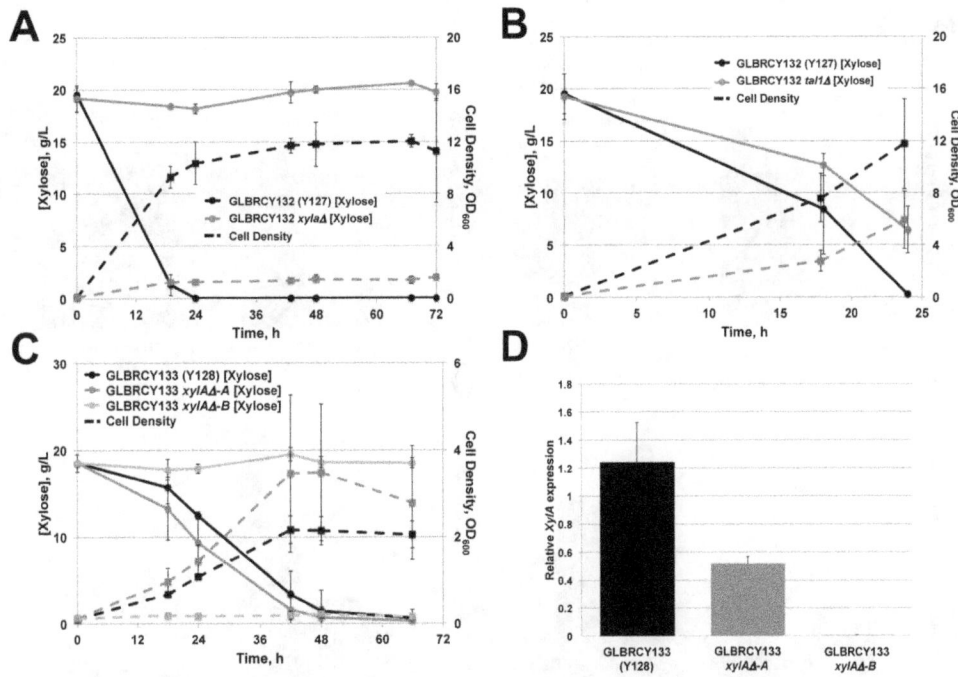

Figure 6. The xylose consumption phenotypes of the evolved Y127 and Y128 strains are dependent upon *CpxylA* and *ScTAL1.*
Extracellular xylose concentrations (solid lines) and cell density (dashed lines) were measured by YSI instrument and OD$_{600}$ readings, respectively, from cultures containing KanMX marker rescued versions of (**A**) GLBRCY127 (Y132) and GLBRCY132 *xylAΔ* or (**B**) Y132 and Y132 *tal1Δ* strains inoculated in aerobic YPX media. In (**C**), extracellular xylose concentrations (solid lines) and cell density (dashed lines) were measured as in (**A**, **B**) for KanMX marker rescued GLBRCY128 (Y133) and two independent GLBRCY133 *xylAΔ* strains inoculated in anaerobic YPX media. These selection marker-rescued Y128 strains were cultured in YPD media and total RNA isolated from a single time point. Expression of *CpxylA* was then quantified and normalized to *ScERV25* RNA levels by qPCR. The bar graph in (**D**) displays the average values and standard deviations for *CpxylA* RNA from three independent biological replicates.

power for 15 minutes. Filtered cells were additionally dried by desiccant for 2–3 days and then weighed. DCW values in grams per 50 mL of culture included subtraction of the filter weight alone. Correlations between g DCW/L and OD$_{600}$ concentrations were calculated to provide cell density measurements based on cell mass/L. Medium glucose and xylose concentrations from aerobic tube and anaerobic flask experiments were determined by YSI instrument. Extracellular glucose, xylose, ethanol, glycerol and xylitol concentrations from aerobic and anaerobic bioreactor experiments were determined by high performance liquid chromatography (HPLC) and refractive index detection (RID) as published elsewhere [35].

Quantification of intracellular pentose metabolic intermediates

To quantify intracellular metabolites, 5–10 mL of cell culture was rapidly removed from bioreactors with a 20 mL sterile syringe and 4 mL aliquots were applied to a filtration manifold unit (Hoefer FH 225 V) outfitted with sterile 25 mm nylon filters (Whatman; Nylon; 0.45 μm pore size), and the cells captured on the filters under vacuum. To reduce the background associated with metabolites present in ACSH, the cells were then rapidly washed with 5 mL of synthetic hydrolysate media [35] at pH 5.0 lacking amino acids and replacing 9% sorbitol in place of 6% glucose and 3% xylose. The filters were then removed and rapidly placed in 15 mL conical tubes containing ice-cold extraction buffer ([43]; acetonitrile-methanol-water, 40:40:20, 0.1% formic acid) and flash frozen in a dry ice ethanol bath.

The concentration of xylulose-5-phosphate was determined using reverse phase ion pairing HPLC [44] and electrospray ionization tandem mass spectrometry (ESI-MS/MS). Reagents and nonlabeled reference compounds were from Sigma Aldrich Co. (Saint Louis, Missouri, USA). Compounds were separated on an Ascentis HSS-T3 C18 column, 150×2.0 mm, 1.8 μm particle size (Waters Acquity). The mobile phase A consisted of 92.5:7.5 water:methanol, 10 mM tributylamine, and 15 mM acetic acid, and mobile phase B was isopropyl alcohol. Xylulose-5-phosphate, whose peak overlapped with ribulose-5-phosphate, was quantified by integrating the portion of the partially resolved peak that was clearly attributable to X5P.

For quantifying intracellular xylose, xylulose, and xylitol, 20 μL aliquots of metabolite extract were transferred to 2.5 mL centrifuge tubes along with 20 μL of a solution containing 100 μM U-^{13}C$_5$-xylitol (Sigma-Aldrich), 100 μM U-^{13}C$_5$-xylose (Sigma-Aldrich) and 50 μM U-^{13}C$_5$ xylulose. U-^{13}C$_5$ xylulose was obtained by enzymatic conversion of U-^{13}C$_5$-xylose by immobilized xylose isomerase (Sigma-Aldrich). Samples were then evaporated to dryness under reduced pressure in a rotary evaporator (Savant SPD131A) with cryogenic cold trap for 3–4 hours. Dried samples were incubated with 30 μL 2% methoxyamine hydrochloride in anhydrous pyridine at 60°C for 45 min, and then derivatized at 60°C for an additional 30 min with 70 μL N-methyl,N-(trimethylsilyl) trifluoroacetamide with 1% trimethylchlorosilane (Fisher Scientific). Derivatized samples were then analyzed by gas chromatography coupled with mass spectrometry (GC-MS) on an Agilent 5975 MSD with a Combi PAL autosampler (CTC analytics), and a 6890A GC oven equipped with a 30 m×0.25 mm ID×0.25 μm film HP5-MS capillary

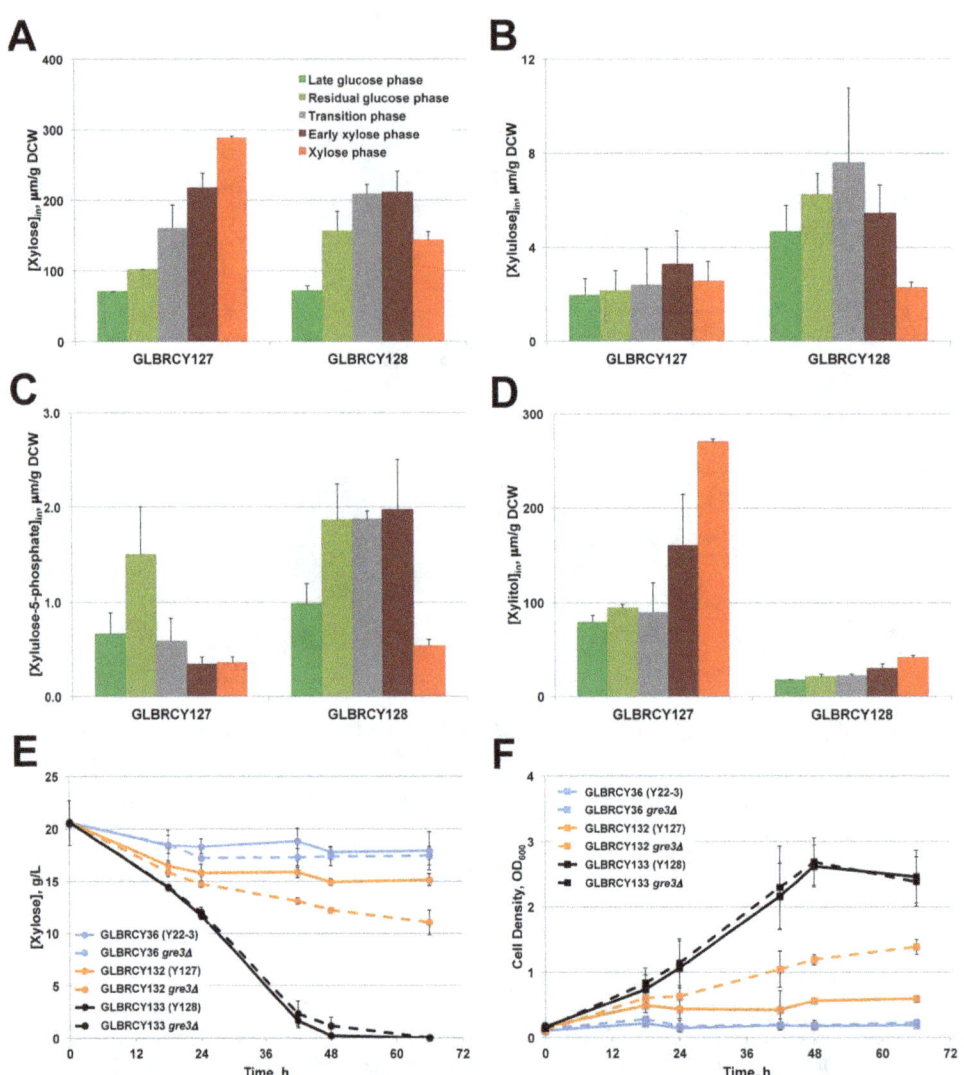

Figure 7. Y128 has a mutation in *GRE3* that reduces xylitol production and contributes towards anaerobic xylose fermentation. Fermentation samples were taken at the indicated time points marked by vertical colored bars in Fig. 5B and C. Cells were then filter-captured, briefly washed and then intracellular metabolites extracted. Average concentrations of xylose (**A**), xylulose (**B**), xylulose-5-phosphate (**C**) and xylitol (**D**) from independent duplicate fermentations were determined by reverse phase ion pairing HPLC-ESI coupled with MS/MS or gas chromatography (see Materials and Methods). Average concentrations and standard deviations are based on two biological replicates. Y22-3, Y127 and Y128 strains with KanMX selection markers excised (Y36, Y132 and Y133, respectively) and with or without deletion of *GRE3* were cultured under anaerobic conditions in YPX media. Samples were taken at the indicated time points to measure xylose concentrations (**E**) or cell density (**F**). Average values and standard deviations were calculated from biological triplicates.

column. The inlet was held at 250°C and operated in split mode with a ratio of 10:1 with a helium carrier gas flow rate maintained at 1.2 mL/min. The oven temperature was held at 125°C for 47 min then increased linearly at 40°C/min to a final temperature of 300°C.

The mass spectrometer was operated in SIM mode divided into time segments so that only the relevant masses were monitored over the times when each target compound eluted, allowing optimal dwell times of 100–150 ms while still recording at least 20 points over the width of a peak. SIM masses were selected corresponding to fragments ($M^{+}-15$) containing all 5 ^{13}C atoms for the labeled internal standards of xylose, xylulose (m/z 457) and xylitol (m/z 427) to allow detection without interference from the isotopic masses of the coeluting natural abundance compounds. The naturally-occurring ^{12}C compounds were measured by peak areas of the corresponding ^{12}C (xylose, m/z 452, xylitol, m/z 422) ions except for xylulose, which was monitored with a much more abundant m/z 263 ion that was found to be free of interference from the labeled internal standard.

Instrument operation, data collection, and calculation of results were conducted by Agilent MassHunter for GC software VB.07.00 and Mass Hunter Workstation Quantitative Analysis v.B.06.00. Results were calculated from relative peak areas of analytes to their corresponding internal standards interpolated with a calibration curve of relative natural standard/^{13}C internal standard peak areas versus relative standard/^{13}C internal standard concentrations.

In vitro xylose reductase activity assays

In vitro xylose reductase (XR) activities were performed as previously described [45–47] with minor modifications. Y132, Y132 *gre3Δ*, Y133 and Y133 *gre3Δ* strains were grown in YPD medium to $OD600 = 0.8$ and then cells from 45 mL of culture were harvested, washed with 10 mL 0.85% NaCl and resuspended in an equal volume of breaking buffer (50 mM potassium phosphate, pH 7.4, 1 mM β-mercaptoethanol and 1x Thermo HALT protease inhibitor cocktail). Resuspended cells were then transferred to a glass tube with glass beads and vortexed. The resulting material was centrifuged at 4°C and the clarified cell lysate used immediately for activity assays. A Tecan M1000 microplate reader was used to measure absorbance at 340 nm in technical triplicates. Protein concentrations from cell extracts were determined using a modified Bradford assay [48] with bovine serum albumin as the standard. Units of enzyme activity were normalized to total protein extract concentration and averaged from two independent biological samples. Specific XR activities were measured in range of 2–40 mU activity/mg total protein in the presence of NADPH co-factor.

Results and Discussion

Phenotypic screening identifies natural *S. cerevisiae* strains tolerant to pretreated lignocellulosic hydrolysates

The primary goal of this study was to create a strain of *S. cerevisiae* that can effectively ferment xylose anaerobically from AFEX-pretreated lignocellulosic biomass. We first evaluated two well-characterized laboratory strains, BY4741 [49] and CEN.PK2 [50] as potential starting points of this research by evaluating their growth abilities in lignocellulosic hydrolysates (LCHs) generated from a variety of established pretreatments and feedstocks (see Materials and Methods). Although both strains reached saturated cell density within 8 h after inoculation in YPD medium, they grew at substantially slower rates and reached low cell densities in the pretreated LCHs compared to standard medium (**Figs. S1A and B**), even though glucose concentrations were significantly higher in LCHs. These results suggest that the BY4741 and CEN.PK2 lab strains would not be able to generate sufficient cell biomass for rapid fermentation of inhibitor-laden hydrolysates. Thus, we sought an alternative strain background with robust cell growth in multiple LCHs in hopes of utilizing a strain with sufficient tolerance to LCH inhibitors. To find such strain, we performed comprehensive phenotyping of a collection of publicly available wild and domesticated *S. cerevisiae* strains obtained from a variety of locations and environments (Supplemental Strain Table 1, [51–53]) for cell growth in multiple pretreated LCHs. Individual strains were inoculated into 96-well plates containing 6% or 9% glucan-loading AFEX pretreated corn stover (ACS), raw or detoxified AHP pretreated corn stover (CS) or switchgrass (SG), [C2mim][OAc]-pretreated SG (IL SGH), or two different proprietary dilute acid pretreated biomass materials, supplemented with yeast extract and peptone (YP). Cell densities were continuously measured for 24–48 h, from which specific growth rates for each strain in every medium condition were calculated and normalized relative to their growth rate in YPD medium.

The collection of strains displayed wide ranges of aerobic growth rates in the various hydrolysate conditions (**Figs. S2–11** and **Table S1**). Supplementation of YP to 6% glucan-loading ACSH (**Figs. S2** and **S6**) improved growth, suggesting that growth defects in hydrolysates were due to lack of specific nutrients or additional nutrients allowed strains to overcome the effect of toxins. Not surprisingly, detoxification of AHP CSH and SGH (**Figs. S4, S5, S7** and **S8**) significantly improved overall rates compared to raw AHP hydrolysates, while growth rates in 9% glucan-loading ACSH, which contained higher concentrations of sugars and inhibitory compounds [34], were lower than in 6% glucan medium (**Figs. S2** and **S3**). In AHP CSH (**Fig. S4**) and the dilute acid hydrolysates (**Figs. S9** and **S10**), many strains did not achieve enough cell growth to calculate exponential growth rates, suggesting that these hydrolysates contained higher inhibitor concentrations, different combinations of inhibitors, or both, compared to the other hydrolysates. Direct growth-rate comparisons between different hydrolysate pretreatments were not made since the hydrolysates were not standardized for glucan-loading or feedstock source (e.g., AHP SGH vs. IL SGH).

To glean insights from the large datasets we amassed, we used hierarchical clustering to organize strains based on their rank in growth rate for each hydrolysate relative their growth rate rank in YPD medium (the reference media), and according to similarities in phenotypes across strains and conditions (**Fig. 1A**). Clustering the hydrolysates based on the distribution of growth phenotypes showed that AHP and ACS hydrolysates grouped together, regardless of plant feedstock, which is consistent with the fact that these LCHs are generated by alkaline-based pretreatments. AHP and AFEX pretreatments have been shown to produce lower levels of inhibitory furans commonly generated from acid dehydration of hexose sugars, particularly furfural, and this absence may be a significant driver in the clustering of alkaline hydrolysates from acid hydrolysates [23–25,29]. In addition, growth in IL and dilute acid-pretreated LCHs differed significantly from growth in other hydrolysates, with most strains unable to grow in dilute acid pretreated LCH #2. One study suggested that the predominant inhibitor in IL-derived LCHs is residual IL itself [26], which likely drives the unique phenotypic profile of yeast strain growth. It is unclear, due to restrictions on proprietary information, how the acid pretreated LCH #1 and #2 differ. These results also suggest that the driving difference between strain profiles is the pretreatment method, with lesser impact of the type of biomass used. Upon further inspection of individual strain performance, we found the BY4741 lab strain and CEN.PK2 (**Fig. 1A**, green arrows) clustered with a group of strains that generally grew slowly in most hydrolysates; this group also included many sake-producing strains, several bread-making strains, and strains isolated from natural fermentations. One cluster contains 11 strains with high growth-rate ranks across most hydrolysate conditions. These include isolates from soil (DBVPG1373), fermentation (Y9, Y12), lab (FL100), banana (YB-210), oak (YPS1009), clinical (YJM440, YJM653, YJM978, YJM981) and unknown (NCYC361) sources. Given our interest in understanding the genetic determinants of xylose fermentation from AFEX-pretreated LCHs, we focused on YB-210 ([33], also referred to as GLBRCY0 [40]), that displayed broad tolerance across the hydrolysates tested but ranked highest in growth for alkaline pretreatments. YB-210 grew robustly in AFEX, AHP and dilute acid pretreated LCHs, and less robustly in IL pretreated LCH (**Figs. 1B and C**) under aerobic conditions. Furthermore, YB-210 displays higher tolerance to elevated temperature [34] and inhibitors found in AHP-pretreated LCHs [25]. Therefore, the YB-210 strain background was selected for metabolic engineering and evolution of anaerobic xylose fermentation. As an added benefit of this work, other researchers can utilize this phenotypic dataset to select from publically available strains tailored for tolerance to their pretreated biomass of interest (*e.g.*, YPS163 or YPS1000 for dilute acid pretreated wheat straw).

Two-stage directed evolution using YB-210 harboring xylose isomerase, transaldolase and xylulokinase permitted anaerobic xylose fermentation

After phenotypic observations revealed its stress tolerant properties, the YB-210 was engineered for xylose metabolism by insertion of an expression cassette containing the *XYL1* (xylose reductase, XR), *XYL2* (xylitol dehydrogenase, XDH) and *XYL3* (xylulokinase) genes from *Sch. stipitis* [40], and then aerobically adapted on xylose. One evolved isolate (GLBRCY73, or Y73) that displayed improved xylose consumption rates in both lab media (**Fig. S12A**) and AHP SG hydrolysate (ASGH) was selected for further study [25]. We first examined the ability of Y73 to ferment xylose under controlled anaerobic conditions in N_2-sparged bioreactors containing YPDX in lab medium (**Fig. S12B**) or ACSH (**Fig. S12C**). Although the Y73 strain aerobically consumed ~40% of the xylose in 48 h, it anaerobically fermented <20% and <5% of the xylose in YPDX and ACSH, respectively, in the same time period. These results indicated that the XR-XDH engineered strain is severely impaired for anaerobic fermentation of xylose, particularly in ACSH, relative to aerobic culturing and suggested that this engineered strain would not be useful for our goals of better understanding anaerobic xylose fermentation. Attempts to adapt Y73 for anaerobic growth on xylose yielded no improved clones.

Similar to our observations with the GLBRCY73 strain, reduced anaerobic xylose consumption rates from other *S. cerevisiae* strains expressing *Sch. stipitis* XR-XDH enzymes have been reported. This limitation is likely due to redox cofactor imbalance. Heterologous engineering of *Sch. stipitis XYL1*, which primarily utilizes NADPH as its reducing cofactor, and *XYL2*, which uses NAD^+ as its oxidizing cofactor, introduces non-regenerative cycles in *S. cerevisiae* that are rapidly imbalanced in the absence of oxygen [4]. A possible alternative to circumvent this problem is to utilize xylose isomerase (XI) [54–57], which catalyzes the conversion of xylose into xylulose without cofactors, in place of *XYL1* and *XYL2*. We therefore re-engineered the diploid YB-210 strain with an expression cassette containing the *ScTDH3* promoter upstream of the *Clostridium phytofermentans* xylose isomerase (*CpxylA*), which has been shown to confer anaerobic xylose fermentation onto *S. cerevisiae* after additional genetic modifications [54]. Our cassette also included *ScTAL1*, a pentose phosphate pathway transaldolase enzyme that can improve xylose metabolism when overexpressed [58,59], and *SsXYL3* driven by the *ScPGK1* and *ScTEF2* promoters, respectively. Finally, to simplify future genomic resequencing of evolved descendants and to rapidly uncover beneficial recessive traits during directed evolution, we isolated one haploid spore, named GLBRCY22-3 (Y22-3), which maintained the *TAL1-xylA-XYL3* gene expression cassette.

To assess whether the engineered Y22-3 strain could metabolize xylose, Y22-3 was cultured aerobically in bioreactors with YPDX medium. Concentrations of extracellular glucose, xylose and dry cell weights were measured over the course of the fermentation (**Fig. 2A**). The Y22-3 strain consumed less than half of the xylose within 64 h, which was significantly less than the Y73 strain. Thus, the Y22-3 strain was subjected to aerobic batch selection in YP medium containing 0.1% glucose and 2% xylose and without exogenous mutagens. Anaerobic batch selection of Y22-3 in YP medium containing 0.1% glucose and 2% xylose was also performed without observing appreciable cell growth and was therefore abandoned, while adaptation in ACSH was not performed because high glucose concentrations (60 g/L in 6% glucan loading ACSH) and diauxic xylose consumption would prevent selection for improved growth on xylose present at 30 g/

L. For the first seven transfer cycles in YP-0.1% glucose and 2% xylose, each of which took place over 3–4 day periods, the culture grew at rates of ~1 generation per day with limited xylose consumption from the medium (**Fig. 2B**), suggesting that most of the growth was on glucose. Over the 8th to 11th transfers, slightly greater xylose consumption was observed, but without substantially faster cell growth rates. By the 12th transfer and beyond, the culture adapted to xylose, consuming all of the sugars within the 2–4 day passaging cycle and reaching saturated growth. After the 34th transfer (~115 generations), the culture was plated and single clones were screened for growth on xylose-containing medium. One clone, GLBRCY127 (Y127), displayed rapid aerobic growth in YPX by 96-well plate assay (data not shown), and was evaluated for aerobic xylose consumption in culture tubes containing YPX medium (**Fig. 2C**) or bioreactors containing YPDX medium (**Fig. 2D**). Compared to the Y22-3 strain, the evolved Y127 isolate displayed 15 to 17-fold faster absolute and specific xylose consumption rates than the Y22-3 parent in aerobic YPDX medium (**Figs. 2A, D** and **Table 2**). These results indicate that the Y127 isolate evolved from Y22-3 with properties allowing faster aerobic xylose consumption.

We next assessed the ability of the Y127 strain to ferment xylose anaerobically in bioreactors sparged with N_2. Similar to the XR-XDH engineered Y73 strain, the aerobically evolved Y127 strain displayed limited xylose fermentation from YPDX medium, consuming less than 30% of the total xylose within 42 h, and did not appear to convert the small amount of consumed xylose into ethanol (**Fig. 3A**). This suggested that, like Y73, the Y127 strain was not capable of effectively fermenting xylose in the absence of oxygen. In an attempt to overcome this barrier, we performed a second round of batch selection of the Y127 strain cultured in YP medium containing 0.1% glucose and 2% xylose under completely anaerobic conditions (**Fig. 3B**). During the first two transfers, 40 µg/L ergosterol and 4 g/L Tween-80 were added to support anaerobic growth, but then omitted for all successive transfers. For the first six transfers, the cell population doubled approximately twice per week. After the 6th transfer, the culture began to grow faster and consumed a greater percentage of the total xylose per day. After reaching 33 generations at the 10th transfer, the culture appeared to plateau in anaerobic growth and xylose consumption rate. After the 14th transfer (~47 generations), the culture was plated and colonies were screened for fastest growth rate in YPX medium by 96-well plate assay. One clone (GLBRCY128, Y128) displaying rapid anaerobic growth on xylose (data not shown) was then evaluated in bioreactors containing YPDX medium sparged with N_2. In contrast to Y127 (**Fig. 3A**), the Y128 strain rapidly fermented xylose in the absence of oxygen, during which time the extracellular ethanol concentration increased (**Fig. 3C**). Consistent with this result, Y128 exhibited higher absolute and specific xylose consumption rates in anaerobic YPDX medium than the Y127 strain (**Table 2**). These results indicate that the two-stage directed evolution yielded Y127 and Y128 strains with enhanced aerobic and anaerobic xylose metabolism, respectively.

The evolved Y128 strain can anaerobically convert xylose into ethanol faster than the parental strains

While the YPDX media matched the predominant sugar concentrations found in 6% glucan loading ACSH, the inclusion of 60 g/L glucose clouded our ability to directly compare the abilities of our engineered and evolved strains in the anaerobic conversion of xylose into ethanol. Therefore, we performed anaerobic fermentations with Y22-3, Y127 and Y128 strains in bioreactors containing YP media with 20 g/L xylose, and quantified xylose consumption and ethanol production from the media over time. For Y128, there was production of ethanol and cell biomass with simultaneous depletion of xylose from the media (**Fig. 4**). In contrast, both Y22-3 and Y127 strains produced less than 1 g/L ethanol (**Table 3**) and consumed less than 1 g/L xylose (**Fig. 4**) by the end of fermentation. Importantly, these differences in xylose consumption and ethanol production showed that Y128 consumed xylose significantly faster and produced ethanol at a higher yield and titer than Y22-3 and Y127 (**Table 3**). These results further support that the evolved Y128 strain displays marked improvement in anaerobic xylose fermentation compared to its parental strains.

The evolved Y128 strain can anaerobically convert xylose from ACSH into ethanol

Within a relatively small number of generations (~162 in total) without exogenous mutagens, our two-stage evolution yielded a *xylA*-engineered *S. cerevisiae* strain with the ability to consume xylose anaerobically in lab medium (summarized in **Fig. 5A**). Although the Y128 genetic background originated from a hydrolysate-tolerant strain, Y128 could have lost stress-tolerance traits during the course of xylose evolution. If so, Y128 might not have been able to convert xylose in lignocellulosic hydrolysates into ethanol anaerobically, which was a chief goal of this research. Previous studies have shown that yeast strains able to ferment xylose rapidly in lab medium are severely impaired in LCHs [34]. Therefore, we assessed the abilities of the Y127 and Y128 strains to ferment sugars anaerobically from ACSH in bioreactors sparged with N_2. Y22-3 was not used in this study, as it clearly does not significantly metabolize xylose aerobically (**Fig. 2**) or anaerobically (**Fig. 4**). Similar to our observations with YPDX lab medium, both Y127 and Y128 strains fermented glucose rapidly (**Figs. 5B and C**). However, the Y128 strain, but not Y127, consumed most of the xylose (~50% within 44 hrs) once glucose was depleted from ACSH. Indeed the absolute and specific xylose consumption rates for Y128 were approximately 10-fold higher than Y127 (**Table 2**). This also resulted in a higher yield in ethanol from xylose for Y128 compared to Y127 and Y73 (**Table 2**). Importantly, the ethanol yield from glucose for Y128 was similar to Y127, suggesting that the anaerobic xylose evolution had little impact on the ability of Y128 to convert glucose into ethanol (**Table 2**). Because Y128 ferments more xylose than Y127 anaerobically, this resulted in a higher ethanol titer for Y128 (**Fig. 5B and C**). Thus, despite the fact that evolution for xylose conversion occurred in lab medium lacking the inhibitors found in LCHs, the Y128 strain could effectively ferment xylose from an industrially relevant pretreated LCH in the absence of oxygen. This ability may be due in part to the innate hydrolysate-tolerant properties of the YB-210 genetic background.

Rapid xylose consumption by GLBRCY127 and Y128 is dependent upon *xylA* and *TAL1* expression

After clearly establishing the faster xylose consumption phenotypes of the Y127 and Y128 strains relative to the parental Y22-3 strain, we next wanted to better understand the potential genetic mechanisms by which these strains could have evolved. One obvious possibility would be through mutations in the engineered genes *xylA*, *TAL1* and *XKS1* that increase their expression or activities. However, when we sequenced the engineered gene cassette, no DNA sequence differences were identified. An alternative possibility is that the Y127 or Y128 strains obtained gain-of-function mutations in native genes that code for xylose metabolism enzymes, which are normally expressed at low levels

or lack sufficient activities for rapid flux into the pentose phosphate pathway [4]. The *S. cerevisiae* genome contains a number of putative enzymes with xylose reductase activities, including Gre3p [60,61], and xylitol dehydrogenases, including an ineffective *XYL2* homolog [62] and *XDH1*, which is present only in some wild *S. cerevisiae* strains and confers detectable xylose consumption [63]. Thus, one possible model for the evolution of Y127 and Y128 is that genetic changes in one or more of these genes allowed for improved xylose consumption independent of engineered *xylA* and *TAL1*. We examined this possibility by first excising the loxP-KanMX-loxP selection marker from the engineered cassette from Y22-3, Y127 and Y128 to generate antibiotic-sensitive GLBRCY36, GLBRCY132 and GLBRCY133, respectively. We then deleted *xylA* from the engineered cassette of the Y127 marker rescued strain (Y132) and assessed its ability to consume xylose aerobically. Indeed, the Y132 *xylAΔ* strain was ablated of its ability to consume or produce cell biomass from xylose (**Fig. 6A**). In contrast, deletion of *TAL1* from the engineered expression cassette, but not endogenous *TAL1*, in Y132 reduced the rate of xylose metabolism but did not impact the final amount of xylose consumed or the cell biomass produced from xylose (**Fig. 6B**), suggesting that the additional copy of engineered *TAL1* was important for determining the rate of xylose consumption but was not essential. In addition, we confirmed two independent marker rescued Y128 (Y133) *xylAΔ* transformants that displayed separate xylose consumption phenotypes. Consistent with the Y132 *xylAΔ* strain, the Y133 *xylAΔ-B* strain could not consume xylose aerobically (data not shown) or anaerobically (**Fig. 6C**). Interestingly, the Y133 *xylAΔ-A* strain fermented xylose at a comparable rate to the Y133 strain. This suggested the possibility that the *xylA* gene was duplicated in *cis*, which could explain why the replacement of *xylA* with the KanMX deletion cassette could be verified by PCR. Indeed, when we compared *xylA* RNA expression in the two Y133 *xylAΔ* strains to the Y133 strain by qPCR, we found that Y133 *xylAΔ-A* expressed half as much *xylA* as Y133, whereas no *xylA* RNA was detected in the Y133 *xylAΔ-B* strain (**Fig. 6D**). Although this result does not rule out possible genetic changes in endogenous xylose metabolizing enzymes, our findings together suggest that the evolved xylose consumption phenotypes in Y127 and Y128 are dependent, at least in part, upon the engineered *xylA* and *TAL1* genes.

Y128 accumulates higher intracellular concentrations of xylose metabolic intermediates but little xylitol

The data presented thus far suggest Y128 has an evolved ability to ferment xylose anaerobically in lab medium (**Figs. 3C and D**) and ACSH (**Fig. 5C**). To investigate how the evolved strains anaerobically ferment xylose, we analyzed the intracellular concentrations of xylose catabolism intermediates: xylose, xylitol, xylulose and xylulose-5-phosphate (X5P). During the anaerobic fermentation of ACSH by the Y127 and Y128 strains (**Figs. 5B and C**), cells were captured from the bioreactors using rapid vacuum filtration at five different time points spanning the glucose and xylose consumption phases; two samples were taken when glucose was detected in the hydrolysate (indicated by vertical lines in shades of green); one sample was taken during the transition to xylose after glucose was undetectable (grey vertical line), and two final samples were taken when xylose consumption (vertical lines in shades of red) was evident. Cell samples for Y127 and Y128 fermentations were taken at comparable sugar concentrations, whereas the last two samples were taken at the equivalent amount of time after the transition time point. During the course of the ACSH fermentation, the Y127 cells accumulated xylose (**Fig. 7A**), whereas xylulose, which is the product of isomerization of xylose by *CpxylA*, did not change (**Fig. 7B**). In contrast, Y128 peaked in intracellular xylose and xylulose levels during the transition phase, after which intracellular xylose decreased slightly at the end of the fermentation coincident with extracellular xylose depletion (**Figs. 7A and B**). The final metabolite of the engineered xylose metabolism pathway, X5P, peaked in intracellular concentration between the residual glucose and early xylose metabolism phases in Y128, whereas for Y127, X5P briefly peaked in the residual glucose phase, then decreased to low levels for the remainder of the fermentations (**Fig. 7C**). These patterns of intracellular xylose, xylulose and X5P accumulation, along with the simultaneous depletion of extracellular xylose and accumulation of ethanol (**Fig. 5C**), suggested the possibility that the xylose consumed by Y128 from ACSH was metabolized through the engineered *xylA* pathway at higher flux than in Y127.

The higher concentrations of catabolized pentose intermediates along with the faster xylose consumption rates (**Tables 2** and **3**) suggest two possible biochemical mechanisms in the evolution of Y128. One possibility is that the apparent increase in xylose consumption from the hydrolysate is due solely to improved xylose transport, without further metabolic conversion of xylose to other products. However, the higher ethanol titer achieved by Y128 coincident with differences in intracellular concentrations of metabolized xylose intermediates supports alternate models in which the evolved Y128 strain has more active xylose catabolism or pentose phosphate enzymes, or both, relative to its Y127 parent, possibly allowing for greater flux of xylose to ethanol.

The accumulation of internalized xylose and unchanging levels of xylulose in Y127 during the ACSH fermentation, along with the paltry change in extracellular xylose levels, suggest a metabolic bottleneck in xylose isomerase activity. Xylitol is a known inhibitor of xylose isomerase [64], and it is know that deletion of *GRE3*, which encodes a reductase enzyme that can convert xylose into xylitol, in *S. cerevisiae* engineered with xylose isomerase improved xylose fermentation [17,55,57,61,65]. Therefore, we quantified and compared the intracellular levels of xylitol in Y127 and Y128 during ACSH fermentation. Strikingly, we observed a severe reduction in intracellular xylitol levels for Y128 at all time points compared to Y127, which accumulated xylitol over time (**Fig. 7D**). This suggested that one difference between the Y127 and Y128 could be an evolved mutation that alters *GRE3* activity or expression. Thus, we sequenced the *GRE3* open reading frames of the Y22-3, Y127 and Y128 strains. We found a single nucleotide polymorphism (SNP) in the Y128 strain producing a G-to-A mutation relative to both Y22-3 and Y127, which changed the alanine at amino acid residue 46 to threonine (A46T) in Gre3p. This residue is conserved in other *S. cerevisiae* strains, as well as in many other yeast species including *Saccharomyces arboricola*, *S. kudriavzevii*, *Candida* and *Kluyveromyces*, and resides within the aldo-ketoreductase catalytic domain, suggesting that the A46T substitution likely impairs Gre3p xylose reductase activity. Although others have rationally deleted *GRE3* from strains to improve XI-mediated xylose metabolism prior to evolution, our observation of a spontaneous *GRE3* mutation acquired from directed evolution of a *xylA*-engineered yeast strain confirms its importance in the xylose metabolism bottleneck.

Taken together, our results suggest that the *gre3^{A46T}* mutation in Y128 may cause a partial or complete loss of Gre3p function, which in turn reduces xylitol production and thus minimizes inhibition of *CpxylA*. To further confirm the possibility that loss of Gre3p activity improves anaerobic xylose fermentation, we deleted *GRE3* in the marker rescued Y22-3 (renamed Y36), Y127 (renamed Y132) and Y128 (renamed Y133) strains and compared their *in vitro* xylose reductase activities and abilities to ferment

xylose in the absence of oxygen. First, xylose reductase activities from extracts generated from selection-marker rescued Y127 (Y132), Y132 *gre3Δ*, marker rescued Y128 (Y133) and Y133 *gre3Δ* cells were determined (**Fig. S13**). Extracts from the Y133 strain, which harbors the *gre3^{A46T}* mutation, and *gre3Δ* strains displayed similar decreased xylose reductase activity compared to Y132, which contains wild-type *GRE3*. This result further supported our *in vivo* observations that strains containing the *gre3^{A46T}* mutation behave biochemically similar to strains lacking *GRE3*. Finally, we found that the Y132 *gre3Δ* strain could anaerobically consume xylose (**Fig. 7E**) and grow faster than the parental Y132 strain (**Fig. 7F**), but not nearly as fast as Y133. This suggests that the Y128 strain contains mutations in addition to *gre3^{A46T}* that aid anaerobic xylose fermentation. Most importantly, there were no differences in the xylose consumption or growth rates between the Y133 and Y133 *gre3Δ* strains, which is consistent with the *gre3^{A46 T}* mutation resulting in a loss of function. Finally, deletion of *GRE3* in the Y36 parental strain had no effect, further indicating that loss of *GRE3* function alone cannot confer anaerobic xylose fermentation. Together, these *in vitro* and *in vivo* results suggest that the Y128 strain evolved a loss-of-function mutation in *GRE3*, which along with other unknown mutations, contributed to improved xylose utilization by reducing the production of inhibitory xylitol.

Conclusions

Here, we report the development of an engineered haploid *S. cerevisiae* strain with the evolved ability to ferment xylose anaerobically in lab media and LCH. Although yeast strains with improved anaerobic xylose fermentation in lab media and pretreated LCHs have been reported, most are derived from polyploid industrial strains with robust tolerance traits [12,17,25,30–32]. Although polyploidy can compensate for detrimental recessive alleles, the multiple gene copies present in these industrial strains make mapping and identifying the causal mutations that accelerate xylose conversion difficult. In contrast, we generated the haploid Y128 strain with the ability to rapidly ferment xylose anaerobically, even in the presence of ACSH inhibitors. Thus, haploid Y22-3 and its evolved Y127 and Y128 descendants are ideally suited for comparative analyses to identify gene sequences that contribute to xylose conversion in the presence and absence of oxygen, and to determine metabolomic and transcriptomic differences that underlie their respective phenotypes. As a proof of concept, we used metabolomic data and targeted gene sequencing to identify a loss-of-function mutation in *GRE3*, which we validated by strain reconstruction (**Fig. 7**). Thus, these strains are exciting tools that will provide future opportunities for multi-omic dissection of the molecular bottlenecks to anaerobic xylose fermentation by *S. cerevisiae* under LCH inhibitor stress.

Supporting Information

Figure S1 Domesticated strains of *S. cerevisiae* grow poorly in lignocellulosic hydrolysates. Representative aerobic growth profiles of lab strains BY4741 (**A**) and CEN.PK2 (**B**) cultured in 96-well plates with hydrolysates made from various pretreated lignocellulose hydrolysates (see Materials and Methods) and supplemented with yeast extract and peptone (YP) are shown by plotting cell density (optical density at 595 nm) every 20 min for 24 h. ACSH; 6% glucan loading AFEX pretreated corn stover hydrolysate, AHP; Alkaline Hydrogen Peroxide pretreatment, IL; Ionic Liquid ([C$_2$mim][OAc]) pretreated, Dil. Acid; Dilute Acid pretreated lignocellulosic hydrolysate, SGH; switchgrass hydrolysate, CSH; corn stover hydrolysate.

Figure S2 Bar graph displaying average growth rates (grey bars) of wild and domesticated *S. cerevisiae* strains in 6% glucan loading ACSH relative to YPD. Averages and standard deviations (black bars) are calculated from at least 3 biological replicates. The row location for NRRL YB-210/GLBRCY0 strain used in this study is identified by opposite coloration (average growth rate in black, standard deviation in grey).

Figure S3 Bar graph displaying average growth rates (grey bars) of wild and domesticated *S. cerevisiae* strains in 9% glucan loading ACSH relative to YPD. Averages and standard deviations (black bars) are calculated from at least biological replicates. The row location for NRRL YB-210/GLBRCY0 strain used in this study is identified by opposite coloration (average growth rate in black, standard deviation in grey).

Figure S4 Bar graph displaying average growth rates (grey bars) of wild and domesticated *S. cerevisiae* strains in YP-AHP CSH relative to YPD. Averages and standard deviations (black bars) are calculated from at least biological replicates. The row location for NRRL YB-210/GLBRCY0 strain used in this study is identified by opposite coloration (average growth rate in black, standard deviation in grey).

Figure S5 Bar graph displaying average growth rates (grey bars) of wild and domesticated *S. cerevisiae* strains in YP-AHP SGH relative to YPD. Averages and standard deviations (black bars) are calculated from at least biological replicates. The row location for NRRL YB-210/GLBRCY0 strain used in this study is identified by opposite coloration (average growth rate in black, standard deviation in grey).

Figure S6 Bar graph displaying average growth rates (grey bars) of wild and domesticated *S. cerevisiae* strains in YP-6% glucan loading ACSH relative to YPD. Averages and standard deviations (black bars) are calculated from at least biological replicates. The row location for NRRL YB-210/GLBRCY0 strain used in this study is identified by opposite coloration (average growth rate in black, standard deviation in grey).

Figure S7 Bar graph displaying average growth rates (grey bars) of wild and domesticated *S. cerevisiae* strains in YP-detoxified AHP CSH relative to YPD. Averages and standard deviations (black bars) are calculated from at least biological replicates. The row location for NRRL YB-210/GLBRCY0 strain used in this study is identified by opposite coloration (average growth rate in black, standard deviation in grey).

Figure S8 Bar graph displaying average growth rates (grey bars) of wild and domesticated *S. cerevisiae* strains in YP-detoxified AHP SGH relative to YPD. Averages and standard deviations (black bars) are calculated from

at least biological replicates. The row location for NRRL YB-210/GLBRCY0 strain used in this study is identified by opposite coloration (average growth rate in black, standard deviation in grey).

Figure S9 Bar graph displaying average growth rates (grey bars) of wild and domesticated *S. cerevisiae* strains in YP-80% dilute acid pretreated hydrolysate #1 relative to YPD. Averages and standard deviations (black bars) are calculated from at least biological replicates. The row location for NRRL YB-210/GLBRCY0 strain used in this study is identified by opposite coloration (average growth rate in black, standard deviation in grey).

Figure S10 Bar graph displaying average growth rates (grey bars) of wild and domesticated *S. cerevisiae* strains in YP-80% dilute acid pretreated hydrolysate #2 relative to YPD. Averages and standard deviations (black bars) are calculated from at least biological replicates. The row location for NRRL YB-210/GLBRCY0 strain used in this study is identified by opposite coloration (average growth rate in black, standard deviation in grey).

Figure S11 Bar graph displaying average growth rates (grey bars) of wild and domesticated *S. cerevisiae* strains in YP-IL SGH relative to YPD. Averages and standard deviations (black bars) are calculated from at least biological replicates. The row location for NRRL YB-210/GLBRCY0 strain used in this study is identified by opposite coloration (average growth rate in black, standard deviation in grey).

Figure S12 Hydrolysate-tolerant YB-210/GLBRCY0 engineered with XR/XDH and evolved for aerobic xylose metabolism does not ferment xylose anaerobically. The YB-210/Y0 strain engineered with *XYL1*, *2* and *3* genes from *S. stipitis* and aerobically-evolved (GLBRCY73) was cultured in bioreactors and evaluated for consumption of xylose in aerobic YPDX (**A**), anaerobic YPDX (**B**) and anaerobic ACSH (**C**) media as described in Materials and Methods. Concentrations (g/L) of glucose (green), xylose (red), dry cell weight (black) and ethanol (blue) are averages and standard deviations from two independent biological replicates.

Figure S13 GLBRCY133 (Y128) cell extracts display reduced *in vitro* xylose reductase activity similar to *GRE3* deletion strains. The indicated strains were cultured aerobically in YPD, harvested and prepared for *in vitro* xylose reductase activity assays as described in Materials and Methods. Xylose and NADPH were added to each extract, and then rates of change in absorbance at 340 nm were measured to determine the Units of enzymatic activity normalized to mg of total protein in the cellular extract. The graph displays the average percent of *in vitro* xylose reductase activities and standard deviations of indicated strains relative to GLBRCY132 (marker rescued GLBRCY127, which contains wild-type *GRE3*) in biological duplicate.

Table S1 Wild and domesticated *S. cerevisiae* strains used in phenotypic growth studies.

Acknowledgments

We thank Drs. Jeff Lewis, Justin Fay, and Clete Kurtzmann for *S. cerevisiae* strains, William Bothfeld, Mick McGee, Brendan Thomson and Robert Zinkel for technical support, Novozymes for providing enzymes, Dr. Brian Burger and James Hose for protocols and technical advice, our industrial collaborator for dilute acid hydrolysates, Drs. Tom Jeffries, Chris Hittinger and Jeff Piotrowski for scientific advice, and Dr. Donna Bates for comments on this manuscript.

Author Contributions

Conceived and designed the experiments: LSP AJH LBW STW BAS DBH JJC BED VB DHK YZ RL APG TKS. Performed the experiments: LSP RJB RAN AJH ALR MT LQ LBW BDB BLB RCP AJB NU TL CL DT HL ELP JS DHK YZ TKS. Analyzed the data: AJH LBW IMO TKS. Contributed reagents/materials/analysis tools: NU TL CL DT BAS DBH MSW JJC BED VB TKS. Contributed to the writing of the manuscript: LSP RJB AJH LBW CL STW BAS DBH JJC BED VB DHK YZ RL APG TKS.

References

1. Sheehan JJ (2009) Biofuels and the conundrum of sustainability. Curr Opin Biotechnol 20: 318–324.
2. Wyman CE (2007) What is (and is not) vital to advancing cellulosic ethanol. Trends Biotechnol 25: 153–157.
3. Pauly M, Keegstra K (2008) Cell-wall carbohydrates and their modification as a resource for biofuels. Plant J 54: 559–568.
4. Jeffries TW, Jin YS (2004) Metabolic engineering for improved fermentation of pentoses by yeasts. Appl Microbiol Biotechnol 63: 495–509.
5. Runquist D, Hahn-Hagerdal B, Radstrom P (2010) Comparison of heterologous xylose transporters in recombinant Saccharomyces cerevisiae. Biotechnol Biofuels 3: 5.
6. Kim SR, Park YC, Jin YS, Seo JH (2013) Strain engineering of Saccharomyces cerevisiae for enhanced xylose metabolism. Biotechnol Adv 31: 851–861.
7. van Maris AJ, Winkler AA, Kuyper M, de Laat WT, van Dijken JP, et al. (2007) Development of efficient xylose fermentation in Saccharomyces cerevisiae: xylose isomerase as a key component. Adv Biochem Eng Biotechnol 108: 179–204.
8. Hahn-Hagerdal B, Karhumaa K, Fonseca C, Spencer-Martins I, Gorwa-Grauslund MF (2007) Towards industrial pentose-fermenting yeast strains. Appl Microbiol Biotechnol 74: 937–953.
9. Laluce C, Schenberg AC, Gallardo JC, Coradello LF, Pombeiro-Sponchiado SR (2012) Advances and developments in strategies to improve strains of Saccharomyces cerevisiae and processes to obtain the lignocellulosic ethanol–a review. Appl Biochem Biotechnol 166: 1908–1926.
10. Kim SR, Skerker JM, Kang W, Lesmana A, Wei N, et al. (2013) Rational and evolutionary engineering approaches uncover a small set of genetic changes efficient for rapid xylose fermentation in Saccharomyces cerevisiae. PLoS One 8: e57048.
11. Madhavan A, Srivastava A, Kondo A, Bisaria VS (2012) Bioconversion of lignocellulose-derived sugars to ethanol by engineered Saccharomyces cerevisiae. Crit Rev Biotechnol 32: 22–48.
12. Demeke MM, Dietz H, Li Y, Foulquie-Moreno MR, Mutturi S, et al. (2013) Development of a D-xylose fermenting and inhibitor tolerant industrial Saccharomyces cerevisiae strain with high performance in lignocellulose hydrolysates using metabolic and evolutionary engineering. Biotechnol Biofuels 6: 89.
13. Bera AK, Ho NW, Khan A, Sedlak M (2011) A genetic overhaul of Saccharomyces cerevisiae 424A(LNH-ST) to improve xylose fermentation. J Ind Microbiol Biotechnol 38: 617–626.
14. Van Vleet JH, Jeffries TW, Olsson L (2008) Deleting the para-nitrophenyl phosphatase (pNPPase), PHO13, in recombinant Saccharomyces cerevisiae improves growth and ethanol production on D-xylose. Metab Eng 10: 360–369.
15. Young EM, Tong A, Bui H, Spofford C, Alper HS (2014) Rewiring yeast sugar transporter preference through modifying a conserved protein motif. Proc Natl Acad Sci U S A 111: 131–136.
16. Barrick JE, Lenski RE (2013) Genome dynamics during experimental evolution. Nat Rev Genet 14: 827–839.
17. Diao L, Liu Y, Qian F, Yang J, Jiang Y, et al. (2013) Construction of fast xylose-fermenting yeast based on industrial ethanol-producing diploid Saccharomyces cerevisiae by rational design and adaptive evolution. BMC Biotechnol 13: 110.
18. Scalcinati G, Otero JM, Van Vleet JR, Jeffries TW, Olsson L, et al. (2012) Evolutionary engineering of Saccharomyces cerevisiae for efficient aerobic xylose consumption. FEMS Yeast Res 12: 582–597.
19. Klimacek M, Kirl E, Krahulec S, Longus K, Novy V, et al. (2014) Stepwise metabolic adaption from pure metabolization to balanced anaerobic growth on xylose explored for recombinant Saccharomyces cerevisiae. Microb Cell Fact 13: 37.
20. Blanch HW, Simmons BA, Klein-Marcuschamer D (2011) Biomass deconstruction to sugars. Biotechnol J 6: 1086–1102.
21. Piotrowski JS, Zhang Y, Bates DM, Keating DH, Sato TK, et al. (2014) Death by a thousand cuts: the challenges and diverse landscape of lignocellulosic hydrolysate inhibitors. Front Microbiol 5: 90.
22. Almeida JR, Runquist D, Sanchez i Nogue V, Liden G, Gorwa-Grauslund MF (2011) Stress-related challenges in pentose fermentation to ethanol by the yeast Saccharomyces cerevisiae. Biotechnol J 6: 286–299.
23. Liu ZL (2006) Genomic adaptation of ethanologenic yeast to biomass conversion inhibitors. Appl Microbiol Biotechnol 73: 27–36.
24. Rasmussen H, Sorensen HR, Meyer AS (2014) Formation of degradation compounds from lignocellulosic biomass in the biorefinery: sugar reaction mechanisms. Carbohydr Res 385: 45–57.
25. Sato TK, Liu T, Parreiras LS, Williams DL, Wohlbach DJ, et al. (2014) Harnessing Genetic Diversity in Saccharomyces cerevisiae for Fermentation of Xylose in Hydrolysates of Alkaline Hydrogen Peroxide-Pretreated Biomass. Appl Environ Microbiol 80: 540–554.
26. Ouellet M, Datta S, Dibble DC, Tamrakar PR, Benke PI, et al. (2011) Impact of ionic liquid pretreated plant biomass on Saccharomyces cerevisiae growth and biofuel production. Green Chemistry 13: 2743–2749.
27. Balan V, Bals B, Chundawat SP, Marshall D, Dale BE (2009) Lignocellulosic biomass pretreatment using AFEX. Methods Mol Biol 581: 61–77.
28. Lau MW, Gunawan C, Balan V, Dale BE (2010) Comparing the fermentation performance of Escherichia coli KO11, Saccharomyces cerevisiae 424A(LNH-ST) and Zymomonas mobilis AX101 for cellulosic ethanol production. Biotechnol Biofuels 3: 11.
29. Chundawat SP, Vismeh R, Sharma LN, Humpula JF, da Costa Sousa L, et al. (2010) Multifaceted characterization of cell wall decomposition products formed during ammonia fiber expansion (AFEX) and dilute acid based pretreatments. Bioresour Technol 101: 8429–8438.
30. Koppram R, Albers E, Olsson L (2012) Evolutionary engineering strategies to enhance tolerance of xylose utilizing recombinant yeast to inhibitors derived from spruce biomass. Biotechnol Biofuels 5: 32.
31. Ismail KS, Sakamoto T, Hatanaka H, Hasunuma T, Kondo A (2013) Gene expression cross-profiling in genetically modified industrial Saccharomyces cerevisiae strains during high-temperature ethanol production from xylose. J Biotechnol 163: 50–60.
32. Garcia Sanchez R, Karhumaa K, Fonseca C, Sanchez Nogue V, Almeida JR, et al. (2010) Improved xylose and arabinose utilization by an industrial recombinant Saccharomyces cerevisiae strain using evolutionary engineering. Biotechnol Biofuels 3: 13.
33. Mortimer RK, Johnston JR (1986) Genealogy of principal strains of the yeast genetic stock center. Genetics 113: 35–43.
34. Jin M, Sarks C, Gunawan C, Bice BD, Simonett SP, et al. (2013) Phenotypic selection of a wild Saccharomyces cerevisiae strain for simultaneous saccharification and co-fermentation of AFEXTM pretreated corn stover. Biotechnol Biofuels 6: 108.
35. Schwalbach MS, Keating DH, Tremaine M, Marner WD, Zhang Y, et al. (2012) Complex physiology and compound stress responses during fermentation of alkali-pretreated corn stover hydrolysate by an Escherichia coli ethanologen. Appl Environ Microbiol 78: 3442–3457.
36. Li M, Foster C, Kelkar S, Pu Y, Holmes D, et al. (2012) Structural characterization of alkaline hydrogen peroxide pretreated grasses exhibiting diverse lignin phenotypes. Biotechnol Biofuels 5: 38.
37. Banerjee G, Car S, Liu T, Williams DL, Meza SL, et al. (2012) Scale-up and integration of alkaline hydrogen peroxide pretreatment, enzymatic hydrolysis, and ethanolic fermentation. Biotechnol Bioeng 109: 922–931.
38. Li C, Tanjore D, He W, Wong J, Gardner JL, et al. (2013) Scale-up and evaluation of high solid ionic liquid pretreatment and enzymatic hydrolysis of switchgrass. Biotechnol Biofuels 6: 154.
39. Sherman F (2002) Getting started with yeast. Methods Enzymol 350: 3–41.
40. Wohlbach DJ, Kuo A, Sato TK, Potts KM, Salamov AA, et al. (2011) Comparative genomics of xylose-fermenting fungi for enhanced biofuel production. Proc Natl Acad Sci U S A 108: 13212–13217.
41. Guldener U, Heck S, Fielder T, Beinhauer J, Hegemann JH (1996) A new efficient gene disruption cassette for repeated use in budding yeast. Nucleic Acids Res 24: 2519–2524.
42. Voth WP, Richards JD, Shaw JM, Stillman DJ (2001) Yeast vectors for integration at the HO locus. Nucleic Acids Res 29: E59–59.
43. Amador-Noguez D, Brasg IA, Feng XJ, Roquet N, Rabinowitz JD (2011) Metabolome remodeling during the acidogenic-solventogenic transition in Clostridium acetobutylicum. Appl Environ Microbiol 77: 7984–7997.
44. Buescher JM, Moco S, Sauer U, Zamboni N (2010) Ultrahigh performance liquid chromatography-tandem mass spectrometry method for fast and robust quantification of anionic and aromatic metabolites. Anal Chem 82: 4403–4412.
45. Witteveen C, Busink R, Van de Vondervoort P, Dijkema C, Swart K, et al. (1989) L-Arabinose and D-xylose catabolism in Aspergillus niger. J Gen Microbiol 135: 2163–2171.

46. Kuhn A, van Zyl C, van Tonder A, Prior BA (1995) Purification and partial characterization of an aldo-keto reductase from Saccharomyces cerevisiae. Appl Environ Microbiol 61: 1580–1585.

47. Jin YS, Jeffries TW (2003) Changing flux of xylose metabolites by altering expression of xylose reductase and xylitol dehydrogenase in recombinant Saccharomyces cerevisiae. Appl Biochem Biotechnol 105–108: 277–286.

48. Zor T, Selinger Z (1996) Linearization of the Bradford protein assay increases its sensitivity: theoretical and experimental studies. Anal Biochem 236: 302–308.

49. Brachmann CB, Davies A, Cost GJ, Caputo E, Li J, et al. (1998) Designer deletion strains derived from Saccharomyces cerevisiae S288C: a useful set of strains and plasmids for PCR-mediated gene disruption and other applications. Yeast 14: 115–132.

50. van Dijken JP, Bauer J, Brambilla L, Duboc P, Francois JM, et al. (2000) An interlaboratory comparison of physiological and genetic properties of four Saccharomyces cerevisiae strains. Enzyme Microb Technol 26: 706–714.

51. Fay JC, Benavides JA (2005) Hypervariable noncoding sequences in Saccharomyces cerevisiae. Genetics 170: 1575–1587.

52. Lewis JA, Elkon IM, McGee MA, Higbee AJ, Gasch AP (2010) Exploiting natural variation in Saccharomyces cerevisiae to identify genes for increased ethanol resistance. Genetics 186: 1197–1205.

53. Liti G, Carter DM, Moses AM, Warringer J, Parts L, et al. (2009) Population genomics of domestic and wild yeasts. Nature 458: 337–341.

54. Brat D, Boles E, Wiedemann B (2009) Functional expression of a bacterial xylose isomerase in Saccharomyces cerevisiae. Appl Environ Microbiol 75: 2304–2311.

55. Karhumaa K, Hahn-Hagerdal B, Gorwa-Grauslund MF (2005) Investigation of limiting metabolic steps in the utilization of xylose by recombinant Saccharomyces cerevisiae using metabolic engineering. Yeast 22: 359–368.

56. Kuyper M, Winkler AA, van Dijken JP, Pronk JT (2004) Minimal metabolic engineering of Saccharomyces cerevisiae for efficient anaerobic xylose fermentation: a proof of principle. FEMS Yeast Res 4: 655–664.

57. Lee SM, Jellison T, Alper HS (2012) Directed evolution of xylose isomerase for improved xylose catabolism and fermentation in the yeast Saccharomyces cerevisiae. Appl Environ Microbiol 78: 5708–5716.

58. Ni H, Laplaza JM, Jeffries TW (2007) Transposon mutagenesis to improve the growth of recombinant Saccharomyces cerevisiae on D-xylose. Appl Environ Microbiol 73: 2061–2066.

59. Walfridsson M, Hallborn J, Penttila M, Keranen S, Hahn-Hagerdal B (1995) Xylose-metabolizing Saccharomyces cerevisiae strains overexpressing the TKL1 and TAL1 genes encoding the pentose phosphate pathway enzymes transketolase and transaldolase. Appl Environ Microbiol 61: 4184–4190.

60. Traff KL, Jonsson LJ, Hahn-Hagerdal B (2002) Putative xylose and arabinose reductases in Saccharomyces cerevisiae. Yeast 19: 1233–1241.

61. Traff KL, Otero Cordero RR, van Zyl WH, Hahn-Hagerdal B (2001) Deletion of the GRE3 aldose reductase gene and its influence on xylose metabolism in recombinant strains of Saccharomyces cerevisiae expressing the xylA and XKS1 genes. Appl Environ Microbiol 67: 5668–5674.

62. Toivari MH, Salusjarvi L, Ruohonen L, Penttila M (2004) Endogenous xylose pathway in Saccharomyces cerevisiae. Appl Environ Microbiol 70: 3681–3686.

63. Wenger JW, Schwartz K, Sherlock G (2010) Bulk segregant analysis by high-throughput sequencing reveals a novel xylose utilization gene from Saccharomyces cerevisiae. PLoS Genet 6: e1000942.

64. Yamanaka K (1969) Inhibition of D-xylose isomerase by pentitols and D-lyxose. Arch Biochem Biophys 131: 502–506.

65. Kuyper M, Hartog MM, Toirkens MJ, Almering MJ, Winkler AA, et al. (2005) Metabolic engineering of a xylose-isomerase-expressing Saccharomyces cerevisiae strain for rapid anaerobic xylose fermentation. FEMS Yeast Res 5: 399–409.

20

Overexpression of the PAP1 Transcription Factor Reveals a Complex Regulation of Flavonoid and Phenylpropanoid Metabolism in *Nicotiana tabacum* Plants Attacked by *Spodoptera litura*

Tomoko Mitsunami[1], Masahiro Nishihara[2], Ivan Galis[3], Kabir Md Alamgir[3], Yuko Hojo[3], Kohei Fujita[2], Nobuhiro Sasaki[2], Keichiro Nemoto[4], Tatsuya Sawasaki[4], Gen-ichiro Arimura[1]*

1 Department of Biological Science & Technology, Faculty of Industrial Science & Technology, Tokyo University of Science, Tokyo, Japan, 2 Iwate Biotechnology Research Center, Kitakami, Japan, 3 Institute of Plant Science and Resources, Okayama University, Kurashiki, Japan, 4 Proteo-Science Center, Ehime University, Matsuyama, Japan

Abstract

Anthocyanin pigments and associated flavonoids have demonstrated antioxidant properties and benefits for human health. Consequently, current plant bioengineers have focused on how to modify flavonoid metabolism in plants. Most of that research, however, does not consider the role of natural biotic stresses (e.g., herbivore attack). To understand the influence of herbivore attack on the metabolic engineering of flavonoids, we examined tobacco plants overexpressing the *Arabidopsis* PAP1 gene (encoding an MYB transcription factor), which accumulated anthocyanin pigments and other flavonoids/phenylpropanoids. In comparison to wild-type and control plants, transgenic plants exhibited greater resistance to *Spodoptera litura*. Moreover, herbivory suppressed the PAP1-induced increase of transcripts of flavonoid/phenylpropanoid biosynthetic genes (e.g., *F3H*) and the subsequent accumulation of these genes' metabolites, despite the unaltered *PAP1* mRNA levels after herbivory. The instances of down-regulation were independent of the signaling pathways mediated by defense-related jasmonates but were relevant to the levels of PAP1-induced and herbivory-suppressed transcription factors, An1a and An1b. Although initially *F3H* transcripts were suppressed by herbivory, after the *S. litura* feeding was interrupted, *F3H* transcripts increased. We hypothesize that in transgenic plants responding to herbivory, there is a complex mechanism regulating enriched flavonoid/phenylpropanoid compounds, via biotic stress signals.

Editor: Jin-Song Zhang, Institute of Genetics and Developmental Biology, Chinese Academy of Sciences, China

Funding: This work was financially supported in part by the MEXT Grants for Excellent Graduate Schools program of Kyoto University; the Research Fund of Tokyo University of Science; and a Grant-in-Aid for Scientific Research from the Japan Society for the Promotion of Science to G.A. (No. 24770019). The LC-MS/MS analyses were supported by the Japan Advanced Plant Science Network. The funders had no role in study design, data collection and analysis, decision to publish, or preparation of the manuscript.

Competing Interests: The authors have declared that no competing interests exist.

* Email: garimura@rs.tus.ac.jp

Introduction

The production of specialized metabolites in plants has attracted great interest as a result of the successful development of mechanisms that enable metabolic enzymes to be produced [1–3]. To synchronously activate metabolic genes, plants use a large variety of transcription factors. These factors may control multiple genes or orchestrate entire metabolic pathways or their specific branches. PRODUCTION OF ANTHOCYANIN PIGMENT1 (PAP1) is a typical R2R3 MYB-type transcription factor in *Arabidopsis thaliana* that is able to ectopically activate the entire array of genes involved in the biosynthesis of phenolic acids and flavonoid compounds (including anthocyanins) in the flavonoid biosynthetic pathway in several plant species, including tobacco, salvia, petunia and rose [4–8]. The overexpression of *Arabidopsis PAP1* in transgenic tobacco plants under the control of a constitutive promoter led to the induction of anthocyanins in the leaves, flowers, stems and roots; the deep red or purple appearance of the pants parts was evidence [5]. Moreover, *PAP1*-overexpressing petunia flowers have been shown to increase not only levels of anthocyanins but also emissions of floral volatiles (benzenoids) [7]. Gene expression profiling and measurements of the levels of pathway intermediates suggest that both increased metabolic flux and the transcriptional activation of scent and color genes underlie the enhancement of petunia flower color and scent production [7]. The coordinated regulation of metabolic steps within or between pathways involved in vital plant functions – for instance, pathways involved in floral display determining plant-pollinator interactions – provides plants with a clear advantage. Indeed, bees and humans can discriminate between the floral scents of *PAP1*-overexpressing and control rose flowers [6]. Therefore, PAP1 overexpression may find applications in horticulture and human health.

Manipulating flavonoid metabolism, especially in response to biotic or abiotic stresses, may have unpredicted side effects, as these compounds belong to the largest group of specialized metabolites produced by plants [9]. Flavonoids serve as essential components of a number of structural polymers that provide protection; lignin, for instance, due to its antioxidant and free radical scavenging properties, protects plants from ultraviolet light and defends plants against herbivores and pathogens [10]. Anthocyanin accumulation in *Arabidopsis* is enhanced in response to a number of environmental stress factors, such as cold, drought, pathogen attack and nutrient depletion [11]. Moreover, it appears that transgenic tobacco expresses structural genes and regulatory genes involved in flavonoid biosynthesis, and these genes defend the plant against herbivores [12,13]. Therefore, *PAP1*-overexpressing plants that typically accumulate large amounts of flavonoids might possess increased plant resistance to biotic stresses, including herbivores, but to date no practical evidence of this prediction has been provided. Whether the *PAP1* transgene interferes with endogenous TFs and processes involved in plants' defense and development has not been extensively investigated.

A separate branch of flavonoid catabolism controlled by other MYB transcription factors (e.g., MYBJS1) was previously shown to control the accumulation of anti-herbivore compounds, phenolamides (PAs), in *Nicotiana* species [14–17]. In addition, CoA-activated phenolic acids with polyamines in PA production is controlled by the MYBJS1 homologue MYB8 in a herbivory-inducible fashion in wild tobacco, *N. attenuata* [18]. MYB8 (and MYBJS1) also likely contributes to lignin accumulation by targeting the control of the hydroxycinnamoyl-CoA:shikimate/quinate hydroxycinnamoyl transferase (HCT) gene in tobacco [19–21].

Because *PAP1*-overexpressing plants constitutively produce large amount of anthocyanins, we asked if the accumulation of other inducible flavonoids and phenylpropanoids was affected. PAP1 and MYBJS1 seem to share common gene targets in the pathways they regulate, and so these two transcription factors are likely to engage in complex interactions when *PAP1* is introduced into plants under a constitutive promoter and also, in response to herbivory, when other MYBs are additionally activated upon. It is likely that when *PAP1*-overexpressing plants are damaged by herbivores, the dynamic activation of the metabolic flow via other MYB members competes with activated coumaric acid units used in the constitutive synthesis of anthocyanins by PAP1-promoted enzymes. Both transcriptional and post-transcriptional regulations of PAP1 have been shown to control anthocyanin levels in ubiquitous light/dark rhythms [22]. Moreover, when anthocyanin production is regulated by independent PAP-transcription and environmental stress, anthocyanin levels are reduced [23]. Given that the stress of herbivory could affect PAP1 function in *PAP1*-overexpressing plants, we asked how efficiently transgenes such as PAP1 could be expected to perform under natural, uncontrolled environmental conditions.

In the current study, we found that the defensive properties of *PAP1*-overexpressing plants increased upon herbivore attack, and addressed the issue of metabolic fluxes and transcriptional regulation in the flavonoid/phenylpropanoid pathways that are co-regulated by powerful PAP1-activators and herbivory. Since jasmonic acid (JA) and its derivatives (referred as jasmonate [JAs]) coordinate the expression of regulatory and structural genes in the flavonoid/phenylpropanoid pathways under stress, we also assessed the role of JAs (and other phytohormones) in the herbivory- and PAP1-mediated regulation of phenylpropanoid metabolism. Finally, we discuss the implications of enhancing the production of specialized metabolites in plants under natural environmental conditions.

Materials and Methods

Production of transgenic tobacco plants

The full-length coding region of *Arabidopsis PAP1* (At1g56650) cDNA was inserted into the binary vector pSMABR35SsGFP by replacing the GFP reporter gene of the vector [24]. The resulting plasmid, pSMABR-35SPAP1, was transformed into *Agrobacterium tumefaciens* strain EHA101 by electroporation. *PAP1*-overexpressing tobacco (*N. tabacum* cv. SR1) was produced by *Agrobacterium*-mediated transformation as described previously [24]. After rooting and acclimatization, the regenerated plants were grown in a controlled greenhouse to set seeds. Nine lines of transgenic T_1 seeds were tested for germination on Murashige and Skoog medium supplemented with 5 mg l^{-1} bialaphos at $25°C$ under a 16-h photoperiod at a light intensity of ca. 30 μE m^{-2} s^{-1}. T_2 seeds harvested from each T_1 individual plant that showed ca. 3:1 segregation ratio were tested for bialaphos-resistance again. Finally, T_2 and T_3 homozygous plant lines were used for further analyses. A homozygous tobacco T_2 line transformed with the binary plasmid pSMABR-35SpGUS (in which the reporter *GFP* gene of pSMABR35SsGFP was replaced by β-glucuronidase [GUS] gene), expressing bialaphos resistance [bar] and GUS genes, was used as a control.

Three representative *PAP1*-overexpressing plants (*PAP1* lines), constitutively expressing *Arabidopsis PAP1* under the control of the cauliflower mosaic virus (CaMV) 35S promoter, were used. The *PAP1* lines exhibited substantial expression of *PAP1* based on their increased pigmentation throughout their development, in comparison to wild-type (WT) and transgenic plants expressing *uid*A (coding for the GUS reporter gene and serving as a control) (Figures S1 and S2). Except for having increased pigmentation, *PAP1* lines developed and matured in a similar manner to their control plants, as shown in previous studies [5,6] (Figure S2).

Plants and caterpillars

Tobacco plants were grown in plastic pots in a growth chamber at $25°C$ (16 h photoperiod at a light intensity of 80 μE m^{-2} s^{-1}) for about 5 weeks. Eggs of *Spodoptera litura* (Fabricius) (Lepidoptera: Noctuidae) were obtained from Sumika Technoservice Co. Ltd. (Takarazuka, Japan), and the hatched larvae were reared on artificial diet (Insecta LF, Nihon Nosan Kogyo Ltd., Tokyo, Japan) in the laboratory at $25°C$.

Chemical and herbivore treatments and leaf sample preparation

For chemical treatment, methyl jasmonate (MeJA, 0.1 mM, Wako Pure Chemical Industrials, Ltd., Osaka, Japan) in 2 mL of 0.1% ethanol solution was sprayed onto intact plants in plastic pots. Ethanol solution (0.1%) was used as the control. For herbivore treatment, four third-instar larvae were incubated with a whole tobacco plant in a pot (Figure S3). All herbivory bioassays were carried out in a climate-controlled chamber at $25°C$ (16 h photoperiod). For the determination of phytochemical concentrations and RNA extractions, leaf tissue samples were harvested into the liquid nitrogen, flash-frozen, and kept at $-80°C$ prior to extraction.

Performance of *S. litura* larvae

Larvae were reared on artificial diet until the assay started, as described above. Second-instar *S. litura* larvae were weighed prior to release onto a potted plant (ranging from 0.7 to 1.2 mg) and released one at a time. Each larva was allowed to move on the whole plant in a climate-controlled room at 25°C under a 16-h photoperiod. Twenty-six to thirty-two larvae were analyzed for each line.

Analysis of phenolic compounds

Phenylpropanoids and flavonoids in tobacco leaves were determined using high-performance liquid chromatography (HPLC) with a slight modification of the method described by Galis et al. (2013) and Matsuba et al. (2010) [25,26]. Replicated analyses were conducted with five independent leaf samples.

For anthocyanin (cyanidin-3-rutinoside) determination, the ground leaf samples (300 mg each) were further homogenized in the liquid nitrogen and extracted with 3 ml of 80% methanol containing 1% trifluoroacetic acid (TFA) at 4°C overnight. After passing through a 0.22 μm syringe filter (Millipore, Billerica, MA, USA), filtrates (10 μl) were subjected to an HPLC (L-6320 Intelligent pump, L-4200 UV-VIS detector, AS-4010 Intelligent Autosampler, L-5025 column oven, Hitachi, Tokyo, Japan) equipped with a COSMOSIL $5C_{18}$-MS-II packed column (4.6 mm×50 mm; Nacalai Tesque, Kyoto, Japan). The separation was performed by the linear gradient elution of 10–70% (v/v) methanol in 1.5% aqueous phosphoric acid for 5 min at a flow rate of 1.5 ml min^{-1} at 35°C. The elution profiles were monitored at 528 nm and recorded using the Chromato-Pro (Runtime Instruments, Kanagawa, Japan). The authentic compound used for quantification was cyanidin-3-rutinoside (Funakoshi Co. Ltd., Tokyo, Japan).

To determine phenylpropanoids (chlorogenic acid) and flavonoids (kaempferol 3-O-rutinoside and rutin), the ground leaf samples (100 mg each) were briefly homogenized in the liquid nitrogen at room temperature in 0.75 ml of 84 mM acetate buffer (adjusted to pH 4.8 with ammonia) in 40% methanol (v/v) using a FastPrep FP120 instrument (Thermo Savant, Thermo Fisher Scientific, Pittsburgh, PA, USA) in the presence of 2.3-mm diameter zirconia beads. Samples were centrifuged for 20 min at 16,100 g, 4°C, and supernatants were collected. Pellets were re-extracted in 0.75 ml buffer containing 80% (v/v) methanol by shaking at RT for 15 min and centrifuged as before; supernatants were pooled with the first extracts. Aliquots (20 μl) of each sample were analyzed on a Shimadzu Prominence HPLC system equipped with a COSMOSIL $5C_{18}$-MS-II packed column (4.6 mm×50 mm; Nacalai Tesque, Kyoto, Japan) in a gradient setup: solvent A (0.1% formic acid, 0.1% ammonia in water, pH 3.6) and solvent B (methanol); time (min)/B (%) gradient conditions: 0/2, 10/80, 15/80, 16/95, 21/95, 24/2, followed by a 6-min equilibration step. Flow rate was set to 1 ml/min and the column was maintained at 35°C. Chromatographic peaks of chlorogenic acid isomers and rutin were detected by an SPD-M20A diode array detector and compared with external calibration curves of authentic chlorogenic acid (Sigma-Aldrich) and rutin (Ishizu Pharmaceuticals Co. Ltd., Osaka, Japan) standards, respectively. Kaempferol 3-O-rutinoside content was estimated based on rutin external calibration.

Identification of phenolic acid and flavonoid compounds

Leaf extracts of *PAP1* lines originally prepared for HPLC analyses were used to identify the major UV-absorbing peaks in chromatograms. A 20-μl aliquot of leaf sample was first separated on HPLC as above, and peak fractions of major UV-absorbing compounds were collected using an automated fraction collector (FRC-10A, Shimadzu, Kyoto, Japan). Fractions from 10 runs were pooled, evaporated, re-dissolved in 70% methanol in water (v/v) and spun shortly before analysis. A 10-μl aliquot of each fraction

was subjected to measurement on a triple quadrupole liquid chromatography-tandem mass spectrometry (MS) system (LC-MS/MS 6410, Agilent Technologies, Santa Clara, CA, USA) equipped with a Zorbax SB-C18 column (2.1 mm id×50 mm, (1.8 μm), Agilent Technologies), maintained at 40°C and coupled to an MS detector operating in positive scan mode (m/z 100–1000; fragmentor at 135 V). Samples were passed through a UV-PDA detector before entering the MS electrospray ion source for reference and later identification of the peaks. Solvent A (0.1% formic acid in water) and solvent B (0.1% formic acid in acetonitrile) were used in time (min)/B (%) gradient mode: 0/5, 0.5/5, 2/40, 6/40, 10/95, 15/95, 16/5, 20/5 at a constant flow rate of 0.4 ml/min. The mass spectra corresponding to UV-absorbing peaks in the chromatograms were analyzed and used to identify each compound. Only the rutin peak could be assigned against an authentic rutin standard (Ishizu Pharmaceutical Co., Ltd., Osaka, Japan), while the remaining peaks were putatively assigned based on their m/z, fragmentation properties and characteristic UV absorbance spectra.

Phytohormone analysis

Leaf hormone concentrations were determined according to the method described by Fukumoto et al. (2013) [27]. Approximately 100 mg of the ground leaf samples was further homogenized in liquid nitrogen and extracted in 1 ml of ethylacetate spiked with internal standards (IS): 25 ng d3-JA; 5 ng d3-JA-Ile; 10 ng d6-ABA and 10 ng d4-SA. After brief homogenization in a FastPrep instrument as above, samples were centrifuged at 4°C, 20 min at 16,100 g. Supernatants were stored and pellets re-extracted in 0.5 ml of ethylacetate without IS. After centrifugation, pooled supernatants were evaporated under reduced pressure in a rotary vacuum concentrator and re-suspended in 250 μl of 70% methanol/water (v/v). Ten-microliter aliquots were analyzed on a triple quadrupole LC-MS/MS 6410 equipped with a Zorbax SB-C18 chromatographic column [2.1 mm id×50 mm, (1.8 μm), Agilent Technologies] protected by a narrow bore guard column Zorbax SB-C8 [2.1 mm id×12.5 mm, (5 μm)]. Samples were chromatographically resolved using solvents A [0.1% formic acid in water (v/v)] and B [0.1% formic acid in acetonitrile (v/v)] used in the following gradient: time (min)/B (%) 0/15, 4.5/98, 12/98, 12.1/15, 18/15. Flow rate was set to 0.4 ml/min and column temperature maintained at 40°C. Mass transitions for each compound [hormone/Q1 precursor ion (m/z)/Q3 product ion (m/z)] were detected in the ESI-negative mode of the mass spectrometer: JA/209/59; JA-Ile/322/130; ABA/263/153; SA/137/93; d3-JA/212/59; d3-JA-Ile/325/130; d6-ABA/269/159; d4-SA/141/97. Hormone amounts were calculated from the ratio of endogenous hormone peak and known amount of IS spike, and adjusted to the exact fresh mass (FM) of the sample used for extraction. Replicated analyses were conducted with five independent leaf samples.

cDNA synthesis and quantitative PCR

Approximately 100 mg of the ground leaf samples was homogenized in liquid nitrogen, and total RNA was isolated from leaf tissues using NucleoSpin RNA Plant (Machery-Nargel, Düren, Germany) following the manufacturer's protocol. First-strand cDNA was synthesized using a ReverTra Ace qPCR RT Master mix (Toyobo, Osaka, Japan) and 0.5 μg of total RNA at 37°C for 15 min. Real-time PCR was performed on a Light Cycler Nano system (Roche Applied Science, Indianapolis, IN, USA) using FastStart Essential DNA Green Master (Roche Applied Science) and gene-specific primers (Table S1). The following protocol was used: initial polymerase activation: 60 s at 95°C; 40 cycles of 15 s at 95°C and 45 s at 60°C; and then melting curve analysis preset by the instrument. Relative transcript abundances were determined after normalization of raw signals with housekeeping transcript abundances of *N. tabacum ELONGATION FACTOR 1α* (*NtEF1α*; GenBank D63396). Replicated analyses were conducted with four to five independent leaf samples.

Statistical analysis

All experiments were repeated at least four times, and statistical analyses were performed using a Tukey's HSD test after a one-way ANOVA. A P value<0.05 was considered statistically significant.

Results

Transgenic tobacco plants constitutively expressing *PAP1* genes are better able to defend themselves against herbivores than are non-transgenic tobacco plants

We evaluated the resistance of *PAP1* lines, constitutively expressing *Arabidopsis PAP1*, to the arthropod herbivore *S. litura*. Larvae were applied onto the leaves of potted WT, GUS and *PAP1* lines for 2 days (Figure 1). Larvae on the PAP-5 and PAP-8 lines had markedly lower weight gain compared to those on WT or GUS lines (*P*<0.05). Larvae on another *PAP1* line (PAP-2) exhibited slightly, but not significantly, less weight gain compared to those on WT or GUS lines (*P*>0.05).

Biosynthesis of flavonoids/phenylpropanoids in *PAP1* lines

Since PAP1 has been shown to activate the phenylpropanoid/flavonoid pathway [28], we measured endogenous levels of their metabolites in *PAP1* lines. According to HPLC analysis supported by identification of the most abundant compounds by LC-MS/MS, there was increased accumulation of an anthocyanin compound (cyanidin-3-*O*-rutinoside) as well as of two other flavonoid compounds (kaempferol 3-*O*-rutinoside and rutin) and of phenylpropanoid compounds (chlorogenic acid isomers 1 and 2) in leaves of *PAP1* lines, relative to WT and GUS lines (Figure 2). PAP-8 leaves accumulated all of the above five compounds at the highest levels, unsurprisingly as this line was best able to defend itself against *S. litura* larvae (see Figure 1). Interestingly, *S. litura* infestation substantially diminished the increase of the concentration of these compounds in *PAP1* lines, whereas no significant

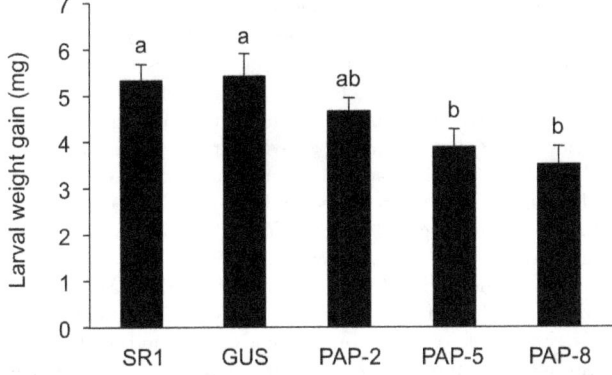

Figure 1. The net body weight that *Spodoptera litura* larvae gained during 4 days on wild-type (WT), GUS control and *PAP1* lines. Data are shown as the mean+standard errors (*n* = 26–32). Means followed by different small letters are significantly different (*P*<0.05).

change (either increase or decreased) of those compounds was observed in both WT and GUS lines (Figure 2).

We therefore assessed the transcriptional pattern of flavonoid/phenylpropanoid pathway genes in leaves of *PAP1* lines not infested or infested with *S. litura* for 2 days (Figure 3). In comparison to GUS lines, PAP-8 showed significantly increased transcript levels of eight genes (*PALs, CHS, CHI, F3H, F3'H, ANS* and *DFR*) involved in phenylpropanoid and flavonoid biosynthesis. The increase of the transcript level was, however, diminished by 34, 60, 29, 39 and 29% of the initial increase in five cases, *CHS, CHI, F3H, ANS* and *DFR*, respectively, in *PAP1* lines in response to herbivory. *PAP1* was expressed similarly between uninfested and infested PAP-8 leaves. Moreover, we also determined the transcript levels of four genes responsible for producing endogenous transcription factors: MYBJS1 (another MYB transcription factor sharing transcriptional targets with PAP1 in the phenylpropanoid pathway [18]), An2 (PAP1 homologue MYB transcription factor [29]), and An1a and An1b (NtAn2-interacting, basic helix-loop-helix (bHLH) transcription factors [30]). An1a, An1b and An2 are known to be responsible for anthocyanin pigmentation in tobacco flowers [29,30]. The transcript of *MYBJS1* was induced similarly between GUS and PAP-8 leaves indicates that the *MYB* gene was not under the direct control of PAP1. *An2* was not expressed in either GUS or PAP-8 leaves, confirming what was previously known, namely, that this transcription factor is abundantly expressed in flowers but not in leaves [29]. On the other hand, although PAP-8 increased transcript levels of *An1a* and *An1b* transcript levels of GUS lines, the increase of the transcript levels was diminished by herbivory; this result was also observed in the structural genes mentioned above (see Figures 3 and S4 for 2 days and 4 days of exposure, respectively).

Upstream and downstream JA signaling for *PAP1*-activated genes

Since JA signaling along with other phytohormone signaling plays a central role in defense responses against chewing herbivores [31], it might be expected that induced JA production and signaling would be relevant to the low concentration of flavonoid/phenylpropanoid pathway products in *PAP1* lines attacked by *S. litura*. To assess whether *PAP1* overexpression affects JA biosynthesis and signaling, we analyzed endogenous levels of jasmonoyl-L-isoleucine (JA-Ile, an active form of JAs [32]) and of two other phytohormones: abscisic acid (ABA, involved in protective wound-healing processes [33]) and salicylic acid (SA, involved in pathogenesis and herbivore responses [34–36]). As shown in Figure 4A, JA and JA-Ile increased and accumulated at similar rates in the infested PAP-8 leaves compared to in the leaves of infested GUS lines. Those data indicate that PAP1 is not an upstream transcription regulator for the biosynthesis of JA. Moreover, neither ABA nor SA was increased in leaves of GUS or PAP-8 lines upon infestation, indicating that neither of those two hormones is directly linked by PAP1 actions.

Next, we investigated the gene expression levels in leaves of GUS and PAP-8 lines by applying an exogenous solution of MeJA (the commonly used active form of JAs [37]: 0.1 mM). Even though the PAP1-regulated transcript levels of *An1a, An1b* and *F3H* are negatively affected by herbivory (as shown in Figure 3), MeJA did not reduce their transcript levels to the same extent that it reduced transcript levels of *PAP1* (Figure 4B). The MeJA concentration used for our treatment was undoubtedly sufficient to induce defense responses in leaf tissues, as the transcript levels of a typical MeJA-inducible protease inhibitor (PIs [38]) and *MYBJS1* genes were substantially up-regulated in the same treated samples.

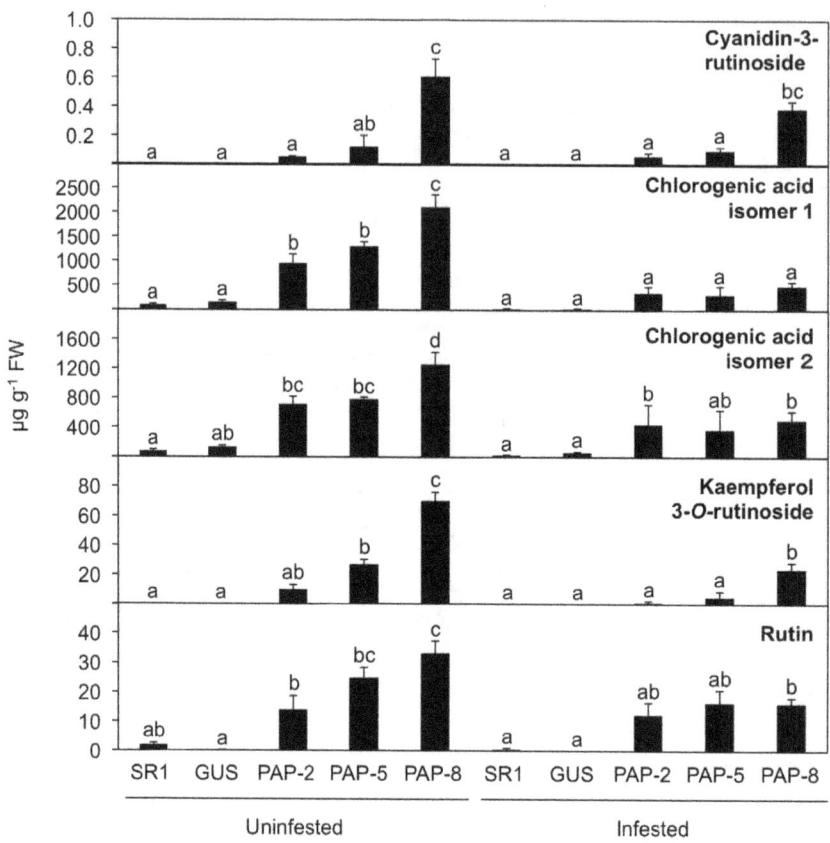

Figure 2. Endogenous accumulation of flavonoids/phenylpropanoids. Values for the major anthocyanin compound (cyanidin-3-*O*-rutinoside), other flavonoid (kaempferol 3-*O*-rutinoside and rutin) and phenylpropanoid (chlorogenic acid) compounds were determined in leaves of wild-type (WT), GUS control and *PAP1* lines not infested or infested with *Spodoptera litura* for 2 days. Data are shown as the mean+standard error ($n = 5$). Means followed by different small letters are significantly different ($P<0.05$).

We therefore conclude that the down-regulation of PAP1-activated genes after herbivory is closely linked not to the JA signaling and biosynthesis cascades but, rather, to other herbivory-associated signals discussed below.

Reversible and transient regulatory patterns of *F3H* and *PI2* during herbivory and post-herbivory periods

To clarify the temporal regulation of PAP1-regulated genes, we analyzed the temporal patterns of the expression levels of *F3H* (PAP1-regulated) and *PI2* (JA-inducible) not only during 48 h after the first exposure to *S. litura* but also for up to 48 h after the larvae were removed (post-herbivory recovery; Figure 5). Consistent with the data shown in Figure 3, the regulation of the expression levels of *An1a*, *An1b* and *F3H* by PAP1 was suppressed 48 h after the first exposure to *S. litura*. After the larvae were removed, the expression levels remained low during the first 8 h but then increased 24 h and 48 h after the damage had ended. In contrast, *PI2* and *PAP1* displayed different expression profiles: the expression level of *PI2* reached a maximum at 48 h after first exposure to *S. litura* but dropped to the basal level 8 h, 24 h and 48 h after the damage had ended, consistent with the defense function of the PI2 protein; the expression level of *PAP1* was unchanged through the experimental period. Collectively, these results suggest that *F3H* might be regulated positively by an additional, JA-independent regulator(s) during the post-herbivory period.

Discussion

To date, light stimuli and abiotic stress have been reported to modulate anthocyanin concentrations in several plant taxa. For instance, in grape berries, and flowers of rose and *Brunsfelsia calycina*, anthocyanin accumulation appears to be suppressed in response to high temperature [39–41]. Most impressively, Rowan et al. (2009) reported that abiotic stress treatments of high temperature and low light cause reductions in the concentrations of anthocyanins and down-regulation of the genes involved in anthocyanin biosynthesis in leaves of *PAP1*-overexpressing *Arabidopsis* plants [23]. The inhibition of transcription occurs independently of *PAP1*, and the loss of anthocyanins is not the result of an obvious catabolic enzyme regulated at the transcription level. The authors addressed whether the decreased flux of anthocyanins into the vacuole occurred as a consequence of a decline in the expression of biosynthetic genes by reducing the expression of three genes (*TT8*, *TTG1* and *EGL3*) of the PAP1 transcriptional complex and enhancing the expression of the potential transcriptional repressors *AtMYB3*, *AtMYB6* and *AtMYBL2*. Those results especially reflect to our finding that transcript levels of both biosynthetic genes and endogenous transcription factor (positive regulator) genes were decreased in infested PAP1 plants compared to those in the uninfested PAP1 plants (Figure 3). *An1a* and *An1b* function by interacting with *An2* (PAP1 homologue) to regulate anthocyanin biosynthesis [30], thus implying that those two factors can also interact with *Arabidopsis* PAP1. In fact, *An1a* and *An1b* transcripts were

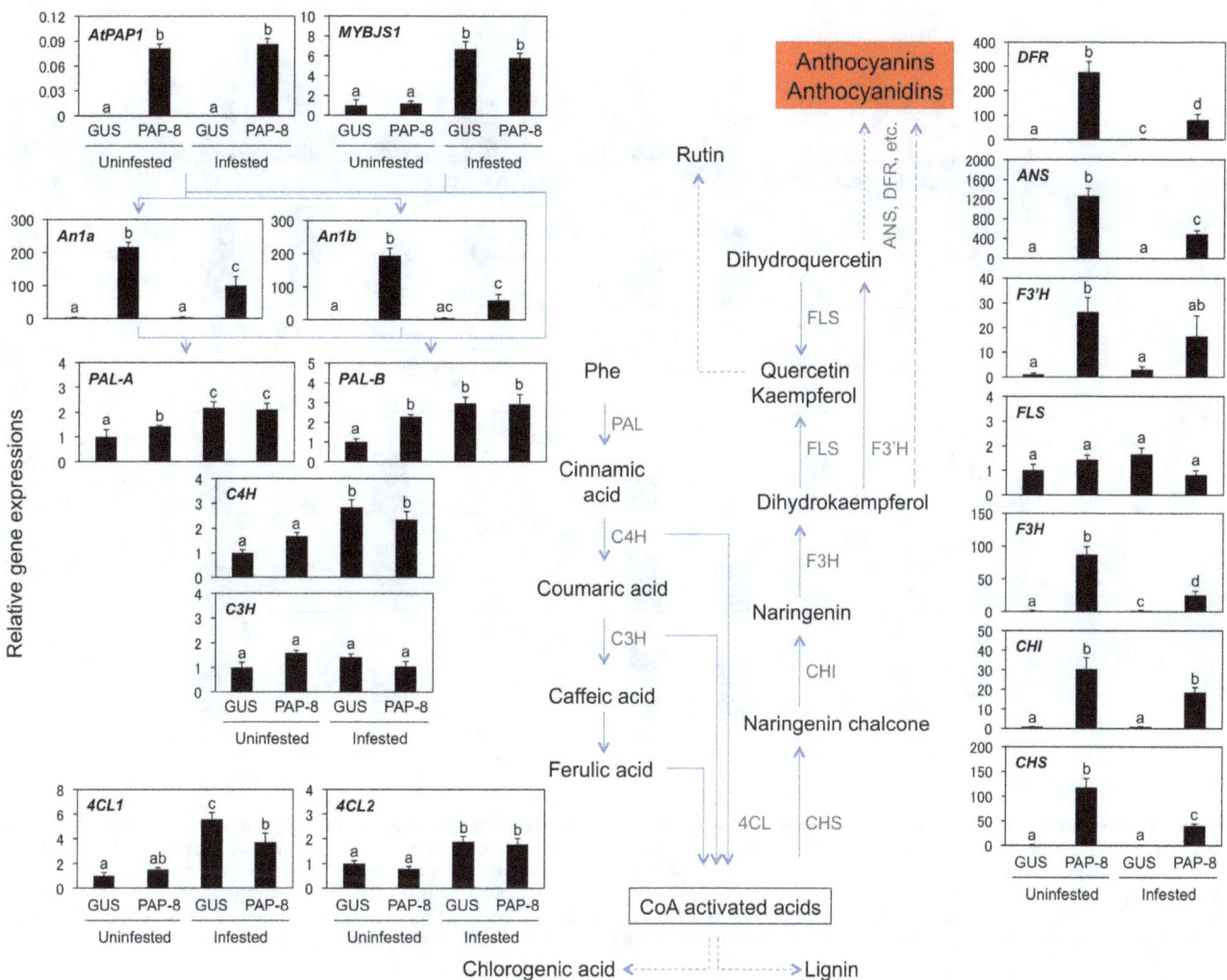

Figure 3. Expression of genes for transcription factors PAP1 and MYBJS1, An1a, An1b and structural genes involved in the flavonoid/phenylpropanoid pathway. Relative transcript levels of genes were determined in leaves of GUS control and *PAP1* lines (PAP-8) not infested or infested with *Spodoptera litura* for 2 days. Transcript levels of genes were normalized by those of *NtEF1α*. Data are shown as the mean+ standard errors ($n = 4$–5). Means followed by different small letters are significantly different ($P < 0.05$). Expression of genes in GUS and PAP-8 leaves after 4 days of damage are presented in Figure S4. ANS, anthocyanidin synthase; C3H, cinnamate 3-hydroxylase; C4H, cinnamate 4-hydroxylase; CHI, chalcone isomerase; CHS, chalcone synthase; 4CL, 4-coumarate coenzyme A ligase; DFR, dihydroflavonol 4-reductase; F3H, flavanone 3-hydroxylase; F3'H, flavonoid 3'-hydroxylase; FLS, flavonol synthase; PAL, phenylalanine ammonia-lyase.

enhanced in the overexpression of *PAP1*, suggesting that those two bHLH proteins are able to interact with foreign PAP1 genes and are involved in the feedback regulation by the MYB/bHLH complexes as reported in *Arabidopsis* [42]. The diminished expression of those two genes suppress the expression of structural genes – genes that are also suppressed by herbivory – and regulates the flux of flavonoid/phenylpropanoid (Figures 2 and 3). Presumably, in return, plants gain energy and the resources required for induced defense responses (so-called homeostasis and feedback regulation), which might eventually make plants better able to withstand to generalist herbivores.

We present here two important findings, namely, that *PAP1* lines that accumulated great amounts of flavonoid/phenylpropanoid pathway products defend against the generalist herbivore *S. litura*, and that the *PAP1*-stimulated accumulations of those compounds are attenuated by herbivory (Figure 6). The first finding is in line with a previous report that transgenic tobacco plants expressing *Arabidopsis* transcription factor *AtMYB12*

exhibit more resistance to *S. litura* and *Helicoverpa armigera* than do non-transgenic plants, most probably due to these plants' enhanced accumulation of rutin [13], a chemical that is toxic for those herbivores [43]. As rutin is one of predominant products in *PAP1* lines (Figure 2), this flavonoid compound is predicted to be a major anti-herbivore agent in our case as well.

The suppression of flavonoid/phenylpropanoid pathway products by herbivory is likely to reflect the phenomena described above, such as the response of *Arabidopsis* plants to abiotic stress, namely, the overexpression of *PAP1* [23]. In addition to the proposed reason for suppression (see above), we should consider an alternative (or additional) explanation, such as that PAP1 protein could be post-transcriptionally degraded under herbivory even though normal transcript levels of *PAP1* in PAP-8 leaves are maintained during herbivory (Figure 3). A similar mechanism is consistent with the repression of anthocyanin in the dark; in this case, the COP1/SPA ubiquitin ligase plays a critical role in the degradation of PAP1 proteins [22]. Although we conducted

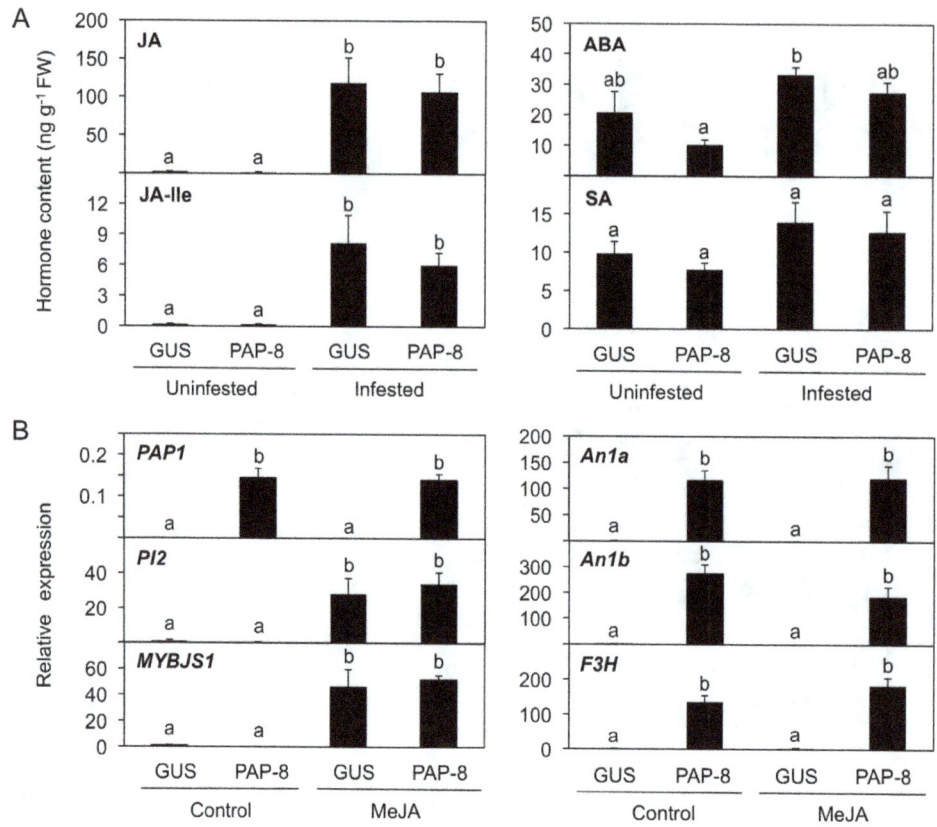

Figure 4. Upstream and downstream transcription regulation of PAP1-activated genes. A, Endogenous phytohormone levels were determined in leaves of GUS control and *PAP1* lines (PAP-8) not infested or infested with *Spodoptera litura* for 2 days. ABA, abscisic acid; JA, jasmonic acid; JA-Ile, jasmonoyl-L-isoleucine; SA, salicylic acid. B, The effect of treatment with methyl jasmonate (MeJA) on gene expression was determined in leaves of GUS and PAP-8 lines treated with 0.1 mM MeJA (in 0.1% ethanol solution) for 24 h. Plants treated with 0.1% ethanol solution for 24 h served as controls. Transcript levels of genes were normalized by comparing them to those of *NtEF1α*. Data are shown as the mean+standard error ($n = 4$–5). Means followed by different small letters are significantly different ($P<0.05$). F3H, flavanone 3-hydroxylase; PI2, proteinase inhibitor 2.

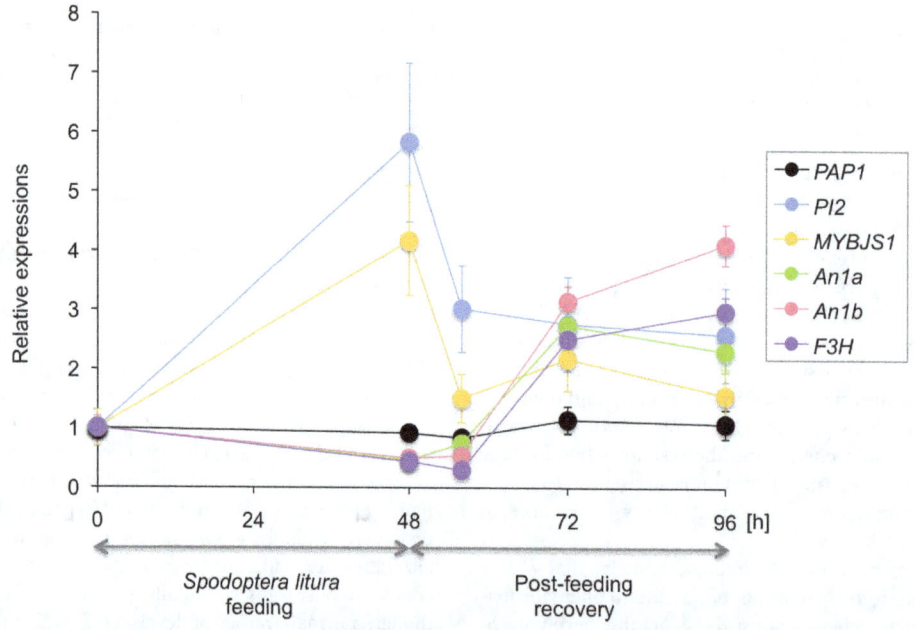

Figure 5. Temporal patterns of *F3H* and *PI2* gene regulation in leaves of *PAP1* lines (PAP-8) plants during exposure to *Spodoptera litura* for 48 h and for up to another 48 h after removal of the larvae (recovery period). Transcript levels of genes were normalized by comparing them to those of *NtEF1α*. Data are shown as the means ± standard errors ($n = 4$–5).

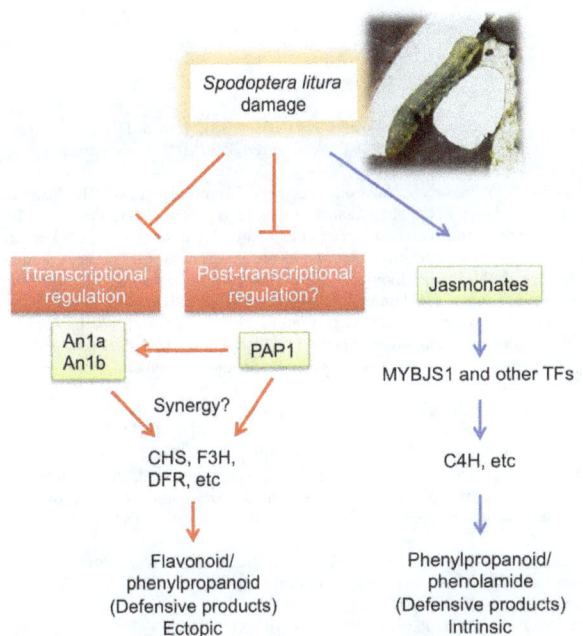

Figure 6. Schematic presentation of effect of *Spodoptera litura* feeding on the PAP1-regulated phenylpropanoid biosynthetic genes in a manner independent of jasmonate signaling. Arrows and bars indicate positive and negative interactions, respectively. TFs, transcription factors.

immunoblot analysis using polyclonal antibodies against partial PAP1 peptide residues ([H]CKIKMKKRDITP[OH]) (produced by Sigma-Aldrich, Ishikari, Japan), our efforts unfortunately failed to detect sufficient PAP1 protein signals extracted from uninfested and infested PAP-8 leaves. Therefore, it remains to be clarified whether PAP1 degradation in response to herbivory is subject to similar mechanisms for light/dark responses observed in *Arabidopsis*.

Alternatively, the possibility that the general decrease observed in the concentration of phenolic compounds after two days of herbivory could be a result of an altered metabolic precursor allocation (competition) cannot be excluded. For example, in transgenic *N. attenuata* plants, the transient silencing of the expression of a hydroxycinnamoyl-CoA:shikimate/quinate hydroxycinnamoyl transferase (HCT) gene involved in lignin biosynthesis dramatically increased herbivory-induced phenolamides produced in the parallel metabolic branch [21]. It was proposed that a strong metabolic tension in the plant, exacerbated during herbivory, exists over the allocation of coumaroyl-CoA units among lignin and other phenolic compounds. This tension could be extended to anthocyanins and flavonoids, as shown in our experiments (see Figure 2), although the branches that are simultaneously enhanced have not yet been identified. In addition, the enhanced catabolism of some phenolics during herbivory stress could be responsible. Unfortunately, very little is known about the anthocyanins, kaempferol-3-*O*-rutinoside and rutin catabolism in plants, and the hypothesis that there is competition among metabolic precursors remains to be tested.

Generally, since JA-mediated defense signaling is predominantly activated when plants are attacked by chewing herbivores [31], it might be expected that induced JA production and signaling would be relevant to the defense responses to *S. litura* attack. Even though *PI2* expression was highly induced during herbivory and

exogenous MeJA application (see Figures 3 and 4), it was likely that the *PAP1*-activated genes were suppressed independently of the induction of JA and JA-Ile (an active form of JA) (Figure 4A). Moreover, since MeJA application also failed to suppress PAP1-activated genes in PAP-8 plants (Figure 4B), JA signaling is not likely a part of the down-regulation pathway. Also, in our case SA and ABA signaling pathways seem to be irrelevant in our case (see Figure 4A). Those facts may reflect a recent report that a mitogen-activated protein kinase (MPK4) suppresses a JA-independent defense pathway in *N. attenuata* in response to *Manduca sexta* feeding [44].

Moreover, the reverse mechanism was also likely used, given that PAP1-regulated *F3H* expression was dramatically increased in PAP-8 leaves during the post-herbivory period, when JA signaling was already switched off (Figure 5). This up-regulation might be caused by the fine-tuning of the PAP1 protein level after the cessation of herbivory, resulting in increased PAP1 levels beyond the basal level in the undamaged condition. Alternatively, additional signal(s) might be facilitated only when JA signaling is turned off, although these speculations require further investigation.

Horticultural ecosystems are very complicated and flexible. Hence, prior to setting genetically modified plants in the field, a wide range of systematic studies should be conducted to understand realistic horticultural ecosystems, where a myriad of plants, animals, and microorganisms interact and coevolve. However, it still remains to be elucidated how other multiple and simultaneous stresses can potentially influence flavonoid/phenylpropanoid catabolism in real agricultural settings.

Supporting Information

Figure S1   Relative mRNA levels of *PAP1* in leaves of wild-type (WT), GUS and *PAP1* lines. Transcript levels of genes were normalized by those of *NtEF1α*. Data are shown as the mean+standard errors ($n = 5$). ND, not detected.

Figure S2   Morphology of *PAP1* lines. A, 4-week-old plants; B, 8-week-old plants; C, flowers.

Figure S3   GUS and PAP-8 plants infested with *Spodoptera litura* for 2 days (A) and 4 days (B).

Figure S4   Expression of genes in GUS and PAP-8 leaves infested with *Spodoptera litura* for 4 days. Relative transcript levels of genes were determined in leaves of GUS control and *PAP1* lines (PAP-8) not infested or infested with *S. litura* for 4 days. Transcript levels of genes were normalized by comparing them to those of *NtEF1α*. Data are shown as the mean+standard errors ($n = 4$–5). Means followed by different small letters are significantly different ($P<0.05$).

Acknowledgments

We thank Ms. Emily Wheeler for editorial assistance.

Author Contributions

Conceived and designed the experiments: TM MN IG KF NS KN GA. Performed the experiments: TM MN KA YH KF NS KN GA. Analyzed the data: TM MN IG KA YH KF NS KN GA. Contributed reagents/materials/analysis tools: MN IG TS GA. Wrote the paper: MN IG GA.

References

1. Dixon RA, Liu C, Jun JH (2013) Metabolic engineering of anthocyanins and condensed tannins in plants. Curr Opin Biotechnol 24: 329–335.
2. Glenn WS, Runguphan W, O'Connor SE (2013) Recent progress in the metabolic engineering of alkaloids in plant systems. Curr Opin Biotechnol 24: 354–365.
3. Wang Y, Chen S, Yu O (2011) Metabolic engineering of flavonoids in plants and microorganisms. Appl Microbiol Biotechnol 91: 949–956.
4. Borevitz JO, Xia Y, Blount J, Dixon RA, Lamb C (2000) Activation tagging identifies a conserved MYB regulator of phenylpropanoid biosynthesis. Plant Cell 12: 2383–2394.
5. Xie DY, Sharma SB, Wright E, Wang ZY, Dixon RA (2006) Metabolic engineering of proanthocyanidins through co-expression of anthocyanidin reductase and the PAP1 MYB transcription factor. Plant J 45: 895–907.
6. Ben Zvi MM, Shklarman E, Masci T, Kalev H, Debener T, et al. (2012) *PAP1* transcription factor enhances production of phenylpropanoid and terpenoid scent compounds in rose flowers. New Phytol 194: 430–439.
7. Ben Zvi MM, Negre-Zakharov F, Masci T, Ovadis M, Shklarman E, et al. (2008) Interlinking showy traits: co-engineering of scent and colour biosynthesis in flowers. Plant Biotechnol J 6: 403–415.
8. Zhang Y, Yan YP, Wang ZZ (2010) The *Arabidopsis* PAP1 transcription factor plays an important role in the enrichment of phenolic acids in *Salvia miltiorrhiza*. J Agric Food Chem 58: 12168–12175.
9. Mellway RD, Tran LT, Prouse MB, Campbell MM, Constabel CP (2009) The wound-, pathogen-, and ultraviolet B-responsive *MYB134* gene encodes an R2R3 MYB transcription factor that regulates proanthocyanidin synthesis in poplar. Plant Physiol 150: 924–941.
10. Korkina LG (2007) Phenylpropanoids as naturally occurring antioxidants: from plant defense to human health. Cell Mol Biol 53: 15–25.
11. Chalker-Scott L (1999) Environmental significance of anthocyanins in plant stress responses. Photochem Photobiol 70: 1–9.
12. Kumar V, Nadda G, Kumar S, Yadav SK (2013) Transgenic tobacco overexpressing tea cDNA encoding dihydroflavonol 4-reductase and anthocyanidin reductase induces early flowering and provides biotic stress tolerance. PLoS ONE 8: e65535.
13. Misra P, Pandey A, Tiwari M, Chandrashekar K, Sidhu OP, et al. (2010) Modulation of transcriptome and metabolome of tobacco by Arabidopsis transcription factor, AtMYB12, leads to insect resistance. Plant Physiol 152: 2258–2268.
14. Gális I, Simek P, Narisawa T, Sasaki M, Horiguchi T, et al. (2006) A novel R2R3 MYB transcription factor NtMYBJS1 is a methyl jasmonate-dependent regulator of phenylpropanoid-conjugate biosynthesis in tobacco. Plant J 46: 573–592.
15. Kaur H, Heinzel N, Schöttner M, Baldwin IT, Gális I (2010) R2R3-NaMYB8 regulates the accumulation of phenylpropanoid-polyamine conjugates, which are essential for local and systemic defense against insect herbivores in *Nicotiana attenuata*. Plant Physiol 152: 1731–1747.
16. Bassard JE, Ullmann P, Bernier F, Werck-Reichhart D (2010) Phenolamides: bridging polyamines to the phenolic metabolism. Phytochemistry 71: 1808–1824.
17. Gulati J, Baldwin IT, Gaquerel E (2014) The roots of plant defenses: Integrative multivariate analyses uncover dynamic behaviors of roots' gene and metabolic networks elicited by leaf herbivory. Plant J 77: 880–892.
18. Onkokesung N, Gaquerel E, Kotkar H, Kaur H, Baldwin IT, et al. (2012) MYB8 controls inducible phenolamide levels by activating three novel hydroxycinnamoyl-coenzyme A:polyamine transferases in *Nicotiana attenuata*. Plant Physiol 158: 389–407.
19. Hoffmann L, Besseau S, Geoffroy P, Ritzenthaler C, Meyer D, et al. (2004) Silencing of hydroxycinnamoyl-coenzyme A shikimate/quinate hydroxycinnamoyltransferase affects phenylpropanoid biosynthesis. Plant Cell 16: 1446–1465.
20. Hoffmann L, Maury S, Martz F, Geoffroy P, Legrand M (2003) Purification, cloning, and properties of an acyltransferase controlling shikimate and quinate ester intermediates in phenylpropanoid metabolism. J Biol Chem 278: 95–103.
21. Gaquerel E, Kotkar H, Onkokesung N, Galis I, Baldwin IT (2013) Silencing an N-acyltransferase-like involved in lignin biosynthesis in *Nicotiana attenuata* dramatically alters herbivory-induced phenolamide metabolism. PLoS ONE 8: e62336.
22. Maier A, Schrader A, Kokkelink L, Falke C, Welter B, et al. (2013) Light and the E3 ubiquitin ligase COP1/SPA control the protein stability of the MYB transcription factors PAP1 and PAP2 involved in anthocyanin accumulation in Arabidopsis. Plant J 74: 638–651.
23. Rowan DD, Cao M, Lin-Wang K, Cooney JM, Jensen DJ, et al. (2009) Environmental regulation of leaf colour in red 35S:*PAP1 Arabidopsis thaliana*. New Phytol 182: 102–115.
24. Mishiba K, Yamasaki S, Nakatsuka T, Abe Y, Daimon H, et al. (2010) Strict *de novo* methylation of the 35S enhancer sequence in gentian. PLoS One 5: e9670.
25. Galis I, Schuman MC, Gase K, Hettenhausen C, Hartl M, et al. (2013) The use of VIGS technology to study plant-herbivore interactions. In: Becker A, editor. Methods in Molecular Biology - Virus-induced gene silencing: Methods and protocols. Totowa, NJ: Humana Press Inc. pp. 109–138.
26. Matsuba Y, Sasaki N, Tera M, Okamura M, Abe Y, et al. (2010) A novel glucosylation reaction on anthocyanins catalyzed by acyl-glucose-dependent glucosyltransferase in the petals of carnation and delphinium. Plant cell 22: 3374–3389.
27. Fukumoto K, Alamgir K, Yamashita Y, Mori IC, Matsuura H, et al. (2013) Response of rice to insect elicitors and the role of OsJAR1 in wound and herbivory-induced JA-Ile accumulation. J Integr Plant Biol 55: 775–784.
28. Tohge T, Nishiyama Y, Hirai MY, Yano M, Nakajima J, et al. (2005) Functional genomics by integrated analysis of metabolome and transcriptome of Arabidopsis plants over-expressing an MYB transcription factor. Plant J 42: 218–235.
29. Pattanaik S, Kong Q, Zaitlin D, Werkman JR, Xie CH, et al. (2010) Isolation and functional characterization of a floral tissue-specific R2R3 MYB regulator from tobacco. Planta 231: 1061–1076.
30. Bai Y, Pattanaik S, Patra B, Werkman JR, Xie CH, et al. (2011) Flavonoid-related basic helix-loop-helix regulators, NtAn1a and NtAn1b, of tobacco have originated from two ancestors and are functionally active. Planta 234: 363–375.
31. Arimura G, Köpke S, Kunert M, Volpe V, David A, et al. (2008) Effects of feeding *Spodoptera littoralis* on Lima bean leaves: IV. Diurnal and nocturnal damage differentially initiate plant volatile emission. Plant Physiol 146: 965–973.
32. Staswick PE, Tiryaki I (2004) The oxylipin signal jasmonic acid is activated by an enzyme that conjugates it to isoleucine in Arabidopsis. Plant Cell 16: 2117–2127.
33. Lulai EC, Suttle JC, Pederson SM (2008) Regulatory involvement of abscisic acid in potato tuber wound-healing. J Exp Bot 59: 1175–1186.
34. Boatwright JL, Pajerowska-Mukhtar K (2013) Salicylic acid: an old hormone up to new tricks. Mol Plant Pathol 14: 623–634.
35. Zhang PJ, Zheng SJ, van Loon JJ, Boland W, David A, et al. (2009) Whiteflies interfere with indirect plant defense against spider mites in Lima bean. Proc Natl Acad Sci USA 106: 21202–21207.
36. Chung SH, Rosa C, Scully ED, Peiffer M, Tooker JF, et al. (2013) Herbivore exploits orally secreted bacteria to suppress plant defenses. Proc Natl Acad Sci USA 110: 15728–15733.
37. Farmer EE, Ryan CA (1990) Interplant communication: airborne methyl jasmonate induces synthesis of proteinase inhibitors in plant leaves. Proc Natl Acad Sci USA 87: 7713–7716.
38. Srinivasan T, Kumar KR, Kirti PB (2009) Constitutive expression of a trypsin protease inhibitor confers multiple stress tolerance in transgenic tobacco. Plant Cell Physiol 50: 541–553.
39. Mori K, Goto-Yamamoto N, Kitayama M, Hashizume K (2007) Loss of anthocyanins in red-wine grape under high temperature. J Exp Bot 58: 1935–1945.
40. Dela G, Or E, Ovadia R, Nissim-Levi A, Weiss D, et al. (2003) Changes in anthocyanin concentration and composition in 'Jaguar' rose flowers due to transient high-temperature conditions. Plant Sci 164: 333–340.
41. Vaknin H, Bar-Akiva A, Ovadia R, Nissim-Levi A, Forer I, et al. (2005) Active anthocyanin degradation in *Brunfelsia calycina* (yesterday-today-tomorrow) flowers. Planta 222: 19–26.
42. Baudry A, Caboche M, Lepiniec L (2006) TT8 controls its own expression in a feedback regulation involving TTG1 and homologous MYB and bHLH factors, allowing a strong and cell-specific accumulation of flavonoids in *Arabidopsis thaliana*. Plant J 46: 768–779.
43. Pandey A, Misra P, Chandrashekar K, Trivedi PK (2012) Development of AtMYB12-expressing transgenic tobacco callus culture for production of rutin with biopesticidal potential. Plant Cell Rep 31: 1867–1876.
44. Hettenhausen C, Baldwin IT, Wu J (2013) *Nicotiana attenuata* MPK4 suppresses a novel jasmonic acid (JA) signaling-independent defense pathway against the specialist insect *Manduca sexta*, but is not required for the resistance to the generalist *Spodoptera littoralis*. New Phytol 199: 787–799.

21

A Meta-Analysis of the Impacts of Genetically Modified Crops

Wilhelm Klümper, Matin Qaim*

Department of Agricultural Economics and Rural Development, Georg-August-University of Goettingen, Goettingen, Germany

Abstract

Background: Despite the rapid adoption of genetically modified (GM) crops by farmers in many countries, controversies about this technology continue. Uncertainty about GM crop impacts is one reason for widespread public suspicion.

Objective: We carry out a meta-analysis of the agronomic and economic impacts of GM crops to consolidate the evidence.

Data Sources: Original studies for inclusion were identified through keyword searches in ISI Web of Knowledge, Google Scholar, EconLit, and AgEcon Search.

Study Eligibility Criteria: Studies were included when they build on primary data from farm surveys or field trials anywhere in the world, and when they report impacts of GM soybean, maize, or cotton on crop yields, pesticide use, and/or farmer profits. In total, 147 original studies were included.

Synthesis Methods: Analysis of mean impacts and meta-regressions to examine factors that influence outcomes.

Results: On average, GM technology adoption has reduced chemical pesticide use by 37%, increased crop yields by 22%, and increased farmer profits by 68%. Yield gains and pesticide reductions are larger for insect-resistant crops than for herbicide-tolerant crops. Yield and profit gains are higher in developing countries than in developed countries.

Limitations: Several of the original studies did not report sample sizes and measures of variance.

Conclusion: The meta-analysis reveals robust evidence of GM crop benefits for farmers in developed and developing countries. Such evidence may help to gradually increase public trust in this technology.

Editor: emidio albertini, University of Perugia, Italy

Funding: This research was financially supported by the German Federal Ministry of Economic Cooperation and Development (BMZ) and the European Union's Seventh Framework Programme (FP7/2007-2011) under Grant Agreement 290693 FOODSECURE. The funders had no role in study design, data collection and analysis, decision to publish, or preparation of the manuscript. Neither BMZ nor FOODSECURE and any of its partner organizations, any organization of the European Union or the European Commission are accountable for the content of this article.

Competing Interests: The authors have declared that no competing interests exist.

* Email: mqaim@uni-goettingen.de

Introduction

Despite the rapid adoption of genetically modified (GM) crops by farmers in many countries, public controversies about the risks and benefits continue [1–4]. Numerous independent science academies and regulatory bodies have reviewed the evidence about risks, concluding that commercialized GM crops are safe for human consumption and the environment [5–7]. There are also plenty of studies showing that GM crops cause benefits in terms of higher yields and cost savings in agricultural production [8–12], and welfare gains among adopting farm households [13–15]. However, some argue that the evidence about impacts is mixed and that studies showing large benefits may have problems with the data and methods used [16–18]. Uncertainty about GM crop impacts is one reason for the widespread public suspicion towards this technology. We have carried out a meta-analysis that may help to consolidate the evidence.

While earlier reviews of GM crop impacts exist [19–22], our approach adds to the knowledge in two important ways. First, we include more recent studies into the meta-analysis. In the emerging literature on GM crop impacts, new studies are published continuously, broadening the geographical area covered, the methods used, and the type of outcome variables considered. For instance, in addition to other impacts we analyze effects of GM crop adoption on pesticide quantity, which previous meta-analyses could not because of the limited number of observations for this particular outcome variable. Second, we go beyond average impacts and use meta-regressions to explain impact heterogeneity and test for possible biases.

Our meta-analysis concentrates on the most important GM crops, including herbicide-tolerant (HT) soybean, maize, and cotton, as well as insect-resistant (IR) maize and cotton. For these crops, a sufficiently large number of original impact studies have been published to estimate meaningful average effect sizes. We estimate mean impacts of GM crop adoption on crop yield, pesticide quantity, pesticide cost, total production cost, and farmer profit. Furthermore, we analyze several factors that may influence outcomes, such as geographic location, modified crop trait, and type of data and methods used in the original studies.

Materials and Methods

Literature search

Original studies for inclusion in this meta-analysis were identified through keyword searches in relevant literature databanks. Studies were searched in the ISI Web of Knowledge, Google Scholar, EconLit, and AgEcon Search. We searched for studies in the English language that were published after 1995. We did not extend the review to earlier years, because the commercial adoption of GM crops started only in the mid-1990s [23]. The search was performed for combinations of keywords related to GM technology and related to the outcome of interest. Concrete keywords used related to GM technology were (an asterisk is a replacement for any ending of the respective term; quotation marks indicate that the term was used as a whole, not each word alone): GM*, "genetically engineered", "genetically modified", transgenic, "agricultural biotechnology", HT, "herbicide tolerant", Roundup, Bt, "insect resistant". Concrete keywords used related to outcome variables were: impact*, effect*, benefit*, yield*, economic*, income*, cost*, soci*, pesticide*, herbicide*, insecticide*, productivity*, margin*, profit*. The search was completed in March 2014.

Most of the publications in the ISI Web of Knowledge are articles in academic journals, while Google Scholar, EconLit, and AgEcon Search also comprise book chapters and grey literature such as conference papers, working papers, and reports in institutional series. Articles published in academic journals have usually passed a rigorous peer-review process. Most papers presented at academic conferences have also passed a peer-review process, which is often less strict than that of good journals though. Some of the other publications are peer reviewed, while many are not. Some of the working papers and reports are published by research institutes or government organizations, while others are NGO publications. Unlike previous reviews of GM crop impacts, we did not limit the sample to peer-reviewed studies but included all publications for two reasons. First, a clear-cut distinction between studies with and without peer review is not always possible, especially when dealing with papers that were not published in a journal or presented at an academic conference [24]. Second, studies without peer review also influence the public and policy debate on GM crops; ignoring them completely would be short-sighted.

Of the studies identified through the keyword searches, not all reported original impact results. We classified studies by screening titles, abstracts, and full texts. Studies had to fulfill the following criteria to be included:

- The study is an empirical investigation of the agronomic and/ or economic impacts of GM soybean, GM maize, or GM cotton using micro-level data from individual plots and/or farms. Other GM crops such as GM rapeseed, GM sugarbeet, and GM papaya were commercialized in selected countries [23], but the number of impact studies available for these other crops is very small.
- The study reports GM crop impacts in terms of one or more of the following outcome variables: yield, pesticide quantity (especially insecticides and herbicides), pesticide costs, total

variable costs, gross margins, farmer profits. If only the number of pesticide sprays was reported, this was used as a proxy for pesticide quantity.

- The study analyzes the performance of GM crops by either reporting mean outcomes for GM and non-GM, absolute or percentage differences, or estimated coefficients of regression models that can be used to calculate percentage differences between GM and non-GM crops.
- The study contains original results and is not only a review of previous studies.

In some cases, the same results were reported in different publications; in these cases, only one of the publications was included to avoid double counting. On the other hand, several publications involve more than one impact observation, even for a single outcome variable, for instance when reporting results for different geographical regions or derived with different methods (e.g., comparison of mean outcomes of GM and non-GM crops plus regression model estimates). In those cases, all observations were included. Moreover, the same primary dataset was sometimes used for different publications without reporting identical results (e.g., analysis of different outcome variables, different waves of panel data, use of different methods). Hence, the number of impact observations in our sample is larger than the number of publications and primary datasets (Data S1). The number of studies selected at various stages is shown in the flow diagram in Figure 1. The number of publications finally included in the meta-analysis is 147 (Table S1).

Effect sizes and influencing factors

Effect sizes are measures of outcome variables. We chose the percentage difference between GM and non-GM crops for five different outcome variables, namely yield, pesticide quantity, pesticide cost, total production cost, and farmer profits per unit area. Most studies that analyze production costs focus on variable costs, which are the costs primarily affected through GM technology adoption. Accordingly, profits are calculated as revenues minus variable production costs (profits calculated in this way are also referred to as gross margins). These production costs also take into account the higher prices charged by private companies for GM seeds. Hence, the percentage differences in profits considered here are net economic benefits for farmers using GM technology. Percentage differences, when not reported in the original studies, were calculated from mean value comparisons between GM and non-GM or from estimated regression coefficients.

Since we look at different types of GM technologies (different modified traits) that are used in different countries and regions, we do not expect that effect sizes are homogenous across studies. Hence, our approach of combining effect sizes corresponds to a random-effects model in meta-analysis [25]. To explain impact heterogeneity and test for possible biases, we also compiled data on a number of study descriptors that may influence the reported effect sizes. These influencing factors include information on the type of GM technology (modified trait), the region studied, the type of data and method used, the source of funding, and the type of publication. All influencing factors are defined as dummy variables. The exact definition of these dummy variables is given in Table 1. Variable distributions of the study descriptors are shown in Table S2.

Statistical analysis

In a first step, we estimate average effect sizes for each outcome variable. To test whether these mean impacts are significantly

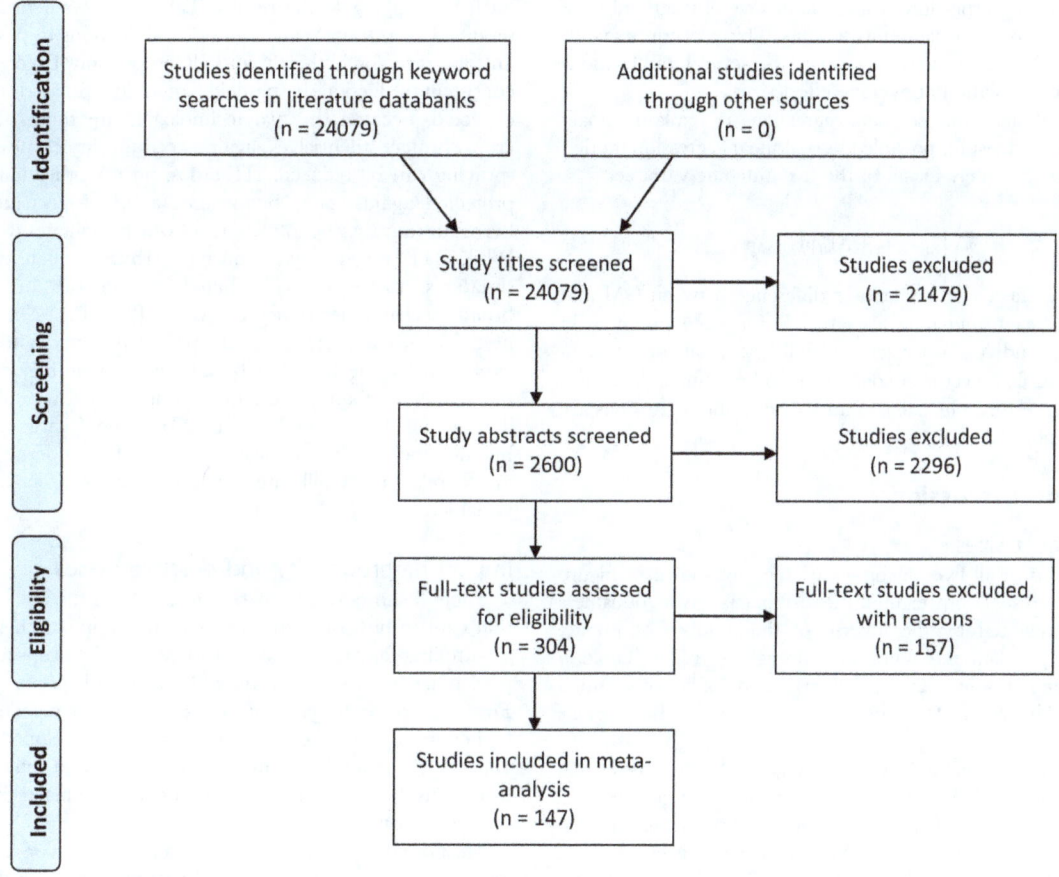

Figure 1. Selection of studies for inclusion in the meta-analysis.

different from zero, we regress each outcome variable on a constant with cluster correction of standard errors by primary dataset. Thus, the test for significance is valid also when observations from the same dataset are correlated. We estimate average effect sizes for all GM crops combined. However, we expect that the results may differ by modified trait, so that we also analyze mean effects for HT crops and IR crops separately.

Meta-analyses often weight impact estimates by their variances; estimates with low variance are considered more reliable and receive a higher weight [26]. In our case, several of the original studies do not report measures of variance, so that weighting by variance is not possible. Alternatively, weighting by sample size is common, but sample sizes are also not reported in all studies considered, especially not in some of the grey literature publications. To test the robustness of the results, we employ a

Table 1. Variables used to analyze influencing factors of GM crop impacts.

Variable name	Variable definition
Insect resistance (IR)	Dummy that takes a value of one for all observations referring to insect-resistant GM crops with genes from *Bacillus thuringiensis* (Bt), and zero for all herbicide-tolerant (HT) GM crops.
Developing country	Dummy that takes a value of one for all GM crop applications in a developing country according to the World Bank classification of countries, and zero for all applications in a developed country.
Field-trial data	Dummy that takes a value of one for all observations building on field-trial data (on-station and on-farm experiments), and zero for all observations building on farm survey data.
Industry-funded study	Dummy that takes a value of one for all studies that mention industry (private sector companies) as source of funding, and zero otherwise.
Regression model result	Dummy that takes a value of one for all impact observations that are derived from regression model estimates, and zero for observations derived from mean value comparisons between GM and non-GM.
Journal publication	Dummy that takes a value of one for all studies published in a peer-reviewed journal, and zero otherwise.
Journal/academic conference	Dummy that takes a value of one for all studies published in a peer-reviewed journal or presented at an academic conference, and zero otherwise.

different weighting procedure, using the inverse of the number of impact observations per dataset as weights. This procedure avoids that individual datasets that were used in several publications dominate the calculation of average effect sizes.

In a second step, we use meta-regressions to explain impact heterogeneity and test for possible biases. Linear regression models are estimated separately for all of the five outcome variables:

$$\%\Delta Y_{hij} = \alpha_h + \mathbf{X}_{hij}\boldsymbol{\beta}_h + \varepsilon_{hij}$$

$\%\Delta Y_{hij}$ is the effect size (percentage difference between GM and non-GM) of each outcome variable h for observation i in publication j, and $\mathbf{X}_{hij}$ is a vector of influencing factors. α_h is a coefficient and $\boldsymbol{\beta}_h$ a vector of coefficients to be estimated; ε_{hij} is a random error term. Influencing factors used in the regressions are defined in Table 1.

Results and Discussion

Average effect sizes

Distributions of all five outcome variables are shown in Figure S1. Table 2 presents unweighted mean impacts. As a robustness check, we weighted by the inverse of the number of impact observations per dataset. Comparing unweighted results (Table 2) with weighted results (Table S3) we find only very small differences. This comparison suggests that the unweighted results are robust.

On average, GM technology has increased crop yields by 21% (Figure 2). These yield increases are not due to higher genetic yield potential, but to more effective pest control and thus lower crop damage [27]. At the same time, GM crops have reduced pesticide quantity by 37% and pesticide cost by 39%. The effect on the cost of production is not significant. GM seeds are more expensive than non-GM seeds, but the additional seed costs are compensated through savings in chemical and mechanical pest control. Average profit gains for GM-adopting farmers are 69%.

Results of Cochran's test [25], which are reported in Figure S1, confirm that there is significant heterogeneity across study observations for all five outcome variables. Hence it is useful to

further disaggregate the results. Table 2 shows a breakdown by modified crop trait. While significant reductions in pesticide costs are observed for both HT and IR crops, only IR crops cause a consistent reduction in pesticide quantity. Such disparities are expected, because the two technologies are quite different. IR crops protect themselves against certain insect pests, so that spraying can be reduced. HT crops, on the other hand, are not protected against pests but against a broad-spectrum chemical herbicide (mostly glyphosate), use of which facilitates weed control. While HT crops have reduced herbicide quantity in some situations, they have contributed to increases in the use of broad-spectrum herbicides elsewhere [2,11,19]. The savings in pesticide costs for HT crops in spite of higher quantities can be explained by the fact that broad-spectrum herbicides are often much cheaper than the selective herbicides that were used before. The average farmer profit effect for HT crops is large and positive, but not statistically significant because of considerable variation and a relatively small number of observations for this outcome variable.

Impact heterogeneity and possible biases

Table 3 shows the estimation results from the meta-regressions that explain how different factors influence impact heterogeneity. Controlling for other factors, yield gains of IR crops are almost 7 percentage points higher than those of HT crops (column 1). Furthermore, yield gains of GM crops are 14 percentage points higher in developing countries than in developed countries. Especially smallholder farmers in the tropics and subtropics suffer from considerable pest damage that can be reduced through GM crop adoption [27].

Most original studies in this meta-analysis build on farm surveys, although some are based on field-trial data. Field-trial results are often criticized to overestimate impacts, because farmers may not be able to replicate experimental conditions. However, results in Table 3 (column 1) show that field-trial data do not overestimate the yield effects of GM crops. Reported yield gains from field trials are even lower than those from farm surveys. This is plausible, because pest damage in non-GM crops is often more severe in farmers' fields than on well-managed experimental plots.

Table 2. Impacts of GM crop adoption by modified trait.

Outcome variable	All GM crops	Insect resistance	Herbicide tolerance
Yield	21.57*** (15.65; 27.48)	24.85*** (18.49; 31.22)	9.29** (1.78; 16.80)
_n/m_	451/100	353/83	94/25
Pesticide quantity	−36.93*** (−48.01; −25.86)	−41.67*** (−51.99; −31.36)	2.43 (−20.26; 25.12)
_n/m_	121/37	108/31	13/7
Pesticide cost	−39.15*** (−46.96; −31.33)	−43.43*** (−51.64; −35.22)	−25.29*** (−33.84; −16.74)
_n/m_	193/57	145/45	48/15
Total production cost	3.25 (−1.76; 8.25)	5.24** (0.25; 10.73)	−6.83 (−16.43; 2.77)
_n/m_	115/46	96/38	19/10
Farmer profit	68.21*** (46.31; 90.12)	68.78*** (46.45; 91.11)	64.29 (−24.73; 153.31)
_n/m_	136/42	119/36	17/9

Average percentage differences between GM and non-GM crops are shown with 95% confidence intervals in parentheses. *, **, *** indicate statistical significance at the 10%, 5%, and 1% level, respectively. _n_ is the number of observations, _m_ the number of different primary datasets from which these observations are derived.

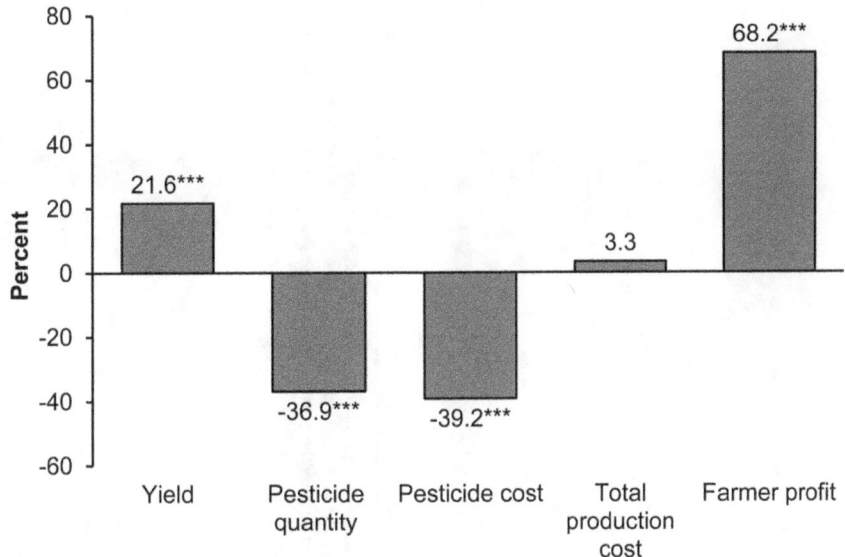

Figure 2. Impacts of GM crop adoption. Average percentage differences between GM and non-GM crops are shown. Results refer to all GM crops, including herbicide-tolerant and insect-resistant traits. The number of observations varies by outcome variable; yield: 451; pesticide quantity: 121; pesticide cost: 193; total production cost: 115; farmer profit: 136. *** indicates statistical significance at the 1% level.

Another concern often voiced in the public debate is that studies funded by industry money might report inflated benefits. Our results show that the source of funding does not significantly influence the impact estimates. We also analyzed whether the statistical method plays a role. Many of the earlier studies just compared yields of GM and non-GM crops without considering possible differences in other inputs and conditions that may also affect the outcome. Net impacts of GM technology can be estimated with regression-based production function models that control for other factors. Interestingly, results derived from regression analysis report higher average yield effects.

Finally, we examined whether the type of publication matters. Controlling for other factors, the regression coefficient for journal publications in column (1) of Table 3 implies that studies published in peer-reviewed journals show 12 percentage points higher yield gains than studies published elsewhere. Indeed, when only including observations from studies that were published in journals, the mean effect size is larger than if all observations are included (Figure S2). On first sight, one might suspect publication bias, meaning that only studies that report substantial effects are accepted for publication in a journal. A common way to assess possible publication bias in meta-analysis is through funnel plots [25], which we show in Figure S3. However, in our case these funnel plots should not be over-interpreted. First, only studies that report variance measures can be included in the funnel plots, which holds true only for a subset of the original studies used here. Second, even if there were publication bias, our mean results would be estimated correctly, because we do include studies that were not published in peer-reviewed journals.

Further analysis suggests that the journal review process does not systematically filter out studies with small effect sizes. The journal articles in the sample report a wide range of yield effects, even including negative estimates in some cases. Moreover, when combining journal articles with papers presented at academic conferences, average yield gains are even higher (Table 3, column 2). Studies that were neither published in a journal nor presented at an academic conference encompass a diverse set of papers, including reports by NGOs and outspoken biotechnology critics.

These reports show lower GM yield effects on average, but not all meet common scientific standards. Hence, rather than indicating publication bias, the positive and significant journal coefficient may be the result of a negative NGO bias in some of the grey literature.

Concerning other outcome variables, IR crops have much stronger reducing effects on pesticide quantity than HT crops (Table 3, column 3), as already discussed above. In terms of pesticide costs, the difference between IR and HT is less pronounced and not statistically significant (column 4). The profit gains of GM crops are 60 percentage points higher in developing countries than in developed countries (column 6). This large difference is due to higher GM yield gains and stronger pesticide cost savings in developing countries. Moreover, most GM crops are not patented in developing countries, so that GM seed prices are lower [19]. Like for yields, studies published in peer-reviewed journals report higher profit gains than studies published elsewhere, but again we do not find evidence of publication bias (column 7).

Conclusion

This meta-analysis confirms that – in spite of impact heterogeneity – the average agronomic and economic benefits of GM crops are large and significant. Impacts vary especially by modified crop trait and geographic region. Yield gains and pesticide reductions are larger for IR crops than for HT crops. Yield and farmer profit gains are higher in developing countries than in developed countries. Recent impact studies used better data and methods than earlier studies, but these improvements in study design did not reduce the estimates of GM crop advantages. Rather, NGO reports and other publications without scientific peer review seem to bias the impact estimates downward. But even with such biased estimates included, mean effects remain sizeable.

One limitation is that not all of the original studies included in this meta-analysis reported sample sizes and measures of variance. This is not untypical for analyses in the social sciences, especially when studies from the grey literature are also included. Future

Table 3. Factors influencing results on GM crop impacts (%).

Variables	(1) Yield	(2) Yield	(3) Pesticide quantity	(4) Pesticide cost	(5) Total cost	(6) Farmer profit	(7) Farmer profit
Insect resistance (IR)	6.58** (2.85)	5.25* (2.82)	-37.38*** (11.81)	-7.28 (5.44)	5.63 (5.60)	-22.33 (21.62)	-33.41 (21.94)
Developing country	14.17*** (2.72)	13.32*** (2.65)	-10.23 (8.99)	-19.16*** (5.35)	3.43 (4.78)	59.52*** (18.02)	60.58*** (17.67)
Field-trial data	-7.14** (3.19)	-7.81** (3.08)	-#	-17.56 (11.45)	-10.69* (5.79)	-#	-#
Industry-funded study	1.68 (5.30)	1.05 (5.21)	37.04 (23.08)	-7.77 (10.22)	-#	-#	-#
Regression model result	7.38* (3.90)	7.29* (3.83)	9.67 (10.40)	-#	-#	-11.44 (24.33)	-9.85 (24.03)
Journal publication	12.00*** (2.52)	-	9.95 (6.79)	-3.71 (4.09)	-3.08 (3.30)	48.27*** (15.48)	-
Journal/academic conference	-	16.48*** (2.64)	-	-	-	-	65.29*** (17.75)
Constant	-0.22 (2.84)	-2.64 (2.86)	-4.44 (10.33)	-16.13 (4.88)	-1.02 (4.86)	8.57 (24.33)	-1.19 (24.53)
Observations	451	451	121	193	115	136	136
R^2	0.23	0.25	0.20	0.14	0.12	0.12	0.14

Coefficient estimates from linear regression models are shown with standard errors in parentheses. Dependent variables are GM crop impacts measured as percentage differences between GM and non-GM. All explanatory variables are 0/1 dummies (for variable definitions see Table 1). The yield models in columns (1) and (2) and the farmer profit models in columns (6) and (7) have the same dependent variables, but they differ in terms of the explanatory variables, as shown. *, **, *** indicate statistical significance at the 10%, 5%, and 1% level, respectively. # indicates that the variable was dropped because the number of observations with a value of one was smaller than 5.

impact studies with primary data should follow more standardized reporting procedures. Nevertheless, our findings reveal that there is robust evidence of GM crop benefits. Such evidence may help to gradually increase public trust in this promising technology.

Acknowledgments

We thank Sinja Buri and Tingting Xu for assistance in compiling the dataset. We also thank Joachim von Braun and three reviewers of this journal for useful comments.

Author Contributions

Conceived and designed the research: WK MQ. Analyzed the data: WK MQ. Contributed to the writing of the manuscript: WK MQ. Compiled the data: WK.

References

1. Gilbert N (2013) A hard look at GM crops. Nature 497: 24–26.
2. Fernandez-Cornejo J, Wechsler JJ, Livingston M, Mitchell L (2014) Genetically Engineered Crops in the United States. Economic Research Report ERR-162 (United Sates Department of Agriculture, Washington, DC).
3. Anonymous (2013) Contrary to popular belief. Nature Biotechnology 31: 767.
4. Andreasen M (2014) GM food in the public mind-facts are not what they used to be. Nature Biotechnology 32: 25.
5. DeFrancesco L (2013) How safe does transgenic food need to be? Nature Biotechnology 31: 794–802.
6. European Academies Science Advisory Council (2013) Planting the Future: Opportunities and Challenges for Using Crop Genetic Improvement Technologies for Sustainable Agriculture (EASAC, Halle, Germany).
7. European Commission (2010) A Decade of EU-Funded GMO Research 2001–2010 (European Commission, Brussels).
8. Pray CE, Huang J, Hu R, Rozelle S (2002) Five years of Bt cotton in China - the benefits continue. The Plant Journal 31: 423–430.
9. Huang J, Hu R, Rozelle S, Pray C (2008) Genetically modified rice, yields and pesticides: assessing farm-level productivity effects in China. Economic Development and Cultural Change 56: 241–263.
10. Morse S, Bennett R, Ismael Y (2004) Why Bt cotton pays for small-scale producers in South Africa. Nature Biotechnology 22: 379–380.
11. Qaim M, Traxler G (2005) Roundup Ready soybeans in Argentina: farm level and aggregate welfare effects. Agricultural Economics 32: 73–86.
12. Sexton S, Zilberman D (2012) Land for food and fuel production: the role of agricultural biotechnology. In: The Intended and Unintended Effects of US Agricultural and Biotechnology Policies (eds. Zivin, G. & Perloff, J.M.), 269–288 (University of Chicago Press, Chicago).
13. Ali A, Abdulai A (2010) The adoption of genetically modified cotton and poverty reduction in Pakistan. Journal of Agricultural Economics 61, 175–192.
14. Kathage J, Qaim M (2012) Economic impacts and impact dynamics of Bt (Bacillus thuringiensis) cotton in India. Proceedings of the National Academy of Sciences USA 109: 11652–11656.
15. Qaim M, Kouser S (2013) Genetically modified crops and food security. PLOS ONE 8: e64879.
16. Stone GD (2012) Constructing facts: Bt cotton narratives in India. Economic & Political Weekly 47(38): 62–70.
17. Smale M, Zambrano P, Gruere G, Falck-Zepeda J, Matuschke I, et al. (2009) Measuring the Economic Impacts of Transgenic Crops in Developing Agriculture During the First Decade: Approaches, Findings, and Future Directions (International Food Policy Research Institute, Washington, DC).
18. Glover D (2010) Is Bt cotton a pro-poor technology? A review and critique of the empirical record. Journal of Agrarian Change 10: 482–509.
19. Qaim M (2009) The economics of genetically modified crops. Annual Review of Resource Economics 1: 665–693.
20. Carpenter JE (2010) Peer-reviewed surveys indicate positive impact of commercialized GM crops. Nature Biotechnology 28: 319–321.
21. Finger R, El Benni N, Kaphengst T, Evans C, Herbert S, et al. (2011) A meta analysis on farm-level costs and benefits of GM crops. Sustainability 3: 743–762.
22. Areal FJ, Riesgo L, Rodríguez-Cerezo E (2013) Economic and agronomic impact of commercialized GM crops: a meta-analysis. Journal of Agricultural Science 151: 7–33.
23. James C (2013) Global Status of Commercialized Biotech/GM Crops: 2013. ISAAA Briefs No.46 (International Service for the Acquisition of Agri-biotech Applications, Ithaca, NY).
24. Rothstein HR, Hopewell S (2009) Grey literature. In: Handbook of Research Synthesis and Meta-Analysis, Second Edition (eds. Cooper, H., Hegdes, L.V. & Valentine, J.C.), 103–125 (Russell Sage Foundation, New York).
25. Borenstein M, Hedges LV, Higgins JPT, Rothstein HR (2009) Introduction to Meta-Analysis (John Wiley and Sons, Chichester, UK).
26. Shadish WR, Haddock CK (2009) Combining estimates of effect size. In: Handbook of Research Synthesis and Meta-Analysis, Second Edition (eds. Cooper, H., Hegdes, L.V. & Valentine, J.C.), 257–277 (Russell Sage Foundation, New York).
27. Qaim M, Zilberman D (2003) Yield effects of genetically modified crops in developing countries. Science 299: 900–902.

Permissions

The contributors of this book come from diverse backgrounds, making this book a truly international effort. This book will bring forth new frontiers with its revolutionizing research information and detailed analysis of the nascent developments around the world.

We would like to thank all the contributing authors for lending their expertise to make the book truly unique. They have played a crucial role in the development of this book. Without their invaluable contributions this book wouldn't have been possible. They have made vital efforts to compile up to date information on the varied aspects of this subject to make this book a valuable addition to the collection of many professionals and students.

This book was conceptualized with the vision of imparting up-to-date information and advanced data in this field. To ensure the same, a matchless editorial board was set up. Every individual on the board went through rigorous rounds of assessment to prove their worth. After which they invested a large part of their time researching and compiling the most relevant data for our readers.

The editorial board has been involved in producing this book since its inception. They have spent rigorous hours researching and exploring the diverse topics which have resulted in the successful publishing of this book. They have passed on their knowledge of decades through this book. To expedite this challenging task, the publisher supported the team at every step. A small team of assistant editors was also appointed to further simplify the editing procedure and attain best results for the readers.

Apart from the editorial board, the designing team has also invested a significant amount of their time in understanding the subject and creating the most relevant covers. They scrutinized every image to scout for the most suitable representation of the subject and create an appropriate cover for the book.

The publishing team has been an ardent support to the editorial, designing and production team. Their endless efforts to recruit the best for this project, has resulted in the accomplishment of this book. They are a veteran in the field of academics and their pool of knowledge is as vast as their experience in printing. Their expertise and guidance has proved useful at every step. Their uncompromising quality standards have made this book an exceptional effort. Their encouragement from time to time has been an inspiration for everyone.

The publisher and the editorial board hope that this book will prove to be a valuable piece of knowledge for researchers, students, practitioners and scholars across the globe.

List of Contributors

Phoebe Lin, Patrick Stauffer, Sean Davin and Yuzhen Pan
Casey Eye Institute, Oregon Health & Science University, Portland, Oregon, United States of America

Mary Bach
Division of Rheumatology, University of Washington, VA Medical Center, Seattle, Washington, United States of America

Mark Asquith
Division of Rheumatology, Oregon Health & Science University, Portland, Oregon, United States of America

Aaron Y. Lee
Moorfield's Eye Institute of London, London, United Kingdom

Lakshmi Akileswaran and Russell N. Van Gelder
Department of Ophthalmology, University of Washington, Seattle, Washington, United States of America

Eric D. Cambronne
Department of Molecular Microbiology & Immunology, Oregon Health & Science University, Portland, Oregon, United States of America

Martha Dorris
Department of Rheumatology, University of Texas Southwestern, Dallas, Texas, United States of America

Justine W. Debelius, Christian L. Lauber, Gail Ackermann and Yoshiki V. Baeza
University of Colorado Boulder, Boulder, Colorado, United States of America

Tejpal Gill and Robert A. Colbert
Pediatric Translational Research Branch, National Institute of Arthritis, Musculoskeletal and Skin Diseases, National Institutes of Health, Baltimore, Maryland, United States of America

Rob Knight
University of Colorado Boulder, Boulder, Colorado, United States of America

Howard Hughes Medical Institute, University of Colorado Boulder, Boulder, Colorado, United States of America

Joel D. Taurog
Department of Rheumatology, University of Texas Southwestern, Dallas, Texas, United States of America

James T. Rosenbaum
Casey Eye Institute, Oregon Health & Science University, Portland, Oregon, United States of America
Division of Rheumatology, Oregon Health & Science University, Portland, Oregon, United States of America,
Dever's Eye Institute, Portland, Oregon, United States of America

Dionne N. Shepherd, Marion E. Bezuidenhout, Jennifer A. Thomson and Francisco M. Lakay
Department of Molecular and Cell Biology, University of Cape Town, Rondebosch, Cape Town, South Africa

Benjamin Dugdale and James Dale
Centre for Tropical Crops and Biocommodities, Queensland University of Technology (QUT), Brisbane, Queensland, Australia

Darren P. Martin
Institute of Infectious Disease and Molecular Medicine, University of Cape Town, Observatory, Cape Town, South Africa
Centre for High-Performance Computing, Rosebank, Cape Town, South Africa

Arvind Varsani
School of Biological Sciences and Biomolecular Interaction Centre, University of Canterbury, Christchurch, New Zealand
Department of Plant Pathology and Emerging Pathogens Institute, University of Florida, Gainesville, Florida, United States of America
Electron Microscope Unit, Division of Medical Biochemistry, Department of Clinical Laboratory Sciences, University of Cape Town, Observatory, Cape Town, South Africa

Adérito L. Monjane and Edward P. Rybicki
Department of Molecular and Cell Biology, University of Cape Town, Rondebosch, Cape Town, South Africa
Institute of Infectious Disease and Molecular Medicine, University of Cape Town, Observatory, Cape Town, South Africa

Guo-Hong Yu, Lin-Lin Jiang, Meng-Meng Liu, Shu-Guang Shan and Xian-Guo Cheng
Key Lab. of Plant Nutrition and Fertilizer, Ministry of Agriculture, Institute of Agricultural Resources and Regional Planning, Chinese Academy of Agricultural Sciences, Beijing, China

Xue-Feng Ma
Key Lab. of Plant Nutrition and Fertilizer, Ministry of Agriculture, Institute of Agricultural Resources and Regional Planning, Chinese Academy of Agricultural Sciences, Beijing, China
Institute of Agro-Products Processing Science and Technology, Chinese Academy of Agricultural Sciences, Beijing, China

Zhao-Shi Xu
Institute of Crop Science, Chinese Academy of Agricultural Sciences (CAAS)/National Key Facility for Crop Gene Resources and Genetic Improvement, Key Laboratory of Biology and Genetic Improvement of Triticeae Crops, Ministry of Agriculture, Beijing, China

Jenny Freitag, Sylvia Heink and Thomas Kamradt
Department of Immunology, University Hospital Jena, Jena, Germany

Edith Roth, Jürgen Wittmann and Hans-Martin Jäck
Division of Molecular Immunology, Department of Internal Medicine III, Nikolaus-Fiebiger- Center, University of Erlangen-Nürnberg, Erlangen, Germany

Fuhui Xu, Zhixue Liu, Hongyan Xie and Jian Zhu
School of Life Sciences and Technology, Tongji University, Shanghai, China

Juren Zhang
School of Life Science, Shandong University, Shandong, China

Josef Kraus, Tasja Blaschnig and Reinhard Nehls
KWS SAAT AG, Einbeck, Germany

Hong Wang
School of Life Sciences and Technology, Tongji University, Shanghai, China
KWS SAAT AG, Einbeck, Germany

Chutharat Chueasiri, Ketsuwan Chunthong, Keasinee Pitnjam, Sriprapai Chakhonkaen, Amorntip Muangprom, Numphet Sangarwut, Kanidta Sangsawang and Malinee Suksangpanomrung
National Center for Genetic Engineering and Biotechnology, Thailand Science Park, Klong Luang, Pathumthani, Thailand

Louise V. Michaelson and Johnathan A. Napier
Biological Chemistry Department, Rothamsted Research, Harpenden, Hertfordshire, United Kingdom

Ellen J. Bennett, Richard J. Mead, Mimoun Azzouz, Pamela J. Shaw and Andrew J. Grierson
Sheffield Institute for Translational Neuroscience, Department of Neuroscience, University of Sheffield, Sheffield, United Kingdom

Amit Sethi, Jennifer Delatte, Lane Foil and Claudia Husseneder
Department of Entomology, Louisiana State University Agricultural Center, Baton Rouge, Louisiana, United States of America

Mintu Desai and Navneet Kaur
Michigan State University-Department of Energy Plant Research Laboratory, Michigan State University, East Lansing, Michigan, United States of America

Jianping Hu
Plant Biology Department, Michigan State University, East Lansing, Michigan, United States of America

Yong Wang, Xiao-Yang Zhou, Peng-Ying Xiang, Lu-Lu Wang, Fei Xie, Liang Li and Hong Wei
Department of Laboratory Animal Science, College of Basic Medical Sciences, Third Military Medical University, Chongqing, China

Huan Tang
Department of Laboratory Animal Science, College of Basic Medical Sciences, Third Military Medical University, Chongqing, China
China Three Gorges Museum, Chongqing, China

Xiaoxian Xie, Yufang Ma, Zhenliang Chen, Rongrong Liao, Xiangzhe Zhang, Qishan Wang and Yuchun Pan
School of Agriculture and Biology, Department of Animal Sciences, Shanghai Jiao Tong University, Shanghai, PR China, Shanghai Key Laboratory of Veterinary Biotechnology, Shanghai, PR China

Hod Dana, Tsai-Wen Chen, Amy Hu, Brenda C. Shields, Caiying Guo, Loren L. Looger, Douglas S. Kim and Karel Svoboda
Janelia Farm Research Campus, Howard Hughes Medical Institute, Ashburn, Virginia, United States of America

Luhua Zhang, Haiwei Chen, Yanlan Li, Yanan Li, Defu Chen and Xiwen Chen
Laboratory of Molecular Genetics, College of Life Sciences, Nankai University, Tianjin, China

Shengjun Wang, Jinping Su and Xuejun Liu
Tianjin Crop Research Institute, Tianjin, China

Geneviéve Jolivet, Bruno DaSilva, Erwana Harscoët, Céline Viglietta, Nathalie Daniel-Carlier and Louis-Marie Houdebine
INRA UMR1198, Biologie du Développement et Reproduction, Jouy en Josas, France

Sandrine Braud and Itzik Harosh
ObeTherapy Biotechnology, Evry, France

Bruno Passet
INRA UMR1313, Génétique Animale et Biologie Intégrative, Jouy-en-Josas, France

Thomas Gautier and Laurent Lagrost
NSERM UMR866, Université de Bourgogne, Dijon, France

Sagarika Mishra and Lingaraj Sahoo
Department of Biotechnology, Indian Institute of Technology Guwahati, Guwahati, India

Hemasundar Alavilli and Byeong-ha Lee
Department of Life Science, Sogang University, Mapo-gu, Seoul, Korea

Sanjib Kumar Panda
Department of Life Sciences and Bioinformatics, Assam University, Silchar, India, Department of Biochemistry & Molecular Biology, Noble Research Centre, Oklahoma State University, Stillwater, OK, United States of America

Balaji Jada, Arto J. Soitamo, Eva-Mari Aro and Kirsi Lehto
Department of Biochemistry, Laboratory of Molecular Plant Biology, University of Turku, Turku, Finland

Shahid Aslam Siddiqui
Department of Agricultural sciences, University of Helsinki, Helsinki, Finland

Gayatri Murukesan and Tapio Salakoski
Department of Information Technology, University of Turku, Turku, Finland

Ido Golan, Zvia Konrad, Doron Shkolnik-Inbar and Dudy Bar-Zvi
Department of Life Sciences and Doris and Bertie Black Center for Bioenergetics in Life Sciences, Ben-Gurion University of the Negev, Beer-Sheva, Israel

Pia Guadalupe Dominguez and Fernando Carrari
Instituto de Biotecnología, Instituto Nacional de Tecnología Agropecuaria, Buenos Aires, Argentina

Takuji Wada and Rumi Tominaga-Wada
Graduate School of Biosphere Sciences, Hiroshima University, Higashi-Hiroshima, Hiroshima, Japan

Asuka Kunihiro
Faculty of Applied Biological Science, Hiroshima University, Higashi-Hiroshima, Hiroshima, Japan

Lucas S. Parreiras, Rebecca J. Breuer, Ragothaman Avanasi Narasimhan, Alex LaReau, Mary Tremaine, Li Qin, Benjamin D. Bice, Irene M. Ong, Haibo Li, Edward L. Pohlmann, Jose Serate, Sydnor T. Withers and Trey K. Sato
DOE Great Lakes Bioenergy Research Center, University of Wisconsin-Madison, Madison, Wisconsin, United States of America

Alan J. Higbee
DOE Great Lakes Bioenergy Research Center, University of Wisconsin-Madison, Madison, Wisconsin, United States of America
Department of Chemistry, University of Wisconsin-Madison, Madison, Wisconsin, United States of America

Laura B. Willis
Department of Bacteriology, University of Wisconsin-Madison, Madison, Wisconsin, United States of America

Brandi L. Bonfert, Rebeca C. Pinhancos, Allison J. Balloon, Michael S. Westphall and Joshua J. Coon
Department of Chemistry, University of Wisconsin-Madison, Madison, Wisconsin, United States of America

Nirmal Uppugundla
DOE Great Lakes Bioenergy Research Center, Michigan State University, East Lansing, Michigan, United States of America
Biomass Conversion Research Laboratory, Department of Chemical Engineering and Materials Science, Michigan State University, Lansing, Michigan, United States of America

Tongjun Liu
DOE Great Lakes Bioenergy Research Center, Michigan State University, East Lansing, Michigan, United States of America
School of Food and Bioengineering, Qilu University of Technology, Jinan, China

Chenlin Li and Deepti Tanjore
Advanced Biofuels Process Demonstration Unit, Lawrence Berkeley National Laboratory, Emeryville, California, United States of America

Blake A. Simmons
Deconstruction Division, Joint BioEnergy Institute, Emeryville, California, United States of America

David B. Hodge
DOE Great Lakes Bioenergy Research Center, Michigan State University, East Lansing, Michigan, United States of America
Department of Chemical Engineering & Materials Science, Michigan State University, East Lansing, Michigan, United States of America
Department of Biosystems & Agricultural Engineering, Michigan State University, East Lansing Michigan, United States of America
Division of Sustainable Process Engineering, Lulea° University of Technology, Lulea°, Sweden

Bruce E. Dale and Venkatesh Balan
DOE Great Lakes Bioenergy Research Center, Michigan State University, East Lansing, Michigan, United States of America

Biomass Conversion Research Laboratory, Department of Chemical Engineering and Materials Science, Michigan State University, Lansing, Michigan, United States of America

Robert Landick
DOE Great Lakes Bioenergy Research Center, University of Wisconsin-Madison, Madison, Wisconsin, United States of America
Department of Bacteriology, University of Wisconsin-Madison, Madison, Wisconsin, United States of America
Department of Biochemistry, University of Wisconsin-Madison, Madison, Wisconsin, United States of America

Audrey P. Gasch
DOE Great Lakes Bioenergy Research Center, University of Wisconsin-Madison, Madison, Wisconsin, United States of America
Laboratory of Genetics, University of Wisconsin-Madison, Madison, Wisconsin, United States of America

Tomoko Mitsunami and Gen-ichiro Arimura
Department of Biological Science & Technology, Faculty of Industrial Science & Technology, Tokyo University of Science, Tokyo, Japan

Masahiro Nishihara, Kohei Fujita and Nobuhiro Sasaki
Iwate Biotechnology Research Center, Kitakami, Japan

Ivan Galis, Kabir Md Alamgir and Yuko Hojo
Institute of Plant Science and Resources, Okayama University, Kurashiki, Japan

Keichiro Nemotoa and Tatsuya Sawasaki
Proteo-Science Center, Ehime University, Matsuyama, Japan

Wilhelm Klümper and Matin Qaim
Department of Agricultural Economics and Rural Development, Georg-August-University of Goettingen, Goettingen, Germany

Index

www.ingramcontent.com/pod-product-compliance
Lightning Source LLC
Chambersburg PA
CBHW080408190526
45161CB00003B/170